# SEISMIC TOMOGRAPHY

SEISMOLOGY AND EXPLORATION GEOPHYSICS

# SEISMIC TOMOGRAPHY

## With Applications in Global Seismology and Exploration Geophysics

*Edited by*

GUUST NOLET

*Department of Theoretical Geophysics, Utrecht University, The Netherlands*

D. REIDEL PUBLISHING COMPANY

A MEMBER OF THE KLUWER ACADEMIC PUBLISHERS GROUP

DORDRECHT / BOSTON / LANCASTER / TOKYO

**Library of Congress Cataloging in Publication Data**

CIP

Seismic tomography.

(Seismology and exploration geophysics)
Bibliography: p.
Includes index.
1. Seismic waves—Data processing. 2. Tomography. I. Nolet, Guust, 1945– . II. Series.
QE538.5.S429 1987 551.2′2 87–9826
ISBN 90-277-2521-7 ISBN 90-277-2583-7 pbk

---

Published by D. Reidel Publishing Company,
P.O. Box 17, 3300 AA Dordrecht, Holland.

Sold and distributed in the U.S.A. and Canada
by Kluwer Academic Publishers,
101 Philip Drive, Assinippi Park, Norwell, MA 02061, U.S.A.

In all other countries, sold and distributed
by Kluwer Academic Publishers Group,
P.O. Box 322, 3300 AH Dordrecht, Holland.





Printed in The Netherlands

## LIST OF CONTENTS


List of Authors vii

Preface ix

**INTRODUCTION**

1 *G. Nolet* - Seismic wave propagation and seismic tomography 1

**PART I: BASIC THEORY**

2 *C.H. Chapman* - The Radon transform and seismic tomography 25

3 *A. van der Sluis and H.A. van der Vorst* - Numerical solution of large, sparse linear algebraic systems arising from tomographic problems 49

4 *E. Wielandt* - On the validity of the ray approximation for interpreting delay times 85

5 *V. Červený* - Ray tracing algorithms in three-dimensional laterally varying layered structures 99

**PART II: APPLICATIONS IN SEISMIC EXPLORATION**

6 *A. Tarantola* - Inversion of travel times and seismic waveforms 135

7 *S. Ivansson* - Crosshole transmission tomography 159

8 *P. Firbas* - Tomography from seismic profiles 189

9 *A. Nur* - Seismic rock properties for reservoir descriptions and monitoring 203

**PART III: APPLICATIONS IN GLOBAL SEISMOLOGY**

10 *G. Poupinet* - Seismic data collection platforms for satellite transmission 239

11 *A. Morelli and A.M. Dziewonski* - The harmonic expansion approach to the retrieval of deep Earth structure 251

12 *N. Jobert and G. Jobert* - Ray tracing for surface waves 275

13 *G. Nolet* - Waveform tomography 301

14 *R. Snieder* - Surface wave holography 323

15 *L.J. Ruff* - Tomographic imaging of seismic sources 339

References 367

Index 383

## LIST OF AUTHORS


| | |
|---|---|
| **V. Červený** | Inst. of Geophysics, Ke Karlovu 3, Praha 2, Czechoslovakia. |
| **C.H. Chapman** | Bullard Laboratories, Dept. of Earth Sciences, University of Cambridge, Madingley Rise, Madingley Road, Cambridge, CB3 0EZ, UK. |
| **A.M. Dziewonski** | Hoffman Laboratory, Harvard University, 20 Oxford Street, Cambridge MA 02138, USA. |
| **P. Firbas** | Geofyzika n.p. Brno, Jecna 29a, 61246 Brno, Czechoslovakia. |
| **S. Ivansson** | Forsvarets Forskningsanstalt, PO Box 27322, S10254 Stockholm, Sweden. |
| **G. Jobert** | Institut de Physique du Globe, Université de Paris 6, 4 Place Jussieu, F75252 Paris, France. |
| **N. Jobert** | Institut de Physique du Globe, Université de Paris 6, 4 Place Jussieu, F75252 Paris, France. |
| **A. Morelli** | Hoffman Laboratory, Harvard University, 20 Oxford Street, Cambridge MA 02138, USA. |
| **G. Nolet** | Department of Theoretical Geophysics, Budapestlaan 4, 3584 CD Utrecht, The Netherlands. |
| **A. Nur** | Rock Physics and Borehole Geophysics Laboratory, Dept. of Geophysics, Stanford University, Stanford, CA 94305, USA. |
| **G. Poupinet** | IRIGM, Observatoire de Grenoble, BP 68, F38402 St Martin d'Heres, France. |
| **L.J. Ruff** | Department of Geological Sciences, University of Michigan, MI 48109 Ann Arbor, USA. |
| **A. van der Sluis** | Mathematical Institute, Budapestlaan 6, 3584CD Utrecht, The Netherlands. |
| **R. Snieder** | Department of Theoretical Geophysics, University of Utrecht, Budapestlaan 4, P.O. Box 80.021, 3508 TA Utrecht, The Netherlands. |
| **A. Tarantola** | Institut de Physique du Globe, Université de Paris 6, 4 Place Jussieu, F75252 Paris, France. |
| **H.A. van der Vorst** | Faculty of Mathematics and Informatics, Delft University of Technology, PO Box 356, 2600 AJ Delft, The Netherlands. |
| **E. Wielandt** | Institute of Geophysics, Swiss Federal Institute of Technology (ETH), CH-8093 Zurich, Switzerland. |

## PREFACE

Methods to construct images of an object from "projections" of x-rays, ultrasound or electromagnetic waves have found wide applications in electron microscopy, diagnostic medicine and radio astronomy. Projections are measurable quantities that are a functional - usually involving a line integral - of physical properties of an object. Convolutional methods, or iterative algorithms to solve large systems of linear equations are used to reconstruct the object. In principle, there is no reason why similar image reconstructions can not be made with seismic waves. In practice, *seismic tomography* meets with a number of difficulties, and it is not until the last decade that imaging of transmitted seismic waves has found application in the Earth sciences.

The most important difference between global seismic tomography and more conventional applications in the laboratory is the fact that the seismologist is confronted with the lack of anything resembling a well-controlled experimental set-up. Apart from a few nuclear tests, it is not in our power to locate or time seismic events. Apart from a few seabottom seismographs, our sensors are located on land - and even there the availability of data depends on cultural and political factors. Even in exploration seismics, practical factors such as the cost of an experiment put strong limitations on the completeness of the data set.

The most commonly used datum is the relative delay of a seismic wave. These delays can - in general - be obtained only with very low precision. Furthermore, unlike x-rays in a human body, seismic waves in the Earth follow strongly curved rays, and the geometry of the ray is an extra unknown in the formulation of the problem.

These difficulties notwithstanding, seismic tomography has already yielded useful results, both in exploration seismics and in global seismology. In fact, some of the images obtained are so spectacular that many Earth scientists may be tempted to forget about the limited resolution or the large errors in the final result.

This book describes the state of the art in seismic tomography, with an emphasis on the methods, rather than on the results. An introductory chapter has been included to make the book accessible to non-specialists or students with only a limited knowledge of seismic wave propagation. The remainder of the book has been divided in three parts. The first part contains a few chapters with theoretical results that are basic to any application of seismic tomography. Even though some of these chapters contain new and important theoretical results, I have asked the authors to give these chapters a slight tutorial flavour, so as to bridge the gap between the introduction and the more specialist contributions. This is followed by a section on applications in exploration seismics. The book is concluded by a number of chapters describing applications in global seismology. The division between "exploration" seismics and "global" seismology is not strict, and it is certainly not my intention to encourage readers to skip part of the book: most of the chapters contain material of very general interest to all Earth scientists working with tomographic methods.

No effort has been made to make the notation uniform, or to eliminate redundancies. Thus, every chapter can be read independently and it should be fairly easy to make a selection for a one- or two-semester course for graduate students.

I wish to thank Michiel ten Raa and Roger Cooper of Reidel Publishing Company for stimulating me to edit this book. Several interested students: Tijmen-Jan Moser, Berend Scheffers, Hong-Kie Thio and Alet Zielhuis assisted in the textprocessing and proofreading.

Utrecht, January 1987

*Guust Nolet*

*Chapter 1*

# Seismic wave propagation and seismic tomography

*G. Nolet*

This chapter develops the basic principles of seismic tomography and serves as a general introduction to this book.

## 1. Introduction

Ever since the first seismometers were placed on the surface of the Earth near the end of the 19th century, seismic waves have been used to locate remote "objects". The first applications involved the location of earthquake epicenters in far-away regions. Efforts during the first World War to locate heavy artillery by seismic and acoustic means evolved later to the first exploration methods for oil and gas (Bates et al., 1982). The imaging technique in exploration seismics - commonly referred to as *migration* - has been improved ever since: at first it merely involved the interpretation of arrival times of observed seismic pulses in terms of the depth and slope of reflecting surfaces; later, complete seismic records were used and imaging methods were developed that are firmly based on the acoustic wave equation (see Claerbout, 1985, for references).

Imaging in global seismology stayed far behind the developments in exploration seismics for several reasons: in contrast to artificial sources, earthquakes are uncontrolled, badly placed sources of wave energy; the Earth is only sparsely covered with seismometers; instrument responses were for a long time widely different and recording was - and often is - not in digital form, although recent developments may soon change this for the better (Nolet et al., 1986). Thus, seismologists are faced with the paradox that the available data contain crucial gaps, despite their enormous volume. The most powerful data source in global seismology is not in fact the collection of individual seismograms



*G. Nolet (ed.), Seismic Tomography, 1–23.*

from earthquakes and nuclear explosions, which is unmanageable both because of its size and its diversity. Of much larger practical importance are the time readings of individual phases on these seismograms, which have been made by local seismograph operators. These data are routinely sent to the International Seismological Centre (ISC) and available on magnetic tape and soon on compact disk as well.

Dziewonski et al. (1977) were the first to recognize the potential of the ISC data set. In a pioneering paper, they used 700,000 P wave travel time residuals to determine some 150 coefficients of a spherical harmonic expansion of velocity perturbations in the Earth's mantle by means of a least-squares analysis. In a more recent attempt, Dziewonski (1984) determined spherical harmonic coefficients up to degree 6 with similar, though very much improved, analysis of ISC delay times.

At the same time, efforts to image the Earth's interior on a more local scale were made by Aki and Lee (1976) and Aki et al. (1976,1977), who used P-wave delay time readings from the Test Ban monitoring arrays like LASA and NORSAR to delineate the seismic velocity structure directly under these arrays. Their method involved the determination of velocity perturbations in individual cells, rather than a harmonic expansion. At first this may seem a trivial difference in methodology. But, as we shall see later on, this usually results in a matrix system that is singular or ill-conditioned (by which we mean that small data errors tend to have a large, disturbing effect on the solution). For small-sized problems we can bring this under control by employing sophisticated algebraic techniques such as singular value decomposition, and this is essentially the method developed by Aki and his colleagues. This method has led to a large number of applications, ranging from large continental areas (Romanowicz, 1979, 1980; Menke, 1977; Hirahara, 1977; Taylor and Toksoz, 1979; Yanovskaya, 1984; Babuska et al., 1984) to very local structures (Ellsworth and Koyanagi, 1977; Mitchell et al., 1977; Reasenberg et al., 1980; Grasso et al., 1983; Nercessian et al., 1984; Burmakov et al., 1984; Maguire et al., 1985; Dorbath et al., 1986).

For larger scale problems (when the number of unknows exceeds, say, $10^3$), computer memories cannot easily accommodate the giant matrices resulting from this parametrization and the *algebraic solution itself* must be approximated using iterative techniques. Errors in the algebraic solution arise from incomplete convergence (forced by limitations in computing time) and add to errors resulting from a propagation of the - often very large - data errors. This problem has so far not yet been fully explored. One reason is probably that the initial results from global tomographic interpretations have yielded spectacular images of the Earth's interior, that have been shown to be compatible with geodynamical interpretations of the long wavelength gravity field of the Earth (Hager et al., 1985). The danger that there are major lapses in tomographic interpretations has however never adequately been staved off, and this is now probably the most important research task in seismic tomography.

The first tomographic results using iterative techniques on a global scale were presented by Clayton and Comer (1984), and some tentative studies of the reliability of the iterative solutions were published by Ivansson (1983) and Nolet (1984,1985). Spakman (1986) used large scale iterative techniques to resolve a very detailed velocity structure near the convergence of the African and Eurasian plate.

Aside from delay times, surface wave phase and group velocities, as well as complete waveforms, have been used to image the S-velocity structure of the Earth. Among the many recent examples we mention Woodhouse and Dziewonski, 1984; Montagner, 1985 and Nataf et al., 1986. In exploration geophysics, tomographic techniques have been employed in seismic soundings between boreholes (see Ivansson, chapter 7, for references) but may also find application in more conventional reflection seismics (Ivansson, 1986; Kennett and Williamson, 1987), or refraction experiments (Firbas, chapter 8). Nur (chapter 9) investigates various applications for oil and gas exploration.

In the next section we shall describe the basic principles of ray theory (readers with sufficient background knowledge in seismic wave propagation may safely skip to section 3).

## 2. Ray theory for seismic waves

Because the mathematics of wave propagation is inherently more simple for the acoustic case, we shall develop some basic principles first for this special case. In section 2.2 we shall than bridge the gap to elastic wave propagation in solids, where both longitudinal (P) and transverse (S) waves are allowed.

### 2.1 Acoustic waves

The balance of forces and tractions working on a small volume element in a solid or fluid leads to:

$$\rho\partial_{tt}u_i = \partial_j\sigma_{ij} + f_i \tag{1}$$

where $u_i$ is the component of a (small) displacement in direction i due to incremental stresses $\sigma_{ij}$ and the body force component $f_i$. By 'incremental' we mean that we are merely interested in changes with respect to the static situation. This keeps the system linear, even if we are dealing with waves that travers deep regions of the Earth, where a very large hydrostatic pressure ( a stress of the form $\sigma_{ij}=-p\,\delta_{ij}$ ) is operating. We use the Einstein convention, which implies summation over all indices that occur twice.

The stress is linearly related to the strain $\varepsilon_{kl}$ by Hooke's law:

$$\sigma_{ij} = c_{ijkl}\varepsilon_{kl} = c_{ijkl}\partial_k u_l \tag{2}$$

where the last term can easily be derived from the definition of the strain ($\varepsilon_{kl}=\frac{1}{2}\partial_k u_l+\frac{1}{2}\partial_l u_k$) and the symmetry properties of the *elasticity tensor*

$$c_{ijkl} = c_{jikl} = c_{ijlk} = c_{lkij}$$

If we insert (2) into (1) we find:

$$\rho\partial_{tt}u_i = \partial_j(c_{ijkl}\partial_k u_l) + f_i \tag{3}$$

In a gas or a fluid we prefer to use 'pressure' $P$ rather than stress field $\sigma$. The relation between the two is

$$\sigma_{ij} = -P\,\delta_{ij} \tag{4}$$

and instead of the elasticity tensor, we deal with a simple scalar quantity $\kappa$ called the 'bulk modulus' or 'incompressibility':

$$-P = \kappa \partial_j u_j \tag{5}$$

Having reduced the elasticity tensor to a simple scalar, we are able to derive the differential equation of acoustic wave propagation, since

$$\rho \partial_{tt} u_i = -\partial_i P + f_i \tag{6}$$

We can eliminate u from this equation when we realize that we can express the divergence of **u** in terms of the pressure. Using $\partial_i u_i = -\kappa^{-1} P$ we find:

$$\partial_{tt} P = \kappa \partial_i \left[ \frac{1}{\rho} \partial_i P \right] - \kappa \partial_i \left[ \frac{1}{\rho} f_i \right] \tag{7}$$

If we take the Fourier transform of (7) we find, in the absence of a body force field **f**:

$$-\omega^2 P = \kappa \partial_i [\frac{1}{\rho} \partial_i P] \tag{8}$$

Suppose that we have a point-source. In a homogeneous medium this would give a solution $P=(1/r)\delta(t-r/c)$, i.e. with an amplitude decay $1/r$ and a time delay $r/c$. Our assumption in ray theory is that we shall have a more general geometrical spreading $A(\mathbf{r})$ and a more general time delay $\theta(\mathbf{r})$, but that the shape of the delta function does not change ( we say that there is "no dispersion"). More advanced ray theories incorporate dispersion as terms of higher order in $\omega$, but these are in our case of little use. Thus we shall substitute into (8) the Fourier transform of a pressure field of the assumed form $P(\mathbf{r},t)=A(\mathbf{r})\delta[t-\theta(\mathbf{r})]$ or:

$$p(\mathbf{r},\omega) = A(\mathbf{r}) e^{i\omega\theta(\mathbf{r})} \tag{9}$$

*Wavefronts* are surfaces of equal phase, hence defined with $\theta(\mathbf{r})=\tau$. Rays are defined as the family of normals to the wavefronts. Thus, $\theta$ defines the rays, and $A$ defines the decrease of wave energy because of geometrical spreading. If we substitute (9) into (8) and retain only terms with $\omega$ and $\omega^2$ (high frequency approximation) we find after division by $\omega^2$:

$$-Ae^{i\omega\theta} = \kappa \left[ \partial_i(\frac{1}{\rho}) \frac{iA\partial_i\theta}{\omega} + \frac{1}{\rho}\frac{iA\partial_{ii}\theta}{\omega} + \frac{2}{\rho}\frac{i\partial_i A \partial_i \theta}{\omega} - \frac{1}{\rho}(\partial_i\theta)^2 A \right] e^{i\omega\theta} \tag{10}$$

plus terms of order $\omega^{-2}$. We assume that the derivative of $\rho$ is bounded, and assume that $\omega \to \infty$. Equating the dominant terms of first order, we find:

$$(\partial_i\theta)^2 = \frac{\rho}{\kappa} = \frac{1}{c^2} \tag{11}$$

Which is known as the *eikonal equation* for the location of the wavefront $\theta$. The eikonal equation implies that $c\,\nabla\theta$ is a unit vector. It is a vector perpendicular to the wavefront, and therefore by definition parallel to the ray. Although (11) gives us the location of the wavefront, it is more useful to have an equation that describes the geometry of the rays. Let

$d\mathbf{r}$ be a tangent along the ray, with length $ds$. Then we can write the same unit vector as $d\mathbf{r}/ds$ (figure 1), or:

$$\nabla\theta = \frac{1}{c}\frac{d\mathbf{r}}{ds} \tag{12}$$

On the other hand we have $\mathbf{n}\cdot\nabla\theta = d\theta/ds = 1/c$, and $d(\nabla\theta)/ds = \nabla(d\theta/ds)$,

$$\nabla(\frac{1}{c}) = \frac{d}{ds}\left[\frac{1}{c}\frac{d\mathbf{r}}{ds}\right] \tag{13}$$

This is a second order differential equation for rays. Computers are much better in solving first-order systems than second order ones. It is not difficult to transform the system to a first order system. Put:

$$p_i = \frac{1}{c}\frac{dx_i}{ds} \quad \text{then} \quad \frac{dp_i}{ds} = \frac{\partial}{\partial x_i}(\frac{1}{c})$$

Starting with $x_i(0)$ and $p_i(0)$ we can trace the ray by numerical integration of this system. $x_i(0)$ tells us where the ray starts, $p_i(0)$ is the *ray parameter* that gives the direction in which it starts out. This does not in general conform to the type of boundary condition that we find in seismology. There we know the location of the source, but instead of the ray specification we only know its end point $x_i(S)$ and even the ray length S is not known. S is actually part of the solution. If we set $s = \eta S$, so that $0 \le \eta \le 1$ then we have

$$\frac{dx_i}{d\eta} = Sc(\mathbf{r})p_i \tag{14}$$

$$\frac{dp_i}{d\eta} = S\frac{\partial}{\partial x_i}(\frac{1}{c})$$

with boundary conditions: $x_i(0) = x_i^0$ and $x_i(1) = x_i^S$. This is a nonlinear eigenvalue problem with Dirichlet boundary conditions at $\eta=0$ and $\eta=1$ (Chin et al., 1984).

Although the derivation of the analytical solutions to (13) for Earth models with spherical symmetry is elementary and can be found in any seismology book, we shall give them here for the sake of completeness.

In a medium with horizontal stratification we have $c(x,y,z) = c(z)$, so that $dp_x/d\eta$ and $dp_y/d\eta$ vanish. From (13) we then have, for a ray in the x-y plane:

$$constant \ p_x \equiv p = \frac{1}{c}\frac{dx}{ds} = \frac{\sin i}{c}$$

which shows that Snell's law is valid for waves satisfying the ray theory $\omega \gg 1$ (see figure 2). If we follow a segment of a ray, it will travers a horizontal distance given by

$$X(p) = \int dx = \int tg\ i\ dz = \int \frac{\sin i}{(1-\sin^2 i)^{1/2}} dz = \int \frac{cp}{(1-c^2p^2)^{1/2}} dz \tag{15}$$

in a time:

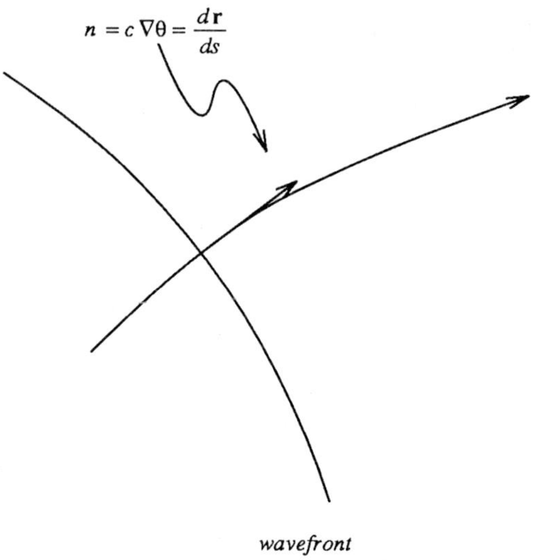


*Figure 1*

$$T(p) = \int dt = \int \frac{ds}{c} = \int \frac{dz}{c\,(1-c^2p^2)^{1/2}} \tag{16}$$

In an Earth with spherical symmetry we find in a similar way:

$$T(p) = \int \frac{\eta^2 dr}{r\,(\eta^2-p^2)^{1/2}} \tag{17}$$

$$\Delta(p) = \int \frac{p dr}{r\,(\eta^2-p^2)^{1/2}} \tag{18}$$

where $\Delta(p)$ is the angle in radians, $\eta = r/c$, and $p = \eta\sin\theta$ ($\theta$ is the angle of the ray with the radius).

For models without symmetry (usually called *laterally heterogeneous* Earth models), the system of differential equations must be solved with numerical methods. In view of the large number of data usually handled in tomographic studies, efficiency of the numerical ray tracing algorithm is of prime importance. Cerveny (chapter 5) gives extensive details on the numerical methods and their efficiency.

### 2.2 Rays in an isotropic, elastic Earth

In an isotropic solid, the elasticity tensor $c_{ijkl}$ simplifies to

$$c_{ijkl} = (\kappa - \frac{2}{3}\mu)\delta_{ij}\delta_{kl} + \mu(\delta_{ik}\delta_{jl} + \delta_{il}\delta_{jk}) \tag{19}$$

where where $\mu$ is the shear modulus and $\kappa - 2\mu/3 = \lambda$, Lamé's parameter. Substitution into (1) gives:

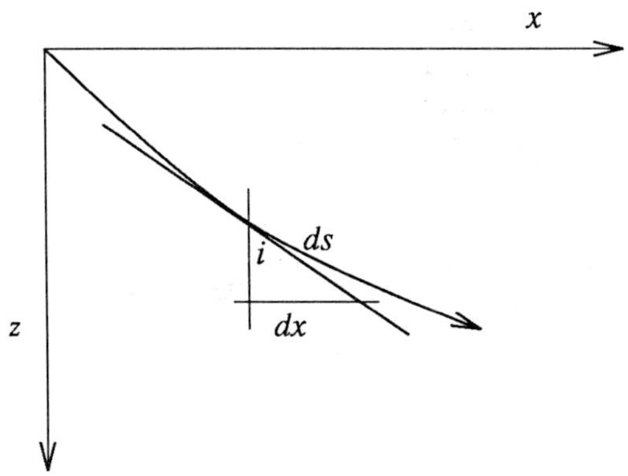


*Figure 2*

$$\rho\partial_{tt}u_i = (\lambda+\mu)\partial_i\partial_k u_k + \mu\partial_j\partial_j u_i + f_i \tag{20}$$

It is fairly easy to show that the equation of elastic waves now gives rise to two different eikonal equations, one for P and one for S waves. As in the acoustic case, we first transform to the frequency domain:

$$-\rho\omega^2 u_i = \partial_i(\lambda\partial_j u_j) + \partial_j[\mu(\partial_i u_j+\partial_j u_i)] \tag{21}$$

we substitute a trial solution $\mathbf{u}(\mathbf{r},t)=\mathbf{A}(\mathbf{r})\delta[t-\theta(\mathbf{r})]$:

$$\mathbf{u}(\mathbf{r},\omega) = \mathbf{A}(\mathbf{r})e^{i\omega\theta(\mathbf{r})} \tag{22}$$

If we substitute this into (21) and collect all terms with $\omega^2$:

$$-\rho A_i = -(\lambda+\mu)\partial_i\theta\partial_j\theta A_j - \mu(\partial_j\theta)^2 A_i \tag{23}$$

or, in vector notation:

$$-\rho\mathbf{A} + (\lambda+\mu)\nabla\theta(\nabla\theta\cdot\mathbf{A}) + \mu\,|\nabla\theta|^2\mathbf{A} = 0 \tag{24}$$

This equation contains 3 terms, two of which are directed along $\mathbf{A}$, the other along $\nabla\theta$. Obviously, the equation can be satisfied only if all nonzero terms are parallel, thus either:

$$\mathbf{A} = constant\times\nabla\theta \;\rightarrow\; |\nabla\theta|^2 = \frac{\rho}{\lambda+2\mu} = 1/\alpha^2 \tag{25}$$

or

$$\mathbf{A}\cdot\nabla\theta = 0 \;\rightarrow\; |\nabla\theta|^2 = \frac{\rho}{\mu} = 1/\beta^2 \tag{26}$$

$\alpha$ is the velocity of P waves, which have their particle motion parallel to the ray and are of compressional typc. S waves, which are shear waves with motion in a plane perpendicular to the ray direction, travel with velocity $\beta$. We see that both P- and S-waves satisfy their own "eikonal" equation. The formal equality of (25),(26) and (12) guarantees that what has been said about ray tracing of acoustic rays is valid in the elastic case as well.

### 2.3 Fermat's principle for seismic rays

An important and very useful principle was formulated by Fermat in the last century, originally for optical rays: the ray path geometry is such that it renders the travel time between two points stationary. We will prove the principle for seismic rays. The travel time $dt$ along a ray segment $d\mathbf{r}$ is given by

$$dt = \frac{|d\mathbf{r}|}{c}$$

where $d\mathbf{r}$ is taken small enough for $c$ to be effectively constant. If we now perturb this ray, such that $d\mathbf{r} \rightarrow d(\mathbf{r} + \delta\mathbf{r})$, the ray will traverse a different velocity $c + \delta c$ and:

$$dt + \delta dt = \frac{|d\mathbf{r} + d\delta\mathbf{r}|}{c + \delta c} = |d\mathbf{r} + d\delta\mathbf{r}|[\frac{1}{c} + \delta(\frac{1}{c})]$$

By writing out, one easily verifies that:

$$|d\mathbf{r}+d\delta\mathbf{r}| = |d\mathbf{r}| + \frac{d\mathbf{r}\cdot\delta d\mathbf{r}}{|d\mathbf{r}|} = \mathbf{n}\cdot(d\mathbf{r}+d\delta\mathbf{r})$$

We therefore have, to first order:

$$\delta dt = \frac{1}{c}\mathbf{n}\cdot d\delta\mathbf{r} + \delta(\frac{1}{c})\mathbf{n}\cdot d\mathbf{r}$$

We find the total travel time perturbation $\delta T$ of a ray between two fixed points A and B by integration:

$$\delta T = \int_A^B \delta(\frac{1}{c})\mathbf{n}\cdot d\mathbf{r} + \int_A^B \frac{1}{c}\mathbf{n}\cdot d\delta\mathbf{r} = \int_A^B \delta(\frac{1}{c})\mathbf{n}\cdot\frac{d\mathbf{r}(s)}{ds}ds + \int_A^B \frac{1}{c}\mathbf{n}\cdot\frac{d\delta\mathbf{r}(s)}{ds}ds \quad (27)$$

In the first integral we may write $\delta(1/c)=\delta\mathbf{r}\cdot\nabla(1/c)$. The second integral can be integrated by parts. This yields, since $\delta\mathbf{r}=0$ in A and B, and $\mathbf{n}=d\mathbf{r}/ds$:

$$\delta T = \int_A^B \delta\mathbf{r}\cdot\left[\nabla(\frac{1}{c}) - \frac{d}{ds}(\frac{1}{c}\mathbf{n})\right]ds$$

For arbitrary $\delta\mathbf{r}$ we have $\delta T=0$ so that:

$$\frac{d}{ds}\left[\frac{1}{c}\frac{d\mathbf{r}}{ds}\right] = \nabla(\frac{1}{c}) \quad (28)$$

which is the same equation as the ray equation (13). Thus, rays have travel times that are stationary for small shifts in the ray location.

## 3. Seismic tomography with delay times

### 3.1 Delay times

Can one derive a velocity model of the interior of the Earth, from travel time measurements at its surface? This is a classical problem in seismology. For a spherically symmetric Earth with velocity increasing monotonically with depth, Herglotz, Wiechert and Bateman showed in the beginning of this century that this is indeed the case. Their classical method of interpretation has survived, in various modifications, until recently. Nowadays, with the increasing computing power available to us, it has become possible to formulate this "inverse" problem for very complicated Earth models using perturbation theory.

For a general, 3-dimensional Earth model, the travel time for a ray is a function of the velocity $c(\mathbf{r})$ and the ray path geometry. Our problem is to derive $c(\mathbf{r})$ from a number of travel time measurements at the surface:

$$T_i = \int_{S_i} \frac{ds}{c(\mathbf{r})} \qquad i=1,...,N \tag{29}$$

(29) is a very complicated problem, since the unknown $c(\mathbf{r})$ is also implicitely present in the ray path $S_i$. This makes the inverse problem highly nonlinear, and nonlinear equations are difficult to solve. Fortunately we have a fairly good idea what $c(\mathbf{r})$ is like, since the spherically symmetric Earth models that have been developed by Jeffreys and Bullen in the 1930's, and by others in more recent times, are able to predict travel times with high accuracy. For teleseismic P waves the deviations rarely exceed 2 seconds on total travel times of 1000 seconds or more. Nevertheless, there are quite systematic deviations from these predictions, and for more nearby events P delays may exceed 5 seconds. This indicates that we can improve on spherically symmetric models, by allowing small deviations from this symmetry, especially in the upper mantle. If we designate the prediction from the starting model by $T_i^0$:

$$T_i^0 = \int_{S_i^0} \frac{ds}{c_0} \tag{30}$$

Where $S_i^0$ is the ray trajectory in the starting model. We define the delay time as

$$\delta T_i = T_i - T_i^0 = \int_{S_i} \frac{ds}{c} - \int_{S_i^0} \frac{ds}{c_0} \approx \int_{S_i^0} (\frac{1}{c} - \frac{1}{c_0}) ds \tag{31}$$

or

$$\delta T_i = -\int_{S_i^0} \frac{\delta c(\mathbf{r})}{c_0(\mathbf{r})^2} ds \tag{32}$$

where $\delta c = c - c_0$. Note that we have used Fermat's principle to substitute the ray path as calculated for the starting model for the (unknown) ray path in the true Earth: we only make a second order error in calculating the time delay. If we choose the starting model symmetric, this makes our calculation very much more efficient.

This principle is widely invoked to justify seismic tomography using the rays of a reference model. One should realize, however, that this is not strictly correct: what Fermat's principle tells us is that we make only a second order error in the direct problem (the calculation of the delay from a model) since the data are insensitive to perturbations in the ray geometry. However, the inverse problem may be ill posed, and there is no guarantee that the second order error in the delay will not blow up to a large error in the solution (see also section 5).

(32) maps a projection of the model into a datum. If a sufficiently good coverage of ray paths is available, analytic solutions of (32) are available (Chapman, chapter 2). In general, this is not so, and we have to take recourse to more general numerical methods, which will be described in section 3.4. A equation similar to (32) turns up in the imaging of the earthquake source process (Ruff, chapter 15).

### 3.2 The width of a ray

So far we have tacitly assumed that the integral in (32) is an integration over a line (ray) in three-dimensional space. This simplifies the mathematics and is so far widely applied in seismic tomography. However, it contradicts our physical intuition, and indeed a small heuristic excursion into the principles of Huygens and Fermat shows that infinitely narrow rays only exist for infinitely small wavelengths $\lambda \to 0$.

Fermat's principle is intimately connected with Huygens' principle and with a principle later deduced by Fresnel: the wave in any point outside a surface $\Sigma$ can be represented as the result of the superposition of coherent secondary waves which are emitted by virtual sources distributed continuously over this surface $\Sigma$. The representation theorems of seismology (Aki and Richards, 1980) are the correct mathematical formulations of this principle, but these shall not be needed here.

We assume a surface $\Sigma$ perpendicular to the ray (figure 3). $SAC$ is the minimum travel time ray path from $S$ to $C$. $SB$ and $BC$ are minimum time paths from $S$ to the virtual source point $B$ and from $B$ to $C$, respectively. As long as the difference in distance $SBC - SAC \leq \lambda/4$ (a quarter wavelength) we can say that the rays interfere constructively, so that the structure in $B$ still influences the wave observed in $C$. We shall adopt this as a measure of the ray width. In *ray coordinates* $s$ (along the ray) and $w$ (normal to the ray), the maximum deviation $w_m(s)$ from the ray of length $L$ is then given by:

$$[w_m(s)^2 + s^2]^{1/2} + [w_m(s)^2 + (L-s)^2]^{1/2} - L = \lambda/4$$

which is an ellipse with $S$ and $C$ as focal points. For $\lambda$=10 km, the maximum ray width now varies from 36 km for a ray length of 1000 km to 112 km for a ray of 10000 km ($w_m \approx \sqrt{\lambda L/8}$).

For large data sets, it is very inefficient to replace the line integral in (32) with a volume integral over some curved ellipsoidal volume. We may however incorporate the finiteness of the ray width in an approximate way, by averaging the solutions $\gamma_k$ over neighbouring cells. This point is further discussed on page 23.

Wielandt (chapter 4) recently discovered a disquieting phenomenon connected with the effects of diffraction: negative delays (faster waves) will easily be detected in the observation point, but the onset of positively delayed waves will drown in the arrival of diffracted waves that are not, or only slightly delayed. This will introduce a bias in tomographic models, in the sense that slow regions may remain undetected. This *Wielandt effect* has so far not been recognized but it is likely to have influenced all published tomographic results, and no way to circumvent the problem has yet been found.

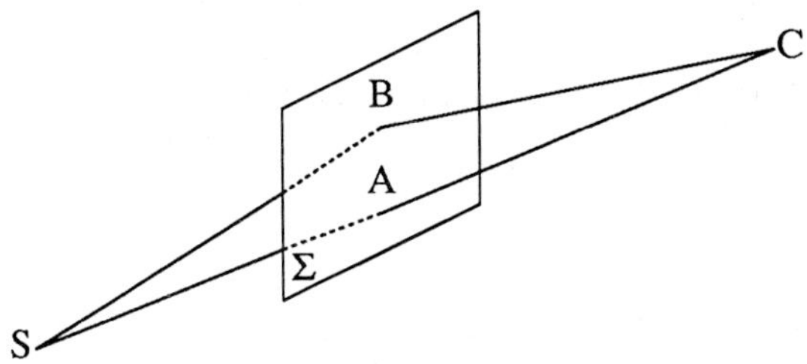


*Figure 3*

### 3.3 The statistics of delay times

As we shall see in section 4, equations (32) can be reduced to a system of linear algebraic equations once a suitable parameterization of the model $\delta c(\mathbf{r})$ is adopted. This system of equations can then be solved, with least-squares fitting or any other criterion.

The least squares estimate is the maximum likelihood estimate for normal (or Gaussian) distributed data errors (see, e.g., Matthews and Walker, 1973). Unfortunately, seismic delay times do not seem to be Gaussian distributed. Reported teleseismic travel times (at a fixed epicentral distance) show a distribution that is both asymmetric and heavily tailed, i.e. there are many more large residuals than one would expect from a Gaussian distribution with its exponential fall-off (Buland, 1984, 1987). Reported travel times are influenced by a combination of random errors arising from the actual observations, systematic biases introduced by the earthquake location algorithm and true delays induced by deviations from spherical symmetry in the Earth. How does one estimate the true delay time from such a set of observations? This problem is an old one in seismology, although the introduction of tomographic interpretation raises some new and interesting questions. Jeffreys (1936) gives the following distribution for the arrival times t at a given epicentral distance:

$$F(t) = \frac{1-\alpha}{(2\pi)^{1/2}\sigma} e^{-(t-t_0)^2/2\sigma^2} + \alpha g(t) \tag{33}$$

where $\sigma$ is the standard deviation, $\alpha$ a (small) constant and $g(t)$ a slowly varying distribution function that is used to explain the large number of statistical outliers. In his pioneering work to determine a spherically symmetric Earth model, and its corresponding travel time tables for seismic waves, Jeffreys developed the method of *uniform reduction*. This method essentially consists of determining the level of the function $g(t)$ in the data, subtracting this level from the histogram and estimating mean and standard deviation from the remaining data. Since $g(t)$ has little effect near $t_0$, where the Gaussian distribution dominates, removal of outliers (very large delay times) from the data set will have a similar effect.

The large number of data in tomographic analysis asks for a simple and practical method of estimation. The simplest way to avoid gross errors to propagate into the solution is to remove outliers. But how large must a delay be to qualify as an outlier? Or, to state it in statistical terms: when is the chance small that the removal of an outlier is not in fact the removal of valid information about the Earth? A thorough statistical analysis of this problem is, to the best of my knowledge, not available at this moment. But we can nevertheless make out a good case for some optimism.

First we note that the true delay is a sum of many individual delays, which the ray acquired as it crossed many different regions in the Earth. The Central Limit theorem of statistics tells us that, if a deviation is the sum of several deviations, then no matter what the probability distribution of the individual deviations may be, their sum will tend more and more to a Gaussian distribution as the number of components increases. Thus we have reason to assume that the *true* delay times are approximately normally distributed. Outliers are then mainly due to observation errors. Truncating at some maximum absolute delay time, or reducing the influence of large delays will therefore increase the "signal-to-noise" ratio of the data.

Essentially, there are two strategies that may be followed that meet the requirement of algorithmic simplicity. The simplest is truncation at some maximum absolute delay time. Ideally, this is done for a delay time $\delta t_{max}$ where the second term in (33) exceeds some predetermined fraction $\gamma$ of the first term, i.e. where

$$\alpha g(t_0+\delta t_{max}) = \gamma \frac{1-\alpha}{(2\pi)^{1/2}\sigma} e^{-\delta t_{max}^2/2\sigma^2} \tag{34}$$

A simple example may illustrate this. If we assume that $g(t)=1/2\tau$ for $-\tau<t<\tau$ and 0 elsewhere, with $\tau$=30 s, $\sigma$=1.2 s, and $\alpha$=0.5 (which is close to parameters used by the International Seismological Centre, as reported by Buland, 1987), we find $|\delta t_{max}|=2.9$ s for $\gamma=1$. Using instead $\sigma=3.2$ s and $\alpha=0.13$, which are values reported by Bolt (1960) we find a much larger $|\delta t_{max}|$ of 8.9 s.

Because the first term in the distribution is exponential, these truncation levels are less sensitive to the choice of $\gamma$ than of $\sigma$. Choosing $\gamma=0.5$ gives $|\delta t_{max}|=2.6$ s, and 8.1 s, respectively. The correct choice of $\sigma$ for determination of $|\delta t_{max}|$ depends on the epicentral distance. $\sigma=1.2$ s is the actual standard deviation of the (carefully selected) ISC data used by Dziewonski (1984) for epicentral distances between 31° and 86°. Both below this range and above this $\sigma$ is higher (e.g. $\sigma=$ 1.6 s at 99°). The model that was derived from these data predicts delay times due to lateral heterogeneity up to 1.8 s, with a true delay $\sigma$ which seems to be close to 0.5 s (Dziewonski, 1984, figure 10). For mantle waves ($\Delta$<25°), standard deviations are probably much larger.

The second way to handle large delay times is to weight the times inversely proportional to their absolute magnitude. It will be shown in the next section that this is equivalent to replacement of the least squares criterion of fit with some other norm.

### 3.4 Model parametrization

Equation (32) constitutes a system of linear equations in $\delta c$. In order to formulate this in a suitable way for computer processing, we divide the Earth into a number $(M)$ of cells, and define:

$$h_i(\mathbf{r}) = v_i^{-1/2} \quad \text{if } \mathbf{r} \textit{ in cell } i$$
$$= 0 \quad \textit{elsewhere} \tag{35a}$$

where $v_i$ is the volume of cell i. The functions $h_i$ are to be seen as a basis, that spans a subspace of the (Hilbert) space of all possible Earth models $c(\mathbf{r})$. If the cells do not overlap, the basis functions are orthonormal:

$$\int_V h_i(\mathbf{r}) h_j(\mathbf{r}) d^3\mathbf{r} = \delta_{ij} \tag{36}$$

An alternative is to expand the Earth's velocity field into a finite number of fully normalized spherical harmonics $Y_l^m(\theta,\phi)$, which form an orthogonal basis as well:

$$h_i(\mathbf{r}) = f_k(r) Y_l^m(\theta,\phi) \tag{35b}$$

where $i = \{klm\}$, a renumbering, and where $f_k(r)$ forms a set of depth functions that is orthonormal over the depth region of interest. This is the approach that has been followed by Dziewonski and others (e.g., Dziewonski, 1984; Morelli and Dziewonski, chapter 11). For low maximum values of $l$ and $k$ this method has the advantage to limit the number of unknown parameters. As a consequence, the matrix may be stored in the computer memory and a full inversion, including the calculation of a posteriori variance in the coefficients of the harmonic expansion, may be performed. A disadvantage is the lack of detail allowed in the Earth model, with a possible decrease of fit to the data since it is very probable that rather detailed heterogeneities exist in the Earth and are reflected in the delay data. Spherical harmonic expansions are probably most useful in the lower mantle and core, where we expect the Earth's heterogeneity to be smoother than in the upper mantle.

Whatever orthogonal basis we choose for the subspace, any function of position in the Earth can be projected on this subspace. For the velocity perturbation we then have:

$$\delta c(\mathbf{r}) = \sum_{k=1}^{M} \gamma_k h_k(\mathbf{r}) \tag{37}$$

where, because of (36):

$$\gamma_k = \int_V \delta c(\mathbf{r}) h_k(\mathbf{r}) d^3\mathbf{r}$$

However, it is not our purpose to decompose a known function into the basis functions $h_i$. Rather, we wish to determine $\delta c(\mathbf{r})$, or $\gamma_k$, from the measurements:

$$\delta T_i = \sum_{k=1}^{M} -\int_{S_i^0} \frac{1}{c_0(\mathbf{r})^2} \gamma_k h_k(\mathbf{r}) ds = \sum_k A_{ik} \gamma_k \tag{38}$$

where

$$A_{ik} = -\int_{S_i^0} \frac{h_k(\mathbf{r})}{c_0(\mathbf{r})^2} ds$$

or, in matrix form:

$$\mathbf{A}\,\boldsymbol{\gamma} = \delta\mathbf{T} \tag{39}$$

The tomographic problem has now been reduced to a discrete system. At this stage, it is trivial to add additional discrete unknowns to the system, such as event origin time and hypocentral coordinate corrections.

### 3.5 Least squares solutions

In general we have many more data (N) than unknowns (M). Since the data have errors, we expect these equations to be inconsistent, so that no exact solution exists. Instead, we shall wish to minimize a measure of the discrepancy in (39):

$$Min \sum_{i=1}^{N} \left| \sum_{j=1}^{M} A_{ij}\gamma_j - \delta T_i \right|^p \qquad (p \geq 1) \tag{40}$$

For the Euclidean norm ($p$=2), we have the familiar *least squares* problem. Differentiating (40) with respect to $\gamma_k$ then yields the *normal equations*:

$$\mathbf{A}^T \mathbf{A}\,\boldsymbol{\gamma} = \mathbf{A}^T \delta\mathbf{T} \tag{41}$$

The least squares criterion is not necessarily the best one to apply for tomographic data. In the following, we shall take a brief excursion into the properties of other norms, and show how these can be reformulated in the same "least squares" form.

Although least-squares methods are widely applied because of their simplicity (the complications that arise from the absolute value disappears when p=2), solutions of (41) tend to be sensitive to the few data with extremely large errors. More generally, the situation can be summarized as follows:

- $p>2$: Dominant influence of outliers. Not desirable.
- $p=2$: Convenient (least squares), but still large influence of outliers.
- $1 \leq p < 2$: Diminishing influence of outliers

Claerbout and Muir (1973) advocate the use of the $p=1$ norm, which leads to the methods of linear programming. Classical linear programming methods are, however, not efficient enough for application in large tomographic inversions.

We can still use least squares and yet circumvent the problems posed by outliers by application of a cut-off criterion such as discussed in section 3.3. In the first iteration we reject those data that differ more than $\delta t_{\max}$ from 0, but in subsequent iterations we may adapt this and reject data that are more than $\delta t_{\max}$ away from the subsequent *model predictions*. If we replace this sharp cut-off by a weighting scheme that decreases the importance of data that are far away from the model predictions it leads to a technique, known as *Iterative Reweighted Least Squares*, or IRLS. IRLS gives approximate solutions to (40) for $p \neq 2$. A lucid review of it is given by Scales et al. (1987). If we differentiate (40) for arbitrary $p$ with respect to $\gamma_k$, we find (using $sgn(r) = r/|r|$):

$$p \sum_{i=1}^{N} r_i A_{ik} \, | r_i |^{p-2} = 0 \tag{42}$$

where $r_i = \sum_j A_{ij} \gamma_j - \delta T_i$, the *residual*. With $\mathbf{R} = diag \, \{ | r_i |^{p-2} \}$ we can write this as:

$$\mathbf{A}^T \mathbf{R} \mathbf{A} \gamma = \mathbf{A}^T \mathbf{R} \delta \mathbf{T} \tag{43}$$

These are the normal equations belonging to the weighted system of linear equations $\mathbf{R}^{1/2} \mathbf{A} \boldsymbol{\gamma} = \mathbf{R}^{1/2} \delta \mathbf{T}$.

In the IRLS method, $\mathbf{R}$ is replaced by $\mathbf{R}_k$, its estimate using the residual vector obtained after k iterations, starting with $\mathbf{R}_0 = \mathbf{I}$. Small residuals $| r_i | < \varepsilon$ are replaced by a cut-off level $\varepsilon$ to avoid instabilities caused by perfectly fitting data. Thus, IRLS allows us to restrict our attention to the solution of least squares systems, and in the following we shall only consider $p = 2$.

Weighting the system (39) is necessary when the data have differing standard deviations. If we denote the covariance matrix of the data by $\mathbf{C}_d$ this can be achieved by multiplying both sides of (39) by $\mathbf{C}_d^{-1/2}$, to avoid that low quality data dominate in the minimization of the residual. In general, $\mathbf{C}_d$ will be diagonal by lack of better, non-diagonal $\mathbf{C}_d$ are awkward to handle in large systems anyhow. The least squares solution then involves the minimization of:

$$Min \, (\mathbf{A} \boldsymbol{\gamma} - \delta \mathbf{T})^T \mathbf{C}_d^{-1} (\mathbf{A} \boldsymbol{\gamma} - \delta \mathbf{T}) \tag{44}$$

### 3.6 Non-unique solutions

A problem that is generally encountered in tomographic studies is that the solution of the system $Min \, | \mathbf{A} \boldsymbol{\gamma} - \delta \mathbf{T} |^2$ is non-unique. Of course we may try to increase the cell size up to the point where the number of cells is equal to the rank of the least squares matrix. Such a coarse model will not be able to give a good representation of the Earth. Rather we prefer to work with small cells and deal with the non-uniqueness in another way. To this end we must formulate additional criteria to select our preferred model among all models that give a best data fit. Nolet (1985) proposes to minimize:

$$Min \int_V \delta c \, (\mathbf{r})^2 \, dV \tag{45}$$

which leads, with the orthogonality (36), to:

$$Min \sum_{k=1}^{M} \gamma_k^2 \tag{46}$$

for an evenly distributed set of rays throughout the model, this will adequately correct for the effects of differing cell volumes. A simple example can illustrate this. Suppose we have one ray that crosses two cells with lengths $l_1$ and $l_2$, respectively (figure 4). For slownesses $s_1$ and $s_2$ we than have just one linear equation: $l_1 s_1 + l_2 s_2 = \delta T$. If we minimize $s_1^2 + s_2^2$ we find, with the method of Lagrange multipliers, that $s_1 = \delta T l_1 / (l_1^2 + l_2^2)$ and $s_2 = \delta T l_2 / (l_1^2 + l_2^2)$. Thus, as an unwanted effect of differing cell lengths we find that the slownesses are unevenly distributed over the two cells: $s_1 / s_2 = l_1 / l_2$. Usually we should

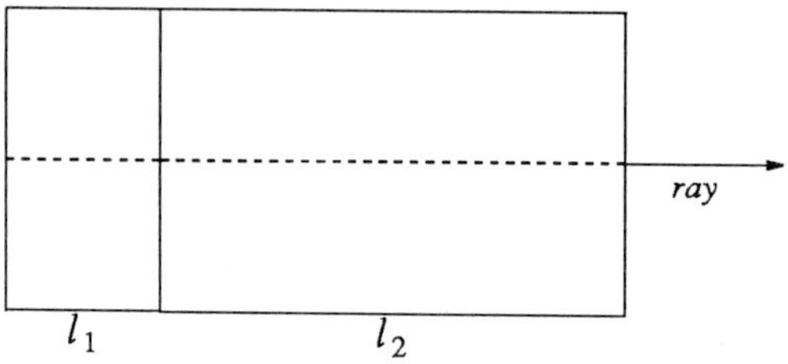


*Figure 4* The influence of cell size on the solution. Without scaling the slownesses in the two cells will be proportional to $l_i$ to satisfy the delay of just one ray, introducing unwarranted heterogeneity.

prefer the two slownesses to be equal, in the absence of any other evidence of lateral heterogeneity. If, instead, we scale $s_i = \gamma_i h_i$ we find that the minimization of $\gamma_1^2+\gamma_2^2$ results in $s_1/s_2 = l_1 h_1^2 / l_2 h_2^2$. If we impose a condition of "minimum heterogeneity" we have $s_1 = s_2$, and we must scale $h_i = l_i^{-1/2}$ to obtain this equality. When there are more rays, the total ray length in a cell will be of the order $Nl_i$, where N is the number of rays. Since N is expected to be proportional to the cross-section $\Delta O$, we see that the total ray length is proportional to the cell *volume* , so that we must scale $h_i = v_i^{-1/2}$. This is exactly what was proposed in (35a).

We may also consider to scale with the *actual* ray length in the cell instead of the expected one. This will indeed further reduce systematic effects due to the uneven distribution of ray paths through the model.

Franklin (1970), Jackson (1979), Tarantola and Nercessian (1984) have - each in their own way - proposed a more elaborate minimum model criterion. Instead of (46) we can minimize:

$$Min \iint \delta c(\mathbf{r}')\mathbf{C}_\gamma^{-1}(\mathbf{r},\mathbf{r}')\delta c(\mathbf{r}) d^3\mathbf{r}' d^3\mathbf{r} \tag{47}$$

where $\mathbf{C}_\gamma^{-1}(\mathbf{r},\mathbf{r}')$ is the a priori covariance function of the model $\delta c(\mathbf{r})$. As an example, if we assume a uniform, isotropic correlation length for the lateral heterogeneity in the model, we could take $\mathbf{C}_\gamma(\mathbf{r},\mathbf{r}')=\sigma^2\exp[-\frac{1}{2}|\mathbf{r}-\mathbf{r}'|^2/L]$. Note that $\mathbf{C}_\gamma$ is a smoothing operator, so that, conversely, $\mathbf{C}_\gamma^{-1}$ will "roughen" the model when it operates on it. (47) imposes a minimum norm condition on this roughened model. The effect of $\mathbf{C}_\gamma^{-1}$ is thus to emphasize the minimization of short wavelength changes.

With cell parametrization, (47) becomes:

$$Min \sum_{ij} \gamma_i \gamma_j \iint h_i(\mathbf{r}')\mathbf{C}_\gamma^{-1}(\mathbf{r},\mathbf{r}')h_j(\mathbf{r}) d^3\mathbf{r}' d^3\mathbf{r}$$

For small cell sizes, we can replace $\mathbf{C}_\gamma^{-1}(\mathbf{r},\mathbf{r}')$ by a constant factor $(\mathbf{C}_\gamma^{-1})_{ij}$ and obtain the condition:

$$Min \sum_{ij} \gamma_i v_i^{-1/2} (\mathbf{C}_\gamma^{-1})_{ij} v_j^{-1/2} \gamma_j$$

If we redefine $\tilde{\mathbf{C}}_\gamma = v_i^{1/2} v_j^{1/2} (\mathbf{C}_\gamma)_{ij}$ we obtain:

$$Min\ \boldsymbol{\gamma}^T \tilde{\mathbf{C}}_\gamma^{-1} \boldsymbol{\gamma} \tag{48}$$

We may now define

$$\tilde{\boldsymbol{\gamma}} = \mathbf{C}_\gamma^{-1/2} \boldsymbol{\gamma} \qquad \tilde{\mathbf{A}} = \mathbf{A}\mathbf{C}_\gamma^{1/2}$$

and solve the transformed system:

$$Min\ |\tilde{\mathbf{A}}\tilde{\boldsymbol{\gamma}} - \delta\mathbf{T}| \tag{49}$$

with a minimum norm condition now imposed on $\tilde{\boldsymbol{\gamma}}$ instead of $\boldsymbol{\gamma}$, and find the unique untransformed solution through $\boldsymbol{\gamma} = \mathbf{C}_\gamma^{1/2}\tilde{\boldsymbol{\gamma}}$. Note that the inverse of $\mathbf{C}_\gamma^{1/2}$ is not needed in this procedure. We may even simply define $\mathbf{C}_\gamma^{1/2}$ as a convenient smoothing matrix.

This approach differs in this final step from the tomographic method proposed by Tarantola and Nercessian (1984). These authors obtain a regularized system of equations by minimizing instead of (44) a weighted sum of data misfit and model norm:

$$Min\ (\mathbf{A}\boldsymbol{\gamma} - \delta\mathbf{T})^T \mathbf{C}_d^{-1} (\mathbf{A}\boldsymbol{\gamma} - \delta\mathbf{T}) + \boldsymbol{\gamma}^T \mathbf{C}_\gamma^{-1} \boldsymbol{\gamma}^T \tag{50}$$

of which the solution is

$$\boldsymbol{\gamma} = (\mathbf{A}^T \mathbf{C}_d^{-1} \mathbf{A} + \mathbf{C}_\gamma^{-1})^{-1} \mathbf{A} \mathbf{C}_d^{-1} \delta\mathbf{T} \tag{51a}$$

Because of the equality $\mathbf{A}^T \mathbf{C}_d^{-1}(\mathbf{A}\mathbf{C}_\gamma \mathbf{A}^T + \mathbf{C}_d) = (\mathbf{A}^T \mathbf{C}_d^{-1} \mathbf{A} + \mathbf{C}_\gamma^{-1}) \mathbf{C}_\gamma \mathbf{A}^T$ this can also be written as:

$$\boldsymbol{\gamma} = \mathbf{C}_\gamma \mathbf{A}^T (\mathbf{A}\mathbf{C}_\gamma \mathbf{A}^T + \mathbf{C}_d)^{-1} \delta\mathbf{T} \tag{51b}$$

which does involve either the calculation of $\mathbf{C}_\gamma^{-1}$ and the inverse of another $M \times M$ matrix, or the computation of an even larger $(N \times N)$ matrix inverse. Neither of these options is attractive in large scale tomography with $M \sim 10^4$ and $N \sim 10^6$, and this strategy is only feasible by making gross approximations to these inverses. But (50) can also be recognized as the minimization problem belonging to the linear system

$$\begin{pmatrix} \mathbf{C}_d^{-1/2}\mathbf{A} \\ \mathbf{C}_\gamma^{-1/2} \end{pmatrix} \boldsymbol{\gamma} = \begin{pmatrix} \mathbf{C}_d^{-1/2}\delta\mathbf{T} \\ \mathbf{0} \end{pmatrix} \tag{52}$$

which can be solved - after we specify $\mathbf{C}_\gamma^{-1/2}$ - with any of the row action methods available for large matrices (see next section). The specification of $\mathbf{C}_\gamma^{-1/2}$ is less of a problem than it seems at first sight, when we realize that it is essentially a roughening operator. The simplest roughening operator takes the difference of a cell with all its neighbours. If we impose the condition:

$$Min \sum_i \frac{1}{2N_i} \sum_{neighb} (\gamma_i - \gamma_j)^2$$

where $N_i$ is the number of neighbours of cell $i$, this results in $M$ extra equations:

$$x_k - \sum_{neighb} \gamma_j = 0 \quad (k = 1, \ldots, M)$$

which form the bottom part of (52), enabling us to identify a convenient form of $\mathbf{C}_\gamma^{-1/2}$.

```
subroutine pstomo(m,n,x,u,v,w,itmax)
#
# subroutine to solve the linear tomographic problem Ax=u using the
# lsqr algorithm (C.C.Paige and M.A.Saunders, ACM Trans.Math.Softw.
# 8, 43 - 71 and 195 - 209, 1982).
#
# Input: m is the number of data, n the number of unknowns, u cont-
#    ains the data (is overwritten), itmax is the nr of iterations.
# Output: x is the solution
# Scratch: arrays v(n) and w(n)
# Subroutines: routines avpu and atupv to be supplied by the user.
#    avpu(m,n,u,v) computes u=u+A*v for given input u,v
#    atupv(m,n,u,v) computes v=v+A(transpose)*u for given u,v
#
dimension x(n),u(m),v(n),w(n)
open (9,file='tomo.int',form='unformatted')
do i=1,n { x(i)=0; v(i)=0 }                         # initialize
call normlz(m,u,beta); b1=beta;
call atupv(m,n,u,v); call normlz(n,v,alfa)
rhobar=alfa; phibar=beta; do i=1,n { w(i)=v(i) }
write(6,*) 0,x(1),beta,1
do iter =1,itmax {                              # repeat
   a= - alfa; do i=1,m { u(i)=a*u(i) }              # bidiagonalization
   call avpu(m,n,u,v); call normlz(m,u,beta);
   b= - beta; do i=1,n { v(i)=b*v(i) }
   call atupv(m,n,u,v); call normlz(n,v,alfa)
   rho=sqrt(rhobar*rhobar+beta*beta)                # modified QR
   c=rhobar/rho; s=beta/rho; teta=s*alfa;
   rhobar= - c*alfa; phi=c*phibar; phibar=s*phibar;
   t1=phi/rho; t2= - teta/rho
   do i=1,n { x(i)=t1*w(i)+x(i); w(i)=t2*w(i)+v(i) } # update
   r=phibar/b1; write(6,*) iter,x(1),phibar,r
   write(9) iter,x,v,alfa,beta,phibar,r             # intermediate output
   }
return; end                                     # return
#
subroutine normlz(n,x,s)                        # normalizes vector x
dimension x(n)
s=0.; do i=1,n { s=s+x(i)**2 }
s=sqrt(s); ss=1./s; do i=1,n { x(i)=x(i)*ss }
return; end
```

*Figure 5* A very simple Ratfor (Rational Fortran) version of the lsqr algorithm. Output variables phibar and r are the absolute and relative lengths of the residual vector Ax-u. Vector v and scalars $\alpha,\beta$ correspond to the quantities described by Van der Sluis and Van der Vorst (chapter 3).

### 3.7 Solving large systems of equations

Seismic tomography would not exist if we had not recourse to numerical methods that solve the very large, sparse systems without the need to store the whole matrix in memory but that rather solve the system using one row at the time. Some of these methods have been briefly compared by Neumann-Denzau and Behrens (1984) and Nolet (1985).

My own preferred method is the LSQR variant of the Conjugate Gradient algorithm (Paige and Saunders, 1982). In (Nolet, 1984, 1985) I showed that the LSQR algorithm was superior to a SIRT type of method of Dines and Lyttle (1979), both in its suppression of the propagation of data errors and in the rate of convergence to the true solution of a 400×200 system of linear tomographic equations. These findings were later confirmed in a theoretical analysis by Van der Sluis and Van der Vorst (chapter 3) and a test on a large tomographic system by Spakman and Nolet (work in prep.). For details about these matrix solvers I refer the reader to chapter 3. My simple Ratfor version of the LSQR algorithm is reproduced in figure 5, and it should give no difficulties to translate this into any other programming language.

Whatever method we use to find an approximate solution $\hat{\boldsymbol{\gamma}}$, it can be formally expressed as a linear combination of the data:

$$\hat{\boldsymbol{\gamma}} = \mathbf{A}^{-}\delta\mathbf{T} \tag{53}$$

where, in general, neither $\mathbf{A}^{-}\mathbf{A}$ nor $\mathbf{A}\mathbf{A}^{-}$ equals the identity matrix, although, of course, we hope that $\mathbf{A}^{-}\mathbf{A} \approx \mathbf{I}$. Here we shall split $\delta\mathbf{T}$ into an uncontaminated data vector $\delta\ddot{\mathbf{T}}$ and a noise, or error, vector $\mathbf{e}$:

$$\delta\mathbf{T} = \delta\ddot{\mathbf{T}} + \mathbf{e} = \mathbf{A}\boldsymbol{\gamma}_{true} + \mathbf{e} \tag{54}$$

where $\boldsymbol{\gamma}_{true}$ represents the "true" model of the Earth. Substitution in (53) and subtraction of the true model gives the error in the solution $\hat{\boldsymbol{\gamma}}$:

$$\hat{\boldsymbol{\gamma}} - \boldsymbol{\gamma}_{true} = (\mathbf{A}^{-}\mathbf{A} - \mathbf{I})\boldsymbol{\gamma}_{true} + \mathbf{A}^{-}\mathbf{e} \tag{55}$$

The result of a tomographic experiment is thus contaminated by two kinds of errors: lack of resolution, and the fact that we will never continue iterations to the very end are the cause that $\mathbf{A}^{-}\mathbf{A} \neq \mathbf{I}$, and errors in the data propagate into the solution as $\mathbf{A}^{-}\mathbf{e}$. The covariance matrix of the model error is given by:

$$\mathbf{C}(\hat{\boldsymbol{\gamma}} - \boldsymbol{\gamma}_{true}) = (\mathbf{A}^{-}\mathbf{A} - \mathbf{I})\mathbf{C}_{\gamma}(\mathbf{A}^{-}\mathbf{A} - \mathbf{I})^{T} + \mathbf{A}^{-}\mathbf{C}_{d}(\mathbf{A}^{-})^{T} \tag{56}$$

Franklin (1970) and Jackson (1979) showed that the formal solution (51) can be obtained by minimizing the diagonal elements of $\mathbf{C}(\hat{\boldsymbol{\gamma}} - \boldsymbol{\gamma}_{true})$ if the data and model are uncorrelated.[1] Wielandt (chapter 4) argues that there may be correlations between the two, in which case (51b) might be replaced by

1. If we assume that both $\boldsymbol{\gamma}$ and $\delta\mathbf{T}$ are Gaussian, (51) is the maximum likelihood estimate

$$\boldsymbol{\gamma} = (\mathbf{C}_{\gamma\gamma}\mathbf{A}^T + \mathbf{C}_{\gamma d})(\mathbf{A}\mathbf{C}_{\gamma\gamma}\mathbf{A}^T + \mathbf{A}\mathbf{C}_{\gamma d} + \mathbf{C}_{\gamma d}^T\mathbf{A}^T + \mathbf{C}_{dd})^{-1}\delta\mathbf{T} \tag{57}$$

with an obvious change of notation.

That $\mathbf{A}^-\mathbf{A}$ is a measure of the resolution can also be seen in another way. If we take $\mathbf{e}$=$\mathbf{0}$ (no errors in the data), any deviation of $\hat{\boldsymbol{\gamma}}$ from $\boldsymbol{\gamma}_{true}$ is due to lack of resolution, due to lack of data, incomplete convergence or both. Since $\hat{\boldsymbol{\gamma}}=\mathbf{A}^-\delta\mathbf{T}=\mathbf{A}^-\mathbf{A}\boldsymbol{\gamma}_{true}$ we see that $\mathbf{A}^-\mathbf{A}$ acts as a "blurring window" through which we view the true Earth.

Note that the iterative solvers such as LSQR do not give us $\mathbf{A}^-$ explicitly, so that we cannot calculate $\mathbf{A}^-\mathbf{A}$ directly. Nakanishi and Suetsugu (1986) propose to solve $\mathbf{A}\mathbf{a}^{(i)}=\hat{\mathbf{n}}^{(i)}$ for successive unit vectors $\hat{\mathbf{n}}^{(i)}$ to construct the M columns $\mathbf{a}^{(i)}$ of $\mathbf{A}^-$. This becomes very unattractive as the number of data grows, since it requires $M$ inversions.

Nolet (1985) develops a resolution analysis that requires only one inversion for every point in the model where we wish to know the precision. He also adds the constraint that the row sum of $\mathbf{A}^-\mathbf{A}$ be equal to 1, so that the resolution matrix gives a true physical average when it operates on $\boldsymbol{\gamma}_{true}$, again to avoid the introduction of unwarranted heterogeneity. This is the discrete analogon of the type of resolution analysis developed by Backus and Gilbert (1970) and Chou and Booker (1979), and can be viewed as the sparse matrix analogon of the method of winnowing proposed by Gilbert (1972). For very large systems, this resolution estimation will also become prohibitively expensive, and a simpler sensitivity analysis, with synthetic data from a known model seems the only possibility.

## 4. Surface wave tomography

As is well known, dispersive solutions to (1) exist if the Earth is horizontally layered and the seismic velocities increase with depth (Aki and Richards, 1980, chapter 7). Because the energy of these waves is concentrated near the free surface of the Earth, these waves are known as *surface waves*. Surface waves exist in two types: Love waves, with a horizontal movement transverse to the direction of wave propagation, and Rayleigh waves with ellipsoidal motion in a vertical plane. The various modes ($n$) of these waves travel with phase velocity $C_n(\omega)=\omega/k_n(\omega)$ and group velocity $U_n(\omega)=[dk_n(\omega)/d\omega]^{-1}$. In this chapter we shall limit ourselves to measurements of just one mode, and remove the index $n$. Extensions to multi-mode data are trivial.

Because lateral homogeneity is a basic assumption in the derivation of the surface wave formalism, the starting model is always laterally homogeneous, and wavepaths are assumed to be straight lines on a flat surface, or great circles on a sphere. Provided a sufficiently dense coverage of wavepaths is available, surface wave arrival times (at a fixed frequency) ar a prime candidate for inversion with more analytical methods, such as described by Chapman (chapter 2).

Assuming that local deviations from this merely affect the propagation velocity of the surface waves, and leave the wave character largely intact, we seek *local* phase and/or group velocity curves that are a function of position, e.g. longitude $\phi$ and colatitude $\theta$.

Since low-frequency surface wave have a larger penetration (or skin) depth than those of high frequency, frequency takes over the role of the depth coordinate. The interpretation of $C(\omega)$, $U(\omega)$, or both, for a set of frequencies in terms of a local velocity structure as a function of depth, is non-unique. Nevertheless the average S-velocity over a depth interval of some 100 km can be established with high precision (Nolet, 1978).

Techniques to measure phase- and group velocities have been described in detail by Dziewonski and Hales (1972) for fundamental modes, and by Nolet and Panza (1976) for higher modes of surface waves. We now assume that the perturbation in the phase velocity acquired over path $i$ is the average of the local phase perturbations $\delta C(\omega,\theta,\phi)$ over the path of length $L_i$:

$$\delta C_i(\omega) = \frac{1}{L_i}\int_{S_i^0} \delta C(\omega,\theta,\phi)\, ds \tag{58}$$

which is very similar to the tomographic equation for delay times (32). For group velocities, one may substitute $U$ for $C$ in (58).

Whatever the data, it is obvious that (58) leads to a tomographic problem very much similar to the one derived for delay times. The difference now is that the surface wave problem (58) is 2-dimensional, but frequency dependent, whereas the delay time problem is simply 3-dimensional. We may however make use of the fact that a fairly accurate perturbation theory exists to map the variation of S-velocity $\beta$ with depth $z$ into a perturbation of the phase velocity (see also chapter 14):

$$\delta C(\omega,\theta,\phi) = \int_0^\infty \left[\frac{\partial C}{\partial \beta}\right]_{\omega,z} \delta\beta(\theta,\phi,z)\, dz \tag{59}$$

which, when substituted into (58), results in a complete 3-D tomographic formulation to retrieve the S-velocity $\beta(\theta,\phi,z)$:

$$\delta C_i(\omega) = \frac{1}{L_i}\int_{S_i^0}\int_0^\infty \left[\frac{\partial C}{\partial \beta}\right]_{\omega,z} \delta\beta(\theta,\phi,z)\, dz\, ds \tag{60}$$

This substitution is not strictly necessary, and one may also first solve the 2-D tomographic problem at a number of frequencies for the local phase of group velocities, and invert these for local structures afterwards (e.g. Nakanishi and Anderson, 1984). Yanovskaya et al. (1987) give a complete method that utilizes this approach.

Alternatively, one may use the linearized relationship between perturbations in the model and in phase velocities, to construct synthetic seismograms for laterally heterogeneous Earth models. Woodhouse and Dziewonski (1984) have done this for very low frequency surface waves, and obtained spectacular images of the Upper Mantle. Because of the limited frequencies involved (periods larger than 135 s) these authors were able to keep the problem linear, at the expense of limited resolution. For higher frequencies, the product $L_i\delta C_i$ exceeds $\pi/2$, and terms $\exp[iL_i\delta C_i]$ cannot be linearized satisfactorily. Nolet et al. (1986) and Nolet (Chapter 13) formulate this tomographic problem in a form suitable for attack by methods of nonlinear optimization.

For a correct inversion of surface wave waveforms, the effects of focusing and defocusing of the surface waves must be taken into account. This is the subject of a paper by Jobert and Jobert (Chapter 12). With sufficient data available, scattered and diffracted surface waves can also be imaged. Snieder (Chapter 14) discusses a small scale experiment in surface wave "holography". Many aspects of surface wave propagation in laterally heterogeneous structures are summarized in Keilis-Borok (1986).

## 5. Concluding remarks

So far we have described the basic principles of seismic tomography. It is clear that there are, as yet, many unsolved problems. The most important problem is the estimation of the reliability of the result, and none of the methods proposed so far can be judged entirely satisfactory in this respect. Geodynamical processes often give rise to minor deviations in the seismic velocities, and it is these small variations we are after. The data have standard errors which are of the same order of magnitude as the time shifts incurred by these small velocity deviations, and the way out is to average over as many data as we can get, that is, to work with giant overdetermined systems. In solving this system, both errors in the data and errors due to insufficient convergence of the iterative solver find their way into the final solution.

The analysis by Van der Sluis and Van der Vorst (chapter 3) shows that it is more useful to increase the precision of the measurements than to increase the multiplicity of the data, also because the precision of the data largely determines the number of iterations needed to solve the system in an adequate way. The satellite beacons described by Poupinet (chapter 10) provide a promising technological advance for the measurement of delay times.

Some more is known about the validity of the application of Fermat's principle. In systems of smaller size, we may repeat the inversion after the first heterogeneous model is obtained, by performing the full ray-tracing calculations through 3-D heterogeneous models. Such an approach has been followed by Thomson and Gubbins (1982), Koch (1985) Williamson (1986) and Nakanishi and Yamaguchi (1986). In general, velocity variations of a few percent can be accommodated in the linear approximation without serious consequences. Koch as well as Williamson report an increase in resolution when ray bending due to lateral heterogeneity is taken into account. Nakanishi and Yamaguchi, however, reported a failure of the (nonlinear) iterations to converge when the data errors are large. Tarantola (chapter 6) describes the tomographic problem in a more general context of nonlinear inverse problems.

Delay times, reported by the ISC, are with respect to the Jeffreys-Bullen model. The fact that this model does not possess a low velocity layer makes this a rather inadequate background model, especially for S waves in the upper mantle. In this case nonlinearities may indeed become important. Strong deviations between the ray geometry of the model and that in the real Earth can be avoided by recalculating the ISC delay times for a more realistic background model such as 1066B (Gilbert and Dziewonski, 1976) or a more local 1-D model if that is known.

In surface wave analyses, phase and group velocities of Love and Rayleigh waves are often considered incompatible, even if allowance is made for detailed inhomogeneous structures (Levshin and Ratnikova, 1984). The common explanation is to invoke seismic anisotropy in the lithosphere, or even the whole upper mantle, to explain this discrepancy. In a way very similar to the imaging of isotropic velocities, the local anisotropy may be inferred from surface wave data - although it is clear that the model will become even more underdetermined. The effects of anisotropy in the Earth have been the subject of tomographic studies by Nataf et al. (1986), Montagner (1985), Montagner and Nataf (1986) and Tanimoto and Anderson (1984).

Very little tomography has yet been attempted on the other important seismic parameter, the Quality factor Q. Here the difficulties in obtaining accurate measurements, as well as the inextricable trade-off with the effects of inhomogeneity (Levshin, 1985) seem to pose insurmountable difficulties. Nevertheless, Hashida and Shimazaki (1985) have obtained a tomographic image of the anelasticity beneath the Kanto district in Japan from intensity reportings.

*Chapter 2*

# The Radon transform and seismic tomography

*C.H. Chapman*

In this chapter, the Radon transform and its inverse are reviewed. Both the filtered back projection and circular harmonic decomposition methods for the inverse transform are discussed. It is shown how the circular harmonic decomposition method can be extended to any laterally homogeneous reference model. The solution is a Volterra integral equation that can be solved efficiently by iteration for low-wavenumber anomalies. The filtered back-projection method can be extended to inhomogeneous models. The asymptotic solution gives the discontinuities or high-wavenumber anomalies.

## 1. Introduction.

The Radon transform (RT) of a function in 2-dimensions consists of its integral along straight lines. This operation corresponds to many physical experiments in medicine, crystallography, geophysics, etc. In general we refer to these as projection experiments. An excellent review of the RT and its applications has recently appeared in the textbook by Deans (1983). The transform is named after Radon (1917) who first derived an inverse transform (IRT). Although the RT describes many physical experiments, it is only recently that the transform and its inverse have been 'discovered' by physicists. The RT is intimately related to the Fourier transform (FT). In many applications, projection data were and are interpreted using FT theory or numerical inversion techniques. The IRT is widely used in medicine where the ray geometry and data distribution are almost perfect but less frequently in other fields. Although the IRT is an elegant mathematical solution, the alternative methods are often preferred when interpreting real projection data as FT theory is well understood and developed, e.g. the fast Fourier transform, and numerical inversion schemes are more easily generalized to curved ray geometries and imperfect data



*G. Nolet (ed.), Seismic Tomography, 25–47.*

distributions. Nevertheless, it is valuable to investigate the IRT and its extensions further. The RT and IRT are real integrals that are physically easy to interpret and intuitively straightforward. Numerical problems that arise with other inversion techniques can often be anticipated and investigated using analytic inverse methods even if the latter are not suitable for inverting realistic datasets.

The RT or its extension to curved line integrals arises in several problems in seismology. This book deals with various applications of tomography to seismology. In tomography, signals propagating along ray paths (at least approximately, in the high-frequency limit) integrate some property of the model, e.g. slowness or slowness anomaly, attenuation, etc. Multiple ray paths in many directions throughout the model provide sufficient information to reconstruct the model. Tomography on all scales has been investigated. In this book, tomography on a global scale using teleseismic, travel-time anomalies and surface wave velocities is discussed as well as on a small scale in cross-borehole and VSP explorations. The RT arises in other problems in seismology. It can be used in the computation of synthetic seismograms as an alternative to FTs for inverting the waveslowness transform integral (Chapman 1978). The inverse operation is called slant (or velocity) stacking (Schultz and Claerbout 1978) and is used to obtain a plane-wave seismogram section in the 'tau-p' domain. In studying earthquake mechanisms with large fault zones, the signal at any one time and location can be expressed as an integral over a line of constant travel-time on the fault surface (Boatwright 1979), i.e. a generalized RT on the fault surface. Finally, we might mention the application of the generalized IRT to seismic migration and inverse scattering (Beylkin 1985, Cohen et al. 1986).

In this chapter we briefly review the standard RT and IRTs. Two alternative methods of inversion are discussed: the filtered back projection (FBP) and the circular harmonic decomposition (CHD) methods. Although these inverse formulae must be identical for exact, complete datasets, for realistic, noisy and incomplete datasets, their (numerical) properties are very different. Note that we use the terminology FBP and CHD to refer to two alternative, exact, analytic IRTs and qualify the names to 'discrete FBP' etc. when referring to the numerical implementation of the methods for discrete datasets (often other authors use the term FBP to refer to the numerical algorithm).

The standard RT applies to the projection experiment with straight rays. Thus the ray geometry is defined by a homogeneous reference model. We refer to this as the tomographic problem with a 0-dimensional (0-D) reference model. Note this only refers to the reference model that defines the ray paths - the inverse transform will reconstruct a 2-D anomaly. We also discuss extensions of the RT to inhomogeneous reference models. We consider 1-D and 2-D reference models. The former is widely used in seismology with velocity a function of only depth or radius. It generalizes the standard RT but retains the advantage that ray paths are translationally invariant. This results in theoretical and computational advantages. In a 2-D reference model, the ray geometries are completely general and it is expensive but straightforward to solve the kinematic ray equations for the ray paths. In this chapter we discuss two extensions of the IRT: a low-wavenumber solution for 1-D reference models, and a high-wavenumber solution for 2-D reference models.

We have not included any discussion of the 3-D tomographic problem. We know of no analytic solutions that are not trivial extensions of the 2-D methods. It is perhaps worth commenting that the 3-D RT and its generalizations are not relevant to 3-D tomography. The 3-D RT is the integral of a function over planes (or 2-D surfaces) whereas 3-D seismic tomography still refers to line integrals but in a 3-D model.

In addition we include two short sections on the Earth flattening transformation and fan-beam geometry. The Earth flattening transformation allows the tomographic problem with a 1-D polar reference model to be transformed into a cartesian model. Fan-beam geometry is a term used in medical tomography to refer to the situation where data at many projection angles are collected with one source. Analytically, it only corresponds to a reparameterization of the projection data with different variables. In seismology it is analogous to parameterizing data as common source (or receiver) point rather than common mid-point.

## 2. The Radon Transform (RT).

The Radon transform of the function $f(\mathbf{x})$, where $\mathbf{x}$ is the position vector in 2-dimensions, is defined as

$$\hat{f}(p,\phi) = \int_L f(\mathbf{x}) ds \tag{1}$$

where the straight line $L$ is defined by the equation

$$p = x\cos\phi + y\sin\phi = r\cos(\theta - \phi) \tag{2}$$

with $\mathbf{x} = (x,y) = (r,\theta)$ in cartesian and polar coordinates. The geometry and coordinates are illustrated in Figure 1. Several seismic observations can be expressed as line integrals of this form (1). We mention three. If $f(\mathbf{x})$ is the slowness, then $\hat{f}(p,\phi)$ is the travel-time, i.e.

$$T(p,\phi) = \int_L u(\mathbf{x}) ds \ . \tag{3}$$

Unfortunately, of course, seismic rays are rarely straight lines and expression (3) is a poor approximation. A better approximation may be to consider the travel-time anomaly, i.e.

$$\delta T(p,\phi) = \int_L \delta u(\mathbf{x}) ds \tag{4}$$

where $T = T_0 + \delta T$ and $u = u_0 + \delta u$. The reference travel times

$$T_0(p,\phi) = \int_{L_0} u_0(\mathbf{x}) ds \tag{5}$$

are calculated for the ray paths appropriate to the reference slowness, $u_0(\mathbf{x})$, and the straight rays are only assumed when interpreting the anomalies (4). Finally, the function $f(\mathbf{x})$ may be related to the attenuation, when $\hat{f}(p,\phi)$ will be the $t^*$ function, i.e.

$$t^*(p,\phi) = \int_L \frac{u(\mathbf{x})}{Q(\mathbf{x})} ds \ . \tag{6}$$

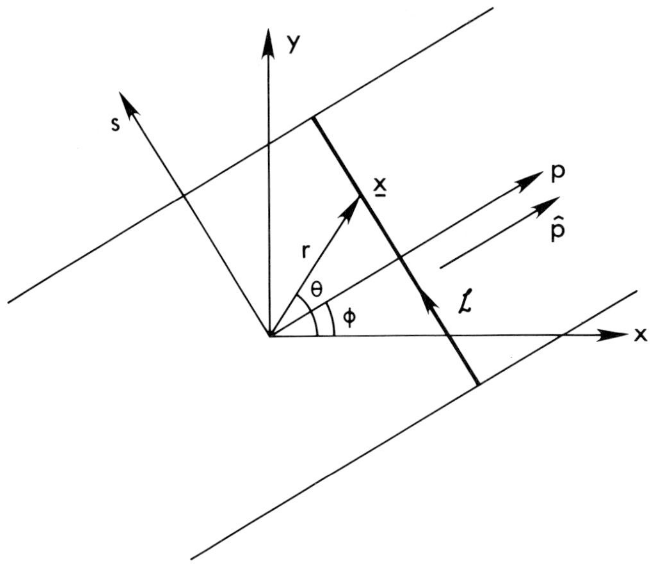


*Figure 1* The geometry and variables in the projection experiment with straight rays. The cartesian coordinates $(x,y)$ and polar coordinates $(r,\theta)$ define the position vector $\mathbf{x}$ in the model. The projection angle $\phi$ defines the coordinates $(p,s)$. The projection experiment integrates the model in the direction of the $s$-coordinate along the line $L$ giving the projection data at coordinates $(p,\phi)$.

In what follows, we refer to the functions, $f(\mathbf{x})$ and $\hat{f}(p,\phi)$, as the model and projection data, respectively. Expression (1) is called the tomography or projection experiment. When $\hat{f}(p,\phi)$ is used to deduce a 'model', i.e. the inverse of (1), we refer to the image or reconstruction.

Several formulae exist for inverting the transform (1), i.e. obtaining $f(\mathbf{x})$ from $\hat{f}(p,\phi)$. Although analytically these are all equivalent, numerically for discrete data they are quite different. These differences are indicative of difficulties that will arise with numerical inversion schemes. The transform (1) is named after Radon (1917) who first derived an inverse transform. His formula is easily obtained using Fourier transform theory.

We define the 2-dimensional FT of the model

$$\tilde{f}(\mathbf{k}) = \iint_{-\infty}^{\infty} f(\mathbf{x}) e^{-2i\pi\mathbf{k}\cdot\mathbf{x}} d\mathbf{x} \tag{7a}$$

and its inverse

$$f(\mathbf{x}) = \iint_{-\infty}^{\infty} \tilde{f}(\mathbf{k}) e^{2i\pi\mathbf{k}\cdot\mathbf{x}} d\mathbf{k} \ . \tag{7b}$$

Similarly, the 1-D FT of the projection data is

$$\check{f}(k,\phi) = \int_{-\infty}^{\infty} \hat{f}(p,\phi) e^{-2i\pi kp}\, dp \tag{8a}$$

and the inverse

$$\hat{f}(p,\phi) = \int_{-\infty}^{\infty} \check{f}(k,\phi) e^{2i\pi kp}\, dk \ . \tag{8b}$$

The Radon transform (1) can be written

$$\hat{f}(p,\phi) = \iint_{-\infty}^{\infty} f(\mathbf{x})\delta(p - \hat{\mathbf{p}}\cdot\mathbf{x})\, d\mathbf{x} \tag{9}$$

where $\hat{\mathbf{p}} = (\cos\phi, \sin\phi)$ is a unit vector in the direction of the $p$-axis (Figure 1). Applying the FT (8a) to this, we obtain

$$\check{f}(k,\phi) = \int_{-\infty}^{\infty}\iint_{-\infty}^{\infty} f(\mathbf{x})\delta(p - \hat{\mathbf{p}}\cdot\mathbf{x}) e^{-2i\pi kp}\, d\mathbf{x}dp$$

$$= \iint_{-\infty}^{\infty} f(\mathbf{x}) e^{-2i\pi k\hat{\mathbf{p}}\cdot\mathbf{x}} d\mathbf{x}$$

by changing the order of integration and evaluating the $p$-integral. The latter can be recognized as the 2-D FT of the model and so

$$\tilde{f}(k\hat{\mathbf{p}}) = \check{f}(k,\phi) \ . \tag{10}$$

Thus the 2-D FT of the model is related to the 1-D FT of the data (Gel'fand et al. 1966, page 4). Early workers, e.g. Bracewell (1956), unaware of Radon's work, used this result to invert the RT (1). It is known as the central slice theorem as for a fixed projection angle $\phi$, the data (RHS of (10)) provide the FT of the model on a slice through the origin of wavenumber space, i.e. $\mathbf{k} = k\hat{\mathbf{p}}$ with $\hat{\mathbf{p}}$ fixed. To invert (10) numerically for discrete data, we must interpolate the data defined on a polar grid to obtain the required results on a rectangular grid in **k**-space.

Inverting the 2-D FT (10), we obtain the model

$$f(\mathbf{x}) = \iint_{-\infty}^{\infty} \tilde{f}(k\hat{\mathbf{p}}) e^{2i\pi\mathbf{k}\cdot\mathbf{x}} d\mathbf{k}$$

$$= \int_{-\infty}^{\infty}\int_{0}^{\pi} \check{f}(k,\phi) e^{2i\pi k\hat{\mathbf{p}}\cdot\mathbf{x}} |k|\, d\phi dk$$

where we have converted to polar coordinates in the **k**-space ( $|k|$ arises as we have written the area integral as $\phi = 0$ to $\pi$ and $k = -\infty$ to $\infty$, rather than the more usual $\phi = 0$ to $2\pi$ and $k = 0$ to $\infty$ ). The final result reduces to

$$f(\mathbf{x}) = \int_0^\pi \hat{f}^*(\hat{\mathbf{p}}\cdot\mathbf{x}, \phi)\, d\phi \ , \tag{11}$$

an inverse Radon transform (IRT). The filtered data, $\hat{f}^*(p,\phi)$ can be obtained from the inverse FT (8b) of $|k|\hat{\tilde{f}}(k,\phi)$ or by a convolution operator in the spatial domain, i.e.

$$\hat{f}^*(p,\phi) = -\frac{1}{2\pi^2}\hat{f}(p,\phi) * \frac{1}{p^2} \tag{12}$$

where the convolution integral is defined by its regularization (Gel'fand and Shilov 1964, p. 52). Apart from the constant factor, the operation corresponds to the derivative of the Hilbert transform of the data, $\hat{f}(p,\phi)$. The operation (11) is referred to as a back-projection as for each projection angle, $\phi$, filtered data are projected back along the ray direction (Figure 1). Then for all projected angles, the functions are integrated (or summed). We call expression (11), filtered back projection (FBP).

## 3. Filtered Back Projection (FBP).

The numerical evaluation of expressions (11) and (12) presents various difficulties. The RT (1) integrates the model and the projection data, $\hat{f}(p,\phi)$, are 'smoother' than the model, $f(\mathbf{x})$. For instance, if the model contains an anomalous region surrounded by a first-order discontinuity, then in general if the boundary is smooth, the projection data have a square root discontinuity, $\sim|p-p_0|^{1/2}$. Similarly, if the boundary has a corner, the projection data have a ramp discontinuity, $\sim|p-p_0|$. If the boundary has a straight edge aligned with the projection direction then the projection data will have a step discontinuity, $\sim H(p-p_0)$. Thus in general, the projection data are 'smoother' than the model. Again in the inverse transform (11), the integral smooths the filtered data but these two 'smoothing' operations are compensated for by taking the derivative of the data (12). Thus, the data are 'roughened', $O(k)$, by the filtering operation (12) and then 'smoothed' by the back projection, $O(k^{-1/2})$. Naturally these operations are numerically difficult as high wavenumber noise in the projection data is amplified by the derivative. The numerical problems of differentiation are well known. The data must be smoothed to prevent amplification of the noise. Equivalently, the filter (12) must be band-limited. Various forms for a discrete, band-limited operator equivalent to (12) have been suggested, e.g. Shepp and Logan (1974). In general, the problems and solutions of filtering and, in particular, differentiation, are well known in seismology and we will not pursue the details here.

In addition to the numerical difficulties associated with differentiation, the filtering operation (12) contains the Hilbert transform. This operation only involves a phase shift. Numerically, difficulties may be encountered because the filter has long tails and the result is non-local and acausal, i.e. the filtered data at one position depend on the projection data at all distances (for one projection angle) including data that have not sampled the region of interest.

Having approximated the filter operator (12), by a discrete, band-limited version, e.g. Shepp and Logan (1974), the back projection (11) is evaluated by summation, i.e.

$$f(\mathbf{x}) \approx \sum_{n=0}^{N-1} \hat{f}^*(\hat{\mathbf{p}}_n \cdot \mathbf{x}, n\Delta\phi)\Delta\phi \tag{13}$$

where $\hat{\mathbf{p}}_n = (\cos n\Delta\phi, \sin n\Delta\phi)$ and we have assumed that the discrete data are collected at uniformly distributed angles with $\Delta\phi = \pi/N$. The filtered data must be interpolated in the radial direction. Various approximations can be used - nearest-neighbour and linear interpolation being common. We refer to this approximation (13) as discrete filtered back projection. It is widely used in medical tomography.

As mentioned above, the filter operator (12) presents numerical difficulties. A cruder approximation to the IRT, that is of some use, is back projection (BP). The filtering is omitted. Thus

$$f(\mathbf{x}) \approx c \sum_{n=0}^{N-1} \hat{f}(\hat{\mathbf{p}}_n \cdot \mathbf{x}, n\Delta\phi)\Delta\phi \tag{14}$$

where the constant $c$ is chosen from the size of the model so that the projection data are distributed uniformly across the model. From the above discussion it is obvious that the BP image will be a smoothed version of the true model. If we consider a point anomaly in the model and evaluate the BP integral (expression (14) as $\Delta\phi \to 0$ ), the image is spread. The point anomaly, $f(\mathbf{x}) = \delta(\mathbf{x} - \mathbf{x}_0)$, is reconstructed as $f(\mathbf{x}) \approx |\mathbf{x} - \mathbf{x}_0|^{-1}$. The filtering operation (12) removes this spread. Although the BP is only a crude approximation to the IRT, it is useful as a preliminary interpretation (Wong et al. 1983) or as a starting model for iterative solutions (Bregman et al. 1986).

Although the discrete FBP (13) is a simple approximation to the IRT (11), it unfortunately produces an image that is inconsistent with the original data. For a discrete dataset we would anticipate that the reconstruction would be non-unique and contain ambiguities due to aliasing, but the discrete FBP is inconsistent. This is most easily seen by considering the minimal dataset, i.e. $N = 1$. The reconstruction is $f(\mathbf{x}) = \pi\hat{f}^*(x,0)$ and, if the projection data are computed in the one projection direction, $\phi = 0$, give data of the form $\hat{f}^*(x,0)$ rather than $\hat{f}(x,0)$. A similar result is obtained for more general datasets (Chapman and Cary 1986). The inconsistent artifacts in the reconstruction consist of the well-known radial streaks emanating from anomalies in the image. Despite the inconsistencies, the discrete FBP is remarkably successful. Fortunately, for a band-limited model and dataset, the artifacts are only significant at large radii. In and near an anomaly, the artifacts are unresolved due to the spatial smoothing. The sampling criterion used in applying the discrete FBP, e.g. Brooks et al. 1978, should be considered as a condition for spatially band-limiting the model and data. If the model and data are not band-limited, then the image will contain inconsistent artifacts, but these may not be resolved by the sampling. In other words, the reconstruction will appear noisy due to sampling an inconsistent image, but the sampling will not be sufficient to illuminate the complete nature of the inconsistent artifacts. In medical tomography, discrete FBP is very successful as the sampling criterion can often be met. Artifacts outside the object of interest, i.e. the patient, are of no interest. Unfortunately, in seismic tomography the anomalous region is not known *a priori* to be restricted or band-limited, and inconsistent artifacts will be more troublesome.

## 4. Circular Harmonic Decomposition (CHD).

Although an exact IRT is known (11), alternative formulae exist. These are equivalent for exact, complete datasets but for incomplete, inaccurate datasets will differ significantly. An alternative formulation that is attractive for datasets with discrete projection angles is the circular harmonic decomposition (CHD) (Cormack 1963, 1964, Kershaw 1970, Perry 1975, Deans 1977, Minerbo and Sanderson 1977, Hansen 1979,1981, Verly 1981, Hawkins 1983, Chapman and Cary 1986). Whereas for discrete projection angles, FBP produces an image that is inconsistent with the data, the discrete CHD reconstruction is consistent with the data. Both will contain ambiguities due to aliasing. The CHD method is also of interest as it has at least three inverse forms: one is numerically unstable, but others use a condition for the data to be consistent with the projection experiment, to produce a stable formula.

The model is expanded as a Fourier series in the polar coordinates:

$$f(\mathbf{x}) = \sum_{n=-\infty}^{\infty} f_n(r) e^{in\theta} \quad . \tag{15}$$

Certain results given below are only valid for $n \geq 0$. Since $f(\mathbf{x})$ is real we obtain the required expression for $n < 0$ using the symmetry $f_{-n}(r) = f_n^*(r)$, where just here the asterisk denotes the complex conjugate. Substituting the series (15) in the RT (9), the $\theta$–integral is easily evaluated. The delta function is non-zero at two points, $\theta = \phi \pm \cos^{-1}(p/r)$, and we obtain (Cormack 1963)

$$\hat{f}(p,\phi) = \sum_{n=-\infty}^{\infty} \hat{f}_n(p) e^{in\phi} \tag{16a}$$

with

$$\hat{f}_n(p) = 2\int_p^{\infty} \frac{r f_n(r) T_n(p/r)}{(r^2-p^2)^{1/2}} dr \tag{17}$$

where $T_n(x) = \cos(n\cos^{-1}x)$ is the Chebyshev polynomial of the first kind. This result relates the harmonic coefficients of the data, which can be obtained by

$$\hat{f}_n(p) = \frac{1}{2\pi}\int_0^{2\pi} \hat{f}(p,\phi) \;, e^{-in\phi} d\phi \tag{16b}$$

to those of the model, $f_n(r)$. To invert (17) we need

$$\int_t^r \frac{r\, t\, T_n(p/r) T_n(p/t)}{p(r^2-p^2)^{1/2}(p^2-t^2)^{1/2}} dp = \frac{\pi}{2} \quad . \tag{18}$$

References for the proof of this result can be found in Deans (1983) or Kershaw (1970). It is of interest to note that for $n = 1$, it reduces to the integral commonly used in the proof of the Herglotz-Wiechert-Bateman (HWB) method for seismic travel-time inversion (Aki and Richards 1980, Section 12.1). Following Cormack (1963,1964), we use (18) to invert (17). After some manipulation we obtain

$$f_n(r) = -\frac{1}{\pi}\frac{d}{dr}\int_r^\infty \frac{r\hat{f}_n(p)T_n(p/r)}{p(p^2-r^2)^{1/2}}dp$$

$$= -\frac{1}{\pi}\int_r^\infty \frac{\hat{f}_n'(p)T_n(p/r)}{(p^2-r^2)^{1/2}}dp \tag{19}$$

where the prime $'$ indicates differentiation with respect to the argument. The first integral in (19) may be preferred for computations as $\hat{f}_n'(p)$ can contain (integrable) singularities. Unfortunately, the numerical evaluation of this integral is unstable for large $n$ as for $x>1$, the Chebyshev polynomial grows rapidly ($\sim x^n$ ). Cormack (1964) solved this problem by expanding the data in Chebyshev polynomials of the second kind for which the integral (19) can be evaluated analytically in terms of the Zernike polynomials. This complication can be avoided as other inversion formulae are possible.

Substituting the expression (16a) in the IRT (11), we obtain

$$f(\mathbf{x}) = \frac{1}{2}\int_0^{2\pi} \sum_{n=-\infty}^{\infty} \hat{f}_n^*(\hat{\mathbf{p}}\cdot\mathbf{x})e^{in\phi}d\phi$$

where we have used the symmetry $\hat{f}(-p,\phi) = \hat{f}(p,\phi+\pi)$. Letting $\chi = \phi-\theta$ so $\hat{\mathbf{p}}\cdot\mathbf{x} = r\cos\chi$, we can obtain the Fourier series (15) and identify (Chapman and Cary 1986)

$$f_n(r) = \frac{1}{2}\int_0^{2\pi} \hat{f}_n^*(r\cos\chi)e^{in\chi}d\chi$$

$$= \int_{-r}^{r} \frac{\hat{f}_n^*(p)T_n(p/r)}{(r^2-p^2)^{1/2}}dp \tag{20}$$

This formula is numerically stable as the Chebyshev polynomial is oscillatory, but requires the filtered projection data (12).

In order to avoid evaluating $\hat{f}_n^*(p)$ explicitly, we use its FT (8b) and evaluate the $\chi$–integral first to obtain

$$f_n(r) = 2\pi i^n \int_0^\infty k\, J_n(2\pi kr)\check{f}_n(k)dk \ . \tag{21}$$

This is a standard Fourier-Bessel transform with inverse

$$\check{f}_n(k) = 2\pi i^{-n} \int_0^\infty r\, J_n(2\pi kr)f_n(r)dr \ . \tag{22}$$

This symmetric transform pair has been obtained and used by many authors (Klug et al. 1958, Cormack 1964, Crowther et al. 1970) but suffers from the disadvantage that typically both terms in the integrand are oscillatory. Yet another inversion formula can be derived from (21). Substituting for the FT (8a) of $\hat{f}_n'(p)$ and evaluating the $k$-integral, we obtain (Chapman and Cary 1986)

$$f_n(r) = \frac{1}{\pi r}\int_0^r U_{n-1}(\frac{p}{r})\hat{f}_n'(p)dp - \frac{1}{\pi r}\int_r^\infty \frac{\exp(-n\cosh^{-1}(p/r))}{(p^2-r^2)^{1/2}}\hat{f}_n'(p)dp \tag{23}$$

where $U_n(x)$ is the Chebyshev polynomial of the second kind. This expression has been derived by various techniques by different authors (Perry 1975, Deans 1977, Minerbo and Sanderson 1977, Hansen 1979,1981, Verly 1981 and Hawkins 1983).

The expression (23) is stable as the final integrand decays for large $n$. It is easy to evaluate numerically and has been investigated by Chapman and Cary (1986). It is preferred over expression (20) as the Hilbert transform of the data is not required (although in common with all methods it requires the derivative). In order to prove the equivalence of the unstable (19) and stable (23) formulae, we use the relationship (for $x>1$)

$$\exp(-n\cosh^{-1}x) = T_n(x) - (x^2-1)^{1/2}U_{n-1}(x)$$

to obtain integral (19) from the second integral of (23). The remaining integral

$$\frac{1}{\pi r}\int_0^\infty U_{n-1}(\frac{p}{r})\hat{f}_n'(p)dp$$

can be proved to be zero using the consistency relations (Cormack 1963, Ein-Gal 1974). The consistency relations are integral constraints that the projection data must satisfy if they are consistent with a projection experiment, i.e. equation (17). Numerically and for realistic noisy data, these constraints will not be satisfied and it is these inconsistencies that make expression (19) unstable. When seismic tomography experiments are inverted numerically (with complicated geometries), we must expect similar instabilities if inconsistencies in the data are interpreted. The solution must be regularized to suppress instabilities. It should be noted that inconsistencies in the data may be due to noise, numerical rounding errors or inadequate parameterization of the model, i.e. the parameterization of the model does not provide an adequate solution of the forward problem (1).

The result (23) breaks down for $r = 0$ when

$$\begin{aligned} f_n(0) &= 0 && \text{for } n\neq 0 \\ &= -\frac{1}{\pi}\int_0^\infty \frac{\hat{f}_0'(p)}{p}dp && \text{for } n=0 \;. \end{aligned} \tag{24}$$

The latter expression is finite as $\hat{f}_0'(0) = 0$ and is equivalent to FBP (11).

## 5. Fan-Beam Geometry.

In the standard RT, we have considered the projection experiment for straight, parallel rays. A geometry which is commonly used in medical tomography, and corresponds more closely to seismic tomography (with a common source or receiver), is the fan-beam geometry (Figure 2). The FBP algorithm can be modified for fan-beam projection data (Lakshminarayanan 1975). In the CHD method, the harmonic coefficients of the projection data are easily obtained by a simple change of variable. The fan-beam projection data are

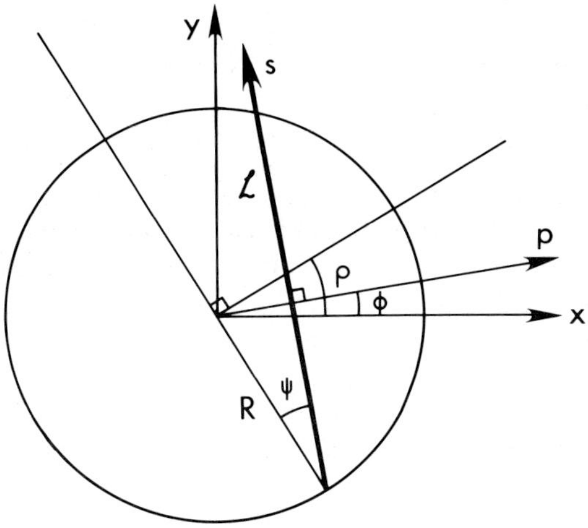


*Figure 2* The geometry and variables in the projection experiment with fan-beam geometry and straight rays. The source lies at a radius $R$, the central ray in the beam is at an angle $\rho$, and rays within the beam make an angle $\psi$. The coordinates equivalent to Figure 1 are also shown.

$\hat{g}(\psi,\rho)$ where $\psi$ is the angle within the beam and $\rho$ is the angle of the central ray in the beam (Figure 2). The equivalent coordinates in the parallel geometry are $(p,\phi)$ where

$$p = R\sin\psi \qquad\qquad \phi = \rho - \psi \tag{25}$$

and $R$ is the reference radius of the source. Substituting in (16b) the equivalent fan-beam data, the harmonic coefficients are

$$\hat{f}_n(p) = \frac{1}{2\pi}\int_0^{2\pi} \hat{g}\left[\sin^{-1}\left(\frac{p}{R}\right), \phi + \sin^{-1}\left(\frac{p}{R}\right)\right] e^{-in\phi} d\phi$$

$$= \hat{g}_n\left(\sin^{-1}\left(\frac{p}{R}\right)\right)e^{in\sin^{-1}(p/R)} \tag{26}$$

where $\hat{g}_n(\psi)$ are the harmonic coefficients with respect to the central beam angle, $\rho$. Thus the harmonic coefficients needed for the CHD can be obtained by a simple phase shift and variable change from the harmonic coefficients of the fan-beam data.

## 6. Earth Flattening Transformation (EFT).

In the preceding sections, we have reviewed the Radon transform and its inverse. These apply to tomographic problems with straight rays. In the following sections we consider some generalizations of the RT that apply to tomography with curved rays. First we discuss the Earth flattening transformation that can be used to map a spherically (or rather, as we are only considering 2-dimensions, cylindrically) symmetric reference model into a plane model. The method is widely used in seismic travel-time and waveform modelling and can also be used in tomography.

We consider a model in polar coordinates $(r,\theta)$ and an equivalent model in cartesian coordinates $(x,z)$. We define a conformal mapping between these models (Gerver and Markuschevich 1966). The horizontal coordinate is mapped as

$$x = R\theta \tag{27a}$$

where $R$ is some reference radius. In order to compensate for the stretching of an arc length $r\theta$ to the horizontal distance $R\theta$, the vertical coordinate is also stretched, i.e. $dz = (R/r)dr$. This maps the coordinate as

$$\frac{z}{R} = \ln(\frac{r}{R}) \ . \tag{27b}$$

If $r$ corresponds to radius in the Earth, then $z$ is measured vertically upwards. In order to compensate for the increased path length in the cartesian model, the velocity must be increased to compensate. We consider a reference model in which the velocity, which defines the ray geometry, is only a function of the radial or vertical coordinate. Thus the velocities in the two models are related by

$$v_c(z) = (\frac{R}{r})v_p(r) \tag{28}$$

where the subscript $c$ refers to the cartesian model and $p$ to the polar model. With these mappings, (27) and (28), the geometry and kinematic ray properties in the two models are exactly equivalent. For instance for a turning ray (Figure 3), the semi-travel time, i.e. the travel time from the turning point, $Z(p)$, to the depth, $z$ (or equivalently $R(p)$ to $r$ in the polar model) is

$$\begin{aligned} T(p,z) &= \int_{L_c} \frac{ds_c}{v_c} = \int_{Z(p)}^{z} \frac{dz}{v_c(z)(1-p^2v_c^2(z))^{1/2}} \\ &= \int_{L_p} \frac{ds_p}{v_p} = \int_{R(p)}^{r} \frac{rdr}{v_p(r)(r^2-(Rp)^2v_p^2(r))^{1/2}} \end{aligned} \tag{29}$$

where $p$ is a ray parameter defining the turning point, i.e.

$$pv_c(z) = \frac{Rp}{r}v_p(r) = 1 \tag{30}$$

at $z = Z(p)$ and $r = R(p)$. Similary, the horizontal semi-range is

$$\begin{aligned} X(p,z) &= \int_{L_c} dx = \int_{Z(p)}^{z} \frac{pv_c(z)dz}{(1-p^2v_c^2(z))^{1/2}} \\ &= R\int_{L_p} d\theta = R\int_{R(p)}^{r} \frac{(Rp)v_p(r)dr}{r(r^2-(Rp)^2v_p^2(r))^{1/2}} \ . \end{aligned} \tag{31}$$

These results, (29) and (31), are the standard ray integrals for rays satisfying Snell's law in 1-D cartesian or polar models.

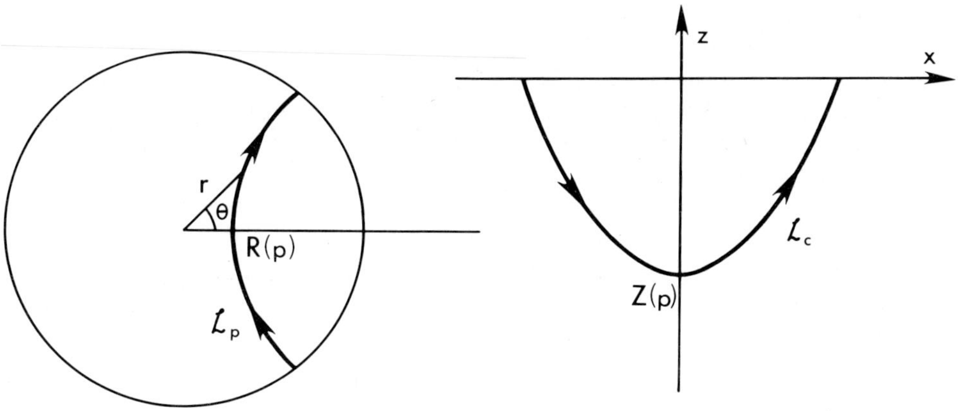


*Figure 3* Curved rays in (a) polar and (b) cartesian models. In the polar model the coordinates are $(r,\theta)$, the ray path is $L_p$ and the turning point at radius $R(p)$. In the cartesian model the coordinates are $(x,z)$, the ray path $L_c$ and the turning point at $Z(p)$.

The conformal mapping between polar and cartesian coordinates, known as the Earth flattening transformation (EFT), allows the kinematic ray properties in the 1-D reference Earth model to be solved exactly in either system. Geometrically and algebraically it is somewhat simpler to use the cartesian model. The forward tomography problem (1) can also be solved in either system. To compensate for the increased path length in the cartesian model, the equivalent anomaly must be reduced. Thus we have

$$\hat{f} = \int_{L_c} f_c(x,z)\,ds_c = \int_{L_p} f_p(r,\theta)\,ds_p$$

provided

$$f_c(x,z) = \frac{r}{R} f_p(r,\theta) \tag{32}$$

(we ignore for the moment the parameterization of the projection data, i.e. the arguments of $\hat{f}$ ).

The standard RT applies to a polar model with constant reference velocity, i.e. $v_p(r) = v_p$, a constant. The equivalent cartesian model has reference velocity increasing exponentially with depth

$$v_c(z) = v_p e^{-z/R} \tag{33}$$

and curved rays defined by

$$X(z,p) = R\cos^{-1}(e^{[Z(p)-z]/R}) \ . \tag{34}$$

In the next section we investigate generalizing the RT to other reference velocities and ray geometries in cartesian models. In what follows we shall only consider results in cartesian models but, using the above transformations, they apply equally to polar models. We drop the subscript $c$.

## 7. Tomography in 1-Dimensional Models: a long wavelength method.

In the previous section we saw how the RT for straight ray tomography in a homogeneous reference model could be transformed into curved ray tomography in an inhomogeneous reference model. Of course, the velocity and ray geometry, (33) and (34), have a special form but this suggests that the tomographic problem might be generalized to any 1-D reference model, $v(z)$. The following development follows Lavrentiev et al. (1970) and Romanov (1974) except that we have included results in the spatial rather than wavenumber domain, used notation more familiar to seismologists, and included terms for low velocity zones (LVZs). We consider turning rays (Figure 4) and later make some comments regarding reflections.

The ray geometry and coordinates are illustrated in Figure 4. We parameterize the projection data by the ray parameter, $p$, and mid-point coordinate, $m$. As the model is 1-D, the ray geometry is translationally invariant, i.e. independent of $m$. The ray parameter defines the turning point, $Z(p)$, as the shallowest solution of the equation, $pv(z)=1$ (30). The turning point is analogous to the projection coordinate in the RT (Figures 1 and 4) which coincidentally was called $p$. The mid-point coordinate, $m$, takes the place of the angular coordinate, $\phi$, i.e. $m = R\,\phi$.

The ray path $L$ is defined by

$$x = m \pm X(z,p) \tag{35}$$

where the semi-range, $X(z,p)$, is defined in (31). Thus the forward projection problem can be written

$$\begin{aligned} \hat{f}(p,m) &= \int_L f(\mathbf{x})ds \\ &= \int_{Z(p)}^{0} \frac{f(m+X(z,p),z)+f(m-X(z,p),z)}{(1-p^2v^2(z))^{1/2}}dz \end{aligned} \tag{36}$$

where we have assumed that the source and receiver lie on the line $z = 0$. Taking the FT (8a) with respect to the horizontal coordinate, $m$, we obtain

$$\tilde{\hat{f}}(p,k) = 2\int_{Z(p)}^{0} \frac{\tilde{f}(k,z)\cos(2\pi kX(z,p))}{(1-p^2v^2(z))^{1/2}}dz \ . \tag{37}$$

This transform and the result (37) are analogous to the harmonic decomposition and result (17) above (effectively, $n \rightarrow 2\pi kR$ ). If the cartesian model has been derived from a polar model then the cartesian model is cyclic, i.e. $f(x+2\pi R,z)=f(x,z)$. The FT should either be replaced by a Fourier series or generalized so that the transform is defined with delta functions at the harmonics ( $n = 2\pi kR$ ). For a non-repetitive cartesian model, the FT is appropriate.

Equation (37) can be recognized as a Volterra integral equation of the first kind. In general it is not known if this type of equation has a solution. We proceed as in Cormack's (1963) solution by applying an integral operator to the projection data

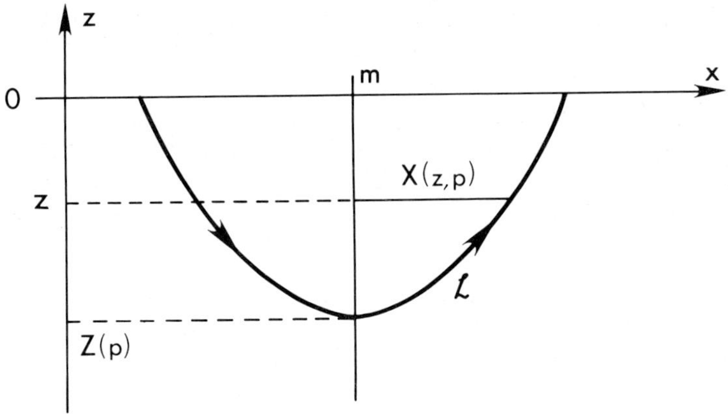


*Figure 4* A curved ray path in a cartesian model illustrating the mid-point coordinate, $m$, the turning point, $Z(p)$, and the semi-range, $X(z,p)$, at $z$.

$$\int_{u(z)}^{u(0)} \frac{\hat{f}(p,m)pdp}{(p^2-u^2(z))^{1/2}} = \int_{u(z)}^{u(0)} \int_{Z(p)}^{0} \frac{f(m+X(\xi,p),\xi)+f(m-X(\xi,p),\xi)}{(1-p^2v^2(\xi))^{1/2}(p^2-u^2(z))^{1/2}} pd\xi dp$$

where for notational simplicity we have defined the reference slowness $u(z) = 1/v(z)$ and used $\xi$ as a dummy depth variable on the RHS. The extent of the double integral is illustrated in Figure 5. Reversing the order of integration we obtain

$$\int_{u(z)}^{u(0)} \frac{\hat{f}(p,m)pdp}{(p^2-u^2(z))^{1/2}} = \left[ \int_{z}^{0} \int_{u(z)}^{u(\xi)} - \sum_i \int_{z_i^L}^{z_i^U} \int_{u_i}^{u(\xi)} \right] \frac{f(m+X(\xi,p),\xi)+f(m-X(\xi,p),\xi)}{(u^2(\xi)-p^2)^{1/2}(p^2-u^2(z))^{1/2}} pdpu(\xi)d\xi$$

where the summation over $i$ includes any LVZs in the range of the integral, i.e. if $u_i > u(z)$. The integral on the RHS can be simplified with the change of variable

$$p^2 = u^2(\xi)\sin^2\chi + u^2(z)\cos^2\chi \tag{38}$$

to obtain

$$\int_{u(z)}^{u(0)} \frac{\hat{f}(p,m)pdp}{(p^2-u^2(z))^{1/2}} = \left[ \int_{z}^{u} \int_{0}^{\pi/2} - \sum_i \int_{z_i^L}^{z_i^U} \int_{\chi_i}^{\pi/2} \right] \left[ f(m+X(\xi,p),\xi)+f(m-X(\xi,p),\xi) \right] d\chi u(\xi)d\xi$$

where

$$\chi_i = \tan^{-1} \left[ \frac{u_i^2-u^2(z)}{u^2(\xi)-u_i^2} \right]^{1/2} .$$

Ignoring lateral variations in the model, or equivalently letting $z \to \xi$, we note

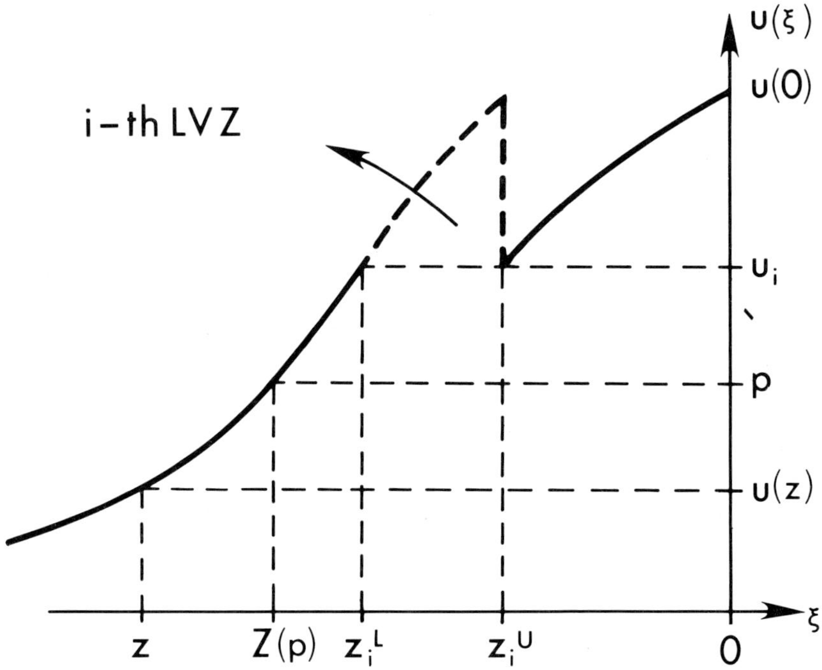


*Figure 5* The range of the double integral over depth (abscissa) and slowness (ordinate). The depth range is $z$ to 0, and the slowness range $u(z)$ to $u(\xi)$. One LVZ is illustrated between depths $z_i^L$ and $z_i^U$. The region in the LVZ where $u(\xi)>u_i$ is excluded from the original integral (depth integral innermost) and must be subtracted when the order of integration is reversed.

$$\int_0^{\pi/2}\left[f(m+X(\xi,p),\xi)+f(m-X(\xi,p),\xi)\right]d\chi \to \pi f(m,\xi) \ .$$

Thus differentiating with respect to $z$ we obtain after rearranging

$$f(m,z)=-\frac{v(z)}{\pi}\frac{d}{dz}\int_{u(z)}^{u(0)}\frac{\hat{f}(p,m)pdp}{(p^2-u^2(z))^{1/2}}$$

$$+\frac{v(z)}{\pi}\int_z^0\frac{\partial}{\partial z}\int_0^{\pi/2}\left[f(m+X(\xi,p),\xi)+f(m-X(\xi,p),\xi)\right]d\chi u(\xi)d\xi$$

$$-\frac{v(z)}{\pi}\sum_i\int_{z_i^L}^{z_i^U}\frac{\partial}{\partial z}\int_{\chi_i}^{\pi/2}\left[f(m+X(\xi,p),\xi)+f(m-X(\xi,p),\xi)\right]d\chi u(\xi)d\xi \ . \quad (39)$$

The result in the transform domain, which without the LVZ term is equivalent to the result in Lavrentiev et al. (1970) and Romanov (1974), is

$$\tilde{f}(k,z) = -\frac{v(z)}{\pi}\frac{d}{dz}\int_{u(z)}^{u(0)} \frac{\tilde{\hat{f}}(p,k)pdp}{(p^2-u^2(z))^{1/2}}$$

$$+\frac{2v(z)}{\pi}\int_z^0 u(\xi)\tilde{f}(k,\xi)\frac{\partial}{\partial z}\int_0^{\pi/2}\cos(2\pi kX(\xi,p))d\chi d\xi$$

$$-\frac{2v(z)}{\pi}\sum_i\int_{z_i^L}^{z_i^U} u(\xi)\tilde{f}(k,\xi)\frac{\partial}{\partial z}\int_{\chi_i}^{\pi/2}\cos(2\pi kX(\xi,p))d\chi d\xi \ . \qquad (40)$$

Ignoring for the moment the LVZ terms or assuming no LVZs exist, this equation can be recognized as a Volterra integral equation of the second kind. Detailed analysis shows that the kernel is well behaved and, at worst, has an integrable singularity on the diagonal, $z=\xi$. It is well known that a Volterra integral equation of the second kind has a solution that can be found, in principle, by the method of successive approximation.

Equations (39) and (40) only apply to depths, $z$, outside LVZs as the lower limit in the integrals must correspond to a turning point. We do not have a formula for determining the anomaly in a LVZ. In common with travel-time inversion (Gerver and Markushevich 1966), the inversion is ambiguous below any LVZs. If the anomaly in a LVZ is known *a priori* (or taken to be zero), then the LVZ terms in (39) and (40) can be evaluated and considered as part of the inhomogeneous term. Then again the integral equation can be solved.

In an iterative solution of the integral equation, the first iteration comes from the inhomogeneous term, i.e.

$$f^{(1)}(m,z) = -\frac{v(z)}{\pi}\frac{d}{dz}\int_{u(z)}^{u(0)}\frac{\hat{f}(p,m)pdp}{(p^2-u^2(z))^{1/2}}$$

$$= -\frac{u'(z)}{\pi}\int_{u(z)}^{u(0)}\frac{\hat{f}'(p,m)dp}{(p^2-u^2(z))^{1/2}} \qquad (41)$$

where the prime $'$ indicates differentiation with respect to the first argument, $p$. As $\hat{f}'(p,m)$ may contain delta function and inverse square root singularities, the first expression in (41) is preferred for numerical evaluation.

If the projection data, $\hat{f}(p,m)$, have no lateral variation, i.e. the model is 1-D but the reference model is incorrect, then the first iteration (41) is the solution (the integral term is zero). If lateral variations exist, then the first iteration (41) is equivalent to a 1-D inversion of common mid-point (CMP) data, i.e. projection data with a CMP, $m$, are interpreted to give the anomaly at the CMP assuming no lateral variation. It is important to note that this solution is not the same as the HWB method for interpreting travel times in 1-D models. The tomographic inversion (with $\hat{f}=T$, the travel time and $f=u$, the slowness (3)) assumes fixed rays in a reference model. The HWB method makes no such assumption and effectively uses the ray paths consistent with the inversion. The tomographic inversion is approximate while the HWB method is exact. As an example, if we consider travel-time data from a linear velocity model, $v(z)=v_0+bz$, and use, in the tomographic inversion, the reference model, $v(z)=V_0+Bz$, it is easily shown that the result of expression (41) is

$v(z) = (b/B)V_0 + bz$, i.e. the tomographic interpretation has the correct gradient but, in general, will be shifted in value.

The integral term in (39) or (40) is straightforward to evaluate numerically. It is over the area contained by the ray with turning point at $(m, z)$ and can be rewritten in various forms with different variables of integration, etc. To date, a thorough investigation of numerical solutions of these integral equations is not available. From the theoretical development and by comparison with the CHD method, it is clear that an iterative solution will be effective if the anomaly is smooth and well described by low-order terms in a Taylor expansion (i.e. only $k$ small is important). If lateral variations are strong and higher derivatives are important (i.e. $k$ large is important), then we would expect the iterative solution to be badly behaved (with large contributions from the integral) and numerically unstable.

The above theoretical development does not include reflections as the depth, $z$, must be a variable, a turning point. If a reflector exists below an inhomogeneous region, then theoretically, turning rays in the inhomogeneous region are needed to find the anomaly in this region. Reflection data from the reflector only provide redundant information on the structure. This is similar to the role reflection data play in the HWB method. In practice, of course, turning ray data may not be available and it would be useful to use the reflection data. We know of no analytic method which solves this problem. Even if turning ray data are present it would be useful to also use the redundant reflection data. If the reflector is replaced by a zone of very high gradient and the above inversion procedure followed, then all the inconsistencies in the reflection data are modelled in the reflector zone which is unsatisfactory. The problem of reflections in a homogeneous reference model is different. The reflecting rays are straight and the problem can be mapped into the normal RT (1) by a simple change of variable (Fawcett 1983). The dataset is incomplete as horizontal rays are never present. This means that laterally homogeneous anomalies cannot be resolved by the reflection data. But apart from this ambiguity, standard inversion techniques can be used.

In this section we have parameterized the projection data by the ray parameter and mid-point, $(p, m)$. It is straightforward to use other variables. Anticipating the next section and reflecting the fan-beam geometry, we might wish to use common source point data (or receiver as by reciprocity the theoretical difference is trivial). Thus defining

$$\rho = m - X(0, p) \tag{42}$$

we collect projection data, $\hat{g}(p, \rho)$. We can derive the required functions

$$\hat{f}(p, m) = \hat{g}(p, m - X(0, p)) \tag{43a}$$

and

$$\tilde{\hat{f}}(p, k) = \tilde{\hat{g}}(p, k)e^{-2i\pi kX(0, p)} . \tag{43b}$$

Finally, we note that if this section's inversion technique is applied to the CHD for the RT in a homogeneous reference model (17), then we obtain

$$f_n(r) = -\frac{1}{\pi}\frac{d}{dr}\int_r^{\infty}\frac{r\hat{f}_n(p)dp}{p(p^2-r^2)^{1/2}}$$

$$+\frac{2}{\pi}\int_r^{\infty} f_n(t)\frac{\partial}{\partial r}\int_r^{t}\frac{rtT_n(p/t)dp}{p(p^2-r^2)^{1/2}(t^2-p^2)^{1/2}}dt \tag{44}$$

The inhomogeneous term is equivalent to the exact result (19) with $n = 0$. The integral in the kernel is an elliptic integral.

## 8. Tomography in 2-Dimensional Models: a short wavelength method.

In this section we consider asymptotic results for the tomographic problem with a 2-D reference model. The method has been developed by Beylkin (1982, 1983, 1984, 1985). We omit the rigorous mathematical development in those papers and follow the approach of Cohen et al. (1986). A similar result has been obtained by Fawcett and Clayton (1984) for 1-D reference models.

In a 2-D model the ray paths are no longer symmetric or translationally invariant. The parameterization of the rays and projection data is therefore less straightforward. We define the ray by its source coordinate on the line $z = 0$, $(\rho,0)$ in cartesian coordinates, and its direction by an angle, $\psi$ (Figure 6). The notation is chosen to emphasize the similarity with the fan-beam geometry. The two parameters $(\psi,\rho)$ uniquely define a ray path. By reciprocity, the roles of the source and receiver can be reversed. To be specific, we refer to $(\rho,0)$ as the source. The sources and receivers need not be on the same line. For turning rays and reflections, we have assumed they lie on the same line, but for cross-borehole tomography they will be on different lines.

The ray path defined by the parameters $(\psi,\rho)$ can be written as

$$\mathbf{x} = \mathbf{x}(\psi,\rho,s) \tag{45}$$

where $s$ is the arc length along the ray. The ray path is found by solving the kinematic ray equations in the 2-D reference model, $v(x,z)$. These equations, and techniques for their numerical solution, are well known and details need not concern us here. Thus the projection experiment is modelled as

$$\hat{g}(\psi,\rho) = \int_L f(\mathbf{x})ds \tag{46}$$

where the ray path $L$ is defined by (45). In order to follow previous analysis, it is necessary to rewrite the ray equation (45):

$$\rho = \rho_j(\mathbf{x},\psi) \ . \tag{47}$$

Although this is less natural than (45), it is equally satisfactory. For any position in the model, $\mathbf{x}$, we trace rays in various directions to the source line, $z = 0$. On this line we find the ray direction, $\psi$, and the position, $x = \rho$. Thus we can consider $\rho$ as a function of $\mathbf{x}$ and $\psi$. Note that for a given $\mathbf{x}$, only a limited range of $\psi$ will be possible. In general, large values of $\psi$ will define shallow rays that may not penetrate to $\mathbf{x}$ (Figure 6). For a given $\mathbf{x}$ and $\psi$ there may be several source positions, $\rho$. These are enumerated by the index, $j$. For

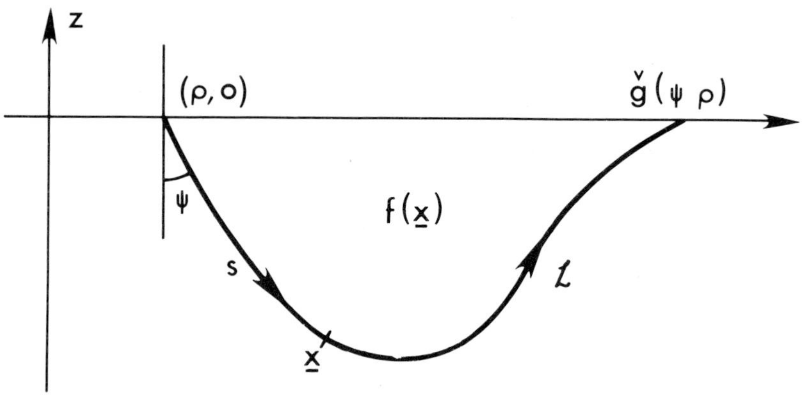


*Figure 6* A ray in a 2-dimensional model. The position, $\mathbf{x}$, on the ray path, $L$, is defined by the source position, $(\rho,0)$, and direction, $\psi$, and the arc length, $s$, along the ray. The projection data, $\check{g}(\psi,\rho)$, are a function of the source position and ray direction.

instance, in 1-D models we have two solutions, $\rho_j = x - X(0,p) \pm X(z,p)$ with $p = u(0)\sin\psi$. In general, there may be more than two solutions. The projection experiment (46) can now be rewritten

$$\check{g}(\psi,\rho) = \iint_{-\infty}^{\infty} f(\mathbf{x}) \sum_{j=1}^{J} |\nabla\rho_j| \delta(\rho - \rho_j(\mathbf{x},\psi)) d\mathbf{x} \tag{48}$$

where $\nabla \equiv \nabla_{\mathbf{x}}$ and normally $J = 1$ (e.g. in cross-borehole tomography with $x$ vertical), or $J = 2$ (e.g. in turning ray and reflection tomography with $x$ horizontal). We take the 1-D FT with respect to the source coordinate to obtain

$$\tilde{\check{g}}(\psi,k) = \iint_{-\infty}^{\infty} f(\mathbf{x}) \sum_{j=1}^{J} |\nabla\rho_j| e^{-2i\pi k \rho_j(\mathbf{x},\psi)} d\mathbf{x} \ . \tag{49}$$

Following Cohen et al. (1986), we guess that the inverse solution of this equation has approximately (asymptotically) the form

$$f(\mathbf{x}) \approx \int_{-\infty}^{\infty} \int \tilde{\check{g}}(\psi,k) \sum_{l=1}^{J} e^{2i\pi k \rho_l(\mathbf{x},\psi)} b_l(\mathbf{x},\psi) |k| \, d\psi dk \ . \tag{50}$$

The range of the $\psi$-integral is that for which $\rho_l(\mathbf{x},\psi)$ is defined. The negative phase is not hard to guess and the unknown amplitude is denoted by $b_l(\mathbf{x},\psi)$. It might depend on $k$ but anticipating the result, we introduce only a factor $|k|$.

If (50) is to be the asymptotic solution, substituting (49) in (50) must reduce approximately to the identity. This requires

$$\iint_{-\infty}^{\infty} \sum_{j=1}^{J} |\nabla\rho_j| b_j(\xi,\psi) |k| e^{2i\pi k[\rho_j(\xi,\psi)-\rho_j(\mathbf{x},\psi)]} d\psi dk \approx \delta(\xi-\mathbf{x}) \tag{51}$$

where we have assumed the terms with $l \neq j$ can be dropped as the integral is highly oscillatory. Using the theory of Fourier integral operators (FIOs), Beylkin (1982) established that asymptotically, if we are only interested in the discontinuity structure of $f(\mathbf{x})$, i.e. $k \to \infty$ in (50), only the lowest order terms in Taylor expansions of the integrand need be retained. Thus we substitute in (51)

$$b_j(\xi,\psi) \approx b_j(\mathbf{x},\psi)$$

and

$$\rho_j(\xi,\psi) - \rho_j(\mathbf{x},\psi) \approx \nabla\rho_j \cdot (\xi - \mathbf{x}) \ .$$

In the resultant integral we make the change of variable

$$\mathbf{k}_j = k\nabla\rho_j \tag{52}$$

and obtain

$$\iint_{-\infty}^{\infty} \sum_{j=1}^{J} \frac{|\nabla\rho_j| b_j(\mathbf{x},\psi)}{|h_j(\mathbf{x},\psi)|} e^{2i\pi \mathbf{k}_j\cdot(\xi-\mathbf{x})} d\mathbf{k}_j \approx \delta(\xi-\mathbf{x})$$

where $h_j(\mathbf{x},\psi)$ arises from the Jacobian of the variable transformation and is given by

$$h_j(\mathbf{x},\psi) = \begin{vmatrix} \dfrac{\partial\rho_j}{\partial x} & \dfrac{\partial\rho_j}{\partial z} \\ \dfrac{\partial^2\rho_j}{\partial x\,\partial\psi} & \dfrac{\partial^2\rho_j}{\partial z\,\partial\psi} \end{vmatrix} \tag{53}$$

This determinant can be derived from the results of kinematic and dynamic ray tracing. Thus we must have

$$b_j(\mathbf{x},\psi) = \frac{|h_j(\mathbf{x},\psi)|}{J\,|\nabla\rho_j|} \tag{54}$$

and the asymptotic inverse formula is

$$\begin{aligned} f(\mathbf{x}) &\approx \int_{-\infty}^{\infty}\int \tilde{g}(\psi,k) \sum_{j=1}^{J} e^{2i\pi k\rho_j(\mathbf{x},\psi)} \frac{|h_j(\mathbf{x},\psi)|}{J\,|\nabla\rho_j|} |k| d\psi dk \\ &= \int \frac{1}{J} \sum_{j=1}^{J} g^*(\psi,\rho_j(\mathbf{x},\psi)) \frac{|h_j(\mathbf{x},\psi)|}{|\nabla\rho_j|} d\psi \end{aligned} \tag{55}$$

i.e. a generalized FBP. A similar result has been obtained by Fawcett (1983) and Fawcett and Clayton (1984) by a different, more intuitive argument.

Beylkin (1982), using the theory of FIOs, has shown how the inverse problem for the generalized RT can be written as a Fredholm equation. Expression (55) is the inhomogeneous term. He also showed how higher-order terms in the asymptotic solution can be derived. Beylkin (1985) and Cohen et al. (1986) have used the method to study the migration problem. A detailed study including numerical examples of the tomographic solution (55) appears not to have been performed yet. We have not included here a complete description of the conditions required of the functions involved in the solution. For example, it is required that the Jacobian (53) is non-singular.

It is interesting to investigate the simplification of expression (55) in more symmetric models. In 1-D models with turning rays and reflections (Figure 4), the weighting function (54) reduces to

$$b_j(\mathbf{x},\psi) = \frac{1}{2}\frac{v(0)v(z)\cos\psi}{v^2(0)-v^2(z)\sin^2\psi} \quad . \tag{56}$$

The inverse formula (55) becomes

$$f(\mathbf{x}) \approx \frac{1}{2}\int_{-u(z)}^{u(z)}\sum_{j=1}^{2} g^*(\psi,\rho_j(\mathbf{x},\psi))\frac{v(z)dp}{1-p^2v^2(z)} \tag{57}$$

where $p = u(0)\sin\psi$ is the ray parameter. The weighting function is simply related to the angle of the ray at depth, $z$ (the squared secant of the angle the ray makes with the vertical). In a homogeneous model with sources and receivers on parallel lines, the inverse formula (55,57) reduces to

$$f(\mathbf{x}) = \int_{-\pi/2}^{\pi/2} g^*(\psi,\rho_1(\mathbf{x},\psi))\sec\psi\, d\psi \tag{58}$$

with $\rho_1(\mathbf{x},\psi) = x + z\tan\psi$, which is identical to the IRT (11) with '$\sec\psi$' compensating for the different variables.

## 9. Conclusions.

In this chapter we have reviewed the Radon transform (RT) and its inverse (IRT). The IRT can be used to solve the 0-D tomographic problem, i.e. tomography with a homogeneous reference model. Two methods of solving the inverse problem are discussed: filtered back projection (FBP) and circular harmonic decomposition (CHD). A discrete version of the former is the obvious implementation of the IRT and is widely used in medical tomography. Unfortunately it introduces some numerical artifacts in the reconstruction which are inconsistent with the data. The CHD method is numerically unstable for high-wavenumber terms. A stable version can be derived using a consistency condition that the data must satisfy if it is derived from a projection experiment.

Two extensions of the IRT are discussed. The first generalizes the CHD method to any laterally homogeneous reference model. The solution is found as a Volterra integral equation. For low-wavenumber terms, an iterative solution will be efficient. The first iteration interprets the CMP data as if the anomaly were laterally homogeneous. Higher-order iterations correct for the lateral variations. The other extension of the IRT, generalizes the FBP method to inhomogeneous reference models. The back projection integral is modified by a weighting factor that compensates for the geometry of the curved rays. The asymptotic solution reconstructs the discontinuous or high-wavenumber anomaly structure.

Neither of these extensions of the IRT has been investigated in detail here or elsewhere, and full numerical implementations and examples are not available. While it may remain more versatile to solve seismic tomographic problems using purely numerical inversion schemes, further study of analytic techniques should be profitable and provide information and insight about difficulties inherent in tomographic inversion.

**Acknowledgments.**

This chapter was written while the author was a Cecil H. and Ida Green Scholar at the Institute of Geophysics and Planetary Physics, Scripps Institution of Oceanography, University of California, San Diego. He gratefully acknowledges this support and the assistance provided. This is Department of Earth Sciences, Cambridge, publication nr 798.

*Chapter 3*

# Numerical solution of large, sparse linear algebraic systems arising from tomographic problems

*A. van der Sluis and H.A. van der Vorst*

In this chapter we will digress on two classes of methods for solving the large sparse matrix systems arising from tomographic problems, viz. *ART-like methods* and *projection methods*.[1] The former class has been in use for several decades, the latter is more recent.

We will discuss the mathematical background and explicate some assumptions which (although perhaps unverifiable) underly the convergence and efficiency of the methods. We will also discuss the phenomenon that in the early stages of such processes the approximate solution suffers less from data errors than is the case later on (a certain regularizing effect in the early stages, if you wish) and give some attention to regularization and smoothing as such.

Some proofs will be given, but on several occasions we will point out that -at least to our knowledge- rigorous results are not available, and that, as a consequence, more mathematical research is in order.

1. In this chapter the term *projection* is used for mathematical projections in vector spaces, and differs from the experimental sense of the term that was used in chapter 2.



*G. Nolet (ed.), Seismic Tomography, 49–83.*

# 1. The problem

## 1.1 The model

In an experimental situation we consider the following *linear* model for the relationship between two vectorial quantities **s** and **d** :

$$\mathbf{As} = \mathbf{d} \tag{1.1}$$

In a (seismic) tomographic context this may have the following interpretation (that we will use repeatedly) :

- **A** is the $m \times n$-matrix whose element $A_{ij}$ denotes the path length of the $i$-th (seismic) ray in the $j$-th cell of a subdivision of the space;
- **s** is the $n$-vector whose coordinate $s_j$ denotes the slowness (inverse (seismic) velocity) in cell $j$ (or possibly the difference of this quantity and the corresponding quantity for an idealized model of the earth);
- **d** is the $m$-vector whose coordinate $d_i$ denotes the total travel time of the $i$-th ray (or possibly the difference of this quantity and the corresponding quantity for an idealized model of the earth).

Occasionally we will call (1.1) the *ideal* or *unperturbed* system. Note that implicit in the above is that (1.1) is exactly satisfied.

## 1.2 The problem

We assume the matrix **A** to be a known non-negative matrix. We assume that instead of **d** a vector **b** is available, which differs from **d** by (considerable) measuring errors, and that an approximation to the vector **s** is to be computed. The problem, then, is to determine a vector **x** from the set of equations

$$\mathbf{Ax} = \mathbf{b} \ . \tag{1.2}$$

Occasionally we will call this the *actual* or *perturbed* system.

Usually, this set of equations

- is sparse, i.e. only relatively few matrix elements are non-zero;
- is strongly overdetermined, i.e. $m >> n$;
- and at the same time is underdetermined, i.e. effectively we have rank(**A**) $< n$;
- is inconsistent, i.e. there exists no vector **x** satisfying (1.2) exactly.

Since (1.2) usually has no exact solution, one often resorts to a *least squares solution* (which always exists), i.e. a vector **x** for which

$$\| \mathbf{Ax} - \mathbf{b} \| \tag{1.3}$$

is minimal, where $\| . \|$ denotes the euclidean norm (cf. sec. 2.1).

If rank(**A**) $< n$ there are an infinity of vectors **x** minimizing (1.3). In this whole set of least squares solutions there is a unique vector whose norm is minimal. This is referred to as the *minimum norm least squares solution* of (1.2) and this is the solution that one is usually after in this case.

There are some complications, however :

- the least squares solvers that work so well for dense matrices become very expensive in terms of computer time and memory space for sparse matrices; therefore there is a demand for more efficient methods for this case;
- the errors in **b** may affect the least squares solutions very badly; in those cases other approximate solutions of (1.2) are desirable which suffer less from the errors in **b**.

## 2. Mathematical preliminaries

### 2.1 Notation

The following notation will be used throughout this paper:

- $\mathbf{A}^T$ will denote the transpose of the matrix **A**;
- $R(\mathbf{A})$ will denote the range of the matrix **A**, i.e. the set of all vectors $\mathbf{Ay}$ ;
- $N(\mathbf{A})$ will denote the nullspace of **A**, i.e. the set of all vectors **y** for which $\mathbf{Ay} = \mathbf{0}$ ;
- sgn($a$) for a real number $a$ will denote 1, 0 or -1 according to $a > 0$, $a = 0$, or $a < 0$ ;
- $\|\mathbf{u}\|$ for a vector **u** will denote the euclidean norm $\left[\sum u_i^2\right]^{1/2}$ ;
- $\|\mathbf{A}\|$ for a matrix **A** will denote $\max\limits_{\mathbf{u}\neq 0} \dfrac{\|\mathbf{Au}\|}{\|\mathbf{u}\|}$ ; we note that $\|\mathbf{A}\| = \sqrt{\mu}$, $\mu$ the largest eigenvalue of $\mathbf{A}^T\mathbf{A}$ ;
- tr(**A**) for a matrix **A** will denote its trace $\sum\limits_i A_{ii}$ .

### 2.2 Basic properties of least squares problems

We note the following basic properties and give some hints how they come about, where it is explicitly allowed that rank( **A**) $< n$ (see also Golub & Van Loan 1983, sec. 6.1).

From a geometrical argument (fig. 2.1) we have

i **x** is a least squares solution of (1.2) if and only if $\mathbf{Ax} - \mathbf{b} \perp R(\mathbf{A})$.
As a consequence **Ax** is the same for all least squares solutions **x** of (1.2).

Rewriting $\mathbf{Ax} - \mathbf{b} \perp R(\mathbf{A})$ as $\mathbf{A}^T(\mathbf{Ax} - \mathbf{b}) = \mathbf{0}$ we get

ii **x** is a least squares solution of (1.2) if and only if

$$\mathbf{A}^T\mathbf{Ax} = \mathbf{A}^T\mathbf{b} \tag{2.1}$$

(the normal equations).

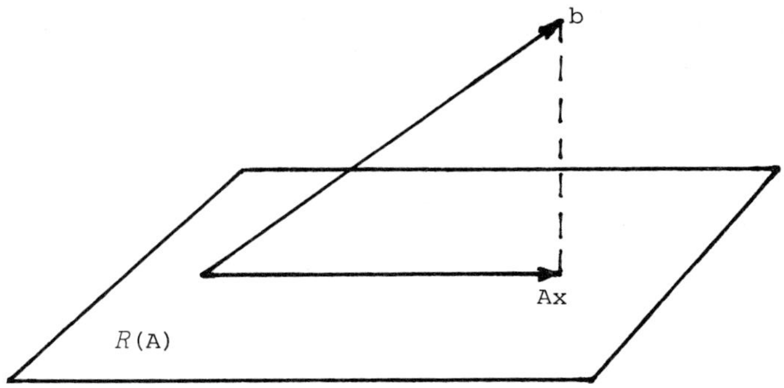


*Figure 2.1*

Finally, if $\mathbf{x}$ is a least squares solution of (1.2) then all other least squares solutions may be written as $\mathbf{x}+\mathbf{y}$, $\mathbf{y} \in N(\mathbf{A})$, and hence, again by a geometrical argument,

iii if $\mathbf{x}$ is a least squares solution of (1.2) then it is the minimum norm least squares solution if and only if $\mathbf{x} \perp N(\mathbf{A})$, or, which amounts to the same, if and only if $\mathbf{x} \in R(\mathbf{A}^T)$.

### 2.3 Effects of row- and column-scaling and shifting

If we multiply each single equation in (1.2) by a constant, i.e. we consider the system $\mathbf{RAx}=\mathbf{Rb}$, $\mathbf{R}$ a diagonal matrix, then the new system will usually not have the same least squares solution(s) as the old one, since we are now, in fact, measuring the residual vector $\mathbf{b}-\mathbf{Ax}$ in a different norm.

Likewise, if rank(**A**) $< n$ and one solves for the minimum norm least squares solution $\mathbf{y}$ of $\mathbf{ACy}=\mathbf{b}$, $\mathbf{C}$ a diagonal matrix (i.e. we multiply each column of $\mathbf{A}$ by a constant), then $\mathbf{Cy}$ will usually be different from the minimum norm least squares solution of $\mathbf{Ax}=\mathbf{b}$ since we are now, in fact, measuring $\mathbf{x}$ in a different norm.

One should, therefore, be careful with scaling rows and columns, and look for that scaling that gives the most meaningful solution. For example, one might choose $\mathbf{R}$ in such a way that the errors in the coordinates of $\mathbf{Rb}$ all have the same variance, since then, at least if rank(**A**) $= n$, on account of the Gauss Markov theorem (cf. Silvey 1970), the least squares solution is the best unbiased linear estimate of $\mathbf{s}$ in the sense that it has minimum variance (see also sec. 2.5).

If rank(**A**) $< n$ it may be desirable that one coordinate of $\mathbf{x}$ weighs heavier than the other when minimizing $\mathbf{x}$. In tomography problems, e.g., with $\mathbf{A}$ and $\mathbf{x}$ as in secs. 1.1, 1.2, one might wish to minimize $\sum n_j v_j x_j^2$, where $n_j$ is the number of rays hitting cell $j$ and $v_j$ is the volume of cell $j$. This can then be accomplished by taking $\mathbf{C} = diag\ [1/\sqrt{n_j v_j}]$.

Another operation to handle with caution if rank(A) $< n$ is that of shifting, i.e. replacing the least squares problem $\mathbf{Ax}=\mathbf{b}$ by $\mathbf{A\tilde{x}}=\mathbf{b}-\mathbf{Ay}$, $\mathbf{y}$ a given vector, and after solving this, taking $\mathbf{x}=\mathbf{y}+\tilde{\mathbf{x}}$. Indeed, if $\tilde{\mathbf{x}}$ is the minimum norm least squares solution of the latter problem, then $\mathbf{x} = \mathbf{y} + \tilde{\mathbf{x}}$ still is a least squares solution of $\mathbf{Ax} = \mathbf{b}$, but it will no longer be a minimum norm least squares solution unless $\mathbf{y} \perp N(\mathbf{A})$. If $\text{rank}(\mathbf{A}) = n$ shifting causes no problems.

Note, however, that in tomography problems, one often has a solution $\mathbf{y}$ for an idealized or model problem, and one may not be so much interested in a minimum norm solution to the real (non-idealized) problem with the same matrix as in a solution with a minimum norm deviation from $\mathbf{y}$. In that case shifting over $\mathbf{y}$ is just the thing to do.

### 2.4 Singular value decomposition

For any $m \times n$-matrix $\mathbf{A}$ there exist an orthogonal $m \times m$-matrix $\mathbf{U}$, an orthogonal $n \times n$-matrix $\mathbf{V}$ and an $m \times n$-diagonal matrix $\mathbf{\Sigma}$ with diagonal elements $\sigma_1 \geq \sigma_2 \geq \cdots \geq \sigma_n \geq 0$ such that

$$\mathbf{A} = \mathbf{U\Sigma V}^T \quad , \tag{2.2}$$

schematically :

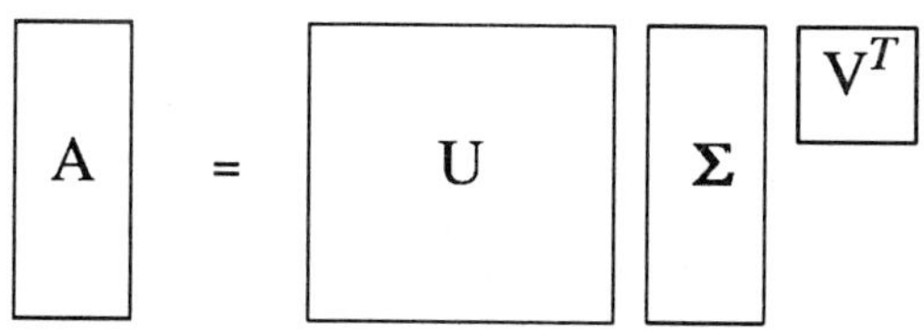


(cf. Golub & Van Loan 1983, sec. 2.3).

This matrix product is called the *singular value decomposition* of $\mathbf{A}$. The columns of $\mathbf{U}$ (or $\mathbf{V}$, respectively) are called the *left (right) singular vectors* of $\mathbf{A}$ *and the* $\sigma_j$ *are called the singular values* of $\mathbf{A}$. The columns of $\mathbf{U}$ (or $\mathbf{V}$) are eigenvectors of $\mathbf{AA}^T$ (or $\mathbf{A}^T\mathbf{A}$) and the corresponding eigenvalues are in either case $\sigma_j^2$.

This singular value decomposition is a very important tool for many kinds of matrix problems, both theoretical and practical. A geometrical interpretation is that for any linear mapping there are orthonormal bases in source and image space (given by the right and left singular vectors, respectively) on which this mapping is represented by a diagonal matrix, and for many purposes diagonal matrices are much simpler, of course, than non-diagonal matrices.

We have $\text{rank}(\mathbf{A}) = p$ if and only if $\sigma_p \neq 0$ and $\sigma_{p+1} = \cdots = \sigma_n = 0$. If $\sigma_{p+1}, \ldots, \sigma_n$ are very small with respect to $\sigma_1, \ldots, \sigma_p$ then for computational purposes $\mathbf{A}$ is hardly distinguishable from a rank $p$ matrix, and we may say that $\mathbf{A}$ has *effectively* rank $p$ (cf. sec. 1.2).

Now suppose we have (1.2) with $\text{rank}(\mathbf{A}) = p$. If we write $\mathbf{b} = \mathbf{Ug}$ (i.e. we express $\mathbf{b}$ as a linear combination of the columns of $\mathbf{U}$; obviously $\mathbf{g} = \mathbf{U}^T\mathbf{b}$) and $\mathbf{x} = \mathbf{Vz}$ (with a similar remark) then $\mathbf{x}$ is a least squares solution of (1.2) if and only if

$$z_j = g_j / \sigma_j, \quad j \le p \tag{2.3}$$

and $\mathbf{x}$ is the minimum norm least squares solution if and only if in addition $z_{p+1} = \cdots = z_n = 0$.

This may be expressed in matrix-vector notation as follows: the minimum norm least squares solution is given by

$$\mathbf{x} = \mathbf{V}\mathbf{\Sigma}^{+}\mathbf{U}^T\mathbf{b} \ , \tag{2.4}$$

where $\mathbf{\Sigma}^{+}$ is the $n \times m$-diagonal matrix with diagonal elements $1/\sigma_1, \ldots, 1/\sigma_p, 0, \ldots, 0$. The matrix $\mathbf{V}\mathbf{\Sigma}^{+}\mathbf{U}^T$ is called the (Moore-Penrose) *generalized inverse* of $\mathbf{A}$, denoted as $\mathbf{A}^{+}$.

From $\mathbf{As} = \mathbf{d}$ (cf. (1.1)) we have, writing $\mathbf{s} = \mathbf{V}\bar{\mathbf{z}}$, $\mathbf{d} = \mathbf{U}\bar{\mathbf{g}}$, right away

$$\bar{z}_j = \bar{g}_j / \sigma_j, \quad j \le p. \tag{2.5}$$

We get no relations for $\bar{z}_{p+1}, \ldots, \bar{z}_n$, however. And indeed, these quantities represent components of $\mathbf{s}$ in the nullspace of $\mathbf{A}$, and therefore don't contribute to $\mathbf{d}$; hence they cannot be reconstructed from $\mathbf{d}$ nor be estimated by solving (1.2). Therefore, in error estimates further on, we will disregard the (possible) components of $\mathbf{s}$ in $N(\mathbf{A})$, or, what amounts to the same, assume $\mathbf{s} \perp N(\mathbf{A})$, i.e. $\bar{z}_{p+1} = \cdots = \bar{z}_n = 0$, or in matrix vector notation

$$\mathbf{s} = \mathbf{V}\mathbf{\Sigma}^{+}\mathbf{U}^T\mathbf{d} \ . \tag{2.6}$$

## 2.5 Effects of data errors

Let us write the measured vector $\mathbf{b}$ as $\mathbf{d} + \boldsymbol{\varepsilon}$ with $\mathbf{d}$ as in (1.1), and suppose that $\boldsymbol{\varepsilon}$ is a vector whose components average 0 and have equal variance $\sigma^2$ but are uncorrelated, i.e. $var(\boldsymbol{\varepsilon}) = \sigma^2\mathbf{I}$.

Then, on account of the Gauss-Markov theorem (cf. Silvey 1970), if $\mathbf{A}$ has rank $n$ the least squares solution $\mathbf{x}$ of (1.2) is the best unbiased linear estimate of the vector s in (1.1) in the sense that it has minimum variance. Likewise, if $\mathbf{A}$ has lower rank, then the minimum norm least squares solution $\mathbf{x}$ is the best unbiased linear estimate of the component of s orthogonal to $N(\mathbf{A})$ (as has already been noted it will never be possible to estimate the component of $\mathbf{s}$ in the nullspace).

The variance of the least squares solution can be very large, however. Indeed , looking at the full rank case we note that the variance matrix $E(\Delta\mathbf{x}\Delta\mathbf{x}^T), E$ denoting the probabilistic expectation and $\Delta\mathbf{x}$ denoting the error vector $\mathbf{x} - \mathbf{s}$ , equals

$$E[(\mathbf{A}^T\mathbf{A})^{-1}\mathbf{A}^T\boldsymbol{\varepsilon}\,\boldsymbol{\varepsilon}^T\mathbf{A}(\mathbf{A}^T\mathbf{A})^{-1}] = \sigma^2(\mathbf{A}^T\mathbf{A})^{-1},$$

and hence

$$E\left[\ \|\Delta\mathbf{x}\|^2\right] = \sigma^2 tr\left[\mathbf{A}^T\mathbf{A}\right]^{-1} = \sigma^2\sum_j \frac{1}{\sigma_j^2} \ . \tag{2.7}$$

Thus, if there are small singular values, large errors in $\mathbf{x}$ should be expected.

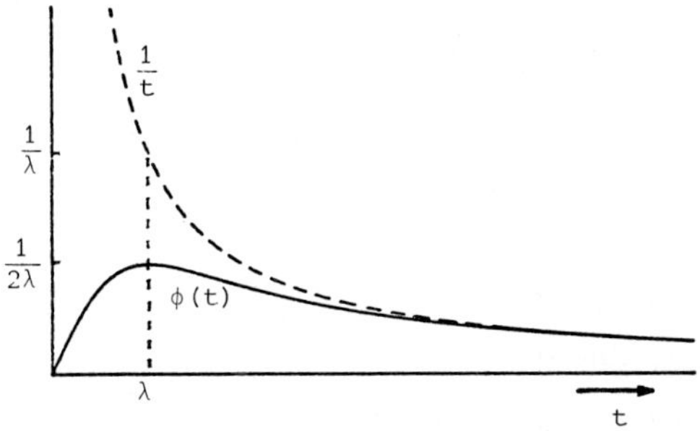


*Figure 2.2*

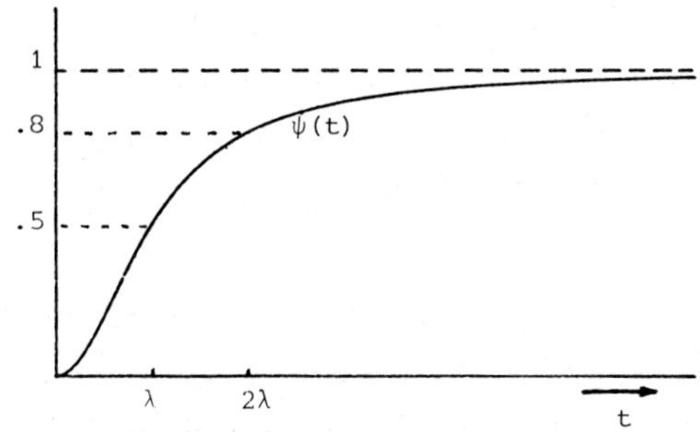


*Figure 2.3*

## 2.6 Regularization

As we saw, when there are very small singular values then already small errors in **b** may have dramatic effects on **x** (see (2.7)). Regularization is a way to circumvent this. Two common ways of doing this are :

*Regularization (i)*

Instead of (1.2) solve the least squares problem

$$\begin{bmatrix} \mathbf{A} \\ \lambda \mathbf{I} \end{bmatrix} \mathbf{x} = \begin{bmatrix} \mathbf{b} \\ \mathbf{0} \end{bmatrix} \tag{2.8}$$

for a certain $\lambda$.

In order to describe what happens we first note that the least squares solution $\mathbf{x}'$ of (2.8) may be expressed as

$$\mathbf{x}' = \left[ \mathbf{A}^T \mathbf{A} + \lambda^2 \mathbf{I} \right]^{-1} \mathbf{A}^T \mathbf{b} \tag{2.9}$$

(cf. sec. 2.2(ii)). Using (2.2) we get $\mathbf{x}' = \mathbf{V}(\mathbf{\Sigma}^T \mathbf{\Sigma} + \lambda^2 \mathbf{I})^{-1} \mathbf{\Sigma} \mathbf{U}^T \mathbf{b}$. If we now express $\mathbf{x}$ (the minimum norm solution of (1.2)), $\mathbf{x}'$ and $\mathbf{b}$ in terms of singular vectors of $\mathbf{A}$: $\mathbf{x} = \mathbf{V}\mathbf{z}$ , $\mathbf{x}' = \mathbf{V}\mathbf{z}'$ , $\mathbf{b} = \mathbf{U}\mathbf{g}$, then we have $\mathbf{z} = (\mathbf{\Sigma}^T \mathbf{\Sigma} + \lambda^2 \mathbf{I})^{-1} \mathbf{\Sigma} \mathbf{g}$ and hence (see also (2.3))

$$z'_j = g_j \phi(\sigma_j) = z_j \psi(\sigma_j) \tag{2.10}$$

$$\phi(t) = \frac{t}{t^2 + \lambda^2} \quad , \quad \psi(t) = \frac{t^2}{t^2 + \lambda^2} = \frac{1}{1 + \lambda^2/t^2} \tag{2.11}$$

with graphs as in figures 2.2 and 2.3.

From the first equality in (2.10) we see that $\phi$ describes how the various components of **b** affect the solution. For this reason we will call $\phi$ the *response function* for (2.8). As we see from figure 2.2, the error components in **b** corresponding to the small singular values now have much smaller effects than in the non-regularized case (where the response function is, in fact, $\phi(t) = 1/t$, as we see from (2.3)).

Applying (2.10) to the unperturbed system we note that $\psi$ describes the effect of regularization on the ideal solution relative to this solution. Therefore $\psi$ will be called the *relative response function.*

A variant of this way of regularization is to replace **I** in (2.8) by another matrix.

*Regularization (ii)*

Compute the singular value decomposition of **A** and then compute the $z_j$ from (2.3), however, taking $z_j = 0$ if $\sigma_j$ is below a certain threshold $\eta$.

We may again define response functions $\phi$ and $\psi$ leading to (2.10), as follows: $\phi(t) = 1/t$ , $\psi(t) = 1$ for $t \geq \eta$ and $\phi(t) = \psi(t) = 0$ for $t < \eta$.

In the next section we shall quantify the effects of data errors and regularization. For the moment we note that a look at the function $\psi$ suffices to see that either way of regularization will only give reasonable results if the unknown vector **s** is such that, writing $\mathbf{s} = \mathbf{V}\bar{\mathbf{z}}$, the sum $\sum \bar{z}_j^2$ for the $\bar{z}_j$ corresponding to the small $\sigma_j$ is not too large. In other words, the total contribution to **s** of the components with respect to the right singular vectors (i.e. the eigenvectors of $\mathbf{A}^T\mathbf{A}$) corresponding to the small singular values should not be too large.

Whether solutions of tomographic problems have this property is an open question. A positive indication to this effect is that positive matrices, like $\mathbf{A}^T\mathbf{A}$, tend to have strongly oscillating eigenvectors corresponding to the small eigenvalues (but there is no proof of this), whereas on the whole one expects the coordinates of **s** (the slownesses) to vary rather slowly as one goes from cell to cell. However, the occurrence of sharp changes in the slownesses will cause the rapidly oscillating components of **s** not to be too small. See also the observation to this effect concerning Nolet's problem at the end of our epilogue (sec. 6).

## 2.7 Quantitative effects of regularization

We will now assess the quantitative effects of regularization. We will do this in a more general setting so as to enable us to use these results for other approximate solution methods as well.

Thus, let some approximate solution method for (1.2) be given, yielding an approximate solution $\mathbf{x}'$. Let there be response functions $\phi$ and $\psi$ so that with $\mathbf{x} = \mathbf{Vz}$ , $\mathbf{x}' = \mathbf{Vz}'$ , $\mathbf{b} = \mathbf{Ug}$ we have (2.10). Then we have $\mathbf{x}' = \mathbf{V\Phi U}^T\mathbf{b}$ with $\mathbf{\Phi} = diag\,(\phi(\sigma_j))$. Since we assumed $\mathbf{s} = \mathbf{V\Sigma}^+\mathbf{U}^T\mathbf{d}$ with $\mathbf{\Sigma}^+ = diag\,(1/\sigma_j)$ (cf. (2.6)) we have for the error $\Delta\mathbf{x} \equiv \mathbf{x}' - \mathbf{s}$:

$$\Delta\mathbf{x} = \mathbf{x}' - \mathbf{s} = \mathbf{V}\left[\mathbf{\Phi} - \mathbf{\Sigma}^+\right]\mathbf{U}^T\mathbf{d} + \mathbf{V\Phi U}^T\left[\mathbf{b} - \mathbf{d}\right] . \qquad (2.12)$$

The first term on the right is the error we would still get if there were no data errors at all. It is caused by the fact that we use an approximate method. This term will therefore, in general, be called the *approximation error* $\Delta\mathbf{x}_{appr}$, in our case the *regularization error,* denoted by $\Delta\mathbf{x}_{reg}$.

The second term on the right in (2.12) is caused by the data errors and may, therefore, be called the *perturbation error* $\Delta\mathbf{x}_{pert}$.

Then we have the following theorem :

**Theorem 2.13**

Suppose $\mathbf{b} = \mathbf{d} + \boldsymbol{\varepsilon}$ with $\boldsymbol{\varepsilon}$ as in the beginning of sec. 2.5. Write $\mathbf{s} = \mathbf{V}\bar{\mathbf{z}}$, $\mathbf{d} = \mathbf{U}\bar{\mathbf{g}}$. Then

$$\| \Delta\mathbf{x}_{appr} \|^2 = \sum \bar{g}_j^2 \left[ \phi(\sigma_j) - \frac{1}{\sigma_j} \right]^2 = \sum \bar{z}_j^2 (\psi(\sigma_j) - 1)^2 \tag{2.14}$$

$$E\left[ \| \Delta\mathbf{x}_{pert} \|^2 \right] = \sigma^2 \sum \phi(\sigma_j)^2 \tag{2.15}$$

$$E\left[ \| \Delta\mathbf{x} \|^2 \right] = \| \Delta\mathbf{x}_{appr} \|^2 + E\left[ \| \Delta\mathbf{x}_{pert} \|^2 \right] . \tag{2.16}$$

**Proof**

(2.14) is straightforward. (2.15) follows from $E(\| \Delta\mathbf{x}_{pert} \|^2) = E(\Delta\mathbf{x}_{pert}^T \Delta\mathbf{x}_{pert}) = E(\boldsymbol{\varepsilon}^T \mathbf{U}\boldsymbol{\Phi}^T \mathbf{V}^T \mathbf{V}\boldsymbol{\Phi}\mathbf{U}^T \boldsymbol{\varepsilon}) = \sigma^2 tr(\mathbf{U}\boldsymbol{\Phi}^T \boldsymbol{\Phi}\mathbf{U}^T) = \sigma^2 tr(\boldsymbol{\Phi}^T \boldsymbol{\Phi})$. Finally (2.16) follows from $E(\Delta\mathbf{x}_{appr}^T \Delta\mathbf{x}_{pert}) = E(\Delta\mathbf{x}_{appr}^T \mathbf{V}\boldsymbol{\Phi}\mathbf{U}^T \boldsymbol{\varepsilon}) = 0$.

□

For regularization (i) we thus get

$$\| \Delta\mathbf{x}_{reg} \|^2 = \lambda^4 \sum \left[ \frac{\bar{g}_j}{(\sigma_j^2 + \lambda^2)\, \sigma_j} \right]^2 = \lambda^4 \sum \left[ \frac{\bar{z}_j}{\sigma_j^2 + \lambda^2} \right]^2 \tag{2.17}$$

$$E\left[ \| \Delta\mathbf{x}_{pert} \|^2 \right] = \sigma^2 \sum \left[ \frac{\sigma_j}{\sigma_j^2 + \lambda^2} \right]^2 . \tag{2.18}$$

For regularization (ii) we get

$$\| \Delta\mathbf{x}_{reg} \|^2 = \sum_{\sigma_j < \eta} \left[ \frac{\bar{g}_j}{\sigma_j} \right]^2 = \sum_{\sigma_j < \eta} \bar{z}_j^2 \tag{2.19}$$

$$E\left[ \| \Delta\mathbf{x}_{pert} \|^2 \right] = \sigma^2 \sum_{\sigma_j \geq \eta} \frac{1}{\sigma_j^2} . \tag{2.20}$$

We note that in either case we get a biased solution, since $E(\Delta\mathbf{x}) = \Delta\mathbf{x}_{reg} \neq 0$. But we also get a reduced variance $E(\|\Delta\mathbf{x}\|^2)$. By a judicious choice of $\lambda$ or $\eta$ we may try to optimize the trade-off between bias and variance.

## 3. SIRT methods

### 3.1 The need for iterative methods

For linear systems with large sparse matrices direct solution methods (such as those based on Householder transformations, cf. Golub & Van Loan 1983, sec. 6.2) , are unattractive since they take too much computer time ($O(mn^2)$) and memory space ($O(mn)$). The reason for this is that, although the original matrix is sparse, it fills up during the process.

Iterative processes just work with the matrix as it stands, and thus do not suffer this fill in. Since, moreover, one will usually not be interested in a high accuracy solution - due to model and data errors - a suitable iterative method might already give an acceptable solution in relatively few steps.

Since for large systems already the storage space for the non-zero matrix elements only may exceed the central storage capacity of the computer, those elements will have to be stored in secondary storage, and this will usually be done in a row-wise fashion. Therefore, methods in which the approximate solution is updated by processing the equations successively are particularly attractive. Such methods are generally referred to as Row Action (RA) methods or Algebraic Reconstruction Techniques (ART). A very early example of this is Kaczmarz's method (cf. Kaczmarz 1937), and many current methods may be considered as outgrowths of this, as we shall now show.

### 3.2 Kaczmarz's method

Starting with some initial approximation $\mathbf{x}^{(0)}$ for the solution of $\mathbf{Ax} = \mathbf{b}$, we define the residual $\mathbf{r}^{(q)}$ for the approximation $\mathbf{x}^{(q)}$ after $q$ iteration steps by

$$\mathbf{r}^{(q)} = \mathbf{b} - \mathbf{A}\mathbf{x}^{(q)} \quad . \tag{3.1}$$

The idea now is to determine a correction $\Delta\mathbf{x}^{(q)}$ to $\mathbf{x}^{(q)}$ such that a certain equation, say the $i$-th one , is satisfied, i.e.

$$r_i^{(q+1)} = 0 \quad . \tag{3.2}$$

Usually one starts with the first equation to find $\Delta\mathbf{x}^{(0)}$, then the second one for $\Delta\mathbf{x}^{(1)}$ and so on. After having used the last equation we start again with the first one, so that we have typically

$$i = q \ mod \ m \tag{3.3}$$

(i.e. $i$ is the remainder if $q$ is divided by $m$). Obviously there are an infinite number of possibilities for choosing $\Delta\mathbf{x}^{(q)}$, but since we should allow only very small corrections $\Delta\mathbf{x}^{(q)}$ when $\mathbf{x}^{(q)}$ is close to its limit, it seems wise to choose the $\Delta\mathbf{x}^{(q)}$ which is minimal in a given norm $\|.\|_p$:

$$\| \Delta \mathbf{x}^{(q)} \|_p \equiv \left[ \sum_j | \Delta x_j^{(q)} |^p \right]^{1/p} , \quad p \geq 1 . \tag{3.4}$$

This then leads to the following correction $\Delta \mathbf{x}^{(q)}$:

$$\Delta x_j^{(q)} = \frac{sgn(A_{ij}) | A_{ij} |^w r_i^{(q)}}{\sum_k | A_{ik} |^{w+1}} \quad , w = \frac{1}{p-1} \tag{3.5}$$

(which is also valid if A has elements of different sign).

The choice $p = 2$ (minimal energy corrections) leads to Kaczmarz's method:

$$x_j^{(q+1)} = x_j^{(q)} + \frac{A_{ij} r_i^{(q)}}{\sum_k A_{ik}^2} . \tag{3.6}$$

This method converges to a solution of $\mathbf{Ax} = \mathbf{b}$ only if $\mathbf{b} \in R(\mathbf{A})$, i.e. if $A\mathbf{x} = \mathbf{b}$ has an exact solution. However, due to data errors, **b** will usually not lie in $R(\mathbf{A})$, and then the iterands necessarily keep fluctuating.

### 3.3 Averaging forms of Kaczmarz's method; SIRT methods

Dines and Lyttle (1979) suggest, among other things, to improve the convergence behaviour by first computing the corrections for all the rows, *while keeping the residual fixed*, and to average these corrections before updating the approximation for **x**. This way, the formula in (3.6) leads to

$$x_j^{(q+1)} = x_j^{(q)} + \frac{1}{M_j} \sum_i \frac{A_{ij} r_i^{(q)}}{\sum_k A_{ik}^2} , \tag{3.7}$$

where $M_j$ denotes the number of non-zero elements in the $j$-th column of **A**, or physically: the number of rays passing through cell $j$. Note that we now no longer have the coupling between $q$ and $i$ as in (3.3), and that one iteration step of (3.7) corresponds more or less to the work of $m$ iteration steps in the original Kaczmarz method.

Methods of this type - and there are many of them - in which the approximate solution is updated only after all equations have been processed are called *simultaneous iterative reconstruction techniques*, SIRT for short.

The iteration method (3.7) is a member of the following two-parameter family:

$$\begin{cases} x_j^{(q+1)} = x_j^{(q)} + \dfrac{\omega}{\gamma_j} \sum_i \dfrac{A_{ij} r_i^{(q)}}{\rho_i} , & 0 < \omega < 2 \\ \gamma_j = \sum_i | A_{ij} |^{\alpha} , \quad \rho_i = \sum_k | A_{ik} |^{2-\alpha} , & 0 \leq \alpha \leq 2 \end{cases} \tag{3.8}$$

and defining $0^0 = 0$.

For $\alpha = 0$, $\omega = 1$ we obtain (3.7), for $\alpha = 1$, $\omega = 1$ we obtain an iteration method suggested by Hager et al. (1985).

For any $\alpha$, $0 \le \alpha \le 2$, and any $\omega$, $0 < \omega < 2$, the sequence $\mathbf{x}^{(q)}$ converges, but not, in general, to the minimum norm solution of the given least squares problem but to the minimum norm solution of a least squares problem obtained from $\mathbf{Ax} = \mathbf{b}$ by a rescaling of rows and columns. For the consequences of this see sec. 2.3. This may be partly undone, however. A proof of this and more detailed information about the convergence behaviour and regularizing properties of this family will be the subject of section 4.

### 3.4 Further generalizations

Beside the averaging process, as in (3.7), Dines and Lyttle (1979) made a number of suggestions to increase the efficiency of iterative methods:

i. Use (3.5) with $p = \infty$, i.e. we minimize in the $\| . \|_\infty$ norm: $\| \mathbf{u} \|_\infty = \max | u_i |$. We then get

$$\Delta x_j^{(q)} = \frac{sgn\,(A_{ij}) r_i^{(q)}}{\sum_k |A_{ik}|} \quad . \tag{3.9}$$

ii. In order to save computations, the residual in (3.9) might be replaced by

$$\tilde{r}_i^{(q)} = b_i - \sum_k sgn\,(A_{ik}) \frac{L_i}{N_i} x_k^{(q)} \tag{3.10}$$

with
$L_i = \sum_k |A_{ik}|$ (in the tomography problem this is the length of the $i$-th ray),
$N_i$ the number of non-zeros on the $i$-th row of $\mathbf{A}$ (i.e. the number of cells through which the $i$-th ray passes).

This actually means changing the equations, of course, but note that the sum of absolute values per row does not change. Hence, when there are very many rays and the rays are sufficiently random, it is quite possible that this does not affect the solution a great deal. Moreover, this could be a useful strategy if one can generate as many rays as one wishes: it could then be decided not to store the equations at all, and just to process each equation as it comes in; it means then a substantial reduction in computing time if one does not have to compute the pathlength of each ray in each cell. Note, however, our concluding remarks in this section.

iii. The averaging algorithm could be taken somewhat more sophisticated than the one in (3.7) (mind the printing errors in Dines & Lyttle 1979)

$$x_j^{(q+1)} = x_j^{(q)} + \sum_i \frac{sgn\,(A_{ij}) r_i^{(q)}}{N_i^4 L_i} \Big/ \sum_i \frac{|sgn\,(A_{ij})|}{N_i^4} \tag{3.11}$$

with $L_i$ and $N_i$ as above. This means that in the averaging process short rays weigh more heavily than long ones.

iv. After each iteration the iterand might be smoothed, i.e. each of its coordinates is replaced by a weighted average of it and its neighbours, neighbour to be understood in the sense of corresponding to a neighbouring cell.

We know of no convergence proofs for algorithms based on these ideas. In this connection it is interesting to note that the variant

$$x_j^{(q+1)} = x_j^{(q)} + \frac{1}{M_j} \sum_i \frac{sgn(A_{ij}) r_i^{(q)}}{\sum_k |A_{ik}|} \tag{3.12}$$

of (3.7) does not converge to anything useful if **A** is a full matrix. Indeed, in that case $x_j^{(q+1)} - x_j^{(q)}$ is independent of j, which means that for the limit vector $\hat{\mathbf{x}}$ we have that $\hat{x}_j - x_j^{(0)}$ is independent of j, and one can actually show that all $\mathbf{x}^{(q)}$ are equal for $q \geq 1$.

## 4. Convergence and regularization properties of a family of SIRT methods

### 4.1 The convergence proof

We consider the family (3.8):

$$\begin{cases} x_j^{(q+1)} = x_j^{(q)} + \dfrac{\omega}{\gamma_j} \displaystyle\sum_i \frac{A_{ij} r_i^{(q)}}{\rho_i} , & 0 < \omega < 2 \\ \gamma_j = \displaystyle\sum_i |A_{ij}|^{\alpha} , \quad \rho_i = \sum_k |A_{ik}|^{2-\alpha} , & 0 \leq \alpha \leq 2 \end{cases} \tag{4.1}$$

Define $\mathbf{C} = diag(\gamma_j)$ , $\mathbf{R} = diag(\rho_j)$.

**Theorem 4.2**

For any of the above values of $\alpha$ and $\omega$ the process converges to a least squares solution $\mathbf{x}$ of $\mathbf{R}^{-1/2}\mathbf{A}\mathbf{x} = \mathbf{R}^{-1/2}\mathbf{b}$. If rank(**A**) $< n$ this solution lies closest to $\mathbf{x}^{(0)}$ in the norm $\|.\|'$ defined by $\|\mathbf{u}\|' = (\mathbf{u}^T\mathbf{C}\mathbf{u})^{1/2}$. If, in particular, $\mathbf{x}^{(0)} = \mathbf{0}$ then $\mathbf{x} = \mathbf{C}^{-1/2}\mathbf{y}$, $\mathbf{y}$ the minimum norm least squares solution of $\mathbf{R}^{-1/2}\mathbf{A}\mathbf{C}^{-1/2}\mathbf{y} = \mathbf{R}^{-1/2}\mathbf{b}$.

**Proof**

(This proof runs mostly along the lines set forth by Ivansson 1983). In matrix-vector notation the process reads

$$\mathbf{x}^{(q+1)} = \mathbf{x}^{(q)} + \omega \mathbf{C}^{-1}\mathbf{A}^T\mathbf{R}^{-1}\mathbf{r}^{(q)} , \quad \mathbf{r}^{(q)} = \mathbf{b} - \mathbf{A}\mathbf{x}^{(q)} . \tag{4.3}$$

Hence

$$\mathbf{x}^{(q+1)} = \left[ \mathbf{I} - \omega \mathbf{C}^{-1}\mathbf{A}^T\mathbf{R}^{-1}\mathbf{A} \right] \mathbf{x}^{(q)} + \omega \mathbf{C}^{-1}\mathbf{A}^T\mathbf{R}^{-1}\mathbf{b} . \tag{4.4}$$

Define $\mathbf{u}^{(q)} = \mathbf{C}^{1/2}\mathbf{x}^{(q)}$ , $\mathbf{W} = \mathbf{R}^{-1/2}\mathbf{A}\mathbf{C}^{-1/2}$ and $\mathbf{f} = \mathbf{R}^{-1/2}\mathbf{b}$. Then

$$\mathbf{u}^{(q+1)} = \left[ \mathbf{I} - \omega \mathbf{W}^T \mathbf{W} \right] \mathbf{u}^{(q)} + \omega \mathbf{W}^T \mathbf{f} \ . \tag{4.5}$$

Let

$$\mathbf{W} = \mathbf{U}\mathbf{\Sigma}\mathbf{V}^T \tag{4.6}$$

be the singular value decomposition of $\mathbf{W}$ (cf. sec. 2.4), and suppose $\sigma_1 \geq \sigma_2 \geq \cdots \geq \sigma_p > 0 = \sigma_{p+1} = \cdots = \sigma_n$ (i.e. $\mathbf{W}$ has rank $p$, and so has $\mathbf{A}$). Then

$$\mathbf{W}^T \mathbf{W} = \mathbf{V} \, diag \left[ \sigma_j^2 \right] \mathbf{V}^T \ . \tag{4.7}$$

Writing $\mathbf{f} = \mathbf{U}\mathbf{g}$ and $\mathbf{u}^{(q)} = \mathbf{V}\mathbf{z}^{(q)}$, i.e. $\mathbf{z}^{(q)}$ denotes the components of $\mathbf{u}^{(q)}$ on an eigenvector basis of $\mathbf{W}^T\mathbf{W}$, we get from (4.5)

$$z_j^{(q+1)} = \left[ 1 - \omega \, \sigma_j^2 \right] z_j^{(q)} + \omega \, \sigma_j g_j \ . \tag{4.8}$$

Hence

$$\begin{cases} z_j^{(q)} = \dfrac{g_j}{\sigma_j} + \left[ z_j^{(0)} - \dfrac{g_j}{\sigma_j} \right] \left[ 1 - \omega\sigma_j^2 \right]^q & , \ j \leq p \\[2ex] z_j^{(q)} = z_j^{(0)} & , \ j > p \ . \end{cases} \tag{4.9}$$

Thus, we will have convergence if $0 < \omega \, \sigma_j^2 < 2$ for all $j$, and this will certainly be the case if all $\sigma_j \leq 1$. The latter will be the case if all eigenvalues of $\mathbf{W}^T\mathbf{W}$ are at most 1 (cf. (4.7)), and we note that the largest eigenvalue of $\mathbf{W}^T\mathbf{W}$ equals $\| \mathbf{W} \|^2$. We have for any vector $\mathbf{v}$

$$\begin{aligned} \| \mathbf{W}\mathbf{v} \|^2 &= \sum_i \left[ \sum_j \frac{A_{ij}}{\sqrt{\rho_i}} \frac{v_j}{\sqrt{\gamma_j}} \right]^2 \leq \sum_i \left[ \sum_j \frac{|A_{ij}|^{1-\alpha/2}}{\sqrt{\rho_i}} \frac{|A_{ij}|^{\alpha/2} v_j}{\sqrt{\gamma_j}} \right]^2 \\ &\leq \underbrace{\sum_i \sum_j \frac{|A_{ij}|^{2-\alpha}}{\rho_i}}_{=1} \sum_j \frac{|A_{ij}|^{\alpha} v_j^2}{\gamma_j} = \sum_j \left[ \sum_i \frac{|A_{ij}|^{\alpha}}{\gamma_j} \right] v_j^2 \\ &= \sum_j v_j^2 \ . \end{aligned} \tag{4.10}$$

Hence $\| \mathbf{W} \| \leq 1$.

This proves that the sequence $\mathbf{z}^{(q)}$ converges to a limit $\hat{\mathbf{z}}$, and that $\hat{\mathbf{z}} - \mathbf{z}^{(0)}$ has its last $n-p$ coordinates 0. Hence $\lim \mathbf{u}^{(q)} = \hat{\mathbf{u}} = \mathbf{V}\hat{\mathbf{z}}$ and

$$\hat{\mathbf{u}} - \mathbf{u}^{(0)} \perp N(\mathbf{W}^T\mathbf{W}) \ . \tag{4.11}$$

From (4.5) we see that $\hat{\mathbf{u}}$ satisfies

$$\mathbf{W}^T\mathbf{W}\hat{\mathbf{u}} = \mathbf{W}^T\mathbf{f} \tag{4.12}$$

which are the normal equations for the least squares problem

$$\mathbf{R}^{-1/2}\mathbf{A}\mathbf{C}^{-1/2}\mathbf{u} = \mathbf{R}^{-1/2}\mathbf{b} \ . \tag{4.13}$$

Via the relation $\mathbf{u} = \mathbf{C}^{1/2}\mathbf{x}$ the solutions of this least squares problem are in 1-1 correspondence with those of

$$\mathbf{R}^{-1/2}\mathbf{A}\mathbf{x} = \mathbf{R}^{-1/2}\mathbf{b} \tag{4.14}$$

This proves the first assertion of the theorem.

Any least squares solution $\tilde{\mathbf{u}}$ of (4.13) is obtained as $\tilde{\mathbf{u}} = \hat{\mathbf{u}} + \mathbf{v}$ , $\mathbf{v} \in N(\mathbf{W}^T\mathbf{W})$, and since $\hat{\mathbf{u}} - \mathbf{u}^{(0)} \perp N(\mathbf{W}^T\mathbf{W})$ we have $\|\tilde{\mathbf{u}} - \mathbf{u}^{(0)}\| \geq \|\hat{\mathbf{u}} - \mathbf{u}^{(0)}\|$. Any least squares solution of (4.14) may be written as $\tilde{\mathbf{x}} = \mathbf{C}^{-1/2}\tilde{\mathbf{u}}$ , $\tilde{\mathbf{u}}$ as just mentioned, and hence $(\tilde{\mathbf{x}} - \mathbf{x}^{(0)})^T\mathbf{C}(\tilde{\mathbf{x}} - \mathbf{x}^{(0)}) = (\tilde{\mathbf{u}} - \mathbf{u}^{(0)})^T(\tilde{\mathbf{u}} - \mathbf{u}^{(0)}) \geq (\hat{\mathbf{u}} - \mathbf{u}^{(0)})^T(\hat{\mathbf{u}} - \mathbf{u}^{(0)}) = (\mathbf{x} - \mathbf{x}^{(0)})^T\mathbf{C}(\mathbf{x} - \mathbf{x}^{(0)})$. This proves the second assertion.

If $\mathbf{x}^{(0)} = \mathbf{0}$ we have $\mathbf{u}^{(0)} = \mathbf{0}$, hence (cf. (4.11)), $\hat{\mathbf{u}}$ is the minimum norm least squares solution of (4.13), i.e. $\mathbf{y} = \hat{\mathbf{u}}$.

□

**Remarks**

i. Regarding the proof that the largest eigenvalue of $\mathbf{W}^T\mathbf{W}$ is at most 1 we note that if $\mathbf{A}$ is nonnegative this eigenvalue is exactly 1 if $\alpha = 1$. Otherwise, in tomographic problems it is less than 1 (actually, if $\mathbf{A}$ is nonnegative and $\alpha \neq 1$, the largest eigenvalue can be 1 only if $\mathbf{A}$ may be written as $\mathbf{u}\mathbf{v}^T$ for two vectors $\mathbf{u}$ and $\mathbf{v}$, which means a full matrix unless there are entire rows or columns 0).

ii. The theorem says that the vector $\mathbf{x}$ to which SIRT converges when started with $\mathbf{x}^{(0)} = \mathbf{0}$ satisfies $\mathbf{x} = \mathbf{C}^{-1/2}\mathbf{y}$ where $\mathbf{y}$ is the minimum norm solution of the least squares problem $\mathbf{W}\mathbf{y} = \mathbf{f}$ with $\mathbf{W} = \mathbf{R}^{-1/2}\mathbf{A}\mathbf{C}^{-1/2}$, $\mathbf{f} = \mathbf{R}^{-1/2}\mathbf{b}$. This means that SIRT has the effect of scaling the rows and columns, and this (cf. sec. 2.3) affects the norms in which the residual vector $\mathbf{b} - \mathbf{A}\mathbf{x}$ and the vector $\mathbf{x}$ (the latter if rank(A)$<n$) are minimized, i.e. *the statistical properties of the solution are affected*. In order to express this change of problem, we will call $\mathbf{W}$ the *associated SIRT matrix.*

   We note that for $\alpha \neq 0$ the effect of row scaling (i.e. the different weighting of the residual vector) may be undone, however, by working with $\tilde{\mathbf{A}} = \mathbf{R}^{1/\alpha}\mathbf{A}$ , $\tilde{\mathbf{b}} = \mathbf{R}^{1/\alpha}\mathbf{b}$. Indeed, writing $\tilde{\mathbf{R}}$ for the "R" corresponding to $\tilde{\mathbf{A}}$, we have $\tilde{\mathbf{R}} = \mathbf{R}^{1+(2-\alpha)/\alpha} = \mathbf{R}^{2/\alpha}$, and hence $\tilde{\mathbf{R}}^{-1/2}\tilde{\mathbf{A}} = \mathbf{A}$, $\tilde{\mathbf{R}}^{-1/2}\tilde{\mathbf{b}} = \mathbf{b}$, so that SIRT applied to $\tilde{\mathbf{A}}\mathbf{x} = \tilde{\mathbf{b}}$ actually solves the least squares problem $\mathbf{A}\mathbf{x} = \mathbf{b}$. Note, however, that if rank(A)$<n$ and we start with $\mathbf{x}^{(0)} = \mathbf{0}$ then the limit is the least squares solution that is minimal in the norm $\|.\|'$ defined by $\|\mathbf{u}\|' = (\mathbf{u}^T\tilde{\mathbf{C}}\mathbf{u})^{1/2}$ , $\tilde{\mathbf{C}}$ the "C" corresponding to $\mathbf{R}^{1/\alpha}\mathbf{A}$, i.e. $\tilde{\gamma}_j = \sum_i \rho_i |A_{ij}|^\alpha$.

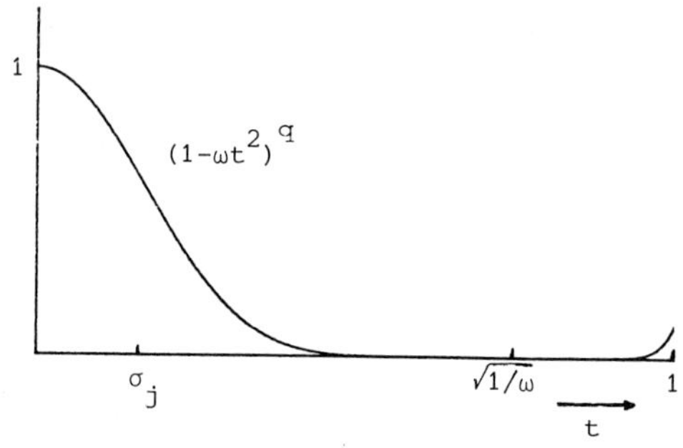


*Figure 4.1*

iii. As regards smoothing (cf. 3.4 (iv)) we note that this amounts to replacing (4.4) by

$$\overline{\mathbf{x}}^{(q+1)} = \mathbf{S}\,(\,\mathbf{I} - \omega\,\mathbf{C}^{-1}\mathbf{A}^T\mathbf{R}^{-1}\mathbf{A}\,)\overline{\mathbf{x}}^{(q)} + \omega\,\mathbf{S}\mathbf{C}^{-1}\mathbf{A}^T\mathbf{R}^{-1}\mathbf{b}\ , \tag{4.15}$$

where **S** denotes the smoothing matrix that does the averaging (a nonnegative matrix with row-sums 1). It is not clear how our proof should be modified to be valid for this case.

### 4.2 The convergence rate of the various components

We see from (4.9) that not all $z_i^{(q)}$, i.e. not all eigenvector components of $\mathbf{u}^{(q)}$, converge equally fast. For simplicity we take $\mathbf{z}^{(0)} = \mathbf{0}$ (i.e. $\mathbf{u}^{(0)} = \mathbf{0}$). Since we have for the limit $\hat{\mathbf{z}}$

$$\hat{z}_j = \frac{g_j}{\sigma_j}\ , \tag{4.16}$$

it follows that

$$z_j^{(q)} - \hat{z}_j = -\,\hat{z}_j\left[1 - \omega\,\sigma_j^2\right]^q\ . \tag{4.17}$$

This shows that the components corresponding to the singular values $\sigma_j$ for which $\omega\,\sigma_j^2$ is close to 0 or 2, converge very slowly, whereas the components for which $\omega\,\sigma_j^2$ is close to 1 converge very fast.

Fig. 4.1 illustrates the error reduction for the components corresponding to the various $\sigma_j$. We see from it that $\omega = 1$ will be rather optimal if the weight of $\hat{\mathbf{z}}$ is concentrated near 1, i.e. if $\sum \hat{z}_j^2$ for the $\hat{z}_j$ corresponding to $\sigma_j$ close to 1 is much larger than $\sum \hat{z}_j^2$ for the remaining $\hat{z}_j$. If, however, a good deal of the weight of $\hat{\mathbf{z}}$ is distributed over a larger part of the interval (0,1) then a larger value of $\omega$ may be profitable. A numerical example will be given in sec. 4.4.

### 4.3 Iteration and perturbation errors; the regularizing effect

As we have mentioned in sec. 4.1, remark ii, SIRT effectively scales the rows and columns. In order to avoid in our discussion the complicating effects of this scaling, we assume that SIRT is applied with the $\tilde{\mathbf{A}}$ and $\tilde{\mathbf{b}}$ as mentioned in that remark. Moreover, rather than studying the errors in the $\mathbf{x}^{(q)}$ themselves, we will study the errors in the transformed iterates $\mathbf{u}^{(q)} = \tilde{\mathbf{C}}^{1/2}\mathbf{x}^{(q)}$ ($\tilde{\mathbf{C}}$ as in the same remark) which converge to the minimum norm solution of the least squares problem $\mathbf{Wu} = \mathbf{b}$, where $\mathbf{W}$ is the SIRT matrix associated with $\tilde{\mathbf{A}}$ (i.e. $\mathbf{W} = \mathbf{A}\tilde{\mathbf{C}}^{-1/2}$). This will give some insight into the behaviour of the process, in particular if the diagonal elements $\tilde{\gamma}_j$ of $\tilde{\mathbf{C}}$ do not differ too much (i.e. the columns of $\tilde{\mathbf{A}}$ have comparable "lengths").

*Note, therefore, that in the remainder of this section* $\mathbf{V}$ *and the* $\sigma_j$ *pertain to the singular value decomposition of this matrix* $\mathbf{W}$.

Like in sec. 2.7 we shall split the errors into approximation errors (here to be called iteration errors) and perturbation errors and we shall discuss the regularizing effect of the present class of methods. This will support the practical observation that in the early stages of the processes in question the iterates seem to suffer less from data errors than is the case later on, i.e. in the beginning the processes seem to have a certain regularizing effect which is lost later on.

We take again $\mathbf{x}^{(0)} = \mathbf{0}$. We assume that the errors in $\mathbf{b}$ are as described in the beginning of section 2.5. Let $\bar{\mathbf{u}}^{(q)}$ and $\bar{\mathbf{u}}$ denote the quantities corresponding to $\mathbf{u}^{(q)}$ and $\mathbf{u}$ if we replace $\mathbf{b}$ by the "ideal" right-hand side $\mathbf{d}$ (note that $\bar{\mathbf{u}} = \tilde{\mathbf{C}}^{1/2}\mathbf{s}$ if $\mathbf{A}$ has rank $n$) and write $\bar{\mathbf{u}} = \mathbf{V}\,\bar{\mathbf{z}}$, $\bar{\mathbf{u}}^{(q)} = \mathbf{V}\,\bar{\mathbf{z}}^{(q)}$.

From (4.9) and (4.17) we have

$$z_j^{(q)} = g_j \phi^{(q)}(\sigma_j) = \hat{z}_j \psi^{(q)}(\sigma_j) \tag{4.18}$$

with $\psi^{(q)}(t) = 1 - (1 - \omega t^2)^q$ , $\phi^{(q)}(t) = \psi^{(q)}(t)\,/\,t$. Then $\phi^{(q)}$ and $\psi^{(q)}$ are (relative) response functions as discussed in sections 2.6 and 2.7. Hence we may apply the theory in sec. 2.7. Thus, if we write

$$\Delta\mathbf{u}^{(q)} \equiv \mathbf{u}^{(q)} - \bar{\mathbf{u}} = \Delta\mathbf{u}_{it}^{(q)} + \Delta\mathbf{u}_{pert}^{(q)} \ , \tag{4.19}$$

where the *iteration error* $\Delta\mathbf{u}_{it}^{(q)} \equiv \bar{\mathbf{u}}^{(q)} - \bar{\mathbf{u}}$ is the error we would get after q iterations using the ideal right-hand side $\mathbf{d}$, and the *perturbation error* $\Delta\mathbf{u}_{pert} = \mathbf{u}^{(q)} - \bar{\mathbf{u}}^{(q)}$ is the change in the iterates caused by using the perturbed right-hand side instead of the ideal one, then we have from theorem 2.13:

**Theorem 4.20**

$$\delta \mathbf{u}_{it}^{(q)} \equiv \| \Delta \mathbf{u}_{it}^{(q)} \| = \sqrt{\sum_j \bar{z}_j^2 \, (1 - \omega \, \sigma_j^2)^{2q}} \tag{4.21}$$

$$\delta \mathbf{u}_{pert}^{(q)} \equiv \sqrt{E\left[ \| \Delta \mathbf{u}_{pert}^{(q)} \|^2 \right]} = \sigma \sqrt{\sum_j \left[ \frac{1 - (1 - \omega \, \sigma_j^2)^q}{\sigma_j} \right]^2} \tag{4.22}$$

$$E\left[ \| \Delta \mathbf{u}^{(q)} \|^2 \right] = \left[ \delta \mathbf{u}_{it}^{(q)} \right]^2 + \left[ \delta \mathbf{u}_{pert}^{(q)} \right]^2 . \tag{4.23}$$

□

From this theorem we see that as $q$ increases the expectation of $\| \Delta \mathbf{u}_{pert}^{(q)} \|^2$ increases, which means that the effect of errors in **b** increases, and this is part of what we wanted to show. We also note that in the limit we get $\sigma \sqrt{\sum \frac{1}{\sigma_j^2}}$, which is indeed how errors in **f** perturb the least squares solution of $\mathbf{Wu} = \mathbf{f}$ (cf. (2.7)).

Although theorem 4.20 gives precise quantitative information on the expected effect of data errors, it does not give much insight. Therefore the following approximate form of it may be useful if $\omega = 1$:

**Property 4.24**

For any $q$ there is a $\theta$, $\frac{1}{2} \leq \theta \leq 1$, such that

$$\begin{aligned} \delta \mathbf{u}_{pert}^{(q)} &= \theta \, \sigma \sqrt{q^2 \sum_{\sigma_j^2 \leq \frac{1}{q}} \sigma_j^2 + \sum_{\sigma_j^2 > \frac{1}{q}} \frac{1}{\sigma_j^2}} \\ &= \theta \, \sigma \sqrt{\sum_j \min(q^2 \sigma_j^2 , \frac{1}{\sigma_j^2})} . \end{aligned} \tag{4.25}$$

**Proof.** This follows right away from the following inequalities for $0 < \gamma \le 1$:

$$\begin{cases} \frac{1}{2} q\gamma \le q\gamma \left[1 - \frac{q\gamma}{2}\right] \le 1 - (1-\gamma)^q \le q\gamma & \text{if } q\gamma \le 1 \\ 1 - \frac{1}{e} \le 1 - (1-\gamma)^q \le 1 & \text{if } q\gamma \ge 1 \end{cases} \tag{4.26}$$

□

We note that $\theta$ will be close to 1 in the case that in neither of the sums in (4.25) the terms with $\sigma_i^2 \approx \frac{1}{q}$ dominate (which, unfortunately, is not the case in the numerical example in the next section).

The last expression in (4.25) leads to the interesting observation that $\delta \mathbf{u}_{pert}^{(q)}$ grows at most proportionally with q (just replace $\min(q^2 \sigma_j^2, \frac{1}{\sigma_j^2})$ by $q^2 \sigma_j^2$ ). If $\sigma_j$ for $j \to \infty$ decreases exponentially, $\sigma_j \le C e^{(k-j)\beta} \sigma_k$ for $j \ge k$ , $\beta > 0$, say, then we can even say that $\delta \mathbf{u}_{pert}^{(q)}$ grows only at most proportionally to $\sqrt{q}$ . Actually, in (4.25) we now have

$$\sum_{\sigma_j^2 \le \frac{1}{q}} \sigma_j^2 \le C \frac{1/q}{1 - e^{-2\beta}} \,, \quad \sum_{\sigma_j^2 > \frac{1}{q}} \frac{1}{\sigma_j^2} \le C \frac{q}{1 - e^{-2\beta}} \tag{4.27}$$

and hence

$$\delta \mathbf{u}_{pert}^{(q)} \le \sigma \sqrt{\frac{2Cq}{1 - e^{-2\beta}}} \,. \tag{4.28}$$

### 4.4 A numerical example

In order to see how the relative iteration and perturbation errors may behave for increasing $q$ we consider a model situation. We make the same assumptions about $\mathbf{W}$, $\sigma_j$ and the errors in $\mathbf{b}$ as in sec. 4.3. We take $\sigma_j = e^{-(j-1)\beta}$ for some $\beta \ll 1$ , where a typical value is $\beta = .025$. This is more or less in line with Nolet's model problem (cf. Nolet 1985) where $\mathbf{A}$ is a $400 \times 200$ matrix which has, beside 3 singular values 0 (to computing accuracy), 197 singular values of which the largest and smallest one have a ratio 200 and any decade within this ratio contains about the same number of singular values. We do not prescribe the size of $\mathbf{A}$, but assume $n$ to be large enough so that $e^{-n\beta} \ll 1$.

In order to compare the variance $\sigma^2$ of the errors in $\mathbf{b}$ with the coordinates of $\mathbf{d}$ (and thus to obtain a scaling invariance) we note that if there are $m$ equations, $\|\mathbf{d}\|/\sqrt{m}$ is the root mean square of the coordinates of $\mathbf{d}$, and we write

$$\sigma = \tilde{\sigma} \, \|\mathbf{d}\| / \sqrt{m} \,. \tag{4.29}$$

Thus $\tilde{\sigma}^2$ denotes a kind of relative variance of the errors in $\mathbf{b}$. We also note for the error vector $\Delta\mathbf{b}$ that $E(\|\Delta\mathbf{b}\|^2) = \tilde{\sigma}^2 \|\mathbf{d}\|^2$.

Now suppose that the solution $\bar{\mathbf{u}}$ of the unperturbed model problem has much larger components with respect to the large singular values than to the small singular values. Let us say $\bar{z}_j = \sigma_j^2 = e^{-2(j-1)\beta}$. Noting that we then have $\|\bar{\mathbf{u}}\| = \|\bar{\mathbf{z}}\| = \sqrt{\sum_j e^{-4j\beta}} \approx \sqrt{1/(4\beta)}$ $= \sqrt{10}$, $\bar{g}_j = \sigma_j^3 = e^{-3(j-1)\beta}$, $\|\mathbf{d}\| = \|\bar{\mathbf{g}}\| \approx \sqrt{1/6\beta} = \sqrt{20/3}$ we get table 4.1.

**Table 4.1**

| $q$ | $\omega$ | $\delta\mathbf{u}_{it}^{(q)}/\|\bar{\mathbf{u}}\|$ | $\delta\mathbf{u}_{pert}^{(q)}/\|\bar{\mathbf{u}}\|$ | $m/\bar{\sigma}^2$ |
|---|---|---|---|---|
| 5 | 1 | .12 | 9.3 $\bar{\sigma}/\sqrt{m}$ | 5700 |
| | 1.6 | .082 | 12 〃 | 22 000 |
| 10 | 1 | .066 | 13 〃 | 41 000 |
| | 1.7 | .040 | 18 〃 | 200 000 |
| 25 | 1 | .027 | 21 〃 | 600 000 |
| | 1.82 | .015 | 30 〃 | 3 700 000 |

The quantities $\delta\mathbf{u}_{it}^{(q)}$ and $\delta\mathbf{u}_{pert}^{(q)}$ have been computed from (4.21) and (4.22). The values of $\omega \neq 1$ have been chosen so as to minimize $\delta\mathbf{u}_{it}^{(q)}$ for the given $q$. The tabulated values for $m/\bar{\sigma}^2$ are such that when $m/\bar{\sigma}^2$ has the tabulated value then $\delta\mathbf{u}_{it}^{(q)}$ and $\delta\mathbf{u}_{pert}^{(q)}$ are equal. Obviously it will not be very useful to continue the iteration when $m/\bar{\sigma}^2$ is less than the given values.

The tabulated values for $\omega = 1$ suggest that, approximately, $\delta\mathbf{u}_{it}^{(q)}$ is inversely proportional to $q$ and that $\delta\mathbf{u}_{pert}^{(q)}$ is proportional to $\sqrt{q}$. Both suggestions are correct and can be proved. The second suggestion corresponds very nicely with the observation at the end of sec. 4.3 (cf. (4.28)). We also note that a proper choice of $\omega$ speeds up the process considerably. Moreover, crude interpolation in the table suggests that, surprisingly, if, for different values of $\omega$ and $q$ about the same iteration errors are obtained, the corresponding perturbation errors are about the same also.

These properties hold for other values of $\beta$ as well, but this may no longer be the case for essentially different distributions of the singular values and the coordinates $\bar{z}_j$. If, e.g., $\sigma_j$ and $\bar{z}_j$ decrease more slowly than exponentially then the rate of convergence and the regularizing effect will be worse. See, however, the epilogue in sec. 6.

### 4.5 The regularizing effect of the method compared with explicit regularization

We now compare the regularizing effect of the method as discussed in secs. 4.3 and 4.4 with the regularization obtained by applying some iterative process to the regularized least squares problem

$$\begin{pmatrix} \mathbf{W} \\ \lambda\mathbf{I} \end{pmatrix} \mathbf{u} = \begin{pmatrix} \mathbf{b} \\ \mathbf{0} \end{pmatrix}, \qquad (4.30)$$

$\mathbf{W}$ the same matrix as in sec. 4.3, and then iterating much longer. Since the purpose of this way of regularization is, in fact, that continued iteration does not deteriorate the result, we just look at the effect of regularization and errors in $\mathbf{b}$ on the true least squares solution of (4.30).

In (2.17) and (2.18) we see what these effects are:

$$\delta\mathbf{u}_{reg} \equiv \sqrt{\|\Delta\mathbf{u}_{reg}\|^2} = \lambda^2 \sqrt{\sum_j \left[\frac{\bar{z}_j}{\sigma_j^2+\lambda^2}\right]^2} \tag{4.31}$$

$$\delta\mathbf{u}_{pert} \equiv \sqrt{E\left[\|\Delta\mathbf{u}_{pert}\|^2\right]} = \sigma \sqrt{\sum_j \left[\frac{\sigma_j}{\sigma_j+\lambda^2}\right]^2} \ . \tag{4.32}$$

Taking the same model case as in sec 4.4 and taking $\sigma = .05$ the value of $\lambda$ for which $(\delta\mathbf{u}_{reg})^2 + (\delta\mathbf{u}_{pert})^2$ is minimal turns out to be .33. We then get $\delta\mathbf{u}_{reg} = .57$ , $\delta\mathbf{u}_{pert} = .65$ and $\sqrt{(\delta\mathbf{u}_{reg})^2 + (\delta\mathbf{u}_{pert})^2} = .87$.

For SIRT with $q = 5, \omega{=}1$, we found (cf. table 4.1) $\delta\mathbf{u}_{it}^{(5)} = .39$ , $\delta\mathbf{u}_{pert}^{(5)} = .57$, and $\sqrt{(\delta\mathbf{u}_{it}^{(5)})^2 + (\delta\mathbf{u}_{pert}^{(5)})^2} = .69$. For $q = 4$ and 6 we find virtually the same total error, for other values of $q$ it is larger. Thus the best attainable total error for SIRT is (in this case) somewhat smaller than for regularization.

Hence, at least for problems with a distribution of singular values and weights like ours (the value of $\beta$ being quite uncritical), stopping an iterative process in time may give somewhat better results than regularizing the problem and at considerably lower cost (if with regularization we iterate much longer).

Finally we look briefly at regularization (ii) (cf. sec. 2.6). Using (2.19) and (2.20) we find $\delta\mathbf{u}_{reg} \approx \eta^2\sqrt{10}$ , $\delta\mathbf{u}_{pert} \approx \dfrac{\sigma}{\eta}\sqrt{20}$, and for $\sigma = .05$ the total error is an optimal .74 for $\eta = .37$, which corresponds to retaining the components belonging to the largest 40 singular values. Thus, at least in this case, the two regularization methods are quite comparable. Note, however, the computational drawbacks of the second method, as outlined, e.g., in Nolet (1985).

## 5. Projection methods

### 5.1 Origin

As we have seen in the proof of theorem 4.2, the SIRT methods considered there could be reformulated as

$$\begin{cases} \mathbf{u}^{(q)} = \mathbf{C}^{1/2}\mathbf{x}^{(q)} \quad , \quad \mathbf{W} = \mathbf{R}^{-1/2}\mathbf{A}\mathbf{C}^{-1/2} \\ \mathbf{u}^{(q+1)} = (\mathbf{I} - \omega\, \mathbf{W}^T\mathbf{W})\mathbf{u}^{(q)} + \omega \mathbf{W}^T\mathbf{R}^{-1/2}\mathbf{b} \ . \end{cases} \tag{5.1}$$

This is the so-called *Richardson iteration* for the (normal) equations

$$\mathbf{W}^T\mathbf{W}\mathbf{u} = \mathbf{W}^T\mathbf{R}^{-1/2}\mathbf{b} \ . \tag{5.2}$$

Indeed, for a set of equations

$$\mathbf{B}\mathbf{u} = \mathbf{f} , \tag{5.3}$$

$\mathbf{B}$ a square $n \times n$-matrix, $\mathbf{f} \in R(\mathbf{B})$, Richardson iteration is defined by

$$\mathbf{u}^{(q+1)} = (\mathbf{I} - \omega\, \mathbf{B})\mathbf{u}^{(q)} + \omega\, \mathbf{f} \ . \tag{5.4}$$

We would get a more general method, of course, if we let the relaxation parameter $\omega$ depend on $q$. Note, however, that for whatever choice of $q$-dependent $\omega$, if we take $\mathbf{u}^{(0)} = \mathbf{0}$ then $\mathbf{u}^{(q+1)}$ may always be written as

$$\mathbf{u}^{(q+1)} = (a_{0q}\mathbf{I} + a_{1q}\mathbf{B} + \cdots + a_{qq}\mathbf{B}^q)\mathbf{f} \tag{5.5}$$

for suitable constants $a_{ij}$. Also note that the assumption $\mathbf{u}^{(0)} = \mathbf{0}$ is no restriction since the sequence $\mathbf{v}^{(q)} = \mathbf{u}^{(q)} - \mathbf{u}^{(0)}$ satisfies (5.4) if we replace $\mathbf{f}$ by $\mathbf{f} - \mathbf{B}\mathbf{u}^{(0)}$, and obviously $\mathbf{v}^{(0)} = \mathbf{0}$ (note, however, the discussion at the end of sec. 2.3).

Consequently, no matter the choice of the relaxation parameters $\omega$, if $\mathbf{u}^{(0)} = \mathbf{0}$ then $\mathbf{u}^{(q)}$ always lies in the so-called *Krylov subspace* $K^{(q)}(\mathbf{B};\mathbf{f})$ defined as the span of the vectors $\mathbf{f}, \mathbf{B}\mathbf{f}, \cdots, \mathbf{B}^{(q-1)}\mathbf{f}$.

This suggests that at the $q$-th iteration we look in the whole of $K = K^{(q)}(\mathbf{B};\mathbf{f})$ for a suitable approximation to the solution $\hat{\mathbf{u}}$ of (5.3). The best approximation would, of course, be the orthogonal projection of $\hat{\mathbf{u}}$ on $K$ but this cannot, unfortunately, be computed in a reasonable amount of time.

Another idea is to project the system (5.3) on $K$ in the following sense: let $\Pi_K$ be the orthogonal projection operater which projects vectors in $R^m$ onto $K$. Then the projected linear system is defined as $\Pi_K \mathbf{B}\mathbf{u} = \Pi_K \mathbf{f}$ and this might then be solved for $\mathbf{u} \in K$. This approach leads to the so-called *projection methods*.

If $\mathbf{B}$ is symmetric (which is the case if $\mathbf{B} = \mathbf{W}^T\mathbf{W}$, cf. (5.2) or $\mathbf{B} = \mathbf{A}^T\mathbf{A}$, cf. (2.1)) then the projected system can be constructed in a reasonable time by the *Lanczos method*, and it is obtained in a form which allows very efficient solving (cf. sec. 5.4). In order to solve our least squares problem (1.2) it is now no longer necessary to transform $\mathbf{A}$ into $\mathbf{W}$ first, but we may operate directly on the normal equations (2.1), thus avoiding solving a differently weighted least squares problem (cf. sec. 4.1, remark ii).

If $\mathbf{B}$ is symmetric and (semi-) positive definite (which is also the case if $\mathbf{B} = \mathbf{A}^T\mathbf{A}$) then the *conjugate gradients method* (cf. sec. 5.2) may be applied, which leads to the same approximate solutions as the Lanczos method, but does so in a cheaper and more elegant way, avoiding the explicit construction of the projected system altogether.

The *LSQR method*, proposed by Paige and Saunders (cf. sec 5.5) is also a projection method. It is suitable for sparse least squares problems and also for sparse systems (5.3) with unsymmetric **B**.

Note that the projection methods, in exact arithmetic, must necessarily terminate within $n$ steps since the dimension of the Krylov space cannot exceed $n$. Therefore they are, strictly speaking, not iterative methods. However, since in practical situations we wish to terminate the iteration process for $q \ll n$, and since rounding errors usually prevent an exact termination, the projection methods are commonly regarded as iterative methods.

Finally we note that the projection methods, when applied to least squares problems, have the important property to give the least residual $\|\mathbf{A}\mathbf{u}^{(q)} - \mathbf{b}\|$ over all methods of the form (5.5), i.e., in particular, to give a smaller residual than the SIRT methods for the same number of iterations. (This minimizing property follows, e.g., from Golub & Van Loan (1983), (10.2-15)).

### 5.2 The conjugate gradients method

For the linear system (5.3) with **B** symmetric positive definite, the conjugate gradients method is given by the following scheme (cf. Hesteness & Stiefel 1952; also Golub & Van Loan 1983, secs. 10.2, 10.3):

$$
\begin{aligned}
&\mathbf{u}^{(0)} = \mathbf{0} \;;\; \mathbf{r}^{(0)} = \mathbf{f} \;;\; \mathbf{p}^{(0)} = \mathbf{r}^{(0)} \\
&\text{for } q = 0,1,2, \cdots \\
&\qquad \alpha_q = (\mathbf{r}^{(q)}, \mathbf{r}^{(q)})/(\mathbf{p}^{(q)}, \mathbf{B}\mathbf{p}^{(q)}) \\
&\qquad \mathbf{u}^{(q+1)} = \mathbf{u}^{(q)} + \alpha_q \mathbf{p}^{(q)} \;;\; \mathbf{r}^{(q+1)} = \mathbf{r}^{(q)} - \alpha_q \mathbf{B}\mathbf{p}^{(q)} \\
&\qquad \text{if } \mathbf{r}^{(q+1)} = \mathbf{0} \text{ then quit} \\
&\qquad \beta_q = (\mathbf{r}^{(q+1)}, \mathbf{r}^{(q+1)})/(\mathbf{r}^{(q)}, \mathbf{r}^{(q)}) \\
&\qquad \mathbf{p}^{(q+1)} = \mathbf{r}^{(q+1)} + \beta_q \mathbf{p}^{(q)}
\end{aligned}
\tag{5.6}
$$

If $\mathbf{r}^{(q+1)} = \mathbf{0}$ then $\mathbf{u}^{(q+1)}$ is the solution of (5.3). Usually we will quit before $\mathbf{r}^{(q+1)} = \mathbf{0}$, and then $\mathbf{u}^{(q+1)}$ will be considered as an approximation to the solution $\hat{\mathbf{u}}$ of (5.3). We note that in exact arithmetic for any $q$

$$\mathbf{r}^{(q)} = \mathbf{f} - \mathbf{B}\mathbf{u}^{(q)} \ , \tag{5.7}$$

i.e. $\mathbf{r}^{(q)}$ is the residual after $q$ iterations, and one may quit when this residual is small enough. Moreover, $\mathbf{r}^{(0)}, \mathbf{r}^{(1)}, \mathbf{r}^{(2)}, \cdots$ are mutually orthogonal, and hence $\mathbf{r}^{(0)}, \cdots, \mathbf{r}^{(q-1)}$ form an orthogonal basis of $K^{(q)}(\mathbf{B};\mathbf{f})$. This is very instrumental, of course, for orthogonal projections (which form the foundation for projection methods, as explained in sec. 5.1).

The process needs only very limited workspace: just space for 4 vectors of $n$ coordinates each. The amount of computing per iteration is very limited also: one matrix-vector multiplication, two inner products and three vector updates (i.e. adding a scalar multiple of a vector to another vector). In tomography problems the matrix-vector multiplication will constitute the major part of the work.

Note that it is not necessary to have the matrix $\mathbf{B}$ explicitly as a, e.g., two dimensional array, it suffices to be able to compute $\mathbf{By}$ for a given vector $\mathbf{y}$. Hence, the programmer may take advantage of any special form of $\mathbf{B}$. Thus, if $\mathbf{B} = \mathbf{A}^T\mathbf{A}$ we need not form $\mathbf{B}$ explicitly, but may just compute $\mathbf{By}$ as $\mathbf{A}^T(\mathbf{Ay})$. However, see sec. 5.3. Also note that the amount of computing per iteration in this case is comparable to that in the SIRT methods where we also had to do two matrix-vector multiplications, one with $\mathbf{A}$ and one with $\mathbf{A}^T$.

Furthermore we note that the requirement that $\mathbf{B}$ be positive definite can be weakened to requiring that $\mathbf{f}$ lies in the span of the eigenvectors corresponding to positive eigenvalues. This is particularly meaningful in our situation, since this weaker requirement is satisfied for the normal equations $\mathbf{A}^T\mathbf{Ax} = \mathbf{A}^T\mathbf{b}$. Moreover the process now converges to the minimum norm solution, and, indeed, any $\mathbf{u}^{(q)}$ is orthogonal to the null space of $\mathbf{A}$ for any $q$. This convergence to the minimum norm solution is, in general, lost, however, if the CG process is used with $\mathbf{u}^{(0)} \neq \mathbf{0}$ and $\mathbf{r}^{(0)} = \mathbf{f} - \mathbf{B}\mathbf{u}^{(0)}$ (which in the full rank case still gives convergence to the solution). For this reason we defined the process with $\mathbf{u}^{(0)} = \mathbf{0}$.

### 5.3 Conjugate gradients for least squares

It has been observed that scheme (5.6) when applied to the normal equations $\mathbf{A}^T\mathbf{Ax} = \mathbf{A}^T\mathbf{b}$ may not lead to numerically satisfactory solutions if $\mathbf{A}^T\mathbf{A}$ is ill-conditioned. According to Paige and Saunders (1982) this is, to a large extent, due to the explicit computation of the vectors $\mathbf{A}^T\mathbf{A}\mathbf{p}^{(q)}$. The basic equality $(\mathbf{p}^{(q)}, \mathbf{A}^T\mathbf{A}\mathbf{p}^{(q)}) = (\mathbf{A}\mathbf{p}^{(q)}, \mathbf{A}\mathbf{p}^{(q)})$ allows to rewrite scheme (5.6) to a scheme in which it is avoided to compute $\mathbf{A}^T\mathbf{A}\mathbf{p}^{(q)}$. This scheme, proposed by Björck and Elfving (1979), is reported to produce better results and reads as follows:

$$
\begin{aligned}
&\mathbf{x}^{(0)} = \mathbf{0} \ ; \ \mathbf{s}^{(0)} = \mathbf{b} \ ; \ \mathbf{r}^{(0)} = \mathbf{A}^T\mathbf{b} \ ; \ \mathbf{p}^{(0)} = \mathbf{r}^{(0)} \\
&\text{for } q = 0,1,2, \cdots \\
&\qquad \mathbf{w}^{(q)} = \mathbf{A}\mathbf{p}^{(q)} \\
&\qquad \alpha_q = (\mathbf{r}^{(q)}, \mathbf{r}^{(q)})/(\mathbf{w}^{(q)}, \mathbf{w}^{(q)}) \\
&\qquad \mathbf{x}^{(q+1)} = \mathbf{x}^{(q)} + \alpha_q \mathbf{p}^{(q)} \ ; \ \mathbf{s}^{(q+1)} = \mathbf{s}^{(q)} - \alpha_q \mathbf{w}^{(q)} \\
&\qquad \mathbf{r}^{(q+1)} = \mathbf{A}^T \mathbf{s}^{(q+1)} \\
&\qquad \text{if } \mathbf{r}^{(q+1)} = \mathbf{0} \text{ then quit} \\
&\qquad \beta_q = (\mathbf{r}^{(q+1)}, \mathbf{r}^{(q+1)})/(\mathbf{r}^{(q)}, \mathbf{r}^{(q)}) \\
&\qquad \mathbf{p}^{(q+1)} = \mathbf{r}^{(q+1)} + \beta_q \mathbf{p}^{(q)}
\end{aligned}
\tag{5.8}
$$

We note that now the $\mathbf{r}^{(q)}$ are the residuals of the normal equations and that

$$\mathbf{s}^{(q)} = \mathbf{b} - \mathbf{A}\mathbf{x}^{(q)} \tag{5.9}$$

is the residual in the least squares system.

### 5.4 The Lanczos method

As we have already observed (sec. 5.2) the vectors $\mathbf{r}^{(0)}, \mathbf{r}^{(1)}, \cdots$ generated by the conjugate gradients method form orthogonal bases for the Krylov subspaces, and this is really what makes projection methods work. If **B** (cf. (5.3)) is symmetric but not (semi-) positive definite and **f** has components in the direction of eigenvectors corresponding to non-positive eigenvalues, then we can still obtain orthogonal bases for the Krylov subspaces $K^{(q)}(\mathbf{B};\mathbf{f})$ by the *Lanczos method* (cf. Lanczos 1950; also Golub & Van Loan 1983, ch. 9):

$$\begin{array}{l} \mathbf{v}^{(0)} = \mathbf{f}/\beta_0 \ ; \ \mathbf{v}^{(-1)} = \mathbf{0} \\ \text{for } q = 0,1,2, \cdots \\ \quad \mathbf{w}^{(q)} = \mathbf{B}\mathbf{v}^{(q)} - \beta_q \mathbf{v}^{(q-1)} \\ \quad \alpha_q = (\mathbf{v}^{(q)}, \mathbf{w}^{(q)}) \\ \quad \mathbf{v}^{(q+1)} = (\mathbf{w}^{(q)} - \alpha_q \mathbf{v}^{(q)})/\beta_{q+1} \text{ with } \beta_{q+1} > 0 \text{ such} \\ \quad \text{that } \| \mathbf{v}^{(q+1)} \| = 1; \text{quit if no} \\ \quad \text{such } \beta_{q+1} \text{ exists.} \end{array} \tag{5.10}$$

Then the vectors $\mathbf{v}^{(0)}, \mathbf{v}^{(1)}, \cdots, \mathbf{v}^{(q-1)}$ (the *Lanczos vectors*) form an orthonormal basis for $K^{(q)}(\mathbf{B};\mathbf{f})$. If $\mathbf{B}$ is positive definite then $\mathbf{v}^{(q)}$ has the same direction as $\mathbf{r}^{(q)}$ in (5.6). The $\alpha_q$ and $\beta_q$ in (5.10) differ from those in (5.6) but are closely related.

The projection of the set of equations $\mathbf{Bu} = \mathbf{f}$ on $K = K^{(q)}(\mathbf{B};\mathbf{f})$ (cf. sec. 5.1) now is obtained as follows.

If $\mathbf{V}_q$ denotes the matrix with the (orthonormal) columns $\mathbf{v}^{(0)}, \mathbf{v}^{(1)}, \ldots, \mathbf{v}^{(q-1)}$ and $\mathbf{T}_q$ denotes the matrix

$$\mathbf{T}_q = \begin{pmatrix} \alpha_0 & \beta_1 & & & \\ \beta_1 & \alpha_1 & \beta_2 & \varnothing & \\ & \cdot & \cdot & \cdot & \\ & & \cdot & \cdot & \cdot \\ & \varnothing & & \cdot & \beta_{q-1} \\ & & & \beta_{q-1} & \alpha_{q-1} \end{pmatrix}, \tag{5.11}$$

then (5.10) amounts to $\mathbf{BV}_q = \mathbf{V}_q \mathbf{T}_q + \beta_q(\mathbf{0}, \mathbf{0}, \ldots, \mathbf{0}, \mathbf{v}^{(q)})$. Since we seek a solution $\mathbf{u}^{(q)} \in K$ of the projected system and any such vector $\mathbf{u}^{(q)}$ may be written as

$$\mathbf{u}^{(q)} = \mathbf{V}_q \mathbf{y}^{(q)}, \tag{5.12}$$

we have $\mathbf{0} = \Pi_K(\mathbf{BV}_q\mathbf{y}^{(q)} - \mathbf{f}) = \mathbf{V}_q\mathbf{T}_q\mathbf{y}^{(q)} - \mathbf{f} = \mathbf{V}_q\mathbf{T}_q\mathbf{y}^{(q)} - \beta_0\mathbf{v}^{(0)}$, since $\mathbf{v}^{(q)} \perp K$ and $\mathbf{f} = \beta_0\mathbf{v}^{(0)} \in K$.

Hence

$$\mathbf{T}_q \mathbf{y}^{(q)} = \beta_0(1, 0, \ldots, 0)^T. \tag{5.13}$$

This system can be solved very cheaply (operation count $O(q)$), and finally $\mathbf{u}^{(q)}$ is obtained from (5.12).

The main advantage of the conjugate gradients method over the Lanczos method (if both are applicable) is that in the former the amount of storage space is fixed, whereas in the latter it increases with $q$, since one has to store the vectors $\mathbf{v}^{(0)}, \cdots, \mathbf{v}^{(q-1)}$ in order to be able to do (5.12). When working in exact arithmetic the results are the same.

Again, if $\mathbf{B}$ is singular and $\mathbf{f} \in R(\mathbf{B})$ the vectors $\mathbf{u}^{(q)}$ converge to the minimum norm solution of (5.3).

### 5.5 The Paige-Saunders method LSQR

A way to get around the disadvantage of the Lanczos method mentioned at the end of sec. 5.4 and at the same time to deal with unsymmetric and least squares problems has been given by Paige and Saunders in their LSQR method (cf. (Paige & Saunders 1982)). We now give a brief sketch of this algorithm.

We first note that the least squares problem $\mathbf{Ax} = \mathbf{b}$ is equivalent to the square linear problem

$$\begin{bmatrix} \mathbf{I} & \mathbf{A} \\ \mathbf{A}^T & \mathbf{0} \end{bmatrix} \begin{bmatrix} \mathbf{r} \\ \mathbf{x} \end{bmatrix} = \begin{bmatrix} \mathbf{b} \\ \mathbf{0} \end{bmatrix} \tag{5.14}$$

(cf. sec. 2.2, (i)).

Applying the Lanczos algorithm to (5.14) yields Lanczos vectors $\mathbf{v}^{(q)}$ (cf. sec. 5.4) which have alternately the form $\begin{bmatrix} \mathbf{y} \\ \mathbf{0} \end{bmatrix}$ and $\begin{bmatrix} \mathbf{0} \\ \mathbf{z} \end{bmatrix}$. Let $\begin{bmatrix} \mathbf{Y}_k \\ \mathbf{0} \end{bmatrix}$ denote the matrix of the first $k$ Lanczos vectors of the form $\begin{bmatrix} \mathbf{y} \\ \mathbf{0} \end{bmatrix}$ in their proper order, and likewise $\begin{bmatrix} \mathbf{0} \\ \mathbf{Z}_k \end{bmatrix}$ for those of the form $\begin{bmatrix} \mathbf{0} \\ \mathbf{z} \end{bmatrix}$. Then on the basis $\begin{bmatrix} \mathbf{Y}_{k+1} & \mathbf{0} \\ \mathbf{0} & \mathbf{Z}_k \end{bmatrix}$ the projection of (5.14) on the Krylov subspace $K^{(2k+1)}$ gets the form

$$\begin{bmatrix} \mathbf{I} & \mathbf{B}_k \\ \mathbf{B}_k^T & \mathbf{0} \end{bmatrix} \begin{bmatrix} \mathbf{t}^{(k+1)} \\ \mathbf{w}^{(k)} \end{bmatrix} = \delta_0 \begin{bmatrix} 1 \\ 0 \\ \cdot \\ \cdot \\ 0 \end{bmatrix} \tag{5.15}$$

and in original coordinates we get

$$\begin{bmatrix} \mathbf{r}^{(k)} \\ \mathbf{x}^{(k)} \end{bmatrix} = \begin{bmatrix} \mathbf{Y}_{k+1} & \mathbf{0} \\ \mathbf{0} & \mathbf{Z}_k \end{bmatrix} \begin{bmatrix} \mathbf{t}^{(k+1)} \\ \mathbf{w}^{(k)} \end{bmatrix} . \tag{5.16}$$

The $(k+1) \times k$ matrix $\mathbf{B}_k$ is lower bidiagonal:

$$\mathbf{B}_k = \begin{bmatrix} \gamma_0 & & & & \\ \delta_1 & \gamma_1 & & & \\ & \delta_2 & \cdot & & \\ & & \cdot & \cdot & \\ & & & \cdot & \gamma_{k-1} \\ & & & & \delta_k \end{bmatrix} . \tag{5.17}$$

Comparing (5.15) with (5.14) we see that $\mathbf{w}^{(k)}$ is least squares solution of the system

$$\mathbf{B}_k \mathbf{w}^{(k)} = \delta_0(1, 0, \cdots, 0)^T \tag{5.18}$$

and this system can be conveniently solved by standard methods based on orthogonal transformations (QR-factorization, cf. Golub & Van Loan 1983, secs. 6.2 and 6.3).

So far it seems that we would still have to store the (expanding) matrices $\mathbf{Y}_k$ and $\mathbf{Z}_k$. For the trick to avoid this we refer the reader to (Paige & Saunders 1982).

We conclude with a few notes. We first note that, when working in exact arithmetic, $\mathbf{x}^{(k)}$ obtained in this way (after $2k + 1$ Lanczos iterations on (5.14)) is the same as $\mathbf{x}^{(k)}$ obtained using conjugate gradients, and, remarkably, that the amounts of computing are about the same.

Secondly, in particular for ill-conditioned systems, LSQR is reported to give better results than the conjugate gradients approach. The explanation for this might be that with conjugate gradients we obtain iterates for the projected system $\mathbf{A}^T\mathbf{A}\mathbf{x} = \mathbf{A}^T\mathbf{b}$, whereas the Paige-Saunders approach leads to the projection of the system $\mathbf{A}\mathbf{x} = \mathbf{b}$ itself.

### 5.6 Projection methods and regularization

The projection methods may, of course, just as well be applied to the regularized system (2.8) or its normal equations as to the original system (1.2) or its normal equations.

In the not-unusual situation that one wishes to solve the regularized system for several values of $\lambda$ the Lanczos and LSQR methods have the pleasant property that this can be done at little extra cost provided the Lanczos vectors are saved. Indeed, from (5.10), applied to the equations $(\mathbf{A}^T\mathbf{A} + \lambda^2\mathbf{I})\mathbf{x} = \mathbf{A}^T\mathbf{b}$ (cf. (2.8) and (2.9)), it follows that the $\mathbf{v}^{(q)}$ and $\beta_q$ do not depend on $\lambda$, and that the $\alpha_q$ are increased by $\lambda^2$. For the LSQR method we refer to Paige and Saunders (1982). See also Björck (1985).

### 5.7 Ritz values and polynomials

A crucial role in discussing the convergence behaviour of conjugate gradients and Lanczos processes (which are, we recall, equivalent if both are applicable) is played by the so-called *Ritz values* (cf. Parlett 1980, van der Sluis & van der Vorst 1986). For any symmetric matrix $\mathbf{B}$ and any vector $\mathbf{f}$ and any $q \geq 1$ they are defined as the eigenvalues

$$\theta_1^{(q)}, \ldots\ldots, \theta_q^{(q)} \tag{5.19}$$

of the matrix in (5.11). Or, in more abstract terms: with $K = K^{(q)}(\mathbf{B};\mathbf{f})$ and $\Pi_K$ the

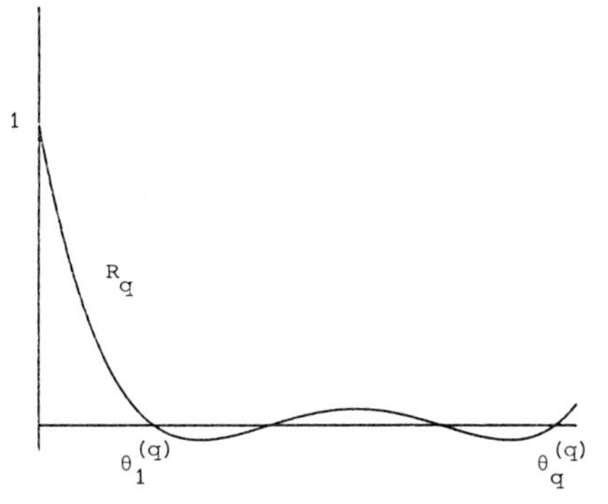


*Figure 5.1*

orthogonal projection onto $K$ the corresponding Ritz values are the eigenvalues of the linear mapping $\Pi_K \mathbf{B} \mid_K$ (the symbol $\mid_K$ denoting that the domain is restricted to K).

We note that the Ritz values lie between the smallest and largest eigenvalue of $\mathbf{B}$ and that they depend continuously (even analytically) on the coordinates of $\mathbf{f}$.

We also note that in the course of a conjugate gradients or Lanczos process the Ritz values may be very cheaply computed should one wish to know them. Indeed, for a matrix like the one in (5.11) all eigenvalues may be computed using the famous QR algorithm (cf. Golub & Van Loan 1983, sec. 8.2) with an operation count of $O(q^2)$.

Using the Ritz values we define the (normalized) *Ritz polynomial* $R_q$ as follows, with a graph as in fig. 5.1:

$$R_q(t) = \prod_j (\theta_j^{(q)} - t) / \prod_j \theta_j^{(q)} \tag{5.20}$$

Now consider the system

$$\mathbf{Bx} = \mathbf{f} \ , \tag{5.21}$$

$\mathbf{B}$ symmetric, semi-positive definite, $\mathbf{f} \in R(\mathbf{B})$, and let $\mathbf{x}$ denote the minimum norm solution. Consider the conjugate gradients process ( with $\mathbf{x}^{(0)} = \mathbf{0}$) or the Lanczos process yielding iterates $\mathbf{x}^{(1)}, \mathbf{x}^{(2)}, \ldots$ (converging to the minimum norm solution of (5.21)).

Write $\mathbf{B} = \mathbf{V \Lambda V}^T$, $\mathbf{V}$ an orthogonal matrix of eigenvectors, $\mathbf{\Lambda}$ the diagonal matrix of corresponding eigenvalues $\lambda_1, \lambda_2, \cdots$. Write $\mathbf{x}^{(q)} = \mathbf{Vz}^{(q)}$, $\mathbf{x} = \mathbf{Vz}$, $\mathbf{f} = \mathbf{Vg}$ (i.e. we express $\mathbf{x}^{(q)}$, $\mathbf{x}$ and $\mathbf{f}$ in terms of the eigenvectors).

Then we have the following fundamental theorem (cf. van der Sluis & van der Vorst 1986, prop. 2.8, but it is not originally ours):

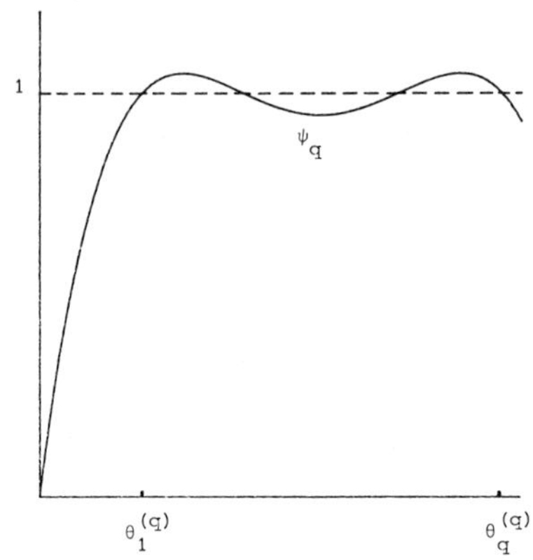


Figure 5.2

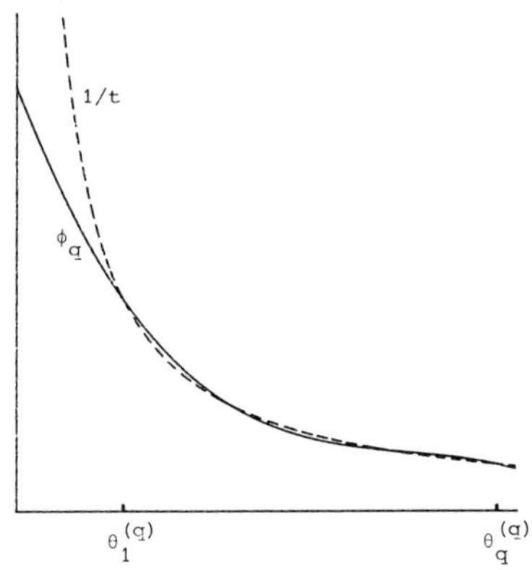


Figure 5.3

**Theorem 5.22**

$$z_j^{(q)} = (1 - R_q(\lambda_j))z_j \tag{5.23}$$

□

Hence, in terms of sec. 2.6, $\psi_q(t) = 1 - R_q(t)$ is the relative response function and $\phi_q(t) = \psi_q(t)/t$ is the response function:

$$z_j^{(q)} = \phi_q(\lambda_j)g_j \ . \tag{5.24}$$

See figs. 5.2 and 5.3.

Note that $\phi_q$ is the Lagrangian interpolation polynomial of the function $\frac{1}{t}$ with the Ritz values as nodes.

Note, however, that, contrary to the situation in previous sections, the (relative) response function depends on **f**.

## 5.8 Convergence and regularization properties of projection methods for least squares problems

We now apply the theory in sec. 5.7 to the normal equations $\mathbf{A}^T\mathbf{A}\mathbf{x} = \mathbf{A}^T\mathbf{b}$ of the least squares problems $\mathbf{A}\mathbf{x} = \mathbf{b}$. Let **x** again denote the minimum norm solution. Write

$$\mathbf{A} = \mathbf{U}\boldsymbol{\Sigma}\mathbf{V}^T \ , \ \mathbf{b} = \mathbf{U}\mathbf{g} \ , \ \mathbf{x} = \mathbf{V}\mathbf{z} \ , \ \mathbf{x}^{(q)} = \mathbf{V}\mathbf{z}^{(q)}$$

$$\mathbf{B} = \mathbf{A}^T\mathbf{A} \ , \ \mathbf{f} = \mathbf{A}^T\mathbf{b} = \mathbf{A}^T\mathbf{U}\mathbf{g} = \mathbf{V}\boldsymbol{\Sigma}^T\mathbf{g} \ .$$

Then we have, defining the ( relative) response functions $\tilde{\psi}_q(t) = \psi_q(t^2)$ , $\tilde{\phi}_q(t) = \psi_q(t^2)/t$ with $\psi_q$ as in sec. 5.7:

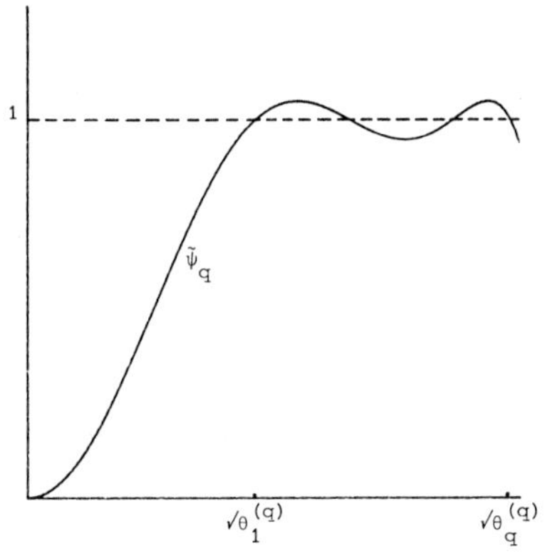


*Figure 5.4*

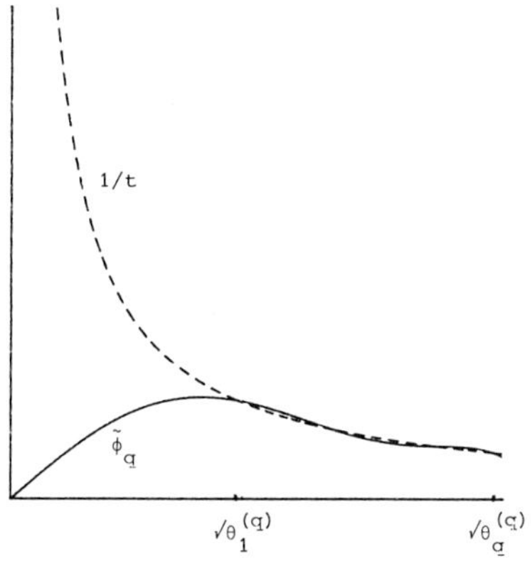


*Figure 5.5*

$$z_j^{(q)} = \tilde{\psi}_q(\sigma_j)\, z_j = \tilde{\phi}_q(\sigma_j)\, g_j \ . \tag{5.25}$$

See figs. 5.4 and 5.5.

We note from the graph of $\tilde{\psi}_q$ that the components $z_j$ corresponding to $\sigma_j \ll \sqrt{\theta_1^{(q)}}$ have large relative errors. The graph for $\tilde{\phi}_q$ shows that data errors affect the components of $\mathbf{x}^{(q)}$ corresponding to small singular values a good deal less than is the case for the corresponding components of $\mathbf{x}$.

As before (cf. secs. 2.7 and 4.3) we wish to decompose the error $\mathbf{x}^{(q)} - \mathbf{s}$ into an approximation error (here again to be called iteration error) $\Delta\mathbf{x}_{it}^{(q)}$, which we would get by applying the process with the ideal right-hand side $\mathbf{d}$, and a perturbation error $\Delta\mathbf{x}_{pert}^{(q)}$ which arises from the errors in $\mathbf{b}$. We assume again that these errors are as described in the beginning of sec. 2.5. Note that we found it convenient to use the function $\psi$ rather than $\tilde{\psi}$.

Writing $\bar{\mathbf{x}}^{(q)}$ , $\bar{\mathbf{z}}^{(q)}$ , $\bar{\mathbf{z}}$ , $\bar{\mathbf{g}}$ , $\bar{R}_q$ , $\bar{\theta}_j^{(q)}$ for the quantities corresponding to those above if we replace $\mathbf{b}$ by $\mathbf{d}$, we then have

$$\Delta\mathbf{x}^{(q)} \equiv \mathbf{x}^{(q)} - \mathbf{s} = \Delta\mathbf{x}_{it}^{(q)} + \Delta\mathbf{x}_{pert}^{(q)} \tag{5.26}$$

with

$$\begin{cases} \Delta\mathbf{x}_{it}^{(q)} \equiv \bar{\mathbf{x}}^{(q)} - \mathbf{s} = \mathbf{V}\Delta\mathbf{z}_{it}^{(q)} \ , \ \Delta z_{it,j}^{(q)} = -\bar{R}_q(\lambda_j)\bar{z}_j \\ \Delta\mathbf{x}_{pert}^{(q)} \equiv \mathbf{x}^q - \bar{\mathbf{x}}^{(q)} = \mathbf{V}\Delta\mathbf{z}_{pert}^{(q)} \ , \ \Delta z_{pert,j}^{(q)} = \psi_q(\lambda_j)z_j - \bar{\psi}_q(\lambda_j)\bar{z}_j \ . \end{cases} \tag{5.27}$$

Since $\psi_q$ depends on $\mathbf{b}$, and, therefore, on the errors in $\mathbf{b}$, in a rather intractable way, we cannot, as before, compute $E(\|\Delta\mathbf{x}_{pert}^{(q)}\|^2)$. If, however, we approximate $\Delta\mathbf{x}_{pert}^{(q)}$ by $\tilde{\Delta}\mathbf{x}_{pert}^{(q)} = \mathbf{V}\tilde{\Delta}\mathbf{z}_{pert}^{(q)}$:

$$\tilde{\Delta} z_{pert,j}^{(q)} = \bar{\psi}_q(\lambda_j)(z_j - \bar{z}_j) \tag{5.28}$$

then we have the following theorem:

**Theorem 5.29**

$$\delta \mathbf{x}_{it}^{(q)} \equiv \| \Delta \mathbf{x}_{it}^{(q)} \| = \sqrt{\sum_j \bar{R}_q(\lambda_j)^2 \bar{z}_j^2} \tag{5.30}$$

$$\tilde{\delta \mathbf{x}}_{pert}^{(q)} \equiv \sqrt{E(\| \tilde{\Delta \mathbf{x}}_{pert}^{(q)} \|^2)} = \sigma \sqrt{\sum_j \frac{\bar{\psi}^{(q)}(\lambda_j)^2}{\lambda_j}} \tag{5.31}$$

$$E\left[ \| \Delta \mathbf{x}_{it}^{(q)} + \tilde{\Delta \mathbf{x}}_{pert}^{(q)} \|^2 \right] = \left[ \delta \mathbf{x}_{it}^{(q)} \right]^2 + \left[ \tilde{\delta \mathbf{x}}_{pert}^{(q)} \right]^2 \tag{5.32}$$

$$E\left[ \| \Delta \mathbf{x}^{(q)} \|^2 \right] \approx \left[ \delta \mathbf{x}_{it}^{(q)} \right]^2 + \left[ \tilde{\delta \mathbf{x}}_{pert}^{(q)} \right]^2 . \tag{5.33}$$

□

Regarding the approximate equality in (5.33) we note that

$$\Delta \mathbf{x}^{(q)} - (\Delta \mathbf{x}_{it}^{(q)} + \tilde{\Delta \mathbf{x}}_{pert}^{(q)}) = \Delta \mathbf{x}_{pert}^{(q)} - \tilde{\Delta \mathbf{x}}_{pert}^{(q)} = \mathbf{V} \Delta \mathbf{z}_{pert}^{(q)} - \mathbf{V} \tilde{\Delta \mathbf{z}}_{pert}^{(q)} \tag{5.34}$$

and in order to estimate this we write

$$\Delta z_{pert,j}^{(q)} - \tilde{\Delta z}_{pert,j}^{(q)} = (\bar{R}_q(\lambda_j) - R_q(\lambda_j))\bar{z}_j + (\psi_q(\lambda_j) - \bar{\psi}_q(\lambda_j))(z_j - \bar{z}_j). \tag{5.35}$$

Assuming the $\theta_k^{(q)}$ to be rather close to the $\bar{\theta}_k^{(q)}$ (which for small $\sigma$ will certainly be the case, cf. sec. 5.7), the quotient $\psi_q(\lambda_j)/\bar{\psi}_q(\lambda_j)$ will be close to 1 for all $j$ (cf. fig. 5.2) and hence $\sum_j [(\psi_q(\lambda_j) - \bar{\psi}_q(\lambda_j))(z_j - \bar{z}_j)]^2 \ll \| \tilde{\Delta \mathbf{x}}_{pert}^{(q)} \|^2$, which is satisfactory. Regarding the first term on the right in (5.35) we note that its form suggests comparison with $\Delta \mathbf{x}_{it}^{(q)}$ rather than with $\tilde{\Delta \mathbf{x}}_{pert}^{(q)}$. However, since we may not expect $R_q(\lambda_j)/\bar{R}_q(\lambda_j)$ to be close to 1 for all $j$ (cf. fig. 5.1) we cannot say that in general $\sum_j [R_q(\lambda_j)\bar{z}_j - \bar{R}_q(\lambda_j)\bar{z}_j]^2 \ll \| \Delta \mathbf{x}_{it}^{(q)} \|^2$, but clearly again, if the $\theta_k^{(q)}$ are rather close to the $\bar{\theta}_k^{(q)}$, this relation can be violated only for rather special choices of the $\lambda_j$ and the $\bar{z}_j$. In view of our numerical example we note that this relation will certainly hold if in (5.30) the sum of the terms with $\lambda_j < \theta_1^{(q)}, \bar{\theta}_1^q$ dominates the rest ( and there are reasons to believe that this is not uncommon either).

Hence, under these circumstances (5.33) will be rather sharp.

### 5.9 Numerical example

Let $\mathbf{A}$ have singular values as used in sec. 4.4, i.e. $\sigma_j = e^{-(j-1)\beta}$, $\beta = .025$, and let again $\bar{z}_j = \sigma_j^2$. We take $n = 200$. Note that $\lambda_j = e^{-2(j-1)\beta}$.

Just to get some feeling for the situation we computed the Ritz values $\bar{\theta}_j^{(q)}$ for some $q$, and display them in table 5.1. The non-displayed Ritz values for the given $q$ are lying more or less equidistantly between $\bar{\theta}_7^{(q)}$ and $\bar{\theta}_{q-2}^{(q)}$.

**Table 5.1**

| $q$ | $\bar\theta_1$ | $\bar\theta_2$ | $\bar\theta_3$ | $\bar\theta_4$ | $\bar\theta_5$ | $\bar\theta_6$ | $\bar\theta_7$ | $\bar\theta_{q-2}$ | $\bar\theta_{q-1}$ | $\bar\theta_q$ |
|---|---|---|---|---|---|---|---|---|---|---|
| 5 | .20 | .42 | .66 | .86 | .99 | | | | | |
| 10 | .069 | .16 | .27 | .39 | .52 | .65 | .77 | .87 | .94 | 1 |
| 15 | .033 | .078 | .13 | .20 | .28 | .37 | .85 | .90 | .95 | 1 |
| 20 | .019 | .045 | .079 | .12 | .17 | .22 | .28 | .90 | .95 | 1 |
| 25 | .012 | .029 | .050 | .077 | .11 | .15 | .19 | .90 | .95 | 1 |
| 30 | .0083 | .019 | .034 | .052 | .074 | .099 | .13 | .90 | .95 | 1 |

We now look in particular at $q = 5$ and $q = 10$. Again to get some feeling we computed the local extrema of the normalized Ritz polynomials $\bar R_q$. In table 5.2 we list the abscissae $t$ and extremal values $\bar R_q(t)$ (remember $\bar R_q(0) = 1$).

**Table 5.2**

| $t$ | .27 | .53 | .76 | .94 | | | | | |
|---|---|---|---|---|---|---|---|---|---|
| $\bar R_5(t)$ | -.038 | .014 | -.0091 | .0088 | | | | | |
| $t$ | .10 | .20 | .32 | .45 | .59 | .71 | .82 | .91 | .98 |
| $\bar R_{10}(t)$ | -.032 | .0099 | -.0046 | .0027 | -.0018 | .0015 | -.0013 | .0015 | -.0031 |

We now consider perturbations with $\sigma = 0.00226$ (we apologize for this funny number which arose from a certain normalization in an early stage of our research). Taking $g_j = \bar g_j - \sigma$ (i.e. we take fixed errors instead of random errors), we find that $\theta_1^{(5)}$ and $\theta_1^{(10)}$ differ only .005 from $\bar\theta_1^{(5)}$ and $\bar\theta_1^{(10)}$, respectively, and that the other Ritz values differ even less from the unperturbed values. For random errors one should expect even smaller perturbations of the Ritz values.

We computed $\delta\mathbf{x}_{it}^{(q)}$, $\tilde\delta\mathbf{x}_{pert}^{(q)}$ for $q = 5$ and $q = 10$ using theorem 5.29. We found the sum in (5.30) to be dominated by the terms with $\lambda_j < \theta_1^{(q)}, \bar\theta_1^{(q)}$ by a large factor. Hence, in view of the discussion after theorem 5.29, the approximate equality in (5.33) should be rather sharp for stochastic perturbations with this kind of $\sigma$.

We obtained the results in table 5.3, with $\tilde\sigma$ as in (4.29):

**Table 5.3**

| $q$ | $\delta\mathbf{x}_{it}^{(q)} / \|\mathbf{s}\|$ | $\tilde\delta\mathbf{x}_{pert}^{(q)} / \|\mathbf{s}\|$ |
|---|---|---|
| 5 | .050 | 15 $\tilde\sigma/\sqrt{m}$ |
| 10 | .016 | 26 " |

Comparing tables 5.1 and 5.3 suggests that $\delta\mathbf{x}_{it}^{(q)}$ is more or less proportional to $\bar\theta_1^{(q)}$ and that $\delta\mathbf{x}_{pert}^{(q)}$ is inversely proportional to $\sqrt{\bar\theta_1^{(q)}}$. This may indeed be shown to be the case for our model problem. We also note from table 5.1 that $\bar\theta_1^{(q)}$ decreases more or less inversely proportionally to $q^2$ (which can be shown also) and hence that $\delta\mathbf{x}_{it}^{(q)}$ does the same.

### 5.10 Comparison of SIRT and projection methods

We recall (cf. sec. 4.1, remark ii) that SIRT, applied to any system $\mathbf{Ax} = \mathbf{b}$, actually solves for the minimum norm least squares solution of $\mathbf{Wy} = \mathbf{f}$, $\mathbf{W} = \mathbf{R}^{-1/2}\mathbf{AC}^{-1/2}$, $\mathbf{f} = \mathbf{R}^{-1/2}\mathbf{b}$ and then takes $\mathbf{x} = \mathbf{C}^{-1/2}\mathbf{y}$.

As an alternative we could compute this $\mathbf{W}$ and $\mathbf{f}$ explicitly and then apply a projection method to $\mathbf{Wy} = \mathbf{f}$. If we do this for the $\mathbf{W}$ and $\mathbf{f}$ considered in sec. 4.4 (where we actually specified $\mathbf{W}$ and $\mathbf{f} = \mathbf{b}$ indirectly, viz. through the singular values of $\mathbf{W}$, the coordinates of the ideal solution with respect to a singular vector basis and the distribution of errors in $\mathbf{b}$) then we get just the numerical example considered in sec. 5.9.

It thus makes sense to compare tables 4.1 and 5.3. Doing this, we note that the projection method with $q = 5$ compares quite well to the SIRT method with $q = 7$ or 8 for both the perturbation and iteration error provided the optimal $\omega$ is chosen; for $\omega = 1$ one should take SIRT with $q = 13$. Likewise the projection method with $q = 10$ compares to SIRT with $q = 25$ and optimal $\omega$ or to some $q > 40$ if $\omega = 1$.

Hence, at least in the range considered, *the projection method has a much faster convergence behaviour than SIRT* (as is also seen from the observations in secs. 5.9 and 4.4 that $\delta\mathbf{x}_{it}^{(q)}$ decreases inversely proportionally to $q^2$ and $q$ respectively), *while having the same regularizing effect* (i.e., that when at some stages of the respective processes they have comparable iteration errors, they also have comparable expectations for the perturbation errors).

Since the amount of work per iteration is about the same for the two methods it is clear that, at least for our model problem, the projection method is preferable.

We can say nothing about how the SIRT and projection methods compare when applied to the same problem $\mathbf{Ax} = \mathbf{b}$. The importance of our findings above is, however, that, at least in our model situation, any difference in *attainable quality* should be attributed to the *scaling* of this problem in relation to the method to be used, and not to the method as such. In the epilogue (sec. 6) we will comment on the generality of this result.

## 6. Epilogue

In this chapter we have discussed and analyzed SIRT and projection methods. As we have seen (sec. 4.1), with SIRT one solves, in effect, a least squares problem that is differently weighted than the given problem and the limit is minimal in a different norm, too. In genera', this implicit scaling of SIRT should be considered a disadvantage. Indeed, no matter how advantageous this may work out in a particular situation, anyone wishing to solve a least squares problem should be allowed to scale his problem in a meaningful way (and many people spend a good deal of thought how to do this appropriately). It is then undesirable that this scaling is changed behind the scenes by the method.

The projection methods do not have this adverse effect. Moreover, at each iteration step they minimize the norm of the residual among all iterative methods of the form (5.5) to which the SIRT methods also belong.

We also noted for a model situation that solutions of comparable quality as those produced by SIRT could be obtained considerably faster by a projection method applied to a suitably scaled version of the problem (cf. sec. 5.10), and that, therefore, any difference in attainable quality should be attributed to the scaling of the problem in relation to the method to be used, and not to the method as such.

It would be interesting to know, of course, whether the latter properties hold for a large, interesting class of problems. In this connection we note that for rather dense spectra and rather modest numbers of iterations it is not so much the particular values of $\sigma_j$ and $z_j$ that are of importance for the convergence behaviour of the cumulative weight function $f(t) \equiv \sum_{\sigma_j \geq t} \bar{z}_j^2$, in the sense that for all problems with similar functions $f$ we will have a similar convergence behaviour (for SIRT methods we see this from (4.21) by writing $\sum_j \bar{z}_j^2(1-\omega\sigma_j^2)^{2q}$ as the Stieltjes integral $\int_1^0 (1-\omega t^2)^{2q}\, df(t)$, for the projection methods this is more complicated-it is related to the orthogonality property of the Ritz polynomials, cf. van der Sluis & van der Vorst 1986). For our model problem we have $f(t) \approx c\,(1-t^4)$, for Nolet's problem (see sec. 4.4), we have $f(t) \approx c \ln \frac{t}{\sigma_1}$ (which means, by the way, that in this case the solution has quite some weight for the smaller singular values).

We can, indeed, prove that the interesting properties above do hold for large classes of functions $f$, including functions of the kinds we just mentioned. This indicates that much more generally a judicious application of a projection method will produce results more efficiently than SIRT without sacrificing accuracy. Proofs and further details will be published elsewhere.

*Chapter 4*

# On the validity of the ray approximation for interpreting delay times

*E. Wielandt*

This chapter deals with the influence of diffracted waves on tomographic observations. Probably most workers in tomography are aware of the existence of such waves and of their potential to interfere with the direct wave on which the tomographic interpretation is based. Strong diffraction effects are expected when the scale of heterogeneity is in the order of the seismic wavelength, but the smallest detail one wants to resolve with seismic tomography is typically one or two orders of magnitude larger. Can we safely neglect diffraction in this case?

The question is important because if diffracted waves have a noticeable amplitude, they will introduce a bias into the travel-time data: diffracted waves arriving before the direct wave would be read as early arrivals while diffracted waves arriving later would in general be ignored. We must remember that in heterogeneous and especially in discontinuous structures the "direct" wave is associated with a path of stationary but not necessarily of absolute minimum travel time. Faster paths may exist which carry no energy in the high-frequency limit and are therefore ignored in ray theory, although they may transmit substantial energy at finite frequencies. There is no physical difference between "direct" and diffracted waves apart from their behaviour in the (unphysical) high-frequency limit, and the classification as one or the other wave type may depend on the parametrization of the model. There is thus no justification for neglecting diffracted waves a priori.

The problem of calculating the amplitudes of diffracted waves could be attacked with finite-element modelling, or in the framework of a scattering theory with Born's approximation. For the present study we have chosen to investigate a geometrically simple case for which a complete analytic solution is available, namely that of a spherical



*G. Nolet (ed.), Seismic Tomography, 85–98.*

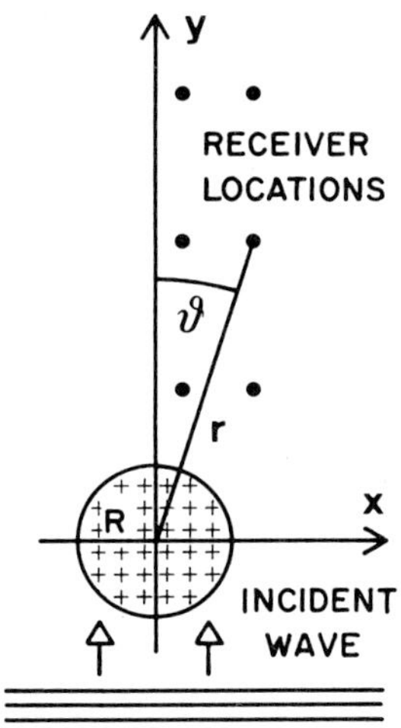


*Figure 1* Geometry of the diffraction problem

inclusion in a homogeneous space; and we will consider only the simplest, acoustic, form of the wave equation. A comprehensive treatment of the elastic case is given by Chapman and Phinney (1972) who model amplitudes and waveforms of P and S waves diffracted at the core-mantle boundary. Related work by Scholte, Nussenzveig, Phinney, Ansell and others is referenced in their paper.

## 1. Statement of the problem and method of solution

We consider a plane acoustic wave of arbitrary time dependence that propagates through a liquid full space with a spherical inclusion centerered at the origin of the coordinate system (fig.1). The problem is of cylindrical symmetry and we will therefore only use the x and y coordinates although we solve the wave equation in three dimensions. The y axis is the axis of symmetry along which the wave propagates. The inclusion is assumed to have a different wave velocity but the same density as the space in which it is embedded. In this case, and for a harmonic wave of frequency $\omega$, the wave equation reduces to

$$(\Delta + k^2)\Phi = 0\,, \quad k = \omega/c \tag{1}$$

$\Delta$ is the Laplace operator and $\Phi$ is a wave function representing the pressure or a related potential. The wave velocity $c$ is a material parameter and not necessarily identical with the phase velocity of a propagating wave, which also depends on the geometry of the wavefront. The physical boundary conditions at the surface of the inclusion require that the wave function and its normal derivative are continuous.

After Sommerfeld (1947), we can construct a solution with the desired properties in the following way:

Considering one harmonic component of the signal at a time, we expand the wavefield outside the anomaly into a series of Bessel functions multiplied with spherical harmonics:

$$\Phi_0(r,\theta) = \sum_{n=0}^{\infty} \left[ a_n \psi_n(k_0 r) + b_n \eta_n(k_0 r) \right] P_n(\cos\theta) \tag{2}$$

where $k_0 = \omega / c_0$ and

$$\psi_n(\rho) = \sqrt{\frac{\pi}{2\rho}} J_{n+1/2}(\rho) , \quad \eta_n(\rho) = \sqrt{\frac{\pi}{2\rho}} Y_{n+1/2}(\rho)$$

The symbols J, Y, and P denote Bessel and Neumann functions and Legendre polynomials, respectively. $\psi_n$ and $\eta_n$ are known as spherical Bessel and Neumann functions. A similar expansion, with the regular solution $\psi_n$ only and with a different wavenumber $k_1$ , describes the wavefield inside the anomaly:

$$\Phi_1(r,\theta) = \sum_{n=0}^{\infty} a'_n \psi_n(k_1 r) P_n(\cos\theta) \tag{3}$$

For each colatitudinal order $n$ we have then three unknown coefficients $a_n$, $b_n$ and $a'_n$. These are determined from the two boundary conditions at the surface of the anomaly, and from the condition that the outer field must represent a plane wave of unit amplitude plus an outgoing wave. The expansions converge late (but then rapidly): at a distance r from the origin we have to retain more than $k_0 r$ terms; by subtracting the plane-wave part in the form

$$\exp(ik_0 y) = \sum_{n=0}^{\infty} (2n+1) i^n \psi_n(k_0 r) P_n(\cos\theta) \tag{4}$$

we reduce the number of significant terms to the order $k_0 R$ where $R$ is the radius of the anomaly. (In place of a direct summation, the expansions could be converted into complex integrals using the Watson transform; see e.g. Chapman and Phinney 1972.) As a last step, we combine harmonic solutions to form a signal of arbitrary, here pulse-like, time dependence. The algorithm has been coded in FORTRAN 77 on a HP-1000 minicomputer and permits seismogram sections such as shown later to be computed within a few minutes. The method has an obvious generalization to spherically layered structures.

## 2. Numerical experiments

To approximate the conditions in upper-mantle tomography, we assume a spherical anomalous body of 100 km diameter with a velocity anomaly between -10% and +10% relative to a background velocity of 8 km/s. The incident signal has either a dominant frequency of 1.5 Hz and a cutoff frequency of 3 Hz, or a dominant frequency of 5 Hz and a cutoff frequency of 10 Hz. The diameter of the anomaly is thus about 20 and 60 times the dominant wavelength, or 40 and 120 times the cutoff wavelength, respectively. (The visually dominant frequency corresponds to the half-width of the signal spectrum, which has the shape of a cosine-square bell around zero frequency.) All seismic traces shown have a length of 6 s. The results can be scaled to other dimensions. Identical waveforms are

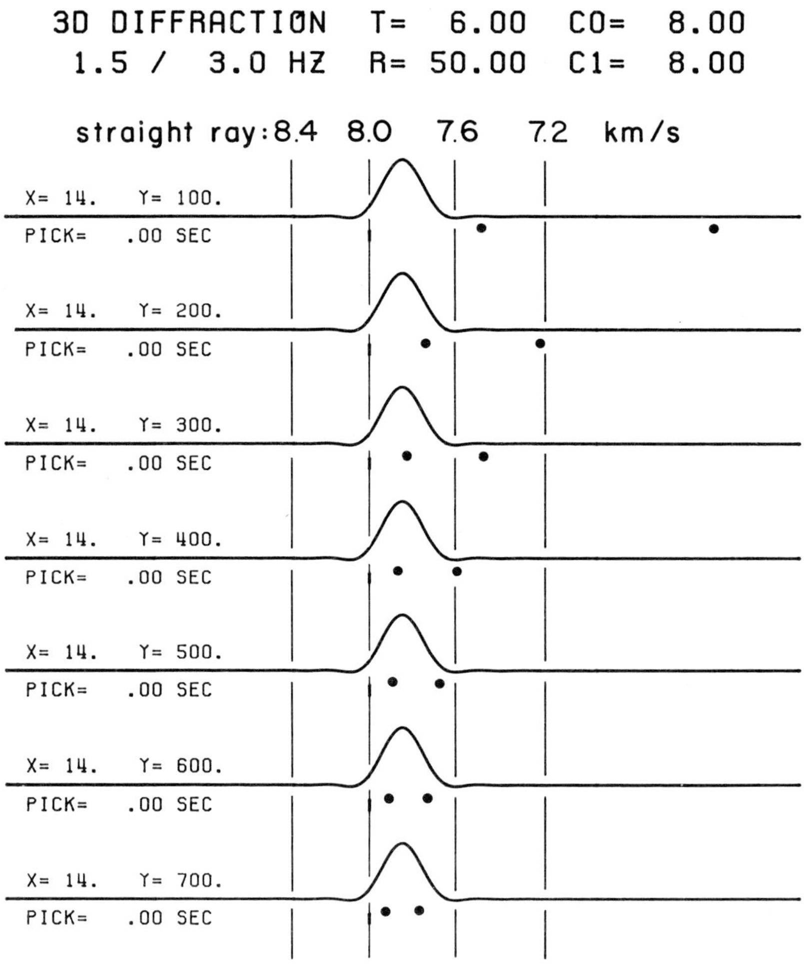


*Figure 2*

obtained with all linear dimensions and all velocities m-times greater; the same applies if we leave the velocities unchanged, take the linear dimensions and the trace length m-times greater, and choose the characteristic frequencies m-times smaller. In particular, after a transformation of the length and time scales the results obtained in the 5/10 Hz band would apply to an inclusion of 333 km diameter observed in the 1.5/3 Hz band which we consider as representative for teleseismic observations.

Seismograms are presented in the form of seismogram sections along linear profiles comprising 7 receiver locations. The most interesting case for a comparison with ray theory is when the receivers are located at different distances behind the anomaly. We have

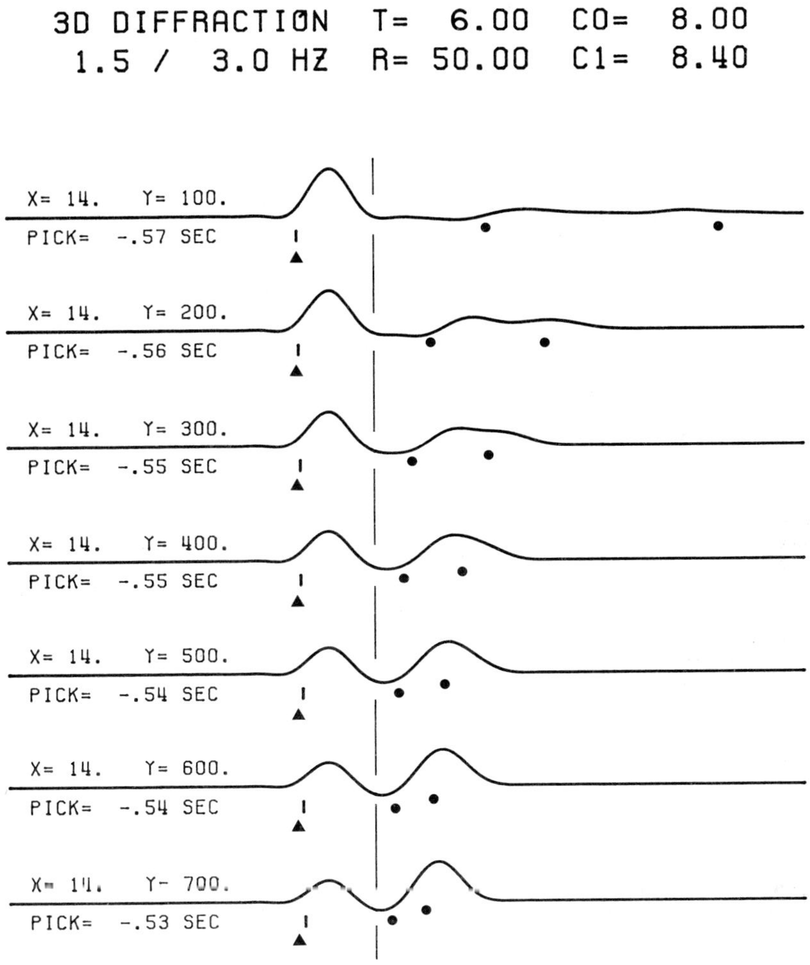


*Figure 3*

chosen a profile that is laterally offset by 14 km from the axis of symmetry and covers the distance range from 100 to 700 km from the center (fig.1). The offset was introduced because a scatterer of cylindrical symmetry generates a diffraction maximum along the axis that would not be observed in a less regular geometry, and would therefore cause untypical results. A profile with a lateral offset of 64 km was used to study effects outside the geometrical "shadow" zone behind the anomaly.

The individual traces in each section are labelled with the receiver coordinates and with automatically picked first-arrival times. We take the time of the first sample that exceeds 15% of the peak amplitude of the incident wave as the time where an onset would be

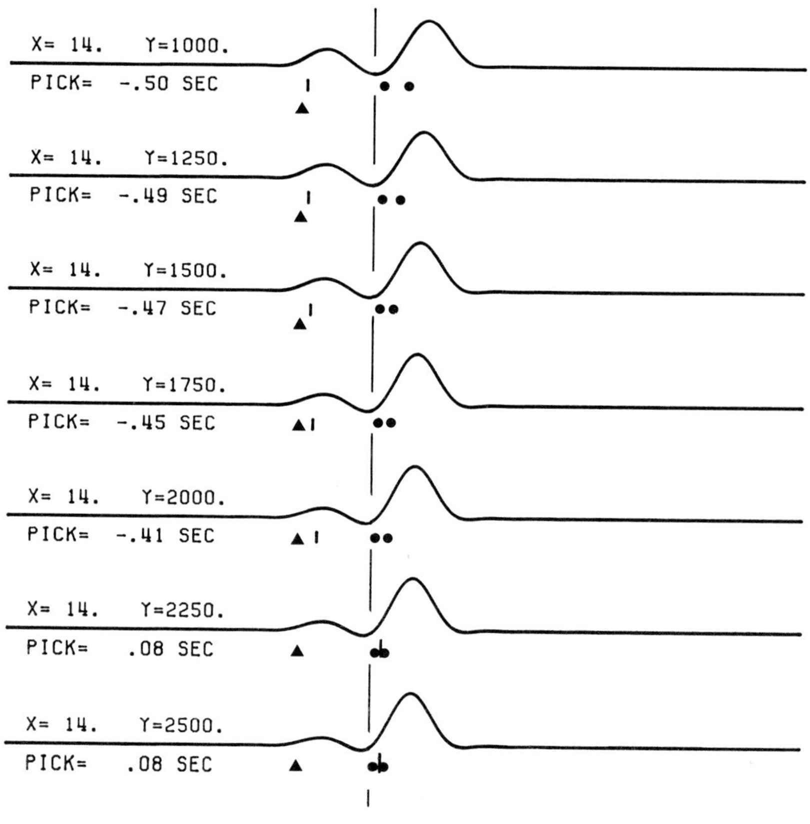


*Figure 4*

observed in the presence of noise. Although quantitative results are of course influenced by this assumption, most of our conclusions remain valid for signals with a better signal-to-noise ratio since diffracted waves are even more likely to be observed in that case.

Four real or hypothetical arrivals are marked in each trace: the undisturbed plane wave with a vertical line across all traces; the transmitted straight ray with a triangular symbol (except in fig.2); two diffracted waves with dots; and the picked first arrival with a small vertical bar. The horizontal distance between the bar and the triangular mark is thus a measure of the discrepancy between full wave theory and the ray approximation. The positions of the triangular and circular symbols indicate geometrically predicted time delays relative to the undisturbed plane wave. The difference between the travel times of the straight ray and of the true transmitted ray that follows Snell's law is negligible in all examples shown here, and we will not distinguish between them.

All plots show true amplitudes; the amplitude of the incident wave is half of the vertical distance between traces (fig.2). Amplitudes larger than twice the incident amplitude have been clipped. The case parameters in the headers are: length of trace (T) in seconds, background velocity (C0) in km/s, dominant and cutoff frequencies in Hz, radius of the anomaly (R) in km, and its velocity (C1) in km/s.

## 3. Results

We now proceed to a discussion of individual seismogram sections, starting from figure 2 which represents the case of no anomaly. This figure is included as a reference for the waveform and the arrival times; it may be copied on a transparency and superimposed on the other seismogram sections. The vertical lines mark the arrivals (at the 15% level) of transmitted rays for velocities of 8.4, 8.0, 7.6 and 7.2 km/s in the inclusion.

Fig.3 shows the case of a positive velocity anomaly of 5%. The first arrival (bar) is early by about 0.55 s in all traces, as expected from the ray approximation (triangle; the predicted value is 0.57 s). We also see two arrivals of diffracted waves (dots) associated with the two shortest paths around the sphere in the X-Y plane. (Paths outside the X-Y plane do also contribute but the resolution of this figure is insufficient to show it; the contribution is however visible in fig.7 between the marked diffracted signals.) The diffracted waves merge at about 400 km distance, from where on they form the strongest arrival. At the same time their delay against the undisturbed plane wave (vertical line) decreases to a fraction of a second. Obviously the exact shape of the inclusion is no longer relevant when the waves diffracted around the inclusion become undistinguishable and essentially restore the incident plane wave. At least from this distance on, our results should apply to all roughly isometric bodies.

Fig.4 is a continuation of fig.3 to longer distances. The amplitude of the transmitted wave decreases as 1/r in the far field while the diffracted wave becomes undistinguishable from the incident wave. Beyond a certain distance, about 2000 km in this specific case (last two traces), the transmitted wave would no longer be recognized, and the anomaly would become invisible; in fact a slightly late arrival is then observed in place of an early one.

Figures 5 and 6 were computed for negative velocity anomalies of 5 and and 10%. They show more complex waveforms than the previous figures. Except for the top trace in each figure, diffracted waves now arrive before the transmitted wave and partly interfere with it; the observed travel time is that of the fastest diffracted wave. There is still a positive delay because the diffracted waves have a longer path, but it is nearly independent of the magnitude of the velocity anomaly. That the diffracted wave is sufficiently strong to be taken for the transmitted wave at a distance as small as twice the diameter of the anomaly is a quite unexpected result.

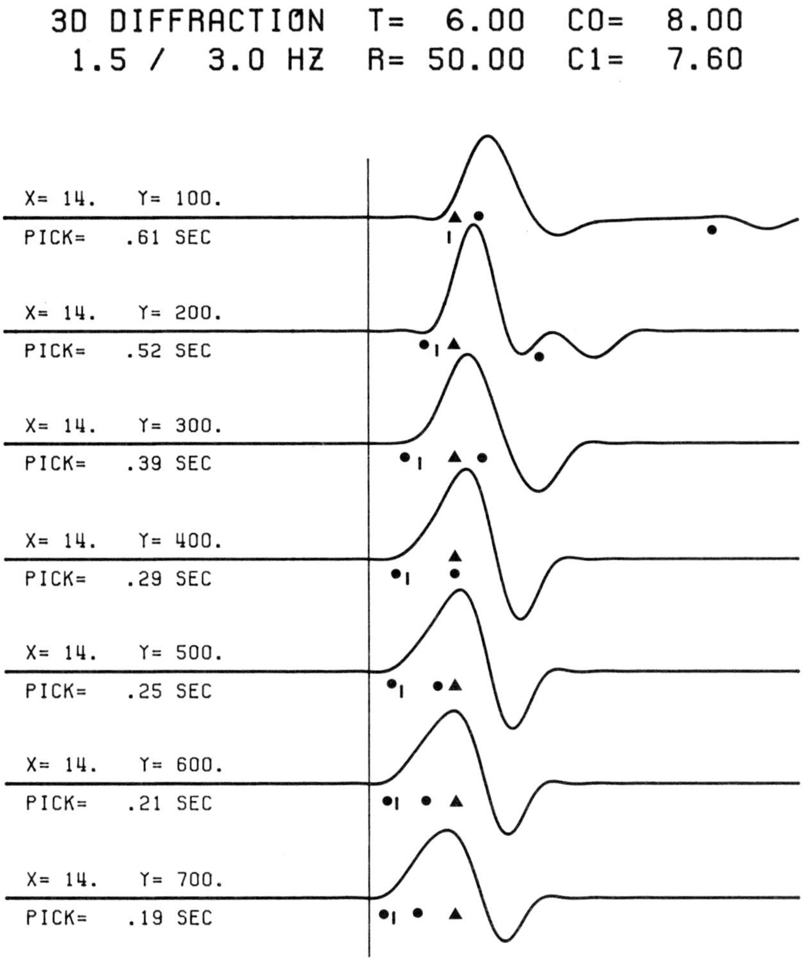


*Figure 5*

Although the relative amplitude of diffracted waves is of course controlled by the wavelength, the observed travel times are remarkably unsensitive to the frequency content of the signal. Seismogram sections corresponding to figures 3 and 5, but using a 5/10 Hz signal, are reproduced in figures 7 and 8. Arrival times picked at the 15% level do not differ substantially from those of the 1.5/3 Hz signal.

We make a few more observations before we discuss the travel-time effects more systematically. The diffracted wave is stronger for a slow than for a fast inclusion, as is apparent from a comparison of figures 3 and 5 or of figures 7 and 8. Also, the diffracted arrivals are slightly earlier than predicted when the inclusion is fast, and vice versa:

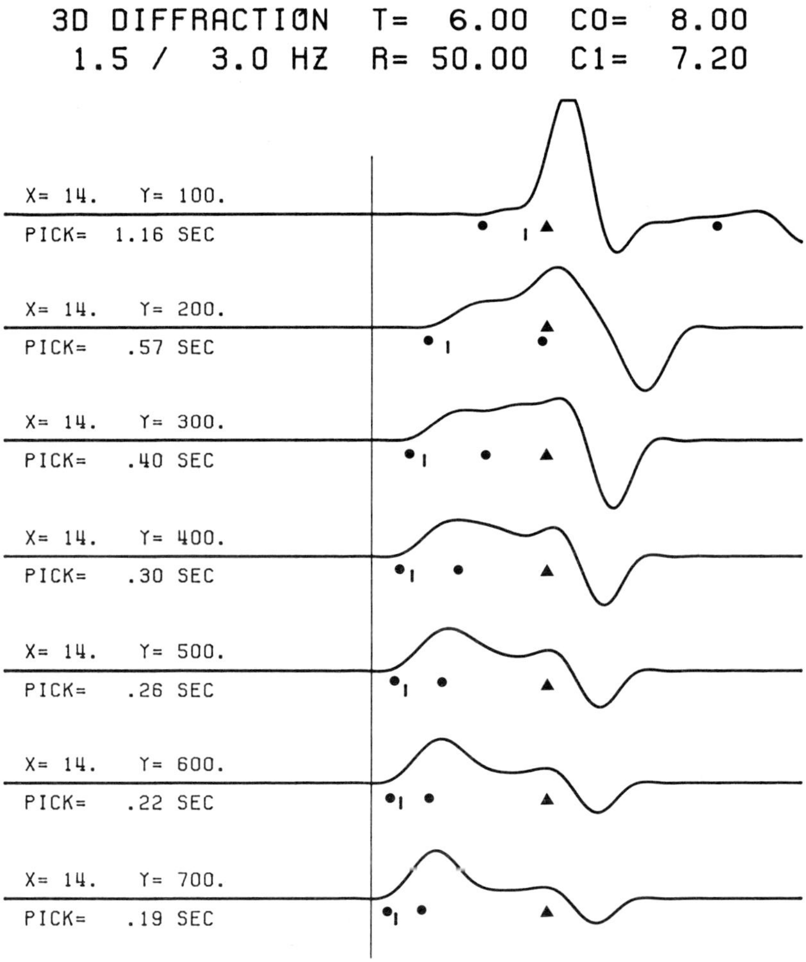


*Figure 6*

diffracted waves do not simply propagate around the inclusion but sample it to some degree. The polarity of the transmitted wave is reversed in the far field after a slow inclusion (figures 5,6 and 8), which must be related to the fact that in a ray approximation the transmitted wave has a caustic. The transmitted wave and the final diffracted wave appear to interchange their character at the crossover point near 400 km (fig.8).

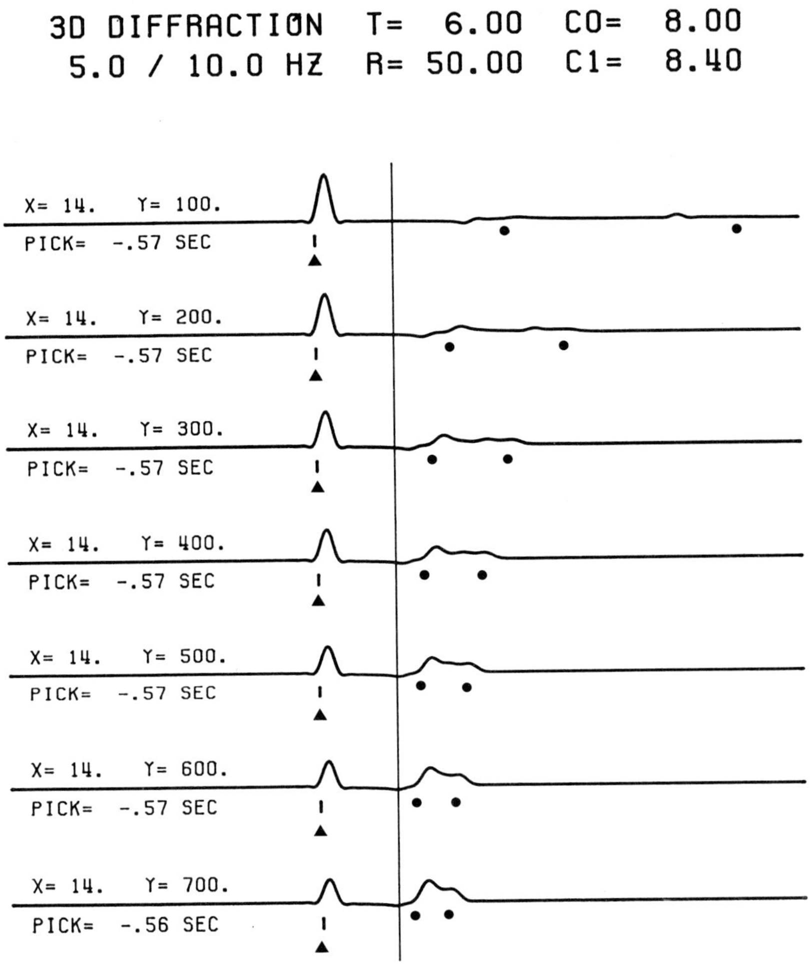


*Figure 7*

## 4. Validity of the ray approximation

Our results reveal a strong asymmetry between positive and negative velocity anomalies. The ray approximation works well for positive anomalies, but in the presence of a negative anomaly, travel-time data are likely to represent the travel time of the fastest path around the anomaly, even at relatively short distances. This would not be a problem if the data were interpreted accordingly, but the ray approximation that is generally used for tomographic modelling is clearly inadequate in this case. It unconditionally predicts a linear relationship between the local slowness anomaly and the observed travel time that in reality exists only for very small anomalies. Figure 10 gives some quantitative information.

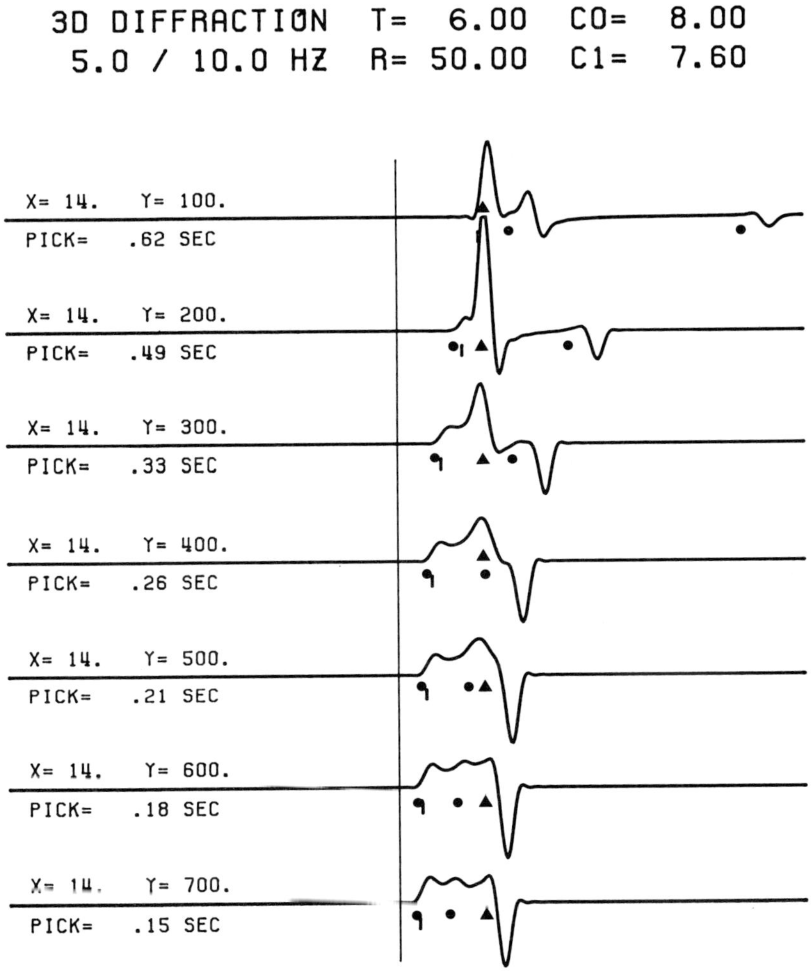


*Figure 8*

It is based on a series of experiments with 5/10 Hz data such as in fig. 7 and 8, and shows the travel-time anomaly observed at various distances as a function of the anomalous velocity. Arrivals were again picked at the 15% level. The asymmetry between positive and negative anomalies is obvious. For negative anomalies, the delays saturate at a distance-dependent value that represents the delay of the fastest diffracted wave. By projecting these limiting delays back onto the delay curve for the straight ray, we find that the ray approximation is inadequate for negative anomalies in excess of 4% at 200 km distance, 2% at 500 km, and 1% at 1000 km (always assuming a diameter of 100 km for the anomaly). Isolated negative anomalies violating these conditions should not appear in a tomographic model because they are unlikely to manifest themselves in travel-time data. If

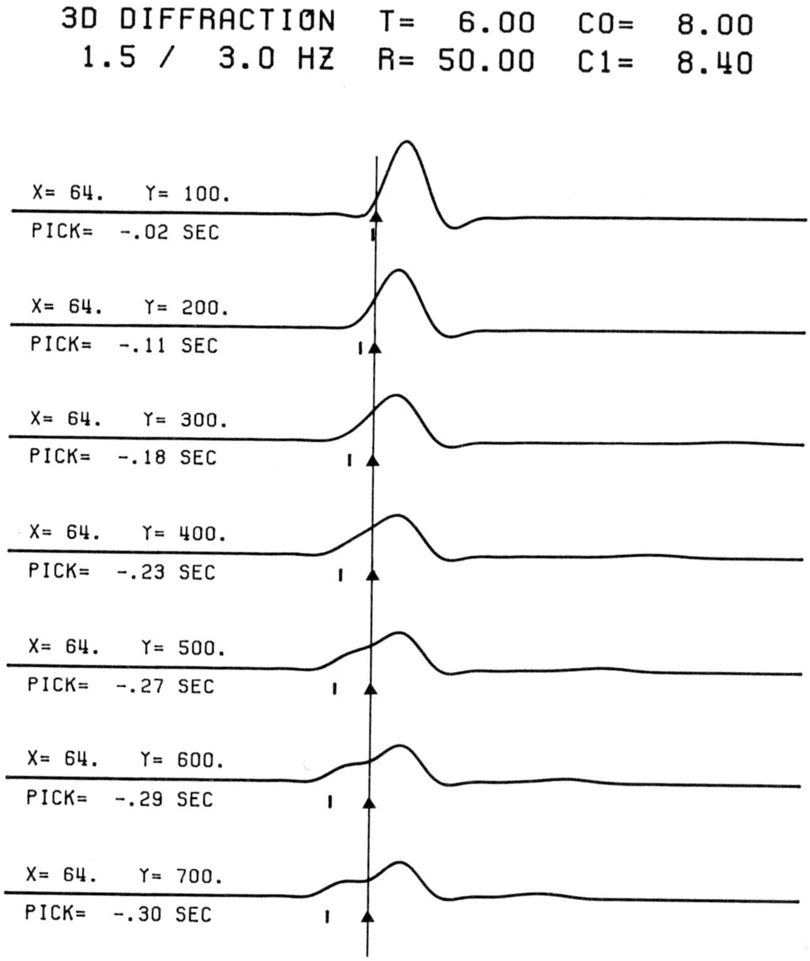


*Figure 9*

such anomalies are present in nature, they are likely to be underestimated in a tomographic inversion.

To maintain symmetry at least in the topics, we show with the last seismogram section that positive anomalies are likely to be overestimated in size.

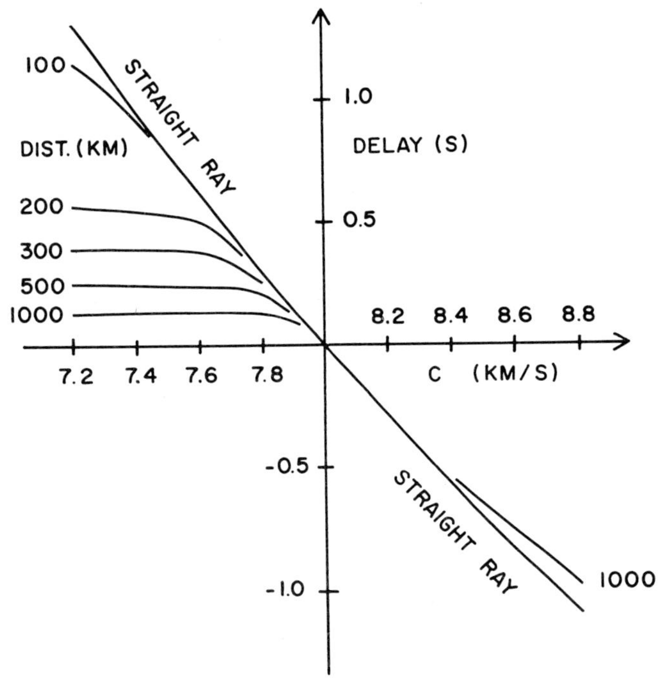


*Figure 10* Delay of a bandlimited impulsive plane wave (0-10 Hz) behind a spherical body of 100 km diameter with a velocity anomaly of -10% to +10% against a background medium with 8.0 km/s. The curves give the delay predicted by the ray approximation, and by full wave theory at various distances between 100 and 1000 km from the center of the anomaly.

## 5. Lateral effects

In all previous experiments we had the receivers located in the geometrical "shadow" zone behind the anomaly. Outside this zone, the straight-ray approximation predicts the absence of travel-time anomalies, and results obtained with raytracing would strongly depend on the shape of the inclusion. Full wave theory is much less critical in this respect and we can easily derive some general results. A wave transmitted through a slow inclusion would not give a first arrival outside the shadow zone, but one transmitted through a fast inclusion would. Fig.9 shows an example, with the same model parameters as fig.3 but receivers located 14 km outside the shadow zone. In all traces from a distance of 200 km on, the transmitted wave produces an early arrival. As in the previous cases, the arrival times do not change much when we extend the frequency band to 5/10 Hz (no figure). We conclude that in a tomographic model a fast inclusion will appear geometrically larger than it actually is. This result becomes a corollary of the previous one when the anomaly and the surrounding medium are interchanged.

## 6. Summary and Conclusion

Diffracted or equivalent laterally refracted waves can have noticeable amplitudes in tomographic observations. The delay vs. slowness relationship of the first arrival then becomes nonlinear and the tomographic ray approximation breaks down. In a model based on this approximation, inclusions with a positive anomaly will appear larger than they are in nature; the amplitude of negative anomalies may be grossly underestimated. Both effects bias the model towards higher average velocities.

Although in the presence of diffracted waves the verification of a tomographic model may require finite-frequency waveform modelling, the arrival times of diffracted waves can approximately be predicted with ray theory, so their incorporation in travel-time tomography should be possible. Forward modelling could be done with a 3-d ray-bending technique and the inverse problem would have to be solved iteratively. The greatest difficulty will probably be the phase identification, i.e. the assignment of a specific ray to an observed signal or (worse) to a bulletin reading. In the absence of other information, the assumption that we see the fastest ray irrespective of its character seems to be more realistic than the assumption that we see only true rays following Snell's law.

*Chapter 5*

# Ray tracing algorithms in three-dimensional laterally varying layered structures

*V. Červený*

## 1. Introduction

This chapter presents a concise review of recent methods of ray tracing and travel time computations of high-frequency seismic body wave propagating in three-dimensional laterally varying isotropic layered structures. Large attention is devoted mainly to *initial value ray tracing* , in which a single ray under consideration is specified by its initial point and by the initial direction of the ray at that point. The numerical, analytic and cell approaches to the solution of initial value ray tracing are discussed in detail. Particularly simple algorithms are obtained for the ray tracing in the *"quadratic slowness"* model, in which the quadratic slowness ($1/V^2$) is used instead of the propagation velocity $V$. In Section 6, the ray tracing system is rewritten into an arbitrary curvilinear orthogonal coordinate system, and in Section 7 into a spherical coordinate system. In all these cases, the ray tracing system is supplemented by the appropriate reflection/transmission laws which allow to perform the ray tracing and the travel time computations even *across arbitrary curved interfaces* .

A short review of the *dynamic ray tracing* and *propagator matrices* is given in Sections 8-10. Dynamic ray tracing and the propagator matrices have recently found important applications in seismology. For example, they can be used to compute easily the second derivatives of the travel time field with respect to the spatial derivatives along the ray and the travel time field in the "paraxial" neighbourhood of the ray. Ray propagator matrices can be also used to compute simply any paraxial ray and to solve *analytically* any boundary value ray tracing problem for paraxial rays. The complete analytic solution



*G. Nolet (ed.), Seismic Tomography, 99–133.*

of the two point ray tracing problem for paraxial rays of any high frequency seismic body wave propagating in a 3-D laterally varying layered structure is given in Section 10.6. Even though this chapter has partly a tutorial character, it does not give the derivation of all presented equations. This would increase the length of the chapter inadmissibly. Moreover, the review does not discuss the linearization approaches to the travel time computations, the ray tracing and travel time computation in anisotropic media, and the evaluation of the ray amplitudes. It also does not give a comprehensive list of references on the ray tracing in three-dimensional models; only selected references needed in the text are presented. The derivation of most of the equations of this chapter can be found in the extensive paper of Červený (1985), including the equations for the ray amplitudes and ray synthetic seismograms. A very concise review of the seismic ray theory is given in Červený (1986a), where also expressions for the Green's function for elastodynamic ray theory with a source and receiver situated arbitrarily in a 3-D laterally varying layered structure can be found.

Boldface letters are used to denote matrices. To distinguish between matrices of different types, the 3×3, 3×2 and 3×1 matrices are written with the circumflex symbol above the letter. If the same bold letter is used for 2×2 and 3×3 matrices with and without the circumflex, e.g. $\mathbf{M}$ and $\hat{\mathbf{M}}$, the matrix without a circumflex ($\mathbf{M}$) denotes a left upper submatrix of the 3×3 matrix with a circumflex ($\hat{\mathbf{M}}$). The relation between the 2×1 matrix $\mathbf{x}$ and 3×1 matrix $\hat{\mathbf{x}}$ is similar, $\mathbf{x} = (x_1, x_2)^T$ and $\hat{\mathbf{x}} = (x_1, x_2, x_3)^T$, where $T$ denotes the transpose. Besides the matrix notation, we also use component notation. Upper case indices $(I, J, K, ...)$ indicate the numbers 1 and 2, lower case indices $(i, j, k, ...)$ the numbers 1,2,3. In this way, $M_{IJ}$ denote components of $\mathbf{M}$, $M_{ij}$ components of $\hat{\mathbf{M}}$. The Einstein summation convention is used throughout the chapter, $M_{ij}x_j = M_{i1}x_1 + M_{i2}x_2 + M_{i3}x_3$, $M_{iJ}x_J = M_{i1}x_1 + M_{i2}x_2$. Unlike the convention followed in other chapters, we denote vectors as, e.g., $\vec{x}$.

## 2. High-frequency solutions of the elastodynamic equation. Eikonal equation

In seismological applications, the high-frequency (HF) solution of the elastodynamic equation is usually assumed to have the following form:

$$u_k(x_j, t) = \mathrm{Re}\left\{ U_k(x_j) F(t - T(x_j)) \right\}. \tag{2.1}$$

Here $x_j$ ($j = 1,2,3$) denote the Cartesian coordinates, $t$ the time, $u_k(x_j, t)$ the Cartesian components of the real-valued displacement vector $\vec{u}(x_j, t)$, $U_k(x_j)$ the Cartesian components of the *vectorial complex-valued amplitude* $\vec{U}(x_j)$, and $T(x_j)$ the *travel time*, also called eikonal. It is assumed that $\vec{U}$ and $T$ depend on spatial coordinates $x_j$ only, not on the time $t$. The scalar complex-valued function $F(\xi)$ is a high-frequency analytical signal. This means that the Fourier spectrum of $F(\xi)$ effectively vanishes for low frequencies and that $F(\xi) = f(\xi) + ig(\xi)$, where $i$ is the imaginary unit, $f$ and $g$ are real-valued and form a Hilbert transform pair. In frequency domain, $F(\xi) = \exp(-i\omega\xi)$, with $\omega$ high.

For constant time $t$, the equation

$$T(x_j) = t \tag{2.2}$$

represents a *wavefront* of the wave under consideration at the time $t$. As $t$ varies, (2.2) represents a moving wavefront. The vector $\vec{p}$, perpendicular to the wavefront,

$$\vec{p} = \nabla T, \tag{2.3}$$

is called the *slowness* vector.

The eikonal $T(x_j)$ and the vectorial complex-valued amplitude $\vec{U}(x_j)$ can be determined by inserting the complex-valued form of (2.1) into the elastodynamic equation. Using the elastodynamic equation, we can show that the HF seismic wave field in a smoothly varying inhomogeneous isotropic perfectly elastic medium separates into two waves, called the compressional (P) and the shear (S) waves.

The displacement vector of the HF *compressional wave* is linearly polarized, perpendicular to the wavefront. The travel time field $T(x_j)$ of the P wave satisfies the equation

$$\nabla T.\nabla T = 1/\alpha^2(x_j), \quad \text{with } \alpha = \left(\frac{\lambda + 2\mu}{\rho}\right)^{1/2} = \left(\frac{k + 4\mu/3}{\rho}\right)^{1/2}. \tag{2.4}$$

Here $\lambda$ and $\mu$ are the Lame's elastic parameters ($\mu$ is the modulus of torsion), $k$ the modulus of incompressibility, $\rho$ the density. $\alpha(x_j)$ represents the local propagation velocity of P waves.

The displacement vector of the *shear wave* is tangent to the wavefront. The S wave is not, in general, linearly polarized, but may be polarized elliptically, even in the HF approximation. The travel time field $T(x_j)$ of the S wave satisfies the equation,

$$\nabla T.\nabla T = 1/\beta^2(x_j), \quad \text{with } \beta = \left(\frac{\mu}{\rho}\right)^{1/2}. \tag{2.5}$$

Here $\beta(x_j)$ represents the local propagation velocity of S waves

Equations (2.4) and (2.5) are usually called the *eikonal equations* or the *equations of the travel time field* . In the isotropic perfectly elastic medium, the local velocities $\alpha(x_j)$ and $\beta(x_j)$ do not depend on the frequency and on the direction of the propagation of the wave. Thus, the waves under consideration are *non-dispersive* in the HF approximation (2.1).

As we can see, the eikonal equations for P and S waves have exactly the same form, only the velocity is different. Therefore, we shall consider only one of the eikonal equations, and denote the velocity of propagation by $V$,

$$\nabla T.\nabla T = 1/V^2(x_j). \tag{2.6}$$

$V$ is either $\alpha$ (for P waves) or $\beta$ (for S waves). Equation (2.6) also immediately implies $|\vec{p}| = 1/V$.

As well known, the decomposition of the complete elastic wave field into two independent waves, P and S, is always possible in a homogeneous medium. Let us, however, emphasize that this decomposition is only approximate in inhomogeneous media, valid for high frequencies only. In regions of strong gradient of velocity, both waves are coupled, at least for lower frequencies. Such situations cannot be investigated by the standard ray method. In the following, we assume that the velocity distribution in the model under consideration is smooth.

The P and S waves become coupled at any interface. At an interface, the incident P (or S) wave generates both P and S, reflected and transmitted, waves.

## 3. Ray tracing systems in cartesian coordinates

The eikonal equation (2.6) is a non-linear partial differential equation of the first order. Such equations are usually solved by means of *characteristics*. From seismological point of view, the characteristics represent *seismic* rays. They satisfy Fermat's principle and determine the direction of the flow of the high frequent part of the energy of the seismic wave field. In other words, this means that the group velocity vector is tangent to the seismic ray at any point of the ray.

Assume that the ray is specified by the parametric equation

$$x_i = x_i(s), \quad i = 1,2,3, \tag{3.1}$$

where $s$ is the arclength along the ray, with $s = 0$ at any reference point of the ray. Then the ray trajectory (3.1) is a solution of the *ray tracing system*,

$$\frac{dx_i}{ds} = Vp_i, \quad \frac{dp_i}{ds} = \frac{\partial}{\partial x_i}(1/V). \tag{3.2}$$

Here $p_i$ are the Cartesian components of the slowness vector $\vec{p} = \nabla T$, see (2.3). The travel time $T(s)$ along the ray is obtained from the relation,

$$\frac{dT}{ds} = 1/V. \tag{3.2'}$$

Equation (3.2') for the travel time can be solved together with the ray tracing system (3.2) or independently from it, after the ray is determined.

The ray tracing system (3.2) can be derived in many ways, see chapter 1 or Červený, Molotkov and Pšenčík (1977), where other references can be found.

To solve the ray tracing system (3.2) and the equation (3.2') for the travel time, the *initial conditions* must be given. The initial conditions for a single ray consist in the coordinates of the initial point $O_0$, in the initial direction of the ray at that point and in the initial travel time. If we denote the Cartesian coordinates of $O_0$ by $x_{i0}$ and specify $s = s_0$ at $O_0$, the initial conditions can be written in the following way:

$$\text{At } O_0: \quad x_i = x_{i0}, \; p_i = p_{i0}, \; T = T_0. \tag{3.3}$$

The components of the initial slowness vector $p_{i0}$ are not independent, they must satisfy the eikonal equation (2.6) at $O_0$,

$$p_{10}^2 + p_{20}^2 + p_{30}^2 = 1/V_0^2, \tag{3.4}$$

where $V_0 = V(x_{i0})$. If the initial direction of the ray at $O_0$ is specified by two take-off angles $\delta_0$ and $\psi_0$, the components of the initial slowness vector $p_{i0}$ can be expressed as follows,

$$p_{10} = V_0^{-1} \sin\delta_0 \cos\psi_0, \quad p_{20} = V_0^{-1} \sin\delta_0 \sin\psi_0, \tag{3.5}$$

$$p_{30} = V_0^{-1} \cos\delta_0.$$

Then (3.4) is automatically satisfied. It is simple to show that the eikonal equation (2.6) is satisfied along the whole ray as soon as it is satisfied at the initial point of the ray.

In certain situations, it may be useful to use, instead of $s$, some other *independent variable* $w$ along the ray . Then the ray trajectory is described by the parametric equation

$$x_i = x_i(w). \tag{3.6}$$

The variable $w$ may be chosen in different ways. It may, e.g., correspond to one of the three Cartesian coordinates. The Cartesian coordinates, however, are not quite suitable, as they do not behave monotonically along the ray. For example, if we choose the coordinate $x_3$ corresponding to the depth as the variable $w$, the ray must be divided into segments going "down" and "up". Some difficulties are also generated at the turning point of the ray. Therefore, it is more suitable to use some *monotonic variable* $w$ .

Often, the travel time $T$ is used as the variable along the ray. As $dT = V^{-1}\,ds$, the ray tracing system (3.2) reads,

$$\frac{dx_i}{dT} = V^2 p_i, \quad \frac{dp_i}{dT} = -\frac{\partial}{\partial x_i}\ln V. \tag{3.7}$$

The initial conditions for $T = T_0$ at $O_0$ are again given by (3.3).

A more general choice of $w$ is as follows:

$$w(T) = w(T_0) + \int_{T_0}^{T} V^N\,dT. \tag{3.8}$$

Then the ray tracing system reads,

$$\frac{dx_i}{dw} = V^{2-N} p_i, \quad \frac{dp_i}{dw} = -V^{-1-N}\frac{\partial V}{\partial x_i} = \frac{1}{N}\frac{\partial}{\partial x_i}(V^{-N}), \tag{3.9}$$

with

$$\frac{dT}{dw} = V^{-N}. \tag{3.9'}$$

The initial conditions for $w = w_0$ at the initial point $O_0$ are again given by (3.3).

We can easily see that the variable $w$ equals to the travel time $T$ for $N$=0. Then (3.9) reduces to (3.7). Similarly, for $N = 1$, we have $w = s$, the arclength along the ray, and (3.9) reduces to (3.2).

From a numerical point of view, no one of the above forms of the ray tracing systems has some distinct advantages with respect to other ray tracing systems. If we wish to evaluate not only rays, but also wavefronts, it may be perhaps useful to use the travel time $T$ as the variable along the ray ($N = 0$). Then the wavefronts are obtained automatically, as a by-product of the ray-tracing. If we use any other variable along the ray, e.g. the arclength $s$, the evaluation of the wavefronts would require an additional integration of (3.2'), or at least an interpolation. In the analytical and cell ray tracing, an especially simple and suitable form of the ray tracing system is obtained for $N = 2$. Therefore, we shall introduce a special symbol $w = \sigma$ for the variable along the ray for $N = 2$,

$$\sigma(T) = \sigma(T_0) + \int_{T_0}^{T} V^2 dT. \tag{3.10}$$

Then, the ray tracing system reads,

$$\frac{dx_i}{d\sigma} = p_i, \quad \frac{dp_i}{d\sigma} = ½\frac{\partial}{\partial x_i}(1/V^2), \tag{3.11}$$

with

$$\frac{dT}{d\sigma} = 1/V^2. \tag{3.11'}$$

The initial conditions for $\sigma = \sigma_0$ at $O_0$ are again by (3.3).

Note that the whole system (3.11) with (3.11') contains only the *quadratic slowness* $1/V^2$, not the velocity $V$ or the slowness $1/V$. This seems formal, but it shows that the quadratic slowness $1/V^2$ may be more useful in certain seismological applications than the velocity $V$ or the slowness $1/V$ (or than any other function of velocity $V$). The main reason of the simplicity of the ray tracing system (3.11) consists in the fact that the system (3.11) contains only the first partial spatial derivatives of $1/V^2$, not the quadratic slowness $1/V^2$ itself. All other ray tracing systems contain not only the derivatives of velocity, but also the velocity $V$ itself. This complicates the analytic ray tracing.

In a medium with a constant gradient of velocity, it is useful to exploit the variable $w = \xi$, given by (3.8) with $N = -1$. This simplifies the analytic solution of the ray tracing system, see (3.9), as the components of the slowness vector $p_i$ become linear functions of $(\xi - \xi_0)$ along the ray. See more details in Section 5.2.4.

All the ray tracing systems presented here consist of ordinary differential equations of the first order. We shall now present the ray tracing system consisting of ordinary differential equations of the second order. Even though such systems have not yet found broader applications in numerical ray tracing, they may find some important applications at least in the analytic and cell ray tracing.

The simplest ray tracing system is obtained if we use the variable $w = \sigma$ along the ray, see (3.10) and (3.11),

$$\frac{d^2x_i}{d\sigma^2} = \frac{1}{2}\frac{\partial}{\partial x_i}(1/V^2), \tag{3.12}$$

with $T$ given by (3.11'). The initial conditions for $\sigma = \sigma_0$ at $O_0$ are given as follows:

$$\text{At } O_0\text{:} \quad x_i = x_{i0},\ x'_i = x'_{i0} = p_{i0},\ T = T_0. \tag{3.13}$$

The quantities $p_{i0}$ must satisfy (3.4) and can be expressed in terms of two take-off angles $\delta_0$ and $\psi_0$.

For completeness, we shall also present the ray tracing system equivalent to (3.9) with a general variable $w$ along the ray,

$$\frac{d}{dw}(V^{N-2}\frac{dx_i}{dw}) = -V^{-1-N}\frac{\partial V}{dx_i}, \tag{3.14}$$

with $T$ given by (3.9'). The initial conditions for (3.14) for $w = w_0$ at the initial point $O_0$ are as follows:

$$\text{At } O_0\text{:} \quad x_i = x_{i0},\ x'_i = x'_{i0} = V_0^{2-N}p_{i0}. \tag{3.15}$$

Note that the term on the R.H.S. of (3.14) can be written as $-\frac{\partial \ln V}{\partial x_i}$ for $N = 0$, and as $N^{-1}\frac{\partial}{\partial x_i}(V^{-N})$ for $N \neq 0$.

## 4. Rays across interfaces

Assume that the ray impinges on a curved interface $\Sigma(x_j) = 0$ at a point $Q$ and denote by $\vec{N}$ the unit normal to $\Sigma$ at $Q$. The unit normal $\vec{N}$ is given by the relation

$$\vec{N} = \frac{\varepsilon_N \nabla\Sigma}{(\nabla\Sigma.\nabla\Sigma)^{1/2}}, \quad i.e., N_k = \frac{\varepsilon_N \frac{\partial \Sigma}{\partial x_k}}{(\frac{\partial \Sigma}{\partial x_j}\frac{\partial \Sigma}{\partial x_j})^{1/2}}. \tag{4.1}$$

Here $\varepsilon_N$ equals either 1 or -1; which we choose is fully arbitrary. Denote the slowness vector of the incident wave at $Q$ by $\vec{p}(Q)$ and call the plane specified by vectors $\vec{N}(Q)$ and $\vec{p}(Q)$ the *plane of incidence*.

Four plane waves are, in general, generated at a plane interface if a plane, P or S, wave is incident at this interface. Two of them are reflected (P and S), and two transmitted (P and S). There are some exceptions (liquid media, free interface, S wave polarized perpendicularly to the plane of incidence, etc.), but we shall consider here the general case. The initial slowness vector of any four generated waves at the point $Q$ on $\Sigma$ can be determined from the *laws of reflection/transmission (R/T laws)* which include the Snell's law. We denote the Cartesian components of the initial slowness vector of any selected

R/T wave by $\tilde{p}_i$. Using Snell's law and the continuity of the projection of $\vec{p}$ on the interface, we can easily derive:

$$\tilde{p}_i = p_i - \left\{ (p_k N_k) \pm \varepsilon [\tilde{V}^{-2} - V^{-2} + (p_k N_k)^2]^{1/2} \right\} N_i . \tag{4.2}$$

Here $V^{-2}$ and $\tilde{V}^{-2}$ are quadratic slownesses of the incident wave and of the selected R/T wave, respectively, $\varepsilon = sign\,(p_i N_i)$ is the so-called orientation index. It equals +1 if the slowness vector $\vec{p}$ of the incident wave makes an acute angle with $\vec{N}$, and equals -1 in the opposite case. The upper sign in (4.2) applies to the reflected waves, the lower sign to the transmitted waves.

The above relation (4.2) is strictly valid only for a plane wave incident on a plane interface between two homogeneous media. For high frequencies, however, equation (4.2) remains approximately valid locally even for slightly curved interfaces between smooth inhomogeneous media and for a slightly curved wavefront of the incident wave. This can be proved by asymptotic methods for $\omega \to \infty$. All the quantities in (4.2) must be taken then at the point of incidence $Q$.

The relation (4.2) is of fundamental importance in the ray tracing and travel time computation of HF seismic body waves propagating in 3-D laterally varying models with curved interfaces. As it expresses the reflection/transmission laws, we shall call it, for simplicity, the R/T law. It is even valid for converted waves, if the quadratic slownesses $1/V^2$ and $1/\tilde{V}^2$ are chosen correspondingly. For monotypic reflected wave, $\tilde{V} = V$ and (4.2) yields $\tilde{p}_i = p_i - 2(p_k N_k) N_i$.

It is interesting to see that the R/T law (4.2) contains again only the quadratic slowness, not the velocity or the slowness.

The R/T law remains approximately valid even for interfaces of higher order. An interface of the (n+1)th order is an interface at which the n-th derivative of the elastic parameters or the density is the lowest derivative that is discontinuous. At an interface of the first order, the elastic parameters and/or the density themselves are discontinuous. In this chapter, we shall usually call the interfaces of the first order simply the interfaces.

## 5. Initial value ray tracing

The solution of a ray tracing system with initial conditions specifying one point of the ray and the direction of the ray at that point is usually called *initial value ray tracing*.

All the equations necessary for initial value ray tracing and for the corresponding travel time computations of any seismic body wave propagating in a 3-D laterally varying layered structure are given in Sections 3 and 4. The relevant equations, however, may be solved in various ways. Many ray tracing procedures have been proposed, mainly in the last few years. The selection of the proper ray tracing procedure is greatly influenced by the practical purpose of the ray tracing, by the dimensionality of the problem, by the preferred approximation of the medium, by the required accuracy and numerical efficiency of computations and by the required completeness of the computations (i.e. do we need rays and travel times only, or ray amplitudes as well).

The simplest and fastest solution of the ray tracing system is usually based on its *analytic solution* , wherever the complexity of the model allows such a solution. It is, however, very difficult to describe the velocity distribution in the whole model (or, in the whole layer or block) by a simple velocity function which would allow the analytic solution of the ray tracing system. Therefore, the whole medium (layer,block) is often divided into suitable cells, in which the velocity may be approximated by simpler velocity laws which permit analytic ray solutions. The ray in the whole model is then obtained as a chain of analytically computed segments. In other cases, the direct numerical solution of the ray tracing system is more suitable. Each of these ray tracing approaches has its advantages and disadvantages. One approach may be suitable in certain applications, and the next method in other applications.

## 5.1 Numerical ray tracing

The most commonly used ray tracing systems from those presented in Section 3 are those expressed in ordinary differential equations of the first order. The numerical procedures for the solution of systems of ordinary differential equations of the first order with specified initial conditions are well known. Various standard numerical techniques are well described in textbooks of numerical mathematics which give the solution of such systems with a required accuracy. The routines designed for the numerical integration of systems of ordinary differential equations of the first order can be found in many subroutine packages. Among the most popular, we can name here the Runge-Kutta method, and the method of the predictor and corrector.

As an independent test of the accuracy of computations, it is also possible to use the eikonal equation which must be satisfied along the whole ray. Thus, we require that $|p_i p_i - 1/V^2| < \delta$, where $\delta$ is some specified small quantity. The accuracy of the ray tracing is increased if $p_i p_i$ is normalized to $1/V^2$ at any step of the numerical ray tracing.

## 5.2 Analytic ray tracing

In certain simpler types of media, the ray tracing may be performed analytically. In this section, we shall present several important examples.

The simplest solution of the ray tracing systems is obtained for *homogeneous* media. Let us consider the ray tracing system (3.2), with $V = const.$ The solution of (3.2) and (3.2'), for initial conditions (3.3) is as follows,

$$x_i = x_{i0} + V p_{i0}(s - s_0), \quad p_i = p_{i0}, \quad T = T_0 + V^{-1}(s - s_0). \tag{5.1}$$

For inhomogeneous media, the simplest analytic solution is obtained for the *"quadratic slowness"* model, if we use the variable $\sigma$ along the ray, see the ray tracing systems (3.11) and (3.12). In Sections 5.2.1 and 5.2.2, we shall consider two such quadratic slowness models. In Sections 5.2.3 and 5.2.4, we shall shortly discuss some other simple velocity distributions, especially the case of a constant velocity gradient.

*5.2.1 Constant gradient of the quadratic slowness.* Assume that the quadratic slowness $1/V^2$ is a linear function of the Cartesian coordinates $x_j$,

$$1/V^2 = A_0 + A_1x_1 + A_2x_2 + A_3x_3. \tag{5.2}$$

Then the ray tracing systems (3.11) or (3.12) with (3.11') and with appropriate initial conditions yield immediately the following *exact* polynomial solution,

$$\begin{aligned} x_i(\sigma) &= x_i(\sigma_0) + p_i(\sigma_0)(\sigma - \sigma_0) + \frac{1}{4}A_i(\sigma - \sigma_0)^2, \\ p_i(\sigma) &= p_i(\sigma_0) + \tfrac{1}{2}A_i(\sigma - \sigma_0), \\ T(\sigma) &= T(\sigma_0) + (A_0 + A_i x_i(\sigma_0))(\sigma - \sigma_0) + \tfrac{1}{2}A_i p_i(\sigma_0)(\sigma - \sigma_0)^2 \\ &\quad + \frac{1}{12}A_i A_i(\sigma - \sigma_0)^3. \end{aligned} \tag{5.3}$$

This polynomial solution is probably the simplest analytical solution for any type of inhomogeneous medium at all.

*5.2.2 Quadratic slowness is a polynomial in Cartesian coordinates.* Assume now a very general 3-D model in which the quadratic slowness $1/V^2$ is a general polynomial in the Cartesian coordinates,

$$1/V^2 = A_0 + A_j x_j + A_{jk} x_j x_k + A_{jkl} x_j x_k x_l + \cdots \tag{5.4}$$

Here summation convention is applied. We shall consider initial conditions for $\sigma = \sigma_0$ at $O_0$,given by (3.3). Then we can rewrite (5.4) in the following form,

$$\begin{aligned} 1/V^2 &= U^0 + U_j^0(x_j - x_{j0}) + \frac{1}{2}U_{jk}^0(x_j - x_{j0})(x_k - x_{k0}) + \\ &\quad + \frac{1}{6}U_{jkl}^0(x_j - x_{j0})(x_k - x_{k0})(x_l - x_{l0}) + \cdots . \end{aligned} \tag{5.5}$$

Here

$$\begin{aligned} U^0 &= [1/V^2]_{x_i = x_{i0}} = A_0 + A_j x_{j0} + A_{jk} x_{j0} x_{k0} + A_{jkl} x_{j0} x_{k0} x_{l0} + \ldots, \\ U_i^0 &= [\frac{\partial}{\partial x_i}(1/V^2)]_{x_i = x_{i0}} = A_i + 2A_{ij} x_{j0} + 3A_{ijk} x_{j0} x_{k0} + \ldots, \\ U_{ij}^0 &= [\frac{\partial^2}{\partial x_i \partial x_j}(1/V^2)]_{x_i = x_{i0}} = 2A_{ij} + 6A_{ijk} x_{k0} + \cdots . \end{aligned} \tag{5.6}$$

Such polynomial model can be also obtained by a Taylor expansion of the quadratic slowness at $O_0$, assuming that the higher derivatives of the quadratic slowness are small and can be neglected.

Then we can assume the solution of the ray tracing system in the following power series form,

$$x_i(\sigma) = x_{i0} + p_{i0}(\sigma - \sigma_0) + \sum_{k=2}^{\infty} X_i^{(k)}(\sigma - \sigma_0)^k, \tag{5.7}$$

where $X_i^{(k)}$, $i = 1,2,3, k = 2,3,...$, are some constants. Inserting (5.7) into (3.12) yields a recurrent system of equations for $X_i^{(2)}, X_i^{(3)}, X_i^{(4)}, ...$. The system can be easily solved. We shall present here the solutions of the first four equations:

$$X_i^{(2)} = \frac{1}{4}U_i^0,\ X_i^{(3)} = \frac{1}{12}U_{ij}^0 p_{j0},\ X_i^{(4)} = \frac{1}{96}U_{ij}^0 U_j^0 + \frac{1}{48}U_{ijk}^0 p_{j0} p_{k0}, \tag{5.8}$$

$$X_i^{(5)} = \frac{1}{480}U_{ij}^0 U_{jk}^0 p_{k0} + \frac{1}{160}U_{ijk}^0 p_{j0} U_k^0 + \frac{1}{240}U_{ijkl}^0 p_{j0} p_{k0} p_{l0}.$$

The solution of the ray tracing system (3.12) with the initial conditions (3.13), up to the fourth order terms in $(\sigma - \sigma_0)$, is then as follows,

$$x_i(\sigma) = x_{i0} + p_{i0}(\sigma - \sigma_0) + \frac{1}{4}U_i^0(\sigma - \sigma_0)^2 + \frac{1}{12}U_{ij}^0 p_{j0}(\sigma - \sigma_0)^3 + X_i^{(4)}(\sigma - \sigma_0)^4, \tag{5.9}$$

$$p_i(\sigma) = p_{i0} + \frac{1}{2}U_i^0(\sigma - \sigma_0) + \frac{1}{4}U_{ij}^0 p_{j0}(\sigma - \sigma_0)^2 +$$

$$+ 4X_i^{(4)}(\sigma - \sigma_0)^3 + 5X_i^{(5)}(\sigma - \sigma_0)^4,$$

$$T(\sigma) = T_0 + U^0(\sigma - \sigma_0) + \frac{1}{2}U_j^0 p_{j0}(\sigma - \sigma_0)^2 + \frac{1}{12}(U_j^0 U_j^0 + 2U_{jk}^0 p_{j0} p_{k0})(\sigma - \sigma_0)^3$$

$$+ \frac{1}{24}(\sigma - \sigma_0)^4(2U_j^0 U_{jk}^0 p_{k0} + U_{jkl}^0 p_{j0} p_{k0} p_{l0}).$$

The solution can be easily extended to higher powers of $(\sigma - \sigma_0)$.

*5.2.3 Constant gradient of $V^{-N}$ or of* $\ln V$. We have shown in Section 5.2.1 that the ray tracing system has a very simple analytical solution for a model with a constant gradient of the quadratic slowness $1/V^2$ (i.e., for $N = 2$). For all other $N$, the solution is more complicated.

Assume that the velocity distribution in the model is specified by the relation,

$$V^{-N} = A_0 + A_1 x_1 + A_2 x_2 + A_3 x_3, \tag{5.10}$$

or,

$$\ln V = A_0 + A_1 x_1 + A_2 x_2 + A_3 x_3. \tag{5.11}$$

In these cases it is useful to use the parameter $w$ given by (3.8) along the ray. The second equation of the ray tracing system (3.9) then yields the following solution for $N \neq 0$,

$$p_i(w) = p_i(w_0) + N^{-1} A_i (w - w_0), \tag{5.12}$$

and for $N = 0$ (i.e., for $w = T$),

$$p_i(T) = p_i(T_0) - A_i(T - T_0). \tag{5.13}$$

Even the differential equations for $x_i(w)$ can be solved analytically. The solutions, however, are not so simple as for $N = 2$. They usually contain some logarithms, square roots, inverse hyperbolic functions, etc. Even algebraically they are more complicated. Therefore, we shall write here only the expressions for the most important case of a constant gradient of velocity; for other solutions the interested reader is referred to a detailed treatment in Červený (1986b).

*5.2.4 Constant gradient of velocity.* The most popular velocity distribution in seismology has been probably the velocity distribution with a constant gradient of velocity ($N = 1$). Therefore, we shall present here the complete solution of this case.

Let us introduce a local Cartesian coordinate system such as that the local $x_3$ axis is oriented along the gradient of velocity $\nabla V$ and the local $x_1$ axis is perpendicular to the local $x_3$ axis and situated in the plane determined by $\vec{p}_0$ and $\nabla V$. The initial conditions for $\vec{p}_0$ then satisfy the relations $p_{20} = 0, p_{10} = p = const.$, where $p$ is the so called ray parameter. The velocity in the local Cartesian coordinate system is specified by the relation

$$V = V_0 + A_3(x_3 - x_{30}). \tag{5.14}$$

Then the analytical parametric solution of the ray tracing system (3.9) is obtained if we consider $N$ = -1 in (3.8), and (3.9), so that the variable $w = \xi$ along the ray is given by the relation,

$$\xi(T) = \xi(T_0) + \int_{T_0}^{T} V^{-1} dT. \tag{5.15}$$

The solution is as follows:

$$\begin{aligned} &p_1(\xi) = p_{10}, \quad p_2(\xi) = 0, \quad p_3(\xi) = p_{30} - A_3(\xi - \xi_0), \\ &x_1(\xi) = x_{10} + (A_3 p)^{-1}[p_{30}V_0 + X^{1/2}(A_3(\xi - \xi_0) - p_{30})], \\ &x_2(\xi) = x_{20}, \quad x_3(\xi) = x_{30} + A_3^{-1}(X^{1/2} - V_0), \\ &T(\xi) = T_0 - A_3^{-1} \ln[(X^{1/2} - A_3(\xi - \xi_0) + p_{30})/(V_0^{-1} + p_{30})], \end{aligned} \tag{5.16}$$

where

$$X = V_0^{-2} - 2p_{30}A_3(\xi - \xi_0) + A_3^2(\xi - \xi_0)^2. \tag{5.17}$$

The square roots $X^{1/2}$ in (5.16) are taken positive.

The above equations are perhaps too cumbersome to be used in practical ray tracing. Using these equations, however, it is not difficult to show that the ray is a circle in the model (5.14),

$$[x_1 - x_{10} - (A_3 p)^{-1} p_{30} V_0]^2 + [x_3 - x_{30} + A_3^{-1} V_0]^2 = (A_3 p)^{-2}. \tag{5.18}$$

Similarly the wavefront in the plane $x_1, x_3$ is also a circle,

$$[x_1 - x_{10}]^2 + [x_3 - x_{30} + A_3^{-1}V_0(1 - \cosh(A_3(T - T_0)))]^2 = \\ = V_0^2 A_3^{-2} \sinh^2(A_3(T - T_0)). \tag{5.19}$$

The circular rays are often used in practical applications. The travel time along any circular segment of the ray can be then determined from (5.19) which can be rewritten in the following form,

$$\cosh[A_3(T - T_0)] = \frac{[x_1 - x_{10}]^2 + [x_3 - x_{30} + V_0 A_3^{-1}]^2 + V_0^2 A_3^{-2}}{2V_0 A_3^{-1}(x_3 - x_{30} + V_0 A_3^{-1})}. \tag{5.19'}$$

These equations can be rewritten in many alternative forms, see e.g., Gjöystdal et al. (1984). No one of them, however, is so simple as the equations (5.3) for the constant gradient of the quadratic slowness.

It is not difficult to find from (5.16) power series expansions for $x_i(\xi)$ and $T(\xi)$ in terms of $(\xi - \xi_0)$, and, consequently, also in $(\sigma - \sigma_0)$, $(T - T_0)$, etc.. Such expansions, for a 2-D case, were given by Langan, Lerche and Cutler (1985) in terms of $(s - s_0)$. Such expansions may be very useful in practical applications. If we take only several terms in these expansions, they will be similar to (5.3). There is, however, one large difference between these expansions: the expansions of (5.16) are only approximate, whereas expansion (5.3) is exact!

## 5.3 Cell ray tracing

In the cell approximation, the whole model (or the whole layer, or block) is subdivided into a network of cells. The velocities, or some functions related to velocities, such as slownesses, quadratic slownesses, etc., may be specified at grid points of the network, alternatively in the centres of the cells. The velocity distribution within individual cells is then approximated by some simple analytical velocity laws. The simplest case is to consider a constant velocity within individual cells. In such a case, interfaces of first order are introduced at boundaries of cells. It is, however, not difficult to choose such velocity laws which do not introduce interfaces of first order at boundaries of cells, but only interfaces of second order. Thus, the velocity distribution is then continuous across the boundaries of cells, only the gradient of velocity is discontinuous. In this section, we shall describe two such possibilities, suitable for tetrahedron cells and for rectangular box cells.

*5.3.1 Tetrahedron cells.* At each tetrahedron, the velocity distribution can be described by one of the analytic velocity distributions (5.10) and (5.11). The four constants in the velocity distributions (5.10) and (5.11) can be determined from the velocity values at four apexes of the tetrahedron. The most popular in seismology is the case $N = -1$, i.e. the case of a constant velocity gradient within a cell. As we know, the ray in such a cell is a circle. The simplest analytical solutions for the ray and for the travel time, however, correspond to the case of the constant gradient of the quadratic slowness, $N = 2$. Therefore, we shall discuss in larger detail the procedure of the cell ray tracing for $N = 2$.

The exact equation for the ray trajectory in such a cell is a quadratic polynomial in $(\sigma-\sigma_0)$. The determination of the intersection of the ray with a plane boundary of the cell leads to a solution of the quadratic equation. As we do not know in advance which wall of the cell will be intersected by the ray, we have to find the intersections of the ray with all the four walls. As a result, we can obtain up to eight possible solutions, from which certain may be complex valued. If the initial point of the ray is situated at one of the walls of the tetrahedron, at most seven solutions exist, as the initial point corresponds to one of these solutions. The actual intersection of the ray with the boundary of the cell corresponds to the smallest positive $(\sigma-\sigma_0)$. We denote the relevant $\sigma$ by $\sigma^*$. As soon as $\sigma^*-\sigma$ is found, we can determine $p_i(\sigma^*), x_i(\sigma^*)$ and $T(\sigma^*)$, using (5.3). The whole procedure requires the solution of four quadratic equations and evaluation of some simple algebraic expressions for each cell. The solution is exact; no computations of trigonometric or transcendental functions and no transformation of coordinates is required. Even the travel time is given merely as the polynomial of the third order in $(\sigma^*-\sigma)$. The initial value ray tracing in a new cell is started with the initial conditions $x_{i0}=x_i(\sigma^*)$, $p_{i0}=p_i(\sigma^*)$, $T_0=T(\sigma^*)$.

Now we shall consider the tetrahedron cell with the constant gradient of velocity (not the constant gradient of the quadratic slowness). Again, the four constants in the approximation (5.10) with $N$ = -1 can be determined from the velocity values at the four apexes of the tetrahedron. As we know, the ray is a circle in this case. To determine the radius of the circle and the position of its centre, it is suitable to perform a transformation from the general Cartesian coordinate system to the local Cartesian coordinate system, with the $x_3$ axis along the gradient of velocity, and $x_1$ axis situated in the plane given by the initial slowness vector and the $x_3$ axis. Then the radius and the centre of the circle follow from (5.18). The intersection of the ray with a plane wall of the tetrahedron can be found by solving four quadratic equations. For each intersection, we have to evaluate the relevant travel time and select the actual intersection as that with the minimum travel time. The travel time is, unfortunately, given by some transcendental functions (logarithmic, inverse hyperbolic, etc., see (5.19')).

There are several alternatives of these approaches, but the basic principles remain the same. Both approaches (for the constant gradient of the quadratic slowness and for the constant gradient of velocity) are similar, but the quadratic slowness approximation is numerically more efficient and simpler for programming.

*5.3.2 Rectangular box cells.* Rectangular box cells with a *constant velocity* inside individual cells have been used in seismology for a long time, e.g. in the method by Aki, Christoffersen and Husebye (1977). As noticed above, such an approximation generates interfaces of the first order at boundaries of the cells. At these interfaces, the R/T laws (4.2) should be satisfied (see Koch,1985). Thus, the ray is a chain of straight line segments, connected by the R/T law at the boundaries of individual cells. Even though this velocity approximation is rather crude, it has yielded a number of important results in seismology.

The simplest analytic velocity distribution which guarantees continuous velocity across the boundaries of rectangular box cells is as follows:

$$V^{-N} = A_0 + A_1x_1 + A_2x_2 + A_3x_3 + B_1x_2x_3 + B_2x_1x_3 + B_3x_1x_2 + Cx_1x_2x_3, \quad (5.20)$$

where $N \neq 0$, but otherwise it may be arbitrary. Instead of $V^{-N}$, we can also use $\ln V$. We have eight constants in (5.20) which can be determined from the $V^{-N}$ values at the eight apexes of the box. The approximation (5.20) is linear along any edge of the box and is continuous across the walls of the cells. In this way, the boundaries of the cells are not interfaces of the first order, but only of the second order.

The exact solution of the ray tracing system for the approximation ( 5.20 ) is not known to the author, even though it is known for $N = 2$ for a 2-D case ( 5.21 ), see Červený (1986b).

$$V^{-2} = A_0 + A_1x_1 + A_3x_3 + B_2x_1x_3. \quad (5.21)$$

Even for a 2-D case, the exact solution contains trigonometric and hyperbolic functions and is not numerically efficient. For practical applications, it is suitable to express the solution in a *power series* of $(\sigma - \sigma_0)$. Such a power series solution can be, however, easily found even for (5.20), if $N$ is 2. In fact, the power series solution for (5.20) is given in Section 5.2.2, as the quadratic slowness distribution (5.20) is a special case of a more general quadratic slowness distribution (5.4).

What would be the algorithm for the cell ray tracing in a rectangular box cell with the velocity distribution given by (5.20) with $N = 2$? First, it would be useful to smooth the quadratic slowness distribution in such a way to decrease the coupling terms with $B_i$ and especially with $C$. Such a smoothing was first proposed by Langan, Lerche and Cutler (1985) for a 2-D model. After this, the algorithm is quite similar as for the tetrahedron cells. We can again use the quadratic power series in $(\sigma - \sigma_0)$ for the ray trajectory, neglecting the higher terms in (5.9). As the rectangular box cells have six walls, we have to solve six quadratic equations for each cell. All other details are the same as for the tetrahedron cells with a constant gradient of the quadratic slowness. The cubic terms in the polynomial expression for $x_i(\sigma)$ in (5.9) can be used to assess the accuracy of computations. Together with it, the components of the slowness vector can be used to assess the accuracy, as they must satisfy the eikonal equation. If the required accuracy is not reached, we have several possibilities: a) To use the third term or even higher terms in the ray expansion (5.9). This would, however, need to solve the cubic or higher order algebraic equations to find the intersection of the ray with the boundaries of the cell. b) To subdivide the rectangular box cell into five or six tetrahedrons and to evaluate the ray segments in each tetrahedron exactly, using (5.3). This yields the exact solution, but numerically and algebraically it is more involved. c) To subdivide the rectangular box cell into several smaller box cells by halving the grid distances, and to evaluate the rays in these smaller boxes using again the same algorithm as in larger boxes. This approach yields again only approximate solutions, but it is algorithmically considerably simpler. If the required accuracy is not reached even with smaller boxes, the halving may be repeated.

## 6. Eikonal equation and ray tracing systems in orthogonal curvilinear coordinates

It is often useful to consider some curvilinear coordinates instead of Cartesian coordinates in the ray tracing. Most common is the application of curvilinear orthogonal coordinates, such as spherical, cylindrical, ellipsoidal, etc.. In seismology, we need mainly the spherical coordinates, or, perhaps, the ellipsoidal coordinates.

In this section, we shall rewrite the eikonal equation, the ray tracing system and the R/T law into general curvilinear orthogonal coordinate system. In the next section, we shall discuss the ray tracing in spherical coordinates in larger detail.

### 6.1 Curvilinear orthogonal coordinates

Consider a right-handed curvilinear orthogonal coordinate system $\xi_1,\xi_2,\xi_3$, with relevant unit basis vectors $\vec{e}_1,\vec{e}_2,\vec{e}_3$, and denote the corresponding scale factors by $h_1,h_2,h_3$, such that $d\vec{x} = h_1 d\xi_1\vec{e}_1 + h_2 d\xi_2\vec{e}_2 + h_3 d\xi_3\vec{e}_3$. The square of the infinitesimal length element $ds^2$ is then given by the formula

$$ds^2 = h_1^2 d\xi_1^2 + h_2^2 d\xi_2^2 + h_3^2 d\xi_3^2. \tag{6.1}$$

It is well known from the vector calculus how to express vectorial differential operators in the curvilinear orthogonal coordinate system. In the following, we shall need only the expressions for the gradient, which is given by a well known formula,

$$\nabla\Phi = \frac{1}{h_1}\frac{\partial\Phi}{\partial\xi_1}\vec{e}_1 + \frac{1}{h_2}\frac{\partial\Phi}{\partial\xi_2}\vec{e}_2 + \frac{1}{h_3}\frac{\partial\Phi}{\partial\xi_3}\vec{e}_3. \tag{6.2}$$

As the slowness vector $\vec{p}$ is defined as $\vec{p} = \nabla T$, we can write the following relation for it :

$$\vec{p} = \frac{1}{h_1}\frac{\partial T}{\partial\xi_1}\vec{e}_1 + \frac{1}{h_2}\frac{\partial T}{\partial\xi_2}\vec{e}_2 + \frac{1}{h_3}\frac{\partial T}{\partial\xi_3}\vec{e}_3. \tag{6.3}$$

We denote

$$T_i = \frac{\partial T}{\partial\xi_i}. \tag{6.4}$$

Then we can write,

$$\vec{p} = \frac{1}{h_1}T_1\vec{e}_1 + \frac{1}{h_2}T_2\vec{e}_2 + \frac{1}{h_3}T_3\vec{e}_3. \tag{6.5}$$

Note that $T_i$'s are not components of a vector. They are related to the slowness vector components $p_i$ by the relation $p_i = \dfrac{T_i}{h_i}$ (no summation over $i$), see (6.5).

### 6.2 The eikonal equation

The eikonal equation $\nabla T.\nabla T = 1/V^2$ in the curvilinear coordinate system $\xi_1,\xi_2,\xi_3$ reads, see (6.3),

$$\frac{1}{h_1^2}T_1^2 + \frac{1}{h_2^2}T_2^2 + \frac{1}{h_3^2}T_3^2 = \frac{1}{V^2(\xi_1,\xi_2,\xi_3)}. \tag{6.6}$$

### 6.3 Ray tracing systems

We shall describe the ray trajectory by a parametric system of equations,

$$\xi_i = \xi_i(w), \tag{6.7}$$

where $w$ is an independent variable along the ray. The variable $w$ may be arbitrary, but we shall again use it in the form of (3.8).

The ray tracing system can be obtained from the eikonal equation (6.6) in the same way as in Cartesian coordinates, e.g. by the method of characteristics. The simplest forms of the ray tracing system are obtained if we use $\sigma$ as the variable along the ray, see (3.10). Then we can write,

$$\frac{d\xi_i}{d\sigma} = h_i^{-2}T_i, \quad \frac{dT_i}{d\sigma} = ½\frac{\partial}{\partial\xi_i}(1/V^2) + \sum_{j=1}^{3} h_j^{-3}T_j^2\frac{\partial h_j}{\partial\xi_i}, \quad \frac{dT}{d\sigma} = 1/V^2 \tag{6.8}$$

(no summation over $i$).

If we use the general variable $w$ given by (3.8) we obtain:

$$\frac{d\xi_i}{dw} = V^{2-N}h_i^{-2}T_i, \quad \frac{dT_i}{dw} = -V^{-N-1}\frac{\partial V}{\partial\xi_i} + V^{2-N}\sum_{j=1}^{3} h_j^{-3}T_j^2\frac{\partial h_j}{\partial\xi_i}, \quad \frac{dT}{dw} = V^{-N}. \tag{6.9}$$

As we can see, for $w = \sigma$ ($N = 2$) Eq. (6.9) reduces to (6.8). For $N = 1$ $w = s$, the arclength along the ray, and for $N = 0$ $w = T$, the travel time along the ray.

Instead of $T_i = \frac{\partial T}{\partial \xi_i}$, we can also use some other related quantities in the ray tracing system, e.g. the components of the slowness vector $p_i = T_i/h_i$. In certain applications, this may simplify the ray tracing system. If we, however, use the components of the slowness vector $p_i$ instead of $T_i$, the ray tracing system becomes in general more complicated.

The *initial conditions* for the ray tracing system (6.8) specify the coordinates of the initial point $O_0$ of the ray and the direction of the ray at the initial point. If we specify the initial point by $w = w_0$ (i.e. $\sigma = \sigma_0$ or $T = T_0$), the initial conditions may be written in the following form,

$$\text{At } O_0\text{:} \quad \xi_i = \xi_{i0}, \quad T_i = T_{i0}, \quad T = T_0. \tag{6.10}$$

Here $T_{i0}$ must satisfy at $O_0$ the eikonal equation (6.6),

$$\frac{1}{h_{10}^2}T_{10}^2 + \frac{1}{h_{20}^2}T_{20}^2 + \frac{1}{h_{30}^2}T_{30}^2 = 1/V_0^2, \tag{6.11}$$

where $V_0$ is the velocity at $O_0$, $h_{i0}$ are the scale factors at $O_0$. Assume now that the initial direction is specified by two take-off angles $\delta_0$ and $\psi_0$ at $O_0$, and denote by $\delta_0$ the angle between the initial slowness vector $\vec{p}_0$ and one selected basis vector, say $\vec{e}_i$. This means that $p_{i0}=V_0^{-1}\cos\delta_0$ and $T_{i0}=h_{i0}V_0^{-1}\cos\delta_0$. In a standard way, we choose also the angle $\psi_0$, considering also the remaining basis vectors $\vec{e}_j,\vec{e}_k (j\neq i, k\neq i)$. The relations for $T_{i0}$, $T_{j0}$ and $T_{k0}$ may be then written as follows,

$$T_{i0}=\frac{h_{i0}}{V_0}\cos\delta_0,\quad T_{j0}=\frac{h_{j0}}{V_0}\sin\delta_0\cos\psi_0,\quad T_{k0}=\frac{h_{k0}}{V_0}\sin\delta_0\sin\psi_0. \tag{6.12}$$

The quantities $T_{10}$,$T_{20}$ and $T_{30}$, determined from (6.12), satisfy automatically (6.11) at $O_0$.

### 6.4 The R/T law

If the ray is incident at a curved interface $\Sigma$ at a point $Q$, the initial conditions for the rays of the generated R/T waves must be determined. As the known coordinates $\xi_i$ of the point of incidence $Q$ and the travel time $T$ at that point represent initial conditions $\xi_{i0}$ and $T_0$ for the R/T waves, we must determine only $T_{i0}$. Assume that the interface is given by the equation $\Sigma(\xi_1,\xi_2,\xi_3)=0$. Then the normal to $\Sigma$ at $Q$ is given by (4.1), or:

$$N_i=\frac{\varepsilon_N}{h_i A}\frac{\partial\Sigma}{\partial\xi_i},\quad A=\left[\sum_{k=1}^{3}\frac{1}{h_k^2}\left(\frac{\partial\Sigma}{\partial\xi_k}\right)^2\right]^{1/2} \tag{6.13}$$

(no summation over $i$). Equation (4.2) yields then the R/T law in the following form,

$$\tilde{T}_i=T_i-h_i\left\{B\pm\varepsilon[(1/\tilde{V})^2-1/V^2+B^2]^{1/2}\right\}N_i \tag{6.14}$$

(no summation over $i$ ), with

$$B=\varepsilon_N\left[\sum_{k=1}^{3}\frac{1}{h_k^2}T_k\frac{\partial\Sigma}{\partial\xi_k}\right]/A. \tag{6.15}$$

All quantities in (6.13)-(6.15) are taken at the point $Q$. The quantities with a tilde, $\tilde{T}_i$ and $\tilde{V}$, correspond to the R/T wave at $Q$. The upper sign corresponds to the reflected wave and the lower to the transmitted wave, $\varepsilon=sign\,(p_iN_i)=signB$.

## 7. Eikonal equation and ray tracing systems in spherical coordinates

In this section, we shall rewrite the eikonal equation, the ray tracing systems and the R/T law in spherical coordinates. We denote the spherical coordinates by $r,\theta,\phi$, where $r$ is the radial distance, $\theta$ is the colatitude and $\phi$ the longitude. The square of the infinitesimal length $ds^2$ in the spherical coordinates is given by the relation,

$$ds^2=dr^2+r^2d\theta^2+r^2\sin^2\theta d\phi^2. \tag{7.1}$$

It follows from (6.1) and (7.1) that the scale factors in spherical coordinates, $h_r$,$h_\phi$ and $h_\theta$, are given by the relations,

$$h_r = 1, \quad h_\theta = r, \quad h_\phi = r \sin\theta. \tag{7.2}$$

We denote

$$T_r = \frac{\partial T}{\partial r}, \quad T_\theta = \frac{\partial T}{\partial \theta}, \quad T_\phi = \frac{\partial T}{\partial \phi}. \tag{7.3}$$

Then the components of the slowness vector $\vec{p}$ can be easily expressed in the following form,

$$p_r = T_r, \; p_\theta = \frac{1}{r} T_\theta, \; p_\phi = \frac{1}{r \sin\theta} T_\phi. \tag{7.4}$$

## 7.1 Eikonal equation

The eikonal equation (6.6) in spherical coordinates reads

$$T_r^2 + \frac{1}{r^2} T_\theta^2 + \frac{1}{r^2 \sin^2\theta} T_\phi^2 = \frac{1}{V^2(r, \theta, \phi)}. \tag{7.5}$$

## 7.2 Ray tracing systems

Ray tracing systems in spherical coordinates can be again written in many forms, see, e.g., Comer (1984). If we use $\sigma$ as an independent variable along the ray, the ray tracing system reads,

$$\begin{aligned} \frac{dr}{d\sigma} &= T_r, & \frac{dT_r}{d\sigma} &= \frac{1}{2} \frac{\partial}{\partial r}(1/V^2) + \frac{1}{r}(1/V^2 - T_r^2), \\ \frac{d\theta}{d\sigma} &= \frac{1}{r^2} T_\theta, & \frac{dT_\theta}{d\sigma} &= \frac{1}{2} \frac{\partial}{\partial \theta}(1/V^2) + \frac{\cos\theta}{r^2 \sin^3\theta} T_\phi^2, \\ \frac{d\phi}{d\sigma} &= \frac{1}{r^2 \sin^2\theta} T_\phi, & \frac{dT_\phi}{d\sigma} &= \frac{1}{2} \frac{\partial}{\partial \phi}(1/V^2). \end{aligned} \tag{7.6}$$

Here we have used the identity $V(T_\theta^2/r^3 + T_\phi^2/r^3 \sin^2\theta) = Vr^{-1}(1/V^2 - T_r^2)$, following from the eikonal equation (7.5). For the travel time $T(\sigma)$ we obtain the equation

$$\frac{dT}{d\sigma} = 1/V^2. \tag{7.7}$$

The *initial conditions* for the ray tracing system for $\sigma = \sigma_0$ specify the coordinates of the initial point $O_0$, the direction of the ray at that point and the travel time,

$$\text{At } O_0\text{: } r = r_0, \theta = \theta_0, \phi = \phi_0, T_r = T_{r0}, T_\theta = T_{\theta 0}\, T_\phi = T_{\phi 0}, T = T_0. \tag{7.8}$$

The quantities $T_{r0}, T_{\theta 0}$ and $T_{\phi 0}$ must satisfy the eikonal equation at $O_0$,

$$T_{r0}^2 + \frac{1}{r_0^2} T_{\theta 0}^2 + \frac{1}{r_0^2 \sin^2\theta_0} T_{\phi 0}^2 = \frac{1}{V_0^2}. \tag{7.9}$$

We can again introduce two take-off angles $\delta_0$ and $\psi_0$ to specify the initial direction of the ray, see (6.12). In seismology, it is common to consider the angle $\delta_0$ with respect to the radial axis, so that the index $i$ in (6.12) corresponds to the $r$ axis. We can then write,

$$T_{r0}=\frac{1}{V_0}\cos\delta_0,\quad T_{\theta 0}=\frac{r_0}{V_0}\sin\delta_0\cos\psi_0,\quad T_{\phi 0}=\frac{r_0\sin\theta_0}{V_0}\sin\delta_0\sin\psi_0. \tag{7.10}$$

Commonly, the angle $\delta_0$ is denoted by $i$.

For completeness, we shall present here also the ray tracing system using the general variable $w$ along the ray, see (6.9). We obtain,

$$\begin{aligned}
&\frac{dr}{dw}=V^{2-N}T_r, && \frac{dT_r}{dw}=-V^{-1-N}\frac{\partial V}{\partial r}+V^{2-N}r^{-1}(V^{-2}-T_r^2),\\
&\frac{d\theta}{dw}=V^{2-N}\frac{1}{r^2}T_\theta, && \frac{dT_\theta}{dw}=-V^{-1-N}\frac{\partial V}{\partial\theta}+V^{2-N}\frac{\cos\theta}{r^2\sin^3\theta}T_\phi^2,\\
&\frac{d\phi}{dw}=V^{2-N}\frac{1}{r^2\sin^2\theta}T_\phi, && \frac{dT_\phi}{dw}=-V^{-1-N}\frac{\partial V}{\partial\phi}.
\end{aligned} \tag{7.11}$$

In the numerical ray tracing, the ray tracing systems (7.6) and (7.11) require an alternative treatment if the computed ray enters a region of very small $\theta$ or $\theta$ very close to $\pi$. Several alternatives are possible; let us name two of them: a) A local transformation of coordinates $r,\theta,\phi\rightarrow r,\theta',\phi'$ removes the problems. b) the velocity distribution in the dangerous region can be locally approximated by a simpler velocity law (homogeneous medium, Bullen's law) which allows local analytic ray tracing.

### 7.3 The R/T law

If the ray strikes an interface $\Sigma(r,\theta,\phi)=0$ at a point $Q$, the initial conditions for rays of R/T waves at $Q$ are again given by (6.13)-(6.15). The components of the unit normal $N_r, N_\theta, N_\phi$ are given by the relations,

$$N_r=\varepsilon_N\frac{\partial\Sigma}{\partial r}/A,\quad N_\theta=\varepsilon_N r^{-1}\frac{\partial\Sigma}{\partial\theta}/A,\quad N_\phi=\frac{\varepsilon_N}{r\sin\theta}\frac{\partial\Sigma}{\partial\phi}/A, \tag{7.12}$$

with

$$A=[(\frac{\partial\Sigma}{\partial r})^2+\frac{1}{r^2}(\frac{\partial\Sigma}{\partial\theta})^2+\frac{1}{r^2\sin^2\theta}(\frac{\partial\Sigma}{\partial\phi})^2]^{1/2}. \tag{7.13}$$

Then the R/T law takes the form,

$$\begin{aligned}
\tilde{T}_r&=T_r-(B\pm\varepsilon[(1/\tilde{V})^2-(1/V)^2+B^2]^{1/2})N_r,\\
\tilde{T}_\theta&=T_\theta-r(B\pm\varepsilon[(1/\tilde{V})^2-(1/V)^2+B^2]^{1/2})N_\theta,\\
\tilde{T}_\phi&=T_\phi-r\sin\theta\,(B\pm\varepsilon[(1/\tilde{V})^2-(1/V)^2+B^2]^{1/2})N_\phi,
\end{aligned} \tag{7.14}$$

where

$$B = \frac{\varepsilon_N}{A}[T_r\frac{\partial\Sigma}{\partial r} + \frac{T_\theta}{r^2}\frac{\partial\Sigma}{\partial\theta} + \frac{1}{r^2\sin^2\theta}T_\phi\frac{\partial\Sigma}{\partial\phi}]. \tag{7.15}$$

Note that $\varepsilon = sign\,(p_i N_i) = signB$.

## 7.4 Modified ray tracing systems

There are several possibilities how to modify the ray tracing systems presented above. We shall show one possibility which leads to more efficient ray tracing systems.

We introduce a new variable u instead of the radial distance $r$ by the relation,

$$u = \ln(r), \quad r = \exp(u), \quad T_u = \frac{\partial T}{\partial u} = r\frac{\partial T}{\partial r} = rT_r. \tag{7.16}$$

Note that $u,\theta,\phi$ form a system of curvilinear *orthogonal* coordinates, with the scale factors $h_u = r = \exp(u), h_\theta = r = \exp(u), h_\phi = r\,\sin\theta = \exp(u)\sin\theta$. The eikonal equation then reads,

$$T_u^2 + T_\theta^2 + \frac{1}{\sin^2\theta}T_\phi^2 = \eta^2, \quad \text{where } \eta(u,\theta,\phi) = (\frac{r}{V(r,\theta,\phi)})_{r=\exp(u)}. \tag{7.17}$$

The ray tracing system can be written in the following simple form,

$$\begin{aligned}
\frac{du}{d\kappa} &= T_u, & \frac{dT_u}{d\kappa} &= \frac{1}{2}\frac{\partial\eta^2}{\partial u},\\
\frac{d\theta}{d\kappa} &= T_\theta, & \frac{dT_\theta}{d\kappa} &= \frac{1}{2}\frac{\partial\eta^2}{\partial\theta} + \frac{\cos\theta}{\sin^3\theta}T_\phi^2,\\
\frac{d\phi}{d\kappa} &= \frac{T_\phi}{\sin^2\theta}, & \frac{dT_\phi}{d\kappa} &= \frac{1}{2}\frac{\partial\eta^2}{\partial\phi},
\end{aligned} \tag{7.18}$$

with

$$\frac{dT}{d\kappa} = \eta^2. \tag{7.18'}$$

The independent variable along the ray is $\kappa$, $d\kappa = r^{-2}d\sigma = dT/\eta^2$. The *initial conditions* for the ray tracing system (7.18) at the point $O_0$ for which $\kappa = \kappa_0$ are as follows,

$$\text{At } O_0\text{:} \quad u = u_0, \theta = \theta_0, \phi = \phi_0, T_u = T_{u0}, T_\theta = T_{\theta 0}, T_\phi = T_{\phi 0}, T = T_0. \tag{7.19}$$

The initial quantities $T_{u0}, T_{\theta 0}$, and $T_{\phi 0}$ satisfy the eikonal equation (7.17) at $O_0$,

$$T_{u0}^2 + T_{\theta 0}^2 + \frac{1}{\sin^2\theta_0}T_{\phi 0}^2 = \eta_0^2. \tag{7.20}$$

If the initial direction of the ray is specified by the take-off angles $\delta_0$ and $\psi_0$, we obtain

$$T_{u0} = \eta_0\cos\delta_0, \; T_{\theta 0} = \eta_0\sin\delta_0\cos\psi_0, \; T_{\phi 0} = \eta_0\sin\delta_0\sin\psi_0\sin\theta_0. \tag{7.21}$$

The system (7.18) can be easily rewritten using the other variables along the ray, instead of $\kappa$. In all such cases, however, the system would be more complicated than (7.18).

It remains to rewrite *the R/T law* in the coordinates $u,\theta,\phi$. We assume that the interface is described by the equation $\Sigma(u,\theta,\phi)=0$. Then we can determine the components of the unit normal $\vec{N}$ at the point of incidence $Q$,

$$N_u=\varepsilon_N\frac{\partial\Sigma}{\partial u}/A\,,\quad N_\theta=\varepsilon_N\frac{\partial\Sigma}{\partial\theta}/A\,,\quad N_\phi=\frac{\varepsilon_N}{\sin\theta}\frac{\partial\Sigma}{\partial\phi}/A\,, \tag{7.22}$$

where

$$A=[(\frac{\partial\Sigma}{\partial u})^2+(\frac{\partial\Sigma}{\partial\theta})^2+\frac{1}{\sin^2\theta}(\frac{\partial\Sigma}{\partial\phi})^2]^{1/2}. \tag{7.23}$$

The R/T law then reads,

$$\tilde{T}_u=T_u-(B\pm\varepsilon[\tilde{\eta}^2-\eta^2+B^2]^{1/2})N_u, \tag{7.24}$$
$$\tilde{T}_\theta=T_\theta-(B\pm\varepsilon[\tilde{\eta}^2-\eta^2+B^2]^{1/2})N_\theta,$$
$$\tilde{T}_\phi=T_\phi-\sin\theta\,(B\pm\varepsilon[\tilde{\eta}^2-\eta^2+B^2]^{1/2})N_\phi,$$

with

$$B=\frac{\varepsilon_N}{A}[T_u\frac{\partial\Sigma}{\partial u}+T_\theta\frac{\partial\Sigma}{\partial\theta}+\frac{1}{\sin^2\theta}T_\phi\frac{\partial\Sigma}{\partial\phi}]. \tag{7.25}$$

Several remarks:

1) The trigonometric functions $\cos\theta$ and $\sin\theta$ may be formally removed from (7.6),(7.11) and (7.18) if we use $y=\cos\theta$ instead of $\theta$.

2) Instead of the variable $u=\ln r$, we can also use the variable $u=-R\ln(r/R)$, where $R$ is the radius of the earth. If we then modify also other quantities in the ray tracing system (7.18) in a similar way, then the ray tracing system will remain the same. For example, the new slowness factor will read $(VR)^{-1}r$.

3) The ray tracing system (7.6) or (7.18) simplifies considerably in two 2-D cases:

a) Ray tracing in a plane $\phi=const$. If $\eta^2$ does not depend on $\phi$ and $T_{\phi 0}=0$, then $T_\phi=0$ along the whole ray, and the system (7.18) for $u$, $\theta$, $T_u$ and $T_\theta$ becomes fully equivalent to 2-D ray tracing system in Cartesian coordinates.

b) Ray tracing along the earth's surface. If $\eta^2$ does not depend on $u$ ( or $r$ ) and $T_{u0}=0$, we can use a new variable $\Theta=\ln tg\,\dfrac{\theta}{2}$ instead of $\theta$. The most common case will be $u=\ln R$ (ray tracing along the earth's surface). Then the ray tracing system (7.18) for $\Theta,\phi,T_\Theta$ and $T_\phi$ becomes exactly the same as the 2-D ray tracing system in Cartesian coordinates, if we replace $\eta$ by $\nu(\Theta,\phi)=[\dfrac{R\sin\theta}{V(R,\theta,\phi)}]_{\theta=2arctan\,\exp(\Theta)}$ and the variable along the ray $\kappa$ by $\Delta$, $d\Delta=\nu^{-2}dT=(\sin\theta)^{-2}d\kappa$.

Thus, any computer program for 2-D ray tracing in Cartesian coordinates can be used for the 2-D ray tracing in spherical coordinates in the two above cases, by simply modifying the input and output data. Similarly, any analytic solution of the 2-D ray tracing system in Cartesian coordinates can be directly taken over from Cartesian to spherical coordinates.

## 8. Ray fields. Dynamic ray tracing in cartesian coordinates

In the preceding sections, we considered only a single ray, specified by the appropriate initial conditions: by the initial point, initial direction and initial time. In actual seismological applications, we are interested in the whole wave field. For each elementary wave, we have to consider a *two-parameter system of rays*. We call the parameters which specify individual rays the *ray parameters* and denote them by $\gamma_1,\gamma_2$. The ray parameters may be, e.g., the take-off angles $\delta_0$ and $\psi_0$ at a point source, or the curvilinear coordinates along an initial surface, exploding reflector, etc..

We introduce the *ray coordinates* $\gamma_1,\gamma_2,\gamma_3$ in such a way that $\gamma_1,\gamma_2$ are the ray parameters and $\gamma_3$ is some monotonic parameter along the ray, e.g. the arclength $s$, the travel time $T$, the parameter $\sigma$, etc.. Here we shall use mostly the general parameter $w$ given by (3.8).

We denote the 3×3 transformation matrix from ray to Cartesian coordinates by $\hat{\mathbf{J}}$, with elements $J_{ij} = \dfrac{\partial x_i}{\partial \gamma_j}$. The partial derivatives $\dfrac{\partial x_i}{\partial \gamma_J}$ ($i$=1,2,3,$J$=1,2) are taken for constant $w$ and characterize the mutual deviations of the individual rays. They can be evaluated by solving an additional system of ordinary differential equations along the ray which is usually called the *dynamic ray tracing system*. In general Cartesian coordinates, the system can be easily obtained from the ray tracing system (3.9),

$$\frac{d}{dw}\left(\frac{\partial x_i}{\partial \gamma_J}\right) = \frac{\partial V^{2-N}}{\partial x_k}\frac{\partial x_k}{\partial \gamma_J}p_i + V^{2-N}\frac{\partial p_i}{\partial \gamma_J}, \tag{8.1}$$

$$\frac{d}{dw}\left(\frac{\partial p_i}{\partial \gamma_J}\right) = \frac{1}{N}\frac{\partial^2}{\partial x_i \partial x_j}(V^{-N})\frac{\partial x_j}{\partial \gamma_J}.$$

Such a system for $\dfrac{\partial x_i}{\partial \gamma_J}$ and $\dfrac{\partial p_i}{\partial \gamma_J}$ consists of 12 equations. It has an especially simple form if we use $N = 2$, with the corresponding variable $w = \sigma$ along the ray, see (3.10),

$$\frac{d}{d\sigma}\left(\frac{\partial x_i}{\partial \gamma_J}\right) = \frac{\partial p_i}{\partial \gamma_J}, \quad \frac{d}{d\sigma}\left(\frac{\partial p_i}{\partial \gamma_J}\right) = \frac{1}{2}\frac{\partial^2}{\partial x_i \partial x_j}(V^{-2})\frac{\partial x_j}{\partial \gamma_J}. \tag{8.2}$$

The system can be rewritten into six ordinary differential equations of the second order for $\dfrac{\partial x_i}{\partial \gamma_J}$,

$$\frac{d^2}{d\sigma^2}\frac{\partial x_i}{\partial \gamma_J} = \frac{1}{2}\frac{\partial^2}{\partial x_i \partial x_j}(V^{-2})\frac{\partial x_j}{\partial \gamma_J}. \tag{8.3}$$

The systems (8.1), (8.2) and (8.3) can be solved easily numerically along the known ray, together with the ray tracing system or after it. We can see from (8.2) and (8.3) that the system can be also simply solved analytically for various quadratic slowness models. For example, for a constant gradient of the quadratic slowness (5.2), the right hand side of (8.3) vanishes, $\dfrac{\partial p_i}{\partial \gamma_J}$ are constant along the ray and $\dfrac{\partial x_i}{\partial \gamma_J}$ are polynomials of the first order in

$(\sigma - \sigma_0)$.

The dynamic ray tracing has found many important applications in the numerical modelling of seismic wave fields in complex structures. We shall not, however, discuss the possibilities and applications of the dynamic ray tracing at this place, we shall pay larger attention to this problem in Sections 9 and 10, in connection with the dynamic ray tracing in ray centred coordinates. It should be noted that the dynamic ray tracing in Cartesian coordinates may find very suitable applications in the analytic and cell ray tracing.

## 9. Ray centred coordinate system. Polarization vectors

Whereas the ray coordinate system $\gamma_1, \gamma_2, \gamma_3$ is connected with the whole ray field, the *ray centred coordinate system* $q_1, q_2, q_3$ is connected with a single, arbitrarily selected ray $\Omega$. The ray $\Omega$ is one of the three coordinate lines of the ray centred coordinate system. As a relevant coordinate $q_3$, we can take any monotonic variable along the ray, e.g. $s, T$ *or* $\sigma$. We shall consider here $q_3 = s$, the arclength along the ray $\Omega$. This choice will allow us to write simple and objective expressions for the transformation matrices from ray centred coordinates to the general Cartesian and local Cartesian coordinate systems. The other two coordinate lines, corresponding to $q_2$ and $q_3$, are straight lines perpendicular to the ray $\Omega$. It would be possible, e.g., to take these straight lines in the direction of the unit normal and unit binormal to the ray at any point of the ray. Such a choice, however, does not yield an orthogonal coordinate system; certain off-diagonal components of the corresponding metric tensor do not vanish. We shall choose the vector basis of the ray centred coordinate system $\vec{e}_1, \vec{e}_2, \vec{e}_3$ in such a way to make the ray centred coordinate system orthogonal. This can be done in the following way: At the initial point, we select $\vec{e}_1, \vec{e}_2, \vec{e}_3$ to be mutually perpendicular and right-handed, with $\vec{e}_3 = \vec{t} = V\vec{p}$ tangent to the ray. (Otherwise $\vec{e}_1$ and $\vec{e}_2$ may be chosen arbitrarily at the initial point.) To guarantee the orthogonality of the ray centred coordinate system, the Cartesian components $e_{1i}, e_{2i}$ of the unit vectors $\vec{e}_1$ and $\vec{e}_2$ must satisfy the following ordinary differential equations along $\Omega$:

$$\frac{de_{1i}}{dw} = -V^{-2}\left(e_{1k}\frac{dp_k}{dw}\right)p_i, \qquad \frac{de_{2i}}{dw} = -V^{-2}\left(e_{2k}\frac{dp_k}{dw}\right)p_i. \tag{9.1}$$

Note that the quantities $dp_k/dw$ are known from the ray tracing system (3.9). We can, of course, replace $dp_k/dw$ by $N^{-1}\partial(V^{-N})/\partial x_k$, see (3.9).

In actual computations, only one of the above vectorial equations (9.1) need to be solved, e.g. the first one for $\vec{e}_1$, as $\vec{e}_2$ can be determined from known $\vec{e}_1$ and $\vec{e}_3$ ( $\vec{e}_3$ is known from the ray tracing). Moreover, even the first vectorial equation for $\vec{e}_1$ can be reduced to one scalar equation only. Such a procedure was proposed by Popov and Pšenčík (1978a,b), see also Červený and Hron (1980), Červený (1985) for details. For spherical coordinates, see Cormier (1984).

The unit vectors $\vec{e}_1$ and $\vec{e}_2$ have one very important property: they determine the *polarization of S waves* . In a smooth medium without interfaces, the displacement vector of the S wave has the direction of the unit vector $\vec{e}_1$ along the whole ray, as soon as it has this direction at any reference point of the ray. The same is valid for the second unit vector

$\vec{e}_2$. In other words, the displacement vector of S waves does not rotate around the ray with respect to $\vec{e}_1$ and $\vec{e}_2$ as the S wave progresses along the ray. It is well known that it rotates around the ray with respect to the unit normal and unit binormal. For this reason, the unit basis vectors $\vec{e}_1, \vec{e}_2, \vec{t}$ are known as the *polarization vectors* ; the unit vector $\vec{t}$ determines the polarization of the P wave.

We now introduce several transformation matrices along the ray $\Omega$. We denote the transformation matrix from the ray centred coordinates to Cartesian coordinates along $\Omega$ by $\hat{\mathbf{H}}$. Its elements are given by partial derivatives $H_{ij} = \dfrac{\partial x_i}{\partial q_j} = \dfrac{\partial q_j}{\partial x_i}$ for $q_1 = q_2 = 0$. The matrix can be simply evaluated: its columns correspond to the Cartesian components of the unit vectors $\vec{e}_1, \vec{e}_2, \vec{t}$. Thus, matrix $\hat{\mathbf{H}}$ is obtained as a by-product of the computation of the polarization vectors. Note that $\det\hat{\mathbf{H}} = 1$, $\hat{\mathbf{H}}^{-1} = \hat{\mathbf{H}}^T$.

A very important role in the dynamic ray tracing is played by the two next 3×3 transformation matrices $\hat{\mathbf{Q}}$ and $\hat{\mathbf{P}}$. Matrix $\hat{\mathbf{Q}}$ represents the transformation matrix from ray coordinates to the ray centred coordinates at $\Omega$, $Q_{ij} = \dfrac{\partial q_i}{\partial \gamma_j}$ for $q_1 = q_2 = 0$. It is obvious that $Q_{13} = Q_{23} = 0$. Later on, we shall mostly use the 2×2 transformation matrix $\mathbf{Q}$ from ray coordinates $\gamma_1, \gamma_2$ to the ray centred coordinates $q_1, q_2$. Matrix $\hat{\mathbf{P}}$ is slightly more complex. We denote the ray centred components of the slowness vector by $p_i^{(q)}$. On the central ray $\Omega$ we have $p_1^{(q)} = p_2^{(q)} = 0$, $p_3^{(q)} = 1/V$. The derivatives $\dfrac{\partial p_1^{(q)}}{\partial \gamma_I}$ and $\dfrac{\partial p_2^{(q)}}{\partial \gamma_I}$, $I = 1,2$, however, do not vanish along $\Omega$. Matrix $\hat{\mathbf{P}}$ represents the transformation matrix from the ray coordinates to the phase space coordinates $p_i^{(q)}$, $P_{ij} = \dfrac{\partial p_i^{(q)}}{\partial \gamma_j} = \dfrac{\partial^2 \tau}{\partial q_i \partial \gamma_j}$ for $q_1 = q_2 = 0$.

It is simple to see that $\mathbf{Q} = \mathbf{O}$ at the point source and at a point caustic, where the neighbouring rays intersect. On the contrary, $\mathbf{P} = \mathbf{O}$ at such a point of the ray $\Omega$, where $p_1^{(q)}$ and $p_2^{(q)}$ vanish in the vicinity of the point, i.e. where the rays are locally parallel and the wavefronts are locally plane. Such points may be called the *telescopic points* or the local *plane wave points* .

## 10. Dynamic ray tracing in ray centred coordinates. Propagator matrices

The dynamic ray tracing system has a very fundamental role in the ray method. It is, therefore, not surprising that it can be derived in many ways. Several of its forms were already given in Section 8. Here we shall present the dynamic ray tracing system in ray centred coordinates, which is in some way the most natural coordinate system connected with the ray. Therefore, the dynamic ray tracing has a very simple form there.

### 10.1 Dynamic ray tracing system

It was first shown by Popov and Pšenčík (1978a,b) that the 2×2 transformation matrices **Q** and **P** satisfy along the ray the following system of ordinary differential equations of the first order,

$$\frac{d\mathbf{Q}}{ds} = V\mathbf{P}, \quad \frac{d\mathbf{P}}{ds} = -V^{-2}\mathbf{V}\mathbf{Q}. \tag{10.1}$$

This system is usually called the dynamic ray tracing system. The elements of the 2×2 matrix **V** are given by relations $V_{IJ} = \partial^2 V/\partial q_I q_J$ for $q_1 = q_2 = 0$. If the second derivatives of velocity are evaluated in Cartesian coordinates, we can write $V_{IJ} = H_{kI} H_{lJ} \partial^2 V/\partial x_k \partial x_l$ at $\Omega$. It is simple to show that the matrices **Q** and **P**, obtained as a solution of (10.1), have the meaning of the transformation matrices along the whole ray as soon as they have that meaning at any initial point on the ray.

If we introduce a general variable $w$ along the ray instead of $s$, we can write the dynamic ray tracing system in the following form,

$$\frac{d\mathbf{Q}}{dw} = V^{2-N}\mathbf{P}, \quad \frac{d\mathbf{P}}{dw} = -V^{-1-N}\mathbf{V}\mathbf{Q}. \tag{10.2}$$

An especially simple dynamic ray tracing system is obtained if we use $N = 2$, i.e. the variable $w = \sigma$ along the ray. Then we obtain

$$\frac{d\mathbf{Q}}{d\sigma} = \mathbf{P}, \quad \frac{d\mathbf{P}}{d\sigma} = -V^{-3}\mathbf{V}\mathbf{Q}. \tag{10.3}$$

This dynamic ray tracing system can be also written in the form of an ordinary differential equation of the second order,

$$\frac{d^2\mathbf{Q}}{d\sigma^2} + \frac{1}{V^3}\mathbf{V}\mathbf{Q} = \mathbf{O}. \tag{10.4}$$

In the following, we shall introduce a 4×2 matrix **X** defined as $\begin{pmatrix}\mathbf{Q}\\ \mathbf{P}\end{pmatrix}$. Then the dynamic ray tracing system (10.1) can be rewritten in the following form,

$$\frac{d\mathbf{X}}{ds} = \mathbf{S}\mathbf{X}, \quad \text{where } \mathbf{X} = \begin{pmatrix}\mathbf{Q}\\ \mathbf{P}\end{pmatrix}, \ \mathbf{S} = \begin{pmatrix}\mathbf{O} & V\mathbf{I}\\ -V^{-2}\mathbf{V} & \mathbf{O}\end{pmatrix}. \tag{10.5}$$

Let us note that $tr(\mathbf{S}) = 0$.

Another very important interpretation of the dynamic ray tracing system is connected with the *ray tracing of paraxial rays*. Under the paraxial rays we understand such rays which are close to the central ray $\Omega$. It is simple to see that the dynamic ray tracing system (10.5) represents also the ray tracing system for paraxial rays. We introduce a column 4×1 matrix $\mathbf{W}(s) = (q_1, q_2, p_1^{(q)}, p_2^{(q)})^T$, which specifies the ray centred coordinates $q_1, q_2$ of a paraxial ray and relevant ray centred components of the slowness vector $p_1^{(q)}, p_2^{(q)}$ at that point. Then the paraxial ray tracing system reads,

$$\frac{d\mathbf{W}}{ds} = \mathbf{SW}. \tag{10.6}$$

It is interesting to see that the paraxial ray tracing system is linear.

## 10.2 Ray propagator matrices

The dynamic ray tracing system (10.6) has four linearly independent solutions. We introduce a 4×4 matrix of linearly independent solutions and denote it by $\mathbf{\Pi}(s, s_0)$. Such a matrix is called the *fundamental matrix* . We specify the initial conditions for this matrix by the relation

$$\mathbf{\Pi}(s, s_0) = \mathbf{I}, \tag{10.7}$$

where $\mathbf{I}$ is a 4×4 identity matrix. The fundamental matrix which satisfies condition (10.7) at $s = s_0$ is usually called the propagator matrix. We shall call $\mathbf{\Pi}(s, s_0)$ the *ray propagator matrix* , or shortly the *ray propagator* . Using the ray propagator matrix, the solutions of (10.5) and (10.6) can be written in the following form,

$$\mathbf{X}(s) = \mathbf{\Pi}(s, s_0)\mathbf{X}(s_0), \quad \mathbf{W}(s) = \mathbf{\Pi}(s, s_0)\mathbf{W}(s_0), \tag{10.8}$$

where $\mathbf{X}(s_0)$ is an arbitrary 4×2 initial matrix and $\mathbf{W}(s_0)$ is an arbitrary 4×1 initial matrix.

The fundamental matrix may be suitably rewritten in the form,

$$\mathbf{\Pi}(s, s_0) = \begin{bmatrix} \mathbf{Q}_1(s, s_0) & \mathbf{Q}_2(s, s_0) \\ \mathbf{P}_1(s, s_0) & \mathbf{P}_2(s, s_0) \end{bmatrix}. \tag{10.9}$$

Here the 2×2 matrices $\mathbf{Q}_1$ and $\mathbf{P}_1$ are solutions of the dynamic ray tracing system (10.1) for the initial conditions $\mathbf{Q}(s_0) = \mathbf{I}$, $\mathbf{P}(s_0) = \mathbf{O}$, where $\mathbf{I}$ is a 2×2 identity matrix and $\mathbf{O}$ is the 2×2 null matrix. In other words, the initial conditions correspond to a telescopic (or local plane wave) point at $s = s_0$. Similarly, $\mathbf{Q}_2$ and $\mathbf{P}_2$ are solutions of (10.1) for initial conditions $\mathbf{Q}(s_0) = \mathbf{O}$, $\mathbf{P}(s_0) = \mathbf{I}$ (point source solution).

The matrices $\mathbf{Q}_1, \mathbf{Q}_2, \mathbf{P}_1, \mathbf{P}_2$ satisfy many important relations along the ray $\Omega$. First four are as follows,

$$\begin{aligned} \mathbf{Q}_1^T\mathbf{P}_1 - \mathbf{P}_1^T\mathbf{Q}_1 = \mathbf{O}, \quad \mathbf{P}_2^T\mathbf{Q}_1 - \mathbf{Q}_2^T\mathbf{P}_1 = \mathbf{I}, \\ \mathbf{Q}_2^T\mathbf{P}_2 - \mathbf{P}_2^T\mathbf{Q}_2 = \mathbf{O}, \quad \mathbf{Q}_1^T\mathbf{P}_2 - \mathbf{P}_1^T\mathbf{Q}_2 = \mathbf{I}. \end{aligned} \tag{10.10}$$

It is not difficult to show that the derivatives of the above expressions (10.10) with respect to $s$ vanish along the ray. From this follows that these expressions remain constant along the ray. Using the initial conditions (10.7) then yields (10.10). In a more compact form, relations (10.10) can be rewritten as follows,

$$\mathbf{\Pi}^T\mathbf{J}\mathbf{\Pi} = \mathbf{J}, \quad \mathbf{J} = \begin{bmatrix} \mathbf{O} & \mathbf{I} \\ -\mathbf{I} & \mathbf{O} \end{bmatrix}. \tag{10.11}$$

The propagator matrix which satisfies (10.10) is called *symplectic*, see Thomson and Chapman (1985).

Another set of equations is as follows,

$$\mathbf{Q}_1\mathbf{Q}_2^T - \mathbf{Q}_2\mathbf{Q}_1^T = \mathbf{O}, \quad \mathbf{Q}_1\mathbf{P}_2^T - \mathbf{Q}_2\mathbf{P}_1^T = \mathbf{I},$$
$$\mathbf{P}_1\mathbf{P}_2^T - \mathbf{P}_2\mathbf{P}_1^T = \mathbf{O}, \quad \mathbf{P}_2\mathbf{Q}_1^T - \mathbf{P}_1\mathbf{Q}_2^T = \mathbf{I}. \tag{10.12}$$

Using the above equations, we can show that the propagator matrix from $s$ to $s_0$ is obtained as an inverse of the propagator matrix from $s_0$ to $s$,

$$\mathbf{\Pi}(s_0,s) = \mathbf{\Pi}^{-1}(s,s_0) = \begin{bmatrix} \mathbf{P}_2^T(s,s_0) & -\mathbf{Q}_2^T(s,s_0) \\ -\mathbf{P}_1^T(s,s_0) & \mathbf{Q}_1^T(s,s_0) \end{bmatrix}. \tag{10.13}$$

It can be proved that the equations (10.12) specify the symplectic properties of the matrix $\mathbf{\Pi}(s_0,s) = \mathbf{\Pi}^{-1}(s,s_0)$. Equation (10.13) is of a basic importance in various practical applications, as it allows us to connect the boundary conditions for the ray given at different points of the ray. The equation may be used e.g. in the boundary value ray tracing of paraxial rays, see Section 10.6.

Two next properties of the propagator matrix are as follows:

$$\det\mathbf{\Pi}(s,s_0) = \det\mathbf{\Pi}(s_0,s_0) = 1. \tag{10.14}$$

This property follows immediately from the fact that the trace of the matrix $\mathbf{S}$ in (10.5) vanishes. Finally, the next very important property of $\mathbf{\Pi}$ is as follows,

$$\mathbf{\Pi}(s,s_0) = \mathbf{\Pi}(s,s')\mathbf{\Pi}(s',s_0), \tag{10.15}$$

where $s'$ is an arbitrary point on $\Omega$, not necessarily between $s$ and $s_0$. This equation can be used to connect the propagator matrices calculated independently along different segments of the ray. For example, the equation can be used in such situations when the ray tracing and the dynamic ray tracing are performed analytically in certain parts of the model and numerically in other parts, see Cormier (1986).

### 10.3 Dynamic ray tracing in layered models

In this section, we shall present, for completeness, all equations necessary for the dynamic ray tracing along a ray of any multiply reflected/transmitted wave in a 3-D laterally varying layered medium.

Let us first consider a single interface and use the notation of Section 4: symbol $\Sigma$ denotes the interface, $Q$ the point of incidence at $\Sigma$, $\vec{N}(Q)$ the unit normal to $\Sigma$ at $Q$, $p_i(Q)$ and $\tilde{p}_i(Q)$ the Cartesian components of the slowness vectors of the incident and of the selected R/T wave at $Q$. In addition, let us introduce a right handed local Cartesian coordinate system $z_1,z_2,z_3$ with its origin at $Q$ and with the basis vectors $\vec{i}_1,\vec{i}_2,\vec{i}_3$. We shall assume that $\vec{i}_3 \equiv \vec{N}(Q)$, and that the $z_1$ axis is situated in the plane of incidence; other details may be arbitrary. Our particular choice of the local Cartesian coordinate system is completely specified by the transformation matrix $\hat{\mathbf{Z}}$ from the local Cartesian coordinate system $z_i$ at $Q$ to the general Cartesian coordinate system, $Z_{ij} = \dfrac{\partial x_i}{\partial z_j}$ at $Q$. The columns of

$\hat{\mathbf{Z}}$ give the Cartesian components of the unit vectors $\vec{i}_1, \vec{i}_2, \vec{i}_3$.

Besides $\hat{\mathbf{Z}}$, it is useful to introduce another transformation matrix $\hat{\mathbf{G}}$ from the ray centred coordinate system $q_i$ to the local Cartesian coordinate system $z_j$ at $Q$, $G_{ij} = \dfrac{\partial z_i}{\partial q_j}$ at $Q$. This matrix can be determined simply from the two above introduced matrices $\hat{\mathbf{H}}$ and $\hat{\mathbf{Z}}$ by the relation

$$\hat{\mathbf{G}} = \hat{\mathbf{Z}}^T \hat{\mathbf{H}}. \tag{10.16}$$

Thus, matrix $\hat{\mathbf{G}}$ can be easily evaluated at $Q$.

Without a derivation, we shall present here the final equations for the change of matrices $\mathbf{P}$ and $\mathbf{Q}$ across the interface. For the derivation and other details see Červený (1985). We denote by $\mathbf{X}(Q)$ the 4×2 matrix $\begin{pmatrix}\mathbf{Q}\\ \mathbf{P}\end{pmatrix}$ corresponding to the incident wave at $Q$ and by $\tilde{\mathbf{X}}(Q)$ the same matrix corresponding to the selected R/T wave at $Q$. The relation between $\tilde{\mathbf{X}}(Q)$ and $\mathbf{X}(Q)$ is as follows,

$$\tilde{\mathbf{X}}(Q) = \mathbf{F}(Q)\mathbf{X}(Q), \quad \mathbf{F}(Q) = \begin{pmatrix}\tilde{\mathbf{G}}^T & \mathbf{O}\\ \mathbf{O} & \tilde{\mathbf{G}}^{-1}\end{pmatrix}\begin{pmatrix}\mathbf{I} & \mathbf{O}\\ u\mathbf{D}+\mathbf{E}-\tilde{\mathbf{E}} & \mathbf{I}\end{pmatrix}\begin{pmatrix}(\mathbf{G}^{-1})^T & \mathbf{O}\\ \mathbf{O} & \mathbf{G}\end{pmatrix}. \tag{10.17}$$

All the elements of the 4×4 matrix $\mathbf{F}(Q)$ are determined at $Q$. The quantities with the tilde above the letter correspond to the R/T wave, those without it to the incident wave at $Q$. $\mathbf{D}$ is the 2×2 matrix of the curvature of the interface $\Sigma$ at $Q$, u and the 2×2 matrices $\mathbf{E}$ and $\tilde{\mathbf{E}}$ are given by the relations,

$$u = V^{-1}G_{33} - \tilde{V}^{-1}\tilde{G}_{33}, \tag{10.18}$$

$$E_{IJ} = -V^{-2}\left(\frac{\partial V}{\partial q_K}G_{JK}G_{I3} + \frac{\partial V}{\partial q_3}G_{I3}G_{J3} + \frac{\partial V}{\partial q_K}G_{IK}G_{J3}\right),$$

$$\tilde{E}_{IJ} = -\tilde{V}^{-2}\left(\frac{\partial \tilde{V}}{\partial q_K}\tilde{G}_{JK}\tilde{G}_{I3} + \frac{\partial \tilde{V}}{\partial q_3}\tilde{G}_{I3}\tilde{G}_{J3} + \frac{\partial \tilde{V}}{\partial q_K}\tilde{G}_{IK}\tilde{G}_{J3}\right),$$

It should be noted that $\dfrac{\partial V}{\partial q_i} = H_{ki}\dfrac{\partial V}{\partial x_k}$, $\dfrac{\partial \tilde{V}}{\partial q_i} = \tilde{H}_{ki}\dfrac{\partial \tilde{V}}{\partial x_k}$.

The above equations can be simply generalized for any multiply reflected, possibly converted, wave propagating in a 3-D laterally varying layered structure, containing interfaces of first and second order $\Sigma_1, \Sigma_2, \ldots, \Sigma_k$. We denote the selected ray by $\Omega$, the initial point of the ray by $O_0$, the end point by $O_S$, and $N$ points of R/T at various interfaces between $O_0$ and $O_S$ by $Q_1, Q_2, \ldots, Q_N$. We also denote $Q_0 = O_0$. Then we obtain,

$$\mathbf{X}(O_S) = \Pi(O_S, Q_N)\prod_{i=N}^{1}[\mathbf{F}(Q_i)\Pi(Q_i, Q_{i-1})]\mathbf{X}(O_0), \tag{10.19}$$

where the symbol $\prod_{i=N}^{1}\mathbf{A}_i$ denotes the matrix product $\mathbf{A}_N.\mathbf{A}_{N-1}.\cdots.\mathbf{A}_1$. In this way, the final form of the ray propagator matrix of an arbitrary multiply reflected wave in a general

3-D layered structure is as follows,

$$\Pi(O_S, O_0) = \Pi(O_S, Q_N)\prod_{i=N}^{1}[\mathbf{F}(Q_i)\Pi(Q_i, Q_{i-1})]. \tag{10.20}$$

## 10.4 Second derivatives of the travel time field along the ray

Using the dynamic ray tracing, we can evaluate the second derivatives of the travel time field with respect to spatial coordinates along the ray.

Let us introduce the 2×2 symmetric matrix $\mathbf{M}(s)$ with elements $M_{IJ}(s) = \partial^2 T(q_1, q_2, s)/\partial q_I q_J$ for $q_1 = q_2 = 0$. It is not difficult to show that the matrix $\mathbf{M}$ satisfies along the ray the nonlinear matrix Riccati equation,

$$\frac{d\mathbf{M}}{ds} + V\mathbf{M}^2 + V^{-2}\mathbf{V} = \mathbf{O}, \tag{10.21}$$

where $\mathbf{V}$ is the matrix of the second derivatives of velocity defined in Section 10.1. Matrix $\mathbf{M}$ is related to matrices $\mathbf{Q}$ and $\mathbf{P}$ as follows:

$$\mathbf{M} = \mathbf{P}\mathbf{Q}^{-1}. \tag{10.22}$$

Matrix $\mathbf{M}$ can be recalculated along the ray using the elements $\mathbf{P}_1, \mathbf{P}_2, \mathbf{Q}_1, \mathbf{Q}_2$ of the propagator matrix, see (10.9),

$$\mathbf{M}(s) = [\mathbf{P}_1(s, s_0) + \mathbf{P}_2(s, s_0)\mathbf{M}(s_0)][\mathbf{Q}_1(s, s_0) + \mathbf{Q}_2(s, s_0)\mathbf{M}(s_0)]^{-1}. \tag{10.23}$$

The 3×3 matrix $\hat{\mathbf{M}}$ of the second derivatives of the travel time field with respect to ray centred coordinates $q_1, q_2, q_3$ along the ray $\Omega$, with elements $M_{ij} = \partial^2 T(q_1, q_2, q_3)/\partial q_i \partial q_j$ at $\Omega$, is simply obtained from the 2×2 matrix $\mathbf{M}$:

$$\hat{\mathbf{M}} = \begin{pmatrix} M_{11} & M_{12} & -V^{-2}\dfrac{\partial V}{\partial q_1} \\ M_{12} & M_{22} & -V^{-2}\dfrac{\partial V}{\partial q_2} \\ -V^{-2}\dfrac{\partial V}{\partial q_1} & -V^{-2}\dfrac{\partial V}{\partial q_2} & -V^{-2}\dfrac{\partial V}{\partial q_3} \end{pmatrix}. \tag{10.24}$$

Here $\partial V/\partial q_i = H_{ki}\partial V/\partial x_k$.

Finally, the 3×3 matrix of second derivatives of the travel time field with respect to Cartesian coordinates $\hat{\mathbf{N}}$, with elements $N_{ij} = \partial^2 T/\partial x_i \partial x_j$ at $\Omega$, is given by the relation,

$$\hat{\mathbf{N}} = \hat{\mathbf{H}}\hat{\mathbf{M}}\hat{\mathbf{H}}^T. \tag{10.25}$$

## 10.5 Paraxial travel times

Using the Taylor series expansion and the above formulae for the second derivatives of the travel time field, we can easily write the expression for the travel time field in the neighbourhood of the ray $\Omega$ (paraxial travel times). Let us consider a point $O_S$ situated on the ray $\Omega$, with Cartesian coordinates $x_i(O_S)$, and a point $S$, situated in the vicinity of the ray $\Omega$ close to the point $O_S$, with coordinates $x_i(S)$. Assume also that the slowness vector $\vec{p}$ and the propagator matrix $\mathbf{\Pi}$ are known at $O_S$ . Then we can write

$$T(S)=T(O_S)+\hat{\mathbf{p}}^T(O_S)\hat{\mathbf{x}}(S,O_S)+\frac{1}{2}\hat{\mathbf{x}}^T(S,O_S)\hat{\mathbf{N}}(O_S)\hat{\mathbf{x}}(S,O_S), \qquad (10.26)$$

where $\hat{\mathbf{x}}(S,O_S)$ is the column 3×1 matrix with elements $x_i(S)-x_i(O_S)$, and $\hat{\mathbf{p}}(O_S)$ is the 3×1 column matrix with elements $p_i(O_S)$.

## 10.6 Paraxial boundary value ray tracing

Propagator matrices can be suitably used not only in the initial value ray tracing, see Eq. (10.8), but also in all types of boundary value ray tracing.

Various types of the boundary value ray tracing are shortly described in Section 12. Any problem of the boundary value ray tracing can be solved analytically for paraxial rays using the propagator matrices. The detailed derivation and discussion of individual cases of the boundary value ray tracing with paraxial rays is given in Červený, Klimeš and Pšenčík (1986). Here we shall present only analytical solution of the paraxial two point ray tracing problem. This solution is also given and shortly discussed in Červený, Klimeš and Pšenčík (1984) and in Červený (1985).

Let us consider a central ray $\Omega$ connecting points $A$ and $B$, specified by Cartesian coordinates $x_i(A)$ and $x_i(B)$. Assume that we know the Cartesian components of the slowness vector $p_i(A)$ and $p_i(B)$, the matrices $H_{ij}(A)$ and $H_{ij}(B)$, the velocities and their first derivatives at $A$ and $B$, the travel time $T(A,B)$ and the propagator matrix $\mathbf{\Pi}(B,A)$ (satisfying the initial conditions $\mathbf{\Pi}(A,A)=\mathbf{I}$)

Now we consider two new points $C$ and $D$; $C$ being close to $A$ and $D$ close to $B$. We denote the Cartesian coordinates of these points by $x_i(C)$ and $x_i(D)$ and determine the 3×1 column matrices $\hat{\mathbf{x}}(C,A)$ and $\hat{\mathbf{x}}(D,B)$ with components $x_i(C,A)$ and $x_i(D,B)$,

$$x_i(C,A)=x_i(C)-x_i(A), \quad x_i(D,B)=x_i(D)-x_i(B). \qquad (10.27)$$

We wish to determine: a) The travel time from $C$ to $D$, $T(C,D)$, b) The Cartesian components of the slowness vector $p_i(C)$ of the ray passing through points $C$ and $D$.

The resulting system of equations, valid approximately for paraxial rays, is as follows,

$$T(C,D)=T(A,B)+\hat{\mathbf{p}}^T(B)\hat{\mathbf{x}}(D,B)-\hat{\mathbf{p}}^T(A)\hat{\mathbf{x}}(C,A) \qquad (10.28)$$

$$+\frac{1}{2}\hat{\mathbf{q}}^T(D,B)\hat{\mathbf{M}}(B,A)\hat{\mathbf{q}}(D,B)-\frac{1}{2}\hat{\mathbf{q}}^T(C,A)\hat{\mathbf{M}}(A,B)\hat{\mathbf{q}}(C,A)$$

$$-\mathbf{q}^T(C,A)\mathbf{Q}_2^{-1}(B,A)\mathbf{q}(D,B),$$

and

$$\hat{\mathbf{p}}(C)=\hat{\mathbf{p}}(A)+\hat{\mathbf{H}}(A)\hat{\mathbf{M}}(A,B)\hat{\mathbf{q}}(C,A)+\hat{\mathbf{H}}_M(A)\mathbf{Q}_2^{-1}(B,A)\mathbf{q}(D,B). \qquad (10.29)$$

Here we have used the following notation:

$$\hat{\mathbf{q}}(C,A)=\hat{\mathbf{H}}^T(A)\hat{\mathbf{x}}(C,A),\quad \hat{\mathbf{q}}(D,B)=\hat{\mathbf{H}}^T(B)\hat{\mathbf{x}}(D,B),$$

$$\mathbf{M}(B,A)=\mathbf{P}_2(B,A)\mathbf{Q}_2^{-1}(B,A),\quad \mathbf{M}(A,B)=-\mathbf{Q}_2^{-1}(B,A)\mathbf{Q}_1(B,A),$$

$$M_{ij}(X,Y)=\begin{bmatrix} M_{11}(X,Y) & M_{12}(X,Y) & -V^{-2}(X)[\partial V/\partial q_1]_X \\ M_{12}(X,Y) & M_{22}(X,Y) & -V^{-2}(X)[\partial V/\partial q_2]_X \\ -V^{-2}(X)[\partial V/\partial q_1]_X & -V^{-2}(X)[\partial V/\partial q_2]_X & -V^{-2}(X)[\partial V/\partial q_3]_X \end{bmatrix}.$$

The symbols $\mathbf{Q}_1(B,A),\mathbf{Q}_2(B,A),\mathbf{P}_1(B,A),\mathbf{P}_2(B,A)$ denote the 2×2 submatrices of the propagator matrix $\mathbf{\Pi}(B,A)$, see (10.9), $\hat{\mathbf{H}}_M$ is a 3×2 left-hand submatrix of the 3×1 matrix $\hat{\mathbf{H}}$, $\partial V/\partial q_i=H_{ki}\partial V/\partial x_k$. The 3×1 column matrix $\hat{\mathbf{p}}$ has elements $p_i$.

Equations (10.28) and (10.29) are quite general, valid for arbitrary multiply reflected, possibly converted, HF seismic body wave in a 3-D laterally varying layered structure. The equations are only approximate, valid for paraxial rays, i.e. for small $x_i(C,A)$ and $x_i(D,B)$. Equation (10.29) solves the two point ray tracing problem, as it determines the initial direction of the paraxial ray from $C$ passing through $D$. Thus, the ray may be now computed by standard initial value ray tracing. As the equation is only approximate, the ray trajectory will not pass exactly through the point $D$. The process, however, may be performed iteratively.

Equations (10.28) and (10.29) simplify considerably in many important practical applications. Let us assume a point source at $A$ and $C$=$A$. Then $\hat{\mathbf{x}}(C,A)=\mathbf{0}$ and Eqs. (10.28) and (10.29) yield

$$T(A,D)=T(A,B)+\hat{\mathbf{p}}^T(B)\hat{\mathbf{x}}(D,B)+\frac{1}{2}\hat{\mathbf{q}}^T(D,B)\hat{\mathbf{M}}(B,A)\hat{\mathbf{q}}(D,B), \qquad (10.30)$$

$$\hat{\mathbf{p}}'(A)=\hat{\mathbf{p}}(A)+\hat{\mathbf{H}}_M(A)\mathbf{Q}_2^{-1}(B,A)\mathbf{q}(D,B).$$

Here $\hat{\mathbf{p}}'(A)$ denotes the initial slowness vector of the ray from $A$ passing through the point $D$.

## 11. Remarks to the approximation of the velocity distribution in the model

The problem of the approximation of the velocity distribution in the 3-D model in a form suitable for the ray tracing and the travel time computation is perhaps more complex than the ray tracing itself. Usually, the whole model is divided into individual layers and/or blocks by interfaces of the first order. The interfaces and the velocity distributions within individual layers should be sufficiently smooth to guarantee the application of the ray method. If we are interested not only in the evaluation of rays and travel times , but also in ray amplitudes, the requirements on the smoothness of the model are strict. In pure kinematic problems, especially in the initial value ray tracing, even more rough models can be used. The computed travel times are not so influenced by the fictitious edges in interfaces of first order and by the fictitious interfaces of second order, caused, e.g., by a piece-wise linear interpolation, as the ray amplitudes. In the two point ray tracing, however, the piece-wise linear approximation may cause some problems. For example, in case of reflected waves, the rays reflected from individual linear segments of the interface of the first order often overlap or form shadow zones. There are, at least, two possibilities how to fight with this problem. The *first possibility* consists in some smoothing of results of computations. Remind that certain recent methods of the computation of HF seismic wave fields in laterally varying layered structures do not require the two point ray tracing and involve the smoothing automatically, or the smoothing may be easily introduced (Maslov method, method of Gaussian beams). The *second possibility* is to smooth the model. This may be done in many ways, either locally or globally. Procedures to smooth locally the model were proposed recently by Langan, Lerche and Cutler (1985). They are very suitable for tomographic applications. Global smoothing may be performed by splines. A very efficient spline package, suitable for the approximation both of interfaces and of the 3-D velocity distribution, was written by Cline (1981).

The application of splines removes both the fictitious edges in interfaces of the first order and the fictitious interfaces of the second order, and suppresses globally the oscillations in the velocity distribution. Unfortunately, in more complex models, the standard spline algorithm does not remove the oscillation fully, and sometimes generates very unpleasant local oscillations in the velocity itself or in its first derivatives. Such oscillations may have similar, or even larger effect on the ray tracing and travel time computations than the fictitious interfaces of the second order. It is more useful to use splines with smoothing, splines with tension, etc. Moreover, the ray tracing in a medium approximated by the spline approximation is more time consuming than for simpler approximations.

Both the first possibility: smoothing of the results, and the second possibility: smoothing of the model, may be, of course, combined.

Another important question is, whether to use the velocity distribution or the quadratic slowness distribution to describe the model. In seismology, the model has been traditionally described by the distribution of the velocity, or by the distribution of the slowness. In general, however, the simplest numerical algorithms are obtained for the quadratic slowness $V^{-2}$ model. Quadratic slowness appears, as the only quantity describing the velocity distribution, in all basic equations of the ray tracing systems, and in the R/T laws at interfaces. The quadratic slowness model is especially suitable in the 3-D cell ray tracing, where it simplifies considerably the algorithms and increases numerical efficiency of the computations. The author is convinced that the quadratic slowness models will play, sooner or later, a very important role in seismology, especially in tomographic methods, where the numerical efficiency of the ray tracing and travel time computation is of basic importance.

## 12. Boundary value ray tracing

In the boundary value ray tracing, the ray is not specified by the initial conditions at one point of the ray, but by more complicated conditions, related to different points of the ray.

The most important case of the boundary value ray tracing is the two point ray tracing. In the two point ray tracing, we seek the ray which connects two points $M_0$ and $M$. The coordinates of $M_0$ and $M$ are given, but the initial direction of the ray is not known at any of these points. The solution of the two point ray tracing is not necessarily unique due to a possible existence of multiple rays between $M_0$ and $M$. The situation with the multiple rays is very common in practical seismological applications, especially in case of refracted waves.

The next examples of the boundary value ray tracing are the ray tracing of normal rays (perpendicular to some surface $\Sigma$, passing through a given point $M$) and the ray tracing of rays generated at an initial surface (which are not necessarily perpendicular to $\Sigma$). For simplicity, we shall discuss here only the two point ray tracing; the situation in other mentioned problems is very similar.

The two point ray tracing is computationally considerably more involved than the initial value ray tracing. We shall describe here very shortly only three methods to solve it. We also remember from Section 10.6 that the two point ray tracing can be solved analytically for paraxial rays, using the dynamic ray tracing and propagator matrices.

### 12.1 The shooting method

The shooting method exploits the standard initial value ray tracing. One point is taken as an initial point, e.g. the point $M_0$. The rays are shoot from $M_0$ under different take-off angles $\delta_0$ and $\psi_0$. An iterative loop is used to find the ray which passes through the second point.

There are two important problems in the shooting method: a) The initial guess of the radiation angles, b) The algorithm which determines the new take-off angles at each step. In the second problem, the algorithms of the paraxial two point ray tracing may find important applications, see Section 10.6. As soon as a computed ray is not too far from the target point $M$, the paraxial method gives the solution very fast.

The shooting method is also suitable to perform *the 3-D 'profile' ray tracing* in which receivers are distributed along some line profile (possibly curved, piece-wise linear, etc.), e.g. in the surface profile measurements, in the borehole-to-borehole configuration, in the VSP configuration. As soon as we succeed to find a ray which crosses the profile, or at least passes in the vicinity of the profile, the rays to all receivers are found very efficiently by the paraxial two point ray tracing method.

### 12.2 The bending method

The bending method does not exploit the standard initial value ray tracing. In the bending method, an initial ray path is first guessed, and then perturbed iteratively so as to satisfy the appropriate differential equations, or directly the Fermat principle. Note that the first guess trajectory does not correspond to any actual ray, it is merely an artificial curve. Often, as a first guess, the straight line connecting $M_0$ and $M$ is used.

In simpler velocity models, the bending method is usually more efficient than the shooting method. It may be easily used even in the case that we wish to determine the rays between $M_0$ and some irregularly distributed points $M$ in a 3-D model, when the distances between points $M$ are large. In such cases, the efficiency of the shooting method is usually lower than for receivers distributed along some profile.

For more complex velocity structures, where the rays are more complicated curves, the efficiency of the bending method is lower. The bending method has also a strong tendency to overlook some multiple rays.

For more details regarding the boundary value ray tracing see Wesson (1971), Chander (1975), Julian and Gubbins (1977), Pereyra et al. (1980), Lee and Stewart (1981).

### 12.3 The continuation method

In both the shooting method and the bending method, the structure of the model is fixed during iterations. In the continuation method, the structure itself is gradually deformed. The two point ray tracing is first solved in a simpler model than the actual model under consideration. Such a model is selected for which the two point ray tracing problem may be solved analytically. After this, the model is step-by-step gradually deformed, until the derived model is achieved. At each deformation step, the ray equations are solved and the ray from the source to the receiver is found. For details see Keller and Perozzi (1983), Docherty (1985).

*Chapter 6*

# Inversion of travel times and seismic waveforms

*A. Tarantola*

The use of gradient methods of optimization allows to solve large-sized least squares inverse problems iteratively. In travel time inversion, each iteration consists on a *back-projection* of the travel time residuals. In waveform inversion, each iteration consists on a *back- propagation* of the waveform residuals. This chapter reviews methods developed by the author. Least-squares concepts are emphasized, at the expense of algorithmic details.

## 1. Functional least squares

Least-squares are so popular for solving inverse problems because they lead to the easiest computations. Their only drawback is their lack of robustness, i.e., their strong sensitivity to a small amount of big errors (outliers ) in a data set.

The least-squares criterion can be justified from the hypothesis that all sources of errors present in the problem can be modeled using Gaussian probability densities. Covariance operators play a central role in the method. The underlying mathematics are simple and beautiful.

Let **d** denote a generic data vector, **m** denote a generic model vector, and **d**=**g**(**m**) denote the theoretical relationship between data and model parameters. A measurement of the true value of **d** gives $\mathbf{d}_{obs}$ , with Gaussian uncertainties described by the covariance operator $\mathbf{C}_D$ . The a priori information on **m** may be described by the a priori value $\mathbf{m}_{prior}$ , with Gaussian uncertainties described by the covariance operator $\mathbf{C}_M$ . As demonstrated by Tarantola and Valette (1982), then the probability density representing the posterior information in the model space is given by



*G. Nolet (ed.), Seismic Tomography, 135–157.*

$$\sigma_M(\mathbf{m}) \propto \exp\left[-\tfrac{1}{2}\left[(\mathbf{g}(\mathbf{m})-\mathbf{d}_{obs})^t\,\mathbf{C}_D^{-1}(\mathbf{g}(\mathbf{m})-\mathbf{d}_{obs}) + (\mathbf{m}-\mathbf{m}_{prior})^t\,\mathbf{C}_M^{-1}(\mathbf{m}-\mathbf{m}_{prior})\right]\right] \quad (1)$$

The equation $\mathbf{d} = \mathbf{g}(\mathbf{m})$ solving the forward problem is usually nonlinear in geophysical problems. Then, the posterior probability density $\sigma_M(\mathbf{m})$ is not Gaussian. If $\sigma_M(\mathbf{m})$ is reasonably well behaved, the posterior information in the model space may be well represented by a central estimator of $\sigma_M(\mathbf{m})$ and a properly defined covariance operator. Among all the central estimators, the easiest to compute is generally the maximum likelihood point $\mathbf{m}_{ML}$ :

$$\sigma_M(\mathbf{m}) \quad MAXIMUM \quad \text{for} \quad \mathbf{m} = \mathbf{m}_{ML} \ , \quad (2)$$

because the obtention of $\mathbf{m}_{ML}$ corresponds to a problem of optimization of a scalar function, and many methods exist allowing an economical resolution of that problem.

Defining the *misfit* function by

$$S(\mathbf{m}) = \tfrac{1}{2}\left[(\mathbf{g}(\mathbf{m})-\mathbf{d}_{obs})^t\,\mathbf{C}_D^{-1}(\mathbf{g}(\mathbf{m})-\mathbf{d}_{obs})+(\mathbf{m}-\mathbf{m}_{prior})^t\,\mathbf{C}_M^{-1}(\mathbf{m}-\mathbf{m}_{prior})\right] \ , \quad (3)$$

the maximum likelihood point is clearly defined by

$$S(\mathbf{m}) \quad MINIMUM \quad \text{for} \quad \mathbf{m} = \mathbf{m}_{ML} \ . \quad (4)$$

The gradient of $S(\mathbf{m})$ at a given point $\mathbf{m}_n$ is denoted $\bar{\boldsymbol{\gamma}}_n$ and is defined by the first order development

$$S(\mathbf{m}_n + \delta\mathbf{m}) = S(\mathbf{m}_n) + \bar{\boldsymbol{\gamma}}_n^t\,\delta\mathbf{m} + O(\|\delta\mathbf{m}\|^2) \ , \quad (5)$$

where $(.)^t\,(.)$ represents the scalar obtained by contraction of all the variables (the gradient is an element of the dual of the model space ). This gives

$$\bar{\boldsymbol{\gamma}}_n = \mathbf{G}_n^t\,\mathbf{C}_d^{-1}(\mathbf{g}(\mathbf{m}_n)-\mathbf{d}_{obs}) + \mathbf{C}_m^{-1}(\mathbf{m}_n - \mathbf{m}_{prior}) \ , \quad (6)$$

where $\mathbf{G}_n$ is the (Fréchet ) derivative of $\mathbf{g}$ :

$$\mathbf{g}(\mathbf{m}_n + \delta\mathbf{m}) = \mathbf{g}(\mathbf{m}_n) + \mathbf{G}_n\,\delta\mathbf{m}_n + O(\|\delta\mathbf{m}\|^2) \ , \quad (7)$$

and where $\mathbf{G}_n^t$ is the transpose operator (sometimes named the adjoint ), and is defined by the condition that for any $\delta\tilde{\mathbf{d}}$ and $\delta\mathbf{m}$ ,

$$(\delta\tilde{\mathbf{d}})^t\,(\mathbf{G}_n^t\,\delta\mathbf{m}) = (\mathbf{G}_n^t\,\delta\tilde{\mathbf{d}})^t\,\delta\mathbf{m}_n \ . \quad (8)$$

To define the *direction of steepest ascent* at $\mathbf{m}_n$ , we need to define a particular distance in the model space. It is natural to choose the one associated with the covariance operator $\mathbf{C}_M$ :

$$\|\mathbf{m}_1 - \mathbf{m}_2\|^2 = (\mathbf{m}_1 - \mathbf{m}_2)^t\,\mathbf{C}_M^{-1}(\mathbf{m}_1 - \mathbf{m}_2) \ . \quad (9)$$

Then, the direction of steepest ascent is (Tarantola, 1987)

$$\boldsymbol{\gamma}_n = \mathbf{C}_M\,\bar{\boldsymbol{\gamma}}_n = \mathbf{C}_m\,\mathbf{G}_n^t\,\mathbf{C}_d^{-1}(\mathbf{g}(\mathbf{m}_n)-\mathbf{d}_{obs}) + (\mathbf{m}_n - \mathbf{m}_{prior}) \ . \quad (10)$$

A crude steepest descent method for the minimization of the misfit functional $S(\mathbf{m})$ defines an "updated point" $\mathbf{m}_{n+1}$ by

$$\mathbf{m}_{n+1} = \mathbf{m}_n - \mu_n \gamma_n \ , \tag{11}$$

where $\mu_n$ is an arbitrary positive real number small enough for ensuring that $S(\mathbf{m}_{n+1})$ will be smaller than $S(\mathbf{m}_n)$ (its existence is guaranteed because, by definition, $-\gamma_n$ is a direction of descent for S at $\mathbf{m}_n$ ).

There are many ways for giving a definite value to $\mu_n$ . Among them:

- by trial and error,
- by interpolation, as for instance when $S(\mathbf{m}_{n+1})$ is computed for three different values of $\mu_n$ , a parabola is fitted to these values, and the value of $\mu_n$ giving the minimum of the parabola is used,
- by linearization of $\mathbf{g}(\mathbf{m})$ around $\mathbf{m}_n$ :

$$\mathbf{g}(\mathbf{m}_n - \mu_n \gamma_n) \approx \mathbf{g}(\mathbf{m}_n) - \mu_n \mathbf{G}_n \gamma_n \ .$$

The problem of estimating $\mu_n$ will not be adressed in this paper. The reader may refer to any textbook in optimization (e.g, Fletcher, 1980; Scales, 1985).

Also, in practical applications, the reader should use conjugate gradients instead of crude steepest descent. The corresponding formulas can be found in Tarantola (1987).

## 2. Inversion of travel times

To infer the velocity structure of a medium, waves are generated by some sources, and the travel times to some receivers are measured. We wish to solve the inverse problem of estimating the velocity structure of the medium.

We assume that the high frequency limit is acceptable, i.e., that ray theory can be used instead of wave theory. If $c(\mathbf{x})$ denotes the celerity of the waves at point $\mathbf{x}$ , let $m(\mathbf{x})$ be the slowness

$$m(\mathbf{x}) = 1/c(\mathbf{x}) \ . \tag{12}$$

The i-th datum is the travel time for the i-th ray:

$$d^i = g^i(\mathbf{m}) = \int_{R^i(\mathbf{m})} ds^i\, m(\mathbf{x}^i) \ , \tag{13}$$

where $R^i(\mathbf{m})$ denotes the i-th ray path. As this ray path depends on $\mathbf{m}$ , $d^i$ is a nonlinear functional of $\mathbf{m}$ . Given a medium $\mathbf{m}$ , the actual ray path is obtained using the Fermat's theorem (or, equivalently, the eikonal equation ) and some numerical method.

First we wish to obtain the derivative of the nonlinear operator $g^i$ at a point $\mathbf{m}_n$ . We have

$$g^i(\mathbf{m}_n + \delta\mathbf{m}) = \int_{R^i(\mathbf{m}_n + \delta\mathbf{m})} ds^i\, (m_n(\mathbf{x}^i) + \delta m(\mathbf{x}^i)) \ . \tag{14}$$

The travel time being stationnary along the actual ray path (Fermat's theorem ),

$$\int_{R^i(\mathbf{m}_n+\delta\mathbf{m})} ds^i(m(\mathbf{x}^i)_n+\delta m(\mathbf{x}^i)) = \int_{R^i(\mathbf{m}_n)} ds^i(m(\mathbf{x}^i)_n+\delta m(\mathbf{x}^i))+O^i(\delta\mathbf{m}^2) \ . \tag{15}$$

This gives

$$g^i(\mathbf{m}_n+\delta\mathbf{m}) = g^i(\mathbf{m}_n)+\int_{R^i(\mathbf{m}_n)} ds^i\,\delta m(\mathbf{x}^i)+O^i(\|\delta\mathbf{m}\|^2) \ . \tag{16}$$

The comparison with the definition of the derivative of a nonlinear operator (equation (7)) gives

$$(\mathbf{G}_n\,\delta\mathbf{m})^i = \int_{R^i(\mathbf{m}_n)} ds^i\,\delta m(\mathbf{x}^i) \ . \tag{17}$$

To the model perturbation $\delta\mathbf{m}$ , the derivative of $\mathbf{g}$ at the point $\mathbf{m}_n$ associates the travel time perturbations (17).

Now, we wish here to obtain the transpose of $\mathbf{G}_n$ . The definition of transpose operator (equation (8)) is written, explicitly,

$$\int_{\mathbf{V}} dV(\mathbf{x})(\mathbf{G}_n^t\delta\tilde{\mathbf{d}})(\mathbf{x})\delta m(\mathbf{x}) = \sum_i \delta\tilde{d}^i\,(\mathbf{G}_n\,\delta\mathbf{m})^i \ , \tag{18}$$

and, using (17),

$$\int_{\mathbf{V}} dV(\mathbf{x})(\mathbf{G}_n^t\delta\tilde{\mathbf{d}})(\mathbf{x})\delta m(\mathbf{x}) = \sum_i \delta\tilde{d}^i \int_{R^i(\mathbf{m}_n)} ds^i\,\delta m(\mathbf{x}) \ . \tag{19}$$

To interprete equations (10)-(11) with some detail, it is useful to introduce the following partial steps:

$$\delta\mathbf{d}_n = \mathbf{G}_n\,\mathbf{m}_n - \mathbf{d}_{obs} \ , \tag{20}$$

$$\delta\tilde{\mathbf{d}}_n = \mathbf{C}_D^{-1}\delta\mathbf{d}_n \ , \tag{21}$$

$$\delta\tilde{\mathbf{m}}_n = \mathbf{G}_n^t\delta\tilde{\mathbf{d}}_n \ , \tag{22}$$

$$\delta\mathbf{m}_n = \mathbf{C}_M\,\delta\tilde{\mathbf{m}}_n \ , \tag{23}$$

$$\gamma_n = \delta\mathbf{m}_n + (\mathbf{m}_n - \mathbf{m}_{prior}) \ , \tag{24}$$

and

$$\mathbf{m}_{(m+1)} = \mathbf{m}_n - \mathbf{m}_n\,\gamma_n \ . \tag{25}$$

Using (19) and introducing the kernels of the covariance operators this gives, explicitly,

$$\delta d_n^i = \int_{R^i(\mathbf{m}_n)} ds^i\,m_n(\mathbf{x}(s^i)) - d_{obs}^i \ , \tag{26}$$

$$\delta\tilde{d}_n^i = \sum_j (\mathbf{C}_D^{-1})^{ij}\,\delta d_n^j \ , \tag{27}$$

$$\delta m_n(\mathbf{x}) = \sum_i \delta\tilde{d}_n^i\,\Psi^i(\mathbf{x}) \ , \tag{28}$$

$$\gamma_n(\mathbf{x}) = \delta m_n(\mathbf{x}) + (m_n(\mathbf{x}) - m_{prior}(\mathbf{x})) \ , \tag{29}$$

and

$$m_{n+1}(\mathbf{x}) = m_n(\mathbf{x}) - \mu_n \gamma_n(\mathbf{x}) \ , \tag{30}$$

where $\Psi^i(\mathbf{x})$ is defined by

$$\Psi^i(\mathbf{x}) = \int_{R^i(\mathbf{m}_n)} ds^i \, C_M(\mathbf{x},\mathbf{x}(s^i)) \ . \tag{31}$$

The interpretation of these formulas is as follows.

- Equation (26): the values $\delta d_n^i$ are simply (minus ) the data residuals corresponding to the current model $m_n(\mathbf{x})$ .
- Equation (27): the values $\delta \tilde{d}_n^i$ are the data residuals weighted with the usual least squares weights. The data covariance matrix is usually diagonal, so that small effort is needed to evaluate the $\delta \tilde{d}_n^i$ .
- Equation (28): this is the most important of the equations. $\Psi^i(\mathbf{x})$ represents the "coefficient of influence" on the point $\mathbf{x}$ of the i-th ray. We see that the contribution of the i-th ray to the value at point $\mathbf{x}$ of the "model perturbation" $\delta \mathbf{m}_n$ is proportional to the weighted residual for the ray and to the coefficient of influence of the ray at point $\mathbf{x}$ . This corresponds to a "back-projection" of the weighted residual along (the immediate vicinity of ) the ray. This concept of back-projection is usual in algebraic reconstruction techniques (ART) (see for instance Herman, 1980). Here they are generalized to the case where the physical space is not discretized. The reader should notice that it is fundamentally the iterative use of simple back-projections which solves the inverse problem.
- Equation (29): this equation corrects the back-projection of the weighted residuals for the information contained in the a priori model.
- Equation (30): it simply corresponds to the model updating.
- Equation (31): usual examples of a priori covariance functions are

$$C_M(\mathbf{x},\mathbf{x}') = \sigma^2 \exp\left\{ -\frac{1}{2} \frac{\|\mathbf{x}-\mathbf{x}'\|^2}{L^2} \right\} \tag{32}$$

or

$$C_M(\mathbf{x},\mathbf{x}') = \sigma^2 \exp\left\{ -\frac{\|\mathbf{x}-\mathbf{x}'\|}{L} \right\} \tag{33}$$

These functions decrease rapidly away from its diagonal ( $\|\mathbf{x}-\mathbf{x}'\| \gg L \rightarrow C(\mathbf{x},\mathbf{x}') \approx 0$ ) . The function $\Psi^i(\mathbf{x})$ takes then significant values only near the i-th ray: it defines a "tube" along the ray. The closer the point $\mathbf{x}$ is to the ray, the greater is the value $\Psi^i(\mathbf{x})$ , which represents a sort of *coefficient of influence* on the point $\mathbf{x}$ of the i-th ray.

Practically, the model $m(\mathbf{x})$ is numerically defined on a grid of points, one point per pixel of the graphic device used to plot the model. A covariance function like (32) or (33) is used, with a correlation length L equal to the caracteristic size of the holes between rays (so that the model will be smooth ). The integrals along rays in (31) are evaluated using any (accurate ) numerical method. The function $\Psi^i(\mathbf{x})$ may simply be grossly approximated by a reasonably chosen function of the distance between the point $\mathbf{x}$ and the (nearest point of the ) i-th ray. For a numerical example with real data, see Nercessian et al., 1984.

Notice that the problem is solved here in a nonlinear way: each iteration needs ray tracing in the current model. If the prior model is close enough to the actual earth, the problem may be linearized and the rays traced only once.

## 3. Inversion of elastic waveforms

At some points $\mathbf{x}_s$ $(s=1,2,...)$ of the Earth's surface, we shoot seismic sources producing elastic waves. For each shot point, we place receivers at some other points $\mathbf{x}_r$ (r=1,2,...) which record the displacement $u^i(\mathbf{x}_r, t; \mathbf{x}_s)$ of the surface. Our aim here is to use the observed displacements to infer the structure of the Earth. What follows is adapted from Tarantola (1986).

*a ) Choice of parameters.* From a seismological point of view, the Earth may be described using the density $\rho(\mathbf{x})$ , the elastic coefficients $c^{ijkl}(\mathbf{x})$ , and some parameters describing attenuation of waves. At first order, the Earth is isotropic (although heterogeneous ) and non attenuating. Only three parameters are then needed. For instance, in addition to the density $\rho(\mathbf{x})$ usual choices are

- the bulk modulus $K(\mathbf{x})$ and the the shear modulus $\mu(\mathbf{x})$ ,
- the Lame's parameters $\lambda(\mathbf{x})$ and $\mu(\mathbf{x})$ ,
- the velocity of compressional (longitudinal, or P) waves, $\alpha(\mathbf{x})$ , and the velocity of shear (transverse, or S) waves, $\beta(\mathbf{x})$ .

I consider here a small scale experiment, as in petroleum exploration. Using a seismic reflection data set (sources and receivers at the surface of the Earth ), all three parameters are not well resolved. To avoid an intrinsic ill-posedness of the problem, it is important to identify independent parameters (i.e., parameters for which correlation between posterior errors will be small ). Using physical arguments, Tarantola (1986) suggests to use density $\rho(\mathbf{x})$ , longitudinal-wave impedance

$$IP(\mathbf{x}) = \rho(\mathbf{x})\alpha(\mathbf{x}) = \sqrt{\rho(\mathbf{x})[\lambda(\mathbf{x}) + 2\mu(\mathbf{x})]} \ , \tag{34a}$$

and transverse-wave impedance

$$IS(\mathbf{x}) = \rho(\mathbf{x})\beta(\mathbf{x}) = \sqrt{\rho(\mathbf{x})\mu(\mathbf{x})} \ . \tag{34b}$$

*b ) The least-squares criterion of goodness of fit.* I define the best Earth's model $(IP(\mathbf{x}), IS(\mathbf{x}), \rho(\mathbf{x}))$ as the model that minimizes the least squares expression

$$S(\mathbf{IP},\mathbf{IS},\mathbf{p}) = \tfrac{1}{2}(\| \mathbf{u}_{cal} - \mathbf{u}_{obs} \|^2 + \| \mathbf{IP} - \mathbf{IP}_{prior} \|^2 + \| \mathbf{IS} - \mathbf{IS}_{prior} \|^2 + \| \rho - \rho_{prior} \|^2) \quad (35)$$

where

$$\| \mathbf{u}_{cal} - \mathbf{u}_{obs} \|^2 = \sum_s \sum_r \int_0^T dt \int_0^T dt' \, [u^i(\mathbf{x}_r, t; \mathbf{x}_s)_{cal} - u^i(\mathbf{x}_r, t; \mathbf{x}_s)_{obs}]$$
$$\times W^{ij}(t, t', \mathbf{x}_r, \mathbf{x}_s)[u^j(\mathbf{x}_r, t'; \mathbf{x}_s)_{cal} - u^j(\mathbf{x}_r, t'; \mathbf{x}_s)_{obs}] \ , \quad (36a)$$

$$\| \mathbf{IP} - \mathbf{IP}_{prior} \|^2 = \int_V dV(\mathbf{x}) \int_V dV(\mathbf{x}')[IP(\mathbf{x}) - IP(\mathbf{x})_{prior}] W_p(\mathbf{x},\mathbf{x}')[IP(\mathbf{x}') - IP(\mathbf{x}')_{prior}] \quad (36b)$$

$$\| \mathbf{IS} - \mathbf{IS}_{prior} \|^2 = \int_V dV(\mathbf{x}) \int_V dV(\mathbf{x}')[IS(\mathbf{x}) - IS(\mathbf{x})_{prior}] W_s(\mathbf{x},\mathbf{x}')[IS(\mathbf{x}') - IS(\mathbf{x}')_{prior}] \ , \quad (36c)$$

and

$$\| \rho - \rho_{prior} \|^2 = \int_V dV(\mathbf{x}) \int_V dV(\mathbf{x}')[\rho(\mathbf{x}) - \rho(\mathbf{x})_{prior}] W_\rho(\mathbf{x},\mathbf{x}')[\rho(\mathbf{x}') - \rho(\mathbf{x}')_{prior}] \ . \quad (36d)$$

Here $u^i(\mathbf{x}_r, t; \mathbf{x}_s)_{cal}$ represents the data predicted by the model $(IP(\mathbf{x}), IS(\mathbf{x}), \rho(\mathbf{x}))$, and $W^{ij}(t, t', \mathbf{x}_r, \mathbf{x}_s)$, $W_p(\mathbf{x},\mathbf{x}')$, $W_s(\mathbf{x},\mathbf{x}')$, and $W_\rho(\mathbf{x},\mathbf{x}')$ represent weighting functions to be discussed below. A priori uncertainties on density and impedance are assumed uncorrelated.

When so defined, the problem is fully nonlinear (the best model is defined without invoking any linear approximation of the basic equations ). In particular, I do not use the Born's approximation. It should be noticed that, as the computed seismograms are *nonlinear* functionals of the model parameters, the functional (3) is a *nonquadratic* function of the parameters.

*c ) Choice of the weighting functions.* In the context of least-squares, a weighting function is the integral kernel of the inverse of a covariance operator. For instance, if $C(\mathbf{x},\mathbf{x}')$ is a covariance function, the associated weighting function verifies

$$\int_V dV(\mathbf{x}') C(\mathbf{x},\mathbf{x}') W(\mathbf{x}',\mathbf{x}'') = \delta(\mathbf{x} - \mathbf{x}'') \quad (37)$$

Taking the covariance function

$$C(\mathbf{x},\mathbf{x}') = C(x, y, z, x', y', z') = K \, \delta(x - x')\delta(y - y') Min(z, z') \quad (38)$$

gives (see Tarantola, 1987)

$$\| \phi - \phi_{prior} \|^2 = \frac{1}{K} \int_V dV(\mathbf{x}) \left\{ \frac{\partial \phi}{\partial z}(x, y, z) - \frac{\partial \phi_{prior}}{\partial z}(x, y, z) \right\}^2 \ , \quad (39)$$

which is an adequate norm to impose to impedances or density on the Earth: we dont wish our final model to be close to the initial model, but we wish the vertical gradient of the model to be close to the vertical gradient of the a priori model. Taking for instance a homogeneous a priori model in impedances and density, the norm (39) will impose to the final model to have small vertical gradients. So, for each of the model parameters, I choose

$$C_p(x,y,z,x',y',z') = K_p\,\delta(x-x')\delta(y-y')Min(z,z') \ , \tag{40a}$$

$$C_s(x,y,z,x',y',z') = K_s\,\delta(x-x')\delta(y-y')Min(z,z') \ , \tag{40b}$$

and

$$C_p(x,y,z,x',y',z') = K_\rho\,\delta(x-x')\delta(y-y')Min(z,z') \ , \tag{40c}$$

which gives respectively

$$\|\mathbf{IP}-\mathbf{IP}_{prior}\|^2 = \frac{1}{K_p}\int_V dV(\mathbf{x})\left\{\frac{\partial IP}{\partial z}(x,y,z) - \frac{\partial IP_{prior}}{\partial z}(x,y,z)\right\}^2 \ , \tag{41a}$$

$$\|\mathbf{IS}-\mathbf{IS}_{prior}\|^2 = \frac{1}{K_s}\int_V dV(\mathbf{x})\left\{\frac{\partial IS}{\partial z}(x,y,z) - \frac{\partial IS_{prior}}{\partial z}(x,y,z)\right\}^2 \ , \tag{41b}$$

and

$$\|\rho-\rho_{prior}\|^2 = \frac{1}{K_\rho}\int_V dV(\mathbf{x})\left\{\frac{\partial \rho}{\partial z}(x,y,z) - \frac{\partial \rho_{prior}}{\partial z}(x,y,z)\right\}^2 \ . \tag{41c}$$

Thus, final models of impedance and density will have small vertical gradients (more precisely, small vertical gradient differences from the a priori model ).

If there are uncorrelated errors in the data set, depending on time or source and receiver position, then

$$C^{ij}(t,t',\mathbf{x}_r,\mathbf{x}_s) = \sigma^2(t,\mathbf{x}_r,\mathbf{x}_s)\delta^{ij}\delta(t-t') \tag{42}$$

and

$$\|\mathbf{u}_{cal}-\mathbf{u}_{obs}\|^2 = \sum_s\sum_r\int_0^T dt\sum_{i=1}^{3}\frac{[u^i(\mathbf{x}_r,t;\mathbf{x}_s)_{cal}-u^i(\mathbf{x}_r,t;\mathbf{x}_s)_{obs}]^2}{\sigma^2(\mathbf{r}_r,t;\mathbf{x}_s)} \ . \tag{43}$$

Usually, only the vertical component $u^3$ is recorded. The sum over i then disappears from (43).

I now turn to the description of the forward problem.

*d ) The elastodynamic wave equation.* Let us consider an isotropic elastic Earth. In what follows, $\mathbf{x}$ represents a point inside (or at the surface of ) the Earth, and $t$ is the time variable, running in an interval $0 \le t \le T$ . If $f^i(\mathbf{x},t;\mathbf{x}_s)$ is the volume density of internal forces for the s-th shot, $T^i(\mathbf{x},t;\mathbf{x}_s)$ the stress vector (traction ) at the Earth's surface S , and $n^i(\mathbf{x})$ the unit normal at the surface, then the displacement $u^i(\mathbf{x},t;\mathbf{x}_s)$ corresponding to that s-th shot is uniquely defined by the differential equations

$$\rho(\mathbf{x})\ddot{u}^i(\mathbf{x},t;\mathbf{x}_s) - \frac{\partial}{\partial x^i}[\lambda(\mathbf{x})u^{kk}(\mathbf{x},t;\mathbf{x}_s)] - 2\frac{\partial}{\partial x^j}[\mu(\mathbf{x})u^{ij}(\mathbf{x},t;\mathbf{x}_s)] = f^i(\mathbf{x},t;\mathbf{x}_s) \tag{44a}$$

$$\lambda(\mathbf{x})u^{kk}(\mathbf{x},t;\mathbf{x}_s)n^i(\mathbf{x}) + 2\mu(\mathbf{x})u^{ij}(\mathbf{x},t;\mathbf{x}_s)n^j(\mathbf{x}) = T^i(\mathbf{x},t;\mathbf{x}_s) \quad \mathbf{x}\in\mathbf{S} \tag{44b}$$

$$u^i(\mathbf{x},0;\mathbf{x}_s) = 0 \tag{44c}$$

$$\dot{u}^i(\mathbf{x},0;\mathbf{x}_s) = 0 \tag{44d}$$

where $u^{ij}(\mathbf{x},t;\mathbf{x}_s)$ represents the strain tensor

$$u^{ij}(\mathbf{x},t;\mathbf{x}_s) = \frac{1}{2}\left\{ \frac{\partial u^i}{\partial x^j}(\mathbf{x},t;\mathbf{x}_s) + \frac{\partial u^j}{\partial x^i}(\mathbf{x},t;\mathbf{x}_s) \right\} , \tag{45}$$

and where an implicit sum over repeated indexes is assumed.

For a given Earth model $\rho(\mathbf{x})$ , $\lambda(\mathbf{x})$ , $\mu(\mathbf{x})$ , and given volume and surface sources $f^i(\mathbf{x},t;\mathbf{x}_s)$ , $T^i(\mathbf{x},t;\mathbf{x}_s)$ , the displacement field $u^i(\mathbf{x},t;\mathbf{x}_s)$ can be evaluated directly from equations (34) using numerical methods such as finite-differences (Altermann and Karal, 1968). The sources of seismic waves can either be tractions at the surface, described by $T^i(\mathbf{x},t;\mathbf{x}_s)$ , or internal sources (as borehole explosions ), described by $f^i(\mathbf{x},t;\mathbf{x}_s)$ .

*e ) The Green's function.* The Green's function (or impulse response ) of the problem is denoted $\Gamma^{ij}(\mathbf{x},t;\mathbf{x}',t')$ and is defined by

$$\rho(\mathbf{x})\frac{\partial^2\Gamma^{ij}}{\partial t^2}(\mathbf{x},t;\mathbf{x}',t') - \frac{\partial}{\partial x^i}(\lambda(\mathbf{x})\Gamma^{kkj}(\mathbf{x},t;\mathbf{x}',t')) - 2\frac{\partial}{\partial x^k}(\mu(\mathbf{x})\Gamma^{ikj}(\mathbf{x},t;\mathbf{x}',t')) =$$
$$= \delta^{ij}\delta(\mathbf{x}-\mathbf{x}')\delta(t-t') \tag{46a}$$

$$\lambda(\mathbf{x})\Gamma^{kkj}(\mathbf{x},t;\mathbf{x}',t')n^i(\mathbf{x}) + 2\mu(\mathbf{x})\Gamma^{ikj}(\mathbf{x},t;\mathbf{x}',t')n^k(\mathbf{x}) = 0 \quad \mathbf{x}\in \mathbf{S} \tag{46b}$$

$$\Gamma^{ij}(\mathbf{x},t;\mathbf{x}',t') = 0 \;\text{ for } t<t' \tag{46c}$$

$$\dot{\Gamma}^{ij}(\mathbf{x},t;\mathbf{x}',t') = 0 \;\text{ for } t<t' , \tag{46d}$$

where $\Gamma^{ijk}(\mathbf{x},t;\mathbf{x}',t')$ is the strain associated with $\Gamma^{ik}(\mathbf{x},t;\mathbf{x}',t')$ :

$$\Gamma^{ijk}(\mathbf{x},t;\mathbf{x}',t') = \frac{1}{2}\left[ \frac{\partial\Gamma^{ik}}{\partial x^j}(\mathbf{x},t;\mathbf{x}',t') + \frac{\partial\Gamma^{jk}}{\partial x^i}(\mathbf{x},t;\mathbf{x}',t') \right] , \tag{47}$$

and where $\delta^{ij}$ , $\delta(\mathbf{x})$ , and $\delta(t)$ represent respectively the Kronecker's symbol and the Dirac's delta functions in space and time. It is usual to say that $\Gamma^{ij}(\mathbf{x},t;\mathbf{x}',t')$ represents "the i-th component of the displacement at point $\mathbf{x}$ and time $t$ corresponding to a unit impulse in the j-th coordinate direction at point $\mathbf{x}'$ and time $t'$ (for homogeneous boundary and initial conditions )".

As the values of the elastic parameters of the medium are assumed not to depend on time, the Green's function is "invariant by time translation" :

$$\Gamma^{ij}(\mathbf{x},t;\mathbf{x}',t') = \Gamma^{ij}(\mathbf{x},t-t';\mathbf{x}',0) . \tag{48}$$

The introduction of the Green's function allows the following representation of the general solution of (46) (see for instance Aki and Richards, 1980):

$$u^i(\mathbf{x},t;\mathbf{x}_s) = \int_{\mathbf{V}} dV(\mathbf{x}')\Gamma^{ij}(\mathbf{x},t;\mathbf{x}',0)*f^j(\mathbf{x}',t;\mathbf{x}_s) + \int_{\mathbf{S}} dS(\mathbf{x}')\Gamma^{ij}(\mathbf{x},t;\mathbf{x}',0)*T^j(\mathbf{x}',t;\mathbf{x}_s) \tag{49}$$

where * denotes time convolution, and **V** and **S** are respectively the Earth's volume and surface. The Green's function introduced here is not arbitrary, nor it is the Green's function corresponding to an unbounded space, but the actual solution of equations (46). As the

Green's function has infinite bandwidth, it is not possible to evaluate it using standard numerical methods. Although its introduction is useful for analytical developments, it will not appear in the final computations, so that the problem of numerical evaluation will not arise.

Another important property of the Green's function is *reciprocity* :

$$\Gamma^{ij}(\mathbf{x},t;\mathbf{x}',0) = \Gamma^{ji}(\mathbf{x}',t;\mathbf{x},0) \ . \tag{50}$$

It means that the response at point $\mathbf{x}$ in the i-th direction due to a source at point $\mathbf{x}'$ in the j-th direction is identical to the response at point $\mathbf{x}'$ in the j-th direction due to a source at point $\mathbf{x}$ . For a theoretical demonstration of this property, see for instance Aki and Richards (1980).

*f ) The linearized forward problem.* Let $\rho(\mathbf{x})_n$ , $\lambda(\mathbf{x})_n$ , $\mu(\mathbf{x})_n$ represent an arbitrary ("current" or "unperturbed") medium and let $u^i(\mathbf{x},t;\mathbf{x}_s)_n$ represent the displacement field which propagates in this reference medium for given surface and volume sources. A model perturbation

$$\rho(\mathbf{x})_n \rightarrow \rho(\mathbf{x})_n + \delta\rho(\mathbf{x}) \tag{51a}$$

$$\lambda(\mathbf{x})_n \rightarrow \lambda(\mathbf{x})_n + \delta\lambda(\mathbf{x}) \tag{51b}$$

$$\mu(\mathbf{x})_n \rightarrow \mu(\mathbf{x})_n + \delta\mu(\mathbf{x}) \tag{51c}$$

will produce a perturbation of the displacement field

$$u^i(\mathbf{x},t;\mathbf{x}_s)_n \rightarrow u^i(\mathbf{x},t;\mathbf{x}_s)_n + \delta u^i(\mathbf{x},t;\mathbf{x}_s) \ . \tag{52}$$

I wish here to obtain the first order approximation to $\delta u^i(\mathbf{x},t;\mathbf{x}_s)$ .

By definition, $u^i(\mathbf{x},t;\mathbf{x}_s)_n$ is the field propagating in the unperturbed medium. It then satifies

$$\rho(\mathbf{x})_n \ddot{u}^i(\mathbf{x},t;\mathbf{x}_s)_n - \frac{\partial}{\partial x^i}(\lambda(\mathbf{x})_n u^{kk}(\mathbf{x},t;\mathbf{x}_s)_n) - 2\frac{\partial}{\partial x^j}(\mu(\mathbf{x})_n u^{ij}(\mathbf{x},t;\mathbf{x}_s)_n) =$$

$$= f^i(\mathbf{x},t;\mathbf{x}_s) \tag{53a}$$

$$\lambda(\mathbf{x})_n u^{kk}(\mathbf{x},t;\mathbf{x}_s)_n n^i(\mathbf{x}) + 2\mu(\mathbf{x})_n u^{ij}(\mathbf{x},t;\mathbf{x}_s)_n n^j(\mathbf{x}) = T^i(\mathbf{x},t;\mathbf{x}_s) \quad \mathbf{x}\in \mathbf{S} \tag{53b}$$

$$u^i(\mathbf{x},0;\mathbf{x}_s)_n = 0 \tag{53c}$$

$$\dot{u}^i(\mathbf{x},0;\mathbf{x}_s)_n = 0 \tag{53d}$$

Then, the field $u^i(\mathbf{x},t;\mathbf{x}_s)_n + \delta u^i(\mathbf{x},t;\mathbf{x}_s)$ verifies

$$\begin{aligned}(\rho_n + \delta\rho)(\mathbf{x})(\ddot{u}_n^i + \delta\ddot{u}^i)(\mathbf{x},t;\mathbf{x}_s) - \frac{\partial}{\partial x^i}((\mu_n + \delta\mu)(\mathbf{x})(u_n^{kk} + \delta u^{kk})(\mathbf{x},t;\mathbf{x}_s))\\ -2\frac{\partial}{\partial x^j}((\mu_n + \delta\mu)(\mathbf{x})(u_n^{ij} + \delta u^{ij})(\mathbf{x},t;\mathbf{x}_s)) = f^i(\mathbf{x},t;\mathbf{x}_s)\end{aligned} \tag{54a}$$

$$\begin{aligned}(\lambda_n + \delta\lambda)(\mathbf{x})(u_n^{kk} + \delta u^{kk})(\mathbf{x},t;\mathbf{x}_s)n^i(\mathbf{x}) + 2(\mu_n + \delta\mu)(\mathbf{x})(u_n^{ij} + \delta u^{ij})(\mathbf{x},t;\mathbf{x}_s)n^j(\mathbf{x})\\ = T^i(\mathbf{x},t;\mathbf{x}_s) \quad \mathbf{x}\in \mathbf{S}\end{aligned} \tag{54b}$$

$$(u_n^i + \delta u^i)(\mathbf{x},0;\mathbf{x}_s) = 0 \tag{54c}$$

$$(\dot{u}_n^i \delta\dot{u}^i)(\mathbf{x},0;\mathbf{x}_s) = 0 \tag{54d}$$

Using (53), equations (54) simplify to

$$\rho(\mathbf{x})_n \delta\ddot{u}^i(\mathbf{x},t;\mathbf{x}_s) - \frac{\partial}{\partial x^i}(\lambda(\mathbf{x})_n \delta u^{kk}(\mathbf{x},t;\mathbf{x}_s)) - 2\frac{\partial}{\partial x^j}(\mu(\mathbf{x})_n \delta u^{ij}(\mathbf{x},t;\mathbf{x}_s)) =$$

$$= \delta f^i(\mathbf{x},t;\mathbf{x}_s) \tag{55a}$$

$$\lambda(\mathbf{x})_n \delta u^{kk}(\mathbf{x},t;\mathbf{x}_s) n^i(\mathbf{x}) + 2\mu(\mathbf{x})_n \delta u^{ij}(\mathbf{x},t;\mathbf{x}_s) n^j(\mathbf{x}) = \delta T^i(\mathbf{x},t;\mathbf{x}_s) \mathbf{x} \in \mathbf{S} \tag{55b}$$

$$\delta u^i(\mathbf{x},0;\mathbf{x}_s) = 0 \tag{55c}$$

$$\delta\dot{u}^i(\mathbf{x},0;\mathbf{x}_s) = 0 \ , \tag{55d}$$

where

$$\delta f^i(\mathbf{x},t;\mathbf{x}_s) = \ddot{u}^i(\mathbf{x},t;\mathbf{x}_s)_n \delta\rho(\mathbf{x}) + \frac{\partial}{\partial x^i}[u^{kk}(\mathbf{x},t;\mathbf{x}_s)_n \delta\lambda(\mathbf{x})] + 2\frac{\partial}{\partial x^j}[u^{ij}(\mathbf{x},t;\mathbf{x}_s)_n \delta\mu(\mathbf{x})]$$
$$- \delta\ddot{u}^i(\mathbf{x},t;\mathbf{x}_s)\delta\rho(\mathbf{x}) + \frac{\partial}{\partial x^i}[\delta u^{kk}(\mathbf{x},t;\mathbf{x}_s)\delta\lambda(\mathbf{x})] + 2\frac{\partial}{\partial x^j}(\delta u^{ij}(\mathbf{x},t;\mathbf{x}_s)\delta\mu(\mathbf{x})) \tag{56a}$$

and

$$\delta T^i(\mathbf{x},t;\mathbf{x}_s) = -u^{kk}(\mathbf{x},t;\mathbf{x}_s)_n \delta\lambda(\mathbf{x}) n^i(\mathbf{x}) - 2u^{ij}(\mathbf{x},t;\mathbf{x}_s)_n \delta\mu(\mathbf{x}) n^j(\mathbf{x})$$
$$- \delta u^{kk}(\mathbf{x},t;\mathbf{x}_s)\delta\lambda(\mathbf{x}) n^i(\mathbf{x}) - 2\delta u^{ij}(\mathbf{x},t;\mathbf{x}_s)\delta\mu(\mathbf{x}) n^j(\mathbf{x}) \ . \tag{56b}$$

As I am seeking for the first order approximation to $\delta u^i(\mathbf{x},t;\mathbf{x}_s)$ , I can drop second order terms in (56). Then, up to the first order,

$$\delta f^i(\mathbf{x},t;\mathbf{x}_s) = -\ddot{u}^i(\mathbf{x},t;\mathbf{x}_s)_n \delta\rho(\mathbf{x}) + \frac{\partial}{\partial x^i}(u^{kk}(\mathbf{x},t;\mathbf{x}_s)_n \delta\lambda(\mathbf{x}))$$
$$+ 2\frac{\partial}{\partial x^j}[u^{ij}(\mathbf{x},t;\mathbf{x}_s)_n \delta\mu(\mathbf{x})] \tag{57a}$$

$$\delta T^i(\mathbf{x},t;\mathbf{x}_s) = -u^{kk}(\mathbf{x},t;\mathbf{x}_s)_n \delta\lambda(\mathbf{x}) n^i(\mathbf{x}) - 2u^{ij}(\mathbf{x},t;\mathbf{x}_s)_n \delta\mu(\mathbf{x}) n^j(\mathbf{x}) \ . \tag{57b}$$

Equations (55) and (57) are interpreted as follows: up to the first order, the perturbation $\delta u^i(\mathbf{x},t;\mathbf{x}_s)$ of the displacement field due to perturbations $\delta\rho(\mathbf{x})$ , $\delta\lambda(\mathbf{x})$ , and $\delta\mu(\mathbf{x})$ can be interpreted as the field propagating in the medium $\rho(\mathbf{x})_n$ , $\lambda(\mathbf{x})_n$ , and $\mu(\mathbf{x})_n$ (equations (55)) and created by the "secondary sources" (57).

Using the Green's function $\Gamma^{ij}(\mathbf{x},t;\mathbf{x}',0)_n$ corresponding to the reference medium $\rho(\mathbf{x})_n$ , $\lambda(\mathbf{x})_n$ , and $\mu(\mathbf{x})_n$ , the solution of (55) at the receiver locations is given by

$$\delta u^i(\mathbf{x}_r,t;\mathbf{x}_s) = \int_{\mathbf{V}} dV(\mathbf{x}) \Gamma^{ij}(\mathbf{x}_r,t;\mathbf{x},0)_n * \delta f^j(\mathbf{x},t;\mathbf{x}_s)$$
$$+ \int_{\mathbf{S}} dS(\mathbf{x}) \Gamma^{ij}(\mathbf{x}_r,t;\mathbf{x},0)_n * \delta T^j(\mathbf{x},t;\mathbf{x}_s) \ , \tag{58}$$

i.e.,

$$\delta u^i(\mathbf{x}_r,t;\mathbf{x}_s) = -\int_V dV(\mathbf{x})\Gamma^{ij}(\mathbf{x}_r,t;\mathbf{x},0)_n * \ddot{u}^j(\mathbf{x}',t;\mathbf{x}_s)_n \delta\rho(\mathbf{x})$$

$$+\int_V dV(\mathbf{x})\Gamma^{ij}(\mathbf{x}_r,t;\mathbf{x},0)_n * \frac{\partial}{\partial x^j}[u^{kk}(\mathbf{x},t;\mathbf{x}_s)_n \delta\lambda(\mathbf{x})]$$

$$+2\int_V dV(\mathbf{x})\Gamma^{ij}(\mathbf{x}_r,t;\mathbf{x},0)_n * \frac{\partial}{\partial x^k}[u^{jk}(\mathbf{x},t;\mathbf{x}_s)_n \delta\mu(\mathbf{x})]$$

$$-\int_S dS(\mathbf{x})\Gamma^{ij}(\mathbf{x}_r,t;\mathbf{x},0)_n * u^{kk}(\mathbf{x},t;\mathbf{x}_s)_r n^j(\mathbf{x})\delta\lambda(\mathbf{x})$$

$$-2\int_S dS(\mathbf{x})\Gamma^{ij}(\mathbf{x}_r,t;\mathbf{x},0)_n * u^{jk}(\mathbf{x},t;\mathbf{x}_s)_n n^k(\mathbf{x})\delta\mu(\mathbf{x}) \ . \tag{59}$$

Using

$$\Gamma^{ij}(\mathbf{x},t;\mathbf{x}',0) * \frac{\partial}{\partial x'^j}[u^{kk}(\mathbf{x}',t;\mathbf{x}_s)\delta\lambda(\mathbf{x}')] =$$

$$-\frac{\partial\Gamma^{ij}}{\partial x'^j}(\mathbf{x},t;\mathbf{x}',0)*u^{kk}(\mathbf{x}',t;\mathbf{x}_s)\delta\lambda(\mathbf{x}') + \frac{\partial}{\partial x'^j}[\Gamma^{ij}(\mathbf{x},t;\mathbf{x}',0)*u^{kk}(\mathbf{x}',t;\mathbf{x}_s)\delta\lambda(\mathbf{x}')] \tag{60}$$

$$\Gamma^{ij}(\mathbf{x},t;\mathbf{x}',0)*\frac{\partial}{\partial x'^k}(u^{jk}(\mathbf{x}',t;\mathbf{x}_s)\delta\mu(\mathbf{x}')) =$$

$$-\frac{\partial\Gamma^{ij}}{\partial x'^k}(\mathbf{x},t;\mathbf{x}',0)*u^{jk}(\mathbf{x}',t;\mathbf{x}_s)\delta\mu(\mathbf{x}') + \frac{\partial}{\partial x'^k}[\Gamma^{ij}(\mathbf{x},t;\mathbf{x}',0)*u^{jk}(\mathbf{x}',t;\mathbf{x}_s)\delta\mu(\mathbf{x}')] \tag{61}$$

and the Green's theorem

$$\int_V dV(\mathbf{x})\frac{\partial\mathbf{Q}}{\partial x^k}(\mathbf{x}) = \int_S dS(\mathbf{x})n^k(\mathbf{x})\mathbf{Q}(\mathbf{x}) \ , \tag{62}$$

equations (59) simplify to

$$\delta u^i(\mathbf{x}_r,t;\mathbf{x}_s) = -\int_V dV(\mathbf{x})\Gamma^{ij}(\mathbf{x}_r,t;\mathbf{x},0)_n * \ddot{u}^j(\mathbf{x},t;\mathbf{x}_s)_n \delta\rho(\mathbf{x})$$

$$-\int_V dV(\mathbf{x})\frac{\partial\Gamma^{ij}}{\partial x^j}(\mathbf{x}_r,t;\mathbf{x},0)_n * u^{mm}(\mathbf{x},t;\mathbf{x}_s)_n \delta\lambda(\mathbf{x})$$

$$-2\int_V dV(\mathbf{x})\frac{\partial\Gamma^{ip}}{\partial x^m}(\mathbf{x}_r,t;\mathbf{x},0)_n * u^{pm}(\mathbf{x},t;\mathbf{x}_s)_n \delta\mu(\mathbf{x}) \ . \tag{63}$$

If the perturbations $\delta\rho(\mathbf{x})$ , $\delta\lambda(\mathbf{x})$ , $\delta\mu(\mathbf{x})$ are sufficiently small (in a sense to be discussed below ), for computing the displacement field in the medium defined by $(\rho(\mathbf{x})_n + \delta\rho(\mathbf{x})$ , $\lambda(\mathbf{x})_n + \delta\lambda(\mathbf{x})$ , $\mu(\mathbf{x})_n + \delta\mu(\mathbf{x}))$ this first order approximation can be used. It is named the (first) *Born approximation* . I will *not* use such an approximation. But we need the Fréchet's derivative of the displacement field with respect to the model parameters, and, clearly, these Fréchet's derivatives are easily obtained from the first order development (63).

Instead of parametring the Earth using the Lamé's parameters $\lambda(\mathbf{x})$ and $\mu(\mathbf{x})$ , I can use the P-wave impedance $IP(\mathbf{x})$ and the S-wave impedances $IS(\mathbf{x})$ (see above ). I then have

$$\lambda(\mathbf{x}) = \frac{1}{\rho(\mathbf{x})}(IP^2(\mathbf{x}) - 2IS^2(\mathbf{x})) \tag{64a}$$

$$\mu(\mathbf{x}) = \frac{1}{\rho(\mathbf{x})} IS^2(\mathbf{x}) \ , \tag{64b}$$

which gives

$$\delta\lambda(\mathbf{x}) = -(\alpha^2(\mathbf{x}) - 2\beta^2(\mathbf{x}))\delta\rho(\mathbf{x}) + 2\alpha(\mathbf{x})\delta IP(\mathbf{x}) - 4\beta(\mathbf{x})\delta IS(\mathbf{x}) \tag{65a}$$

$$\delta\mu(\mathbf{x}) = -\beta^2(\mathbf{x})\delta\rho(\mathbf{x}) + 2\beta(\mathbf{x})\delta IS(\mathbf{x}) \ . \tag{65b}$$

Equation (63) then becomes

$$\begin{aligned}
\delta u^i(\mathbf{x}_r,t;\mathbf{x}_s) = &-\int_V dV(\mathbf{x})\Big[\Gamma^{ij}(\mathbf{x}_r,t;\mathbf{x},0)_n * \ddot{u}^j(\mathbf{x},t;\mathbf{x}_s)_n \\
&-\Big[\alpha^2(\mathbf{x})_n - 2\beta^2(\mathbf{x})_n\Big]\frac{\partial\Gamma^{ij}}{\partial x^j}(\mathbf{x}_r,t;\mathbf{x},0)_n * u^{mm}(\mathbf{x},t;\mathbf{x}_s)_n \\
&-2\beta^2(\mathbf{x})_n\frac{\partial\Gamma^{ij}}{\partial x^m}(\mathbf{x}_r,t;\mathbf{x},0)_n * u^{jm}(\mathbf{x},t;\mathbf{x}_s)_n\Big]\,\delta\rho(\mathbf{x}) \\
&-2\int_V dV(\mathbf{x})\alpha(\mathbf{x})_n\frac{\partial\Gamma^{ij}}{\partial x^j}(\mathbf{x}_r,t;\mathbf{x},0) * u^{mm}(\mathbf{x},t;\mathbf{x}_s)_n\,\delta IP(\mathbf{x}) \\
&-4\int_V dV(\mathbf{x})\beta(\mathbf{x})_n\Big[\frac{\partial\Gamma^{ij}}{\partial x^m}(\mathbf{x}_r,t;\mathbf{x},0)_n * u^{jm}(\mathbf{x},t;\mathbf{x}_s)_n \\
&-\frac{\partial\Gamma^{ij}}{\partial x^j}(\mathbf{x}_r,t;\mathbf{x},0)_n * u^{mm}(\mathbf{x},t;\mathbf{x}_s)_n\Big]\,\delta IS(\mathbf{x}) \ .
\end{aligned} \tag{66}$$

*g ) The validity of the Born's approximation.* As previously stated, Born's approximation consists in using the first order approximation (66) for estimating the displacement field $u_0^i(\mathbf{x},t;\mathbf{x}_s) + \delta u^i(\mathbf{x},t;\mathbf{x}_s)$ corresponding to the medium $(\rho(\mathbf{x})_n + \delta\rho(\mathbf{x})$ , $IP_n(\mathbf{x}) + \delta IP(\mathbf{x})$ , $IS_n(\mathbf{x}) + \delta IS(\mathbf{x}))$ . Although it is possible to obtain *rigorous* conditions for the validity of such an approximation (see for instance Hudson and Heritage, 1981), it is not so easy to obtain *useful* conditions. Common physical sense suggests that a necessary condition for Born's approximation to be adequate is that *travel times* in the perturbed medium are adequately modeled by the travel times in the unperturbed medium, i.e., that the unperturbed medium *contains the low spatial frequency part of the P-wave and S-wave velocities* .

*h ) The Fréchet derivatives of the displacements.* Using equation (66), we see that the Fréchet derivative (at the point $\rho(\mathbf{x})_n$ , $IP(\mathbf{x})_n$ , $IS(\mathbf{x})_n$) of the displacements $u^i(\mathbf{x},t;\mathbf{x}_s)$ with respect to the P- wave impedance $IP(\mathbf{x})$ is the linear operator that to an arbitrary perturbation $\delta IP(\mathbf{x})$ associates the displacement perturbation corresponding to the first order development

$$\delta u^i(\mathbf{x}_r,t;\mathbf{x}_s) = -2\int_{\mathbf{V}} dV(\mathbf{x})\alpha(\mathbf{x})_n \frac{\partial \Gamma^{ij}}{\partial x^j}(\mathbf{x}_r,t;\mathbf{x},0)_n * u^{mm}(\mathbf{x},t;\mathbf{x}_s)_n \, \delta IP(\mathbf{x}) \tag{67a}$$

Introducing the kernel $A^i(\mathbf{x}_r,t,\mathbf{x}_s \,|\, \mathbf{x})_n$ of this linear operator by

$$\delta u^i(\mathbf{x}_r,t;\mathbf{x}_s) = \int_{\mathbf{V}} dV(\mathbf{x}) A^i(\mathbf{x}_r,t,\mathbf{x}_s \,|\, \mathbf{x})_n \, \delta IP(\mathbf{x}) \tag{67b}$$

gives

$$A^i(\mathbf{x}_r,t,\mathbf{x}_s \,|\, \mathbf{x})_n = -2\alpha(\mathbf{x})_n \frac{\partial \Gamma^{ij}}{\partial x^j}(\mathbf{x}_r,t;\mathbf{x},0)_n * u^{mm}(\mathbf{x},t;\mathbf{x}_s)_n \ . \tag{68}$$

Similarly, introducing the kernels $B^i(\mathbf{x}_r,t,\mathbf{x}_s \,|\, \mathbf{x})_n$ and $C^i(\mathbf{x}_r,t,\mathbf{x}_s \,|\, \mathbf{x})_n$ corresponding respectively to the Fréchet derivatives of the displacements with respect to the S-wave impedance and to the density,

$$\delta u^i(\mathbf{x}_r,t;\mathbf{x}_s) = \int_{\mathbf{V}} dV(\mathbf{x}) B^i(\mathbf{x}_r,t,\mathbf{x}_s \,|\, \mathbf{x})_n \, \delta IS(\mathbf{x}) \tag{69a}$$

$$\delta u^i(\mathbf{x}_r,t;\mathbf{x}_s) = \int_{\mathbf{V}} dV(\mathbf{x}) C^i(\mathbf{x}_r,t,\mathbf{x}_s \,|\, \mathbf{x})_n \, \delta\rho(\mathbf{x}) \ , \tag{69b}$$

gives respectively

$$B^i(\mathbf{x}_r,t,\mathbf{x}_s \,|\, \mathbf{x})_n = -4\beta(\mathbf{x})_n \Big( \frac{\partial \Gamma^{ij}}{\partial x^m}(\mathbf{x}_r,t;\mathbf{x},0)_n * u^{jm}(\mathbf{x},t;\mathbf{x}_s)_n - \frac{\partial \Gamma^{ij}}{\partial x^j}(\mathbf{x}_r,t;\mathbf{x},0)_n * u^{mm}(\mathbf{x},t;\mathbf{x}_s)_n \Big) \ , \tag{70}$$

and

$$\begin{aligned} C^i(\mathbf{x}_r,t,\mathbf{x}_s \,|\, \mathbf{x})_n = {} & -\Gamma^{ij}(\mathbf{x}_r,t;\mathbf{x},0)_n * \ddot{u}^j(\mathbf{x},t;\mathbf{x}_s)_n \\ & + (\alpha^2(\mathbf{x})_n - 2\beta^2(\mathbf{x})_n) \frac{\partial \Gamma^{ij}}{\partial x^j}(\mathbf{x}_r,t;\mathbf{x},0)_n * u^{mm}(\mathbf{x},t;\mathbf{x}_s)_n \\ & + 2\beta^2(\mathbf{x})_n \frac{\partial \Gamma^{ij}}{\partial x^m}(\mathbf{x}_r,t;\mathbf{x},0)_n * u^{jm}(\mathbf{x},t;\mathbf{x}_s)_n \ . \end{aligned} \tag{71}$$

*i ) Transpose operators.*

*i-1) P-wave impedance.* The Fréchet derivative (at the point $\rho_n$ , $\mathbf{IP}_n$ , $\mathbf{IS}_n$) of displacements with respect to the P-wave impedance is the linear operator $\mathbf{A}_n$ that to an arbitrary perturbation $\delta IP$ associates the displacement

$$\delta\mathbf{u} = \mathbf{A}_n \delta\mathbf{IP} \tag{72}$$

i.e.

$$\delta u^i(\mathbf{x}_r,t;\mathbf{x}_s) = \int_{\mathbf{V}} dV(\mathbf{x}) A^i(\mathbf{x}_r,t,\mathbf{x}_s \,|\, \mathbf{x})_n \, \delta IP(\mathbf{x}) \ , \tag{67}$$

where $A^i(\mathbf{x}_r,t,\mathbf{x}_s\,|\,\mathbf{x})_n$ is given by equation (68). By definition of the transpose of an operator (see equation (8)), the operator $\mathbf{A}_n^t$ to any $\delta\tilde{u}^i(\mathbf{x}_r,t;\mathbf{x}_s)$ will associate a $\delta I\tilde{P}(\mathbf{x})$ given by

$$\delta\mathbf{I}\tilde{\mathbf{P}} = \mathbf{A}_n^t \delta\tilde{\mathbf{u}} \ , \tag{73}$$

i.e.,

$$\delta I\tilde{P}(\mathbf{x}) = \sum_r \int_0^T dt A^i(\mathbf{x}_r,t,\mathbf{x}_s\,|\,\mathbf{x})_n \delta\tilde{u}^i(\mathbf{x}_r,t;\mathbf{x}_s) \tag{74}$$

(remember that implicit sum is assumed over repeated indexes ). This gives

$$\delta I\tilde{P}(\mathbf{x}) = -2\alpha(\mathbf{x})_n \sum_r \int_0^T dt \frac{\partial \Gamma^{ij}}{\partial x^j}(\mathbf{x}_r,t;\mathbf{x},0)_n * u^{mm}(\mathbf{x},t;\mathbf{x}_s)_n \delta\tilde{u}^i(\mathbf{x}_r,t;\mathbf{x}_s) \tag{75}$$

Defining

$$\delta\Psi^i(\mathbf{x},t;\mathbf{x}_s) = \sum_r \Gamma^{ij}(\mathbf{x},0;\mathbf{x}_r,t)_n * \delta\tilde{u}^j(\mathbf{x}_r,t;\mathbf{x}_s) \tag{76}$$

and

$$\delta\Psi^{ij}(\mathbf{x},t;\mathbf{x}_s) = \frac{1}{2}\left\{ \frac{\partial\delta\Psi^i}{\partial x^j}(\mathbf{x},t;\mathbf{x}_s) + \frac{\partial\delta\Psi^j}{\partial x^i}(\mathbf{x},t;\mathbf{x}_s) \right\} \ , \tag{77}$$

and using Identity 1 of Appendix 1 this gives

$$\delta I\tilde{P}(\mathbf{x}) = -2\alpha(\mathbf{x})_n \int_0^T dt \ u^{mm}(\mathbf{x},t;\mathbf{x}_s)_n \delta\Psi^{kk}(\mathbf{x},t;\mathbf{x}_s) \ . \tag{78}$$

Given the vector $\delta\tilde{\mathbf{u}}$ , to compute $\mathbf{A}_n^t\delta\tilde{\mathbf{u}}$ , we can then use (78), where $\delta\Psi$ is defined by (76). But equation (76) should not be used as it stands. I will now show that the field $\delta\Psi$ satisfies the equations

$$\rho(\mathbf{x})_n \frac{\partial^2\delta\Psi^i}{\partial t^2}(\mathbf{x},t;\mathbf{x}_s) - \frac{\partial}{\partial x^i}[\lambda(\mathbf{x})_n \delta\Psi^{kk}(\mathbf{x},t;\mathbf{x}_s)]$$

$$-2\frac{\partial}{\partial x^j}[\mu(\mathbf{x})_n \delta\Psi^{ij}(\mathbf{x},t;\mathbf{x}_s)] = 0 \tag{79a}$$

$$\lambda(\mathbf{x})_n \delta\Psi^{kk}(\mathbf{x},t;\mathbf{x}_s) n^i(\mathbf{x}) + 2\mu(\mathbf{x})_n \delta\Psi^{ij}(\mathbf{x},t;\mathbf{x}_s) n^j(\mathbf{x}) = \tag{79b}$$

$$= \sum_r \delta(\mathbf{x}-\mathbf{x}_r)\delta\tilde{u}^i(\mathbf{x}_r,t;\mathbf{x}_s) \qquad \mathbf{x}\in \mathbf{S}$$

$$\delta\Psi^i(\mathbf{x},T;\mathbf{x}_s) = 0 \tag{79c}$$

$$\delta\dot{\Psi}^i(\mathbf{x},T;\mathbf{x}_s) = 0 \ , \tag{79d}$$

where it should be noticed that $\delta\Psi^i(\mathbf{x},t;\mathbf{x}_s)$ satisfies homogeneous *final* conditions (79c-d), instead of initial conditions. The "source" of the field $\delta\Psi^i(\mathbf{x},t,\mathbf{x}_s)$ is $\delta\tilde{u}^i(\mathbf{x}_r,t;\mathbf{x}_s)$ , acting as if it was a traction (79b). Using the representation theorem (49), whith reversed time, shows directly that $\delta\Psi^i(\mathbf{x},t;\mathbf{x}_s)$ satisfies equation (76).

The field $\delta\Psi^i(\mathbf{x},t;\mathbf{x}_s)$ can for instance be numerically obtained using a finite-difference code, whith the time running backwards from $t = T$ to $t = 0$ , and where, for a given shot point $\mathbf{x}_s$ , we consider virtual sources, one at each receiver, radiating the weighted residuals backwards in time. See Gauthier et al. (1986) for a numerical implementation in an acoustic example.

*i-2) S-wave impedance.* Analogously, the Fréchet derivative of displacements with respect to the S-wave impedance is the linear operator **B** that to an arbitrary perturbation $\delta\mathbf{IS}$ associates the displacement

$$\delta\mathbf{u} = \mathbf{B}\delta\mathbf{IS} \ , \tag{80}$$

i.e.,

$$\delta u^i(\mathbf{x}_r,t;\mathbf{x}_s) = \int_{\mathbf{V}} dV(\mathbf{x}) B^i_0(\mathbf{x}_r,t,\mathbf{x}_s \,|\, \mathbf{x}) \delta IS(\mathbf{x}) \ , \tag{69}$$

where $B^i(\mathbf{x}_r,t,\mathbf{x}_s \,|\, \mathbf{x})_n$ is given by equation (70). The operator $\mathbf{B}^t_n$ to any $\delta\tilde{\mathbf{u}}$ will associate a $\delta I\tilde{S}$ given by

$$\delta\tilde{\mathbf{IS}} = \mathbf{B}^t \delta\tilde{\mathbf{u}} \ , \tag{81}$$

i.e.,

$$\delta I\tilde{S}(\mathbf{x}) = \sum_r \int_0^T dt B^i(\mathbf{x}_r,t,\mathbf{x}_s \,|\, \mathbf{x})_n \delta\tilde{u}^i(\mathbf{x}_r,t;\mathbf{x}_s) \ . \tag{82}$$

This gives

$$\delta I\tilde{S}(\mathbf{x}) = -4\beta(\mathbf{x})_n \sum_r \int_0^T dt \left[ \frac{\partial\Gamma^{ij}}{\partial x^m}(\mathbf{x}_r,t;\mathbf{x},0)_n * u^{jm}(\mathbf{x},t;\mathbf{x}_s)_n \right.$$
$$\left. - \frac{\partial\Gamma^{ij}}{\partial x^j}(\mathbf{x}_r,t;\mathbf{x},0)_n * u^{mm}_0(\mathbf{x},t;\mathbf{x}_s)_n \right] \delta\tilde{u}^i(\mathbf{x}_r,t;\mathbf{x}_s) \ . \tag{83}$$

Using identities 1 and 2 of appendix 1, this gives

$$\delta I\tilde{S}(\mathbf{x}) = -4\beta(\mathbf{x})_n \int_0^T dt \ [u^{km}(\mathbf{x},t;\mathbf{x}_s)_n \delta\Psi^{km}(\mathbf{x},t;\mathbf{x}_s) - u^{mm}(\mathbf{x})_n \delta\Psi^{kk}(\mathbf{x},t;\mathbf{x}_s)] \tag{84}$$

where the field $\delta\Psi$ has been defined by equations (78).

*i-3) Density.* Finally, the Fréchet derivative of displacements with respect to the density is the linear operator **C** that to an arbitrary perturbation $\delta\rho$ associates the displacement

$$\delta\mathbf{u} = \mathbf{C}\delta\rho \ , \tag{85}$$

i.e.,

$$\delta u^i(\mathbf{x}_r,t;\mathbf{x}_s) = \int_V dV(\mathbf{x}) C_0^i(\mathbf{x}_r,t,\mathbf{x}_s \mid \mathbf{x}) \delta\rho(\mathbf{x}) \ , \tag{69}$$

where $C_0^i(\mathbf{x}_r,t,\mathbf{x}_s \mid \mathbf{x})$ is given by equation (71). The operator $\mathbf{C}_n^t$ to any $\delta\tilde{\mathbf{u}}$ will associate a $\delta\tilde{\rho}$ given by

$$\delta\tilde{\rho} = \mathbf{C}^t \delta\tilde{\mathbf{u}} \ , \tag{86}$$

i.e.,

$$\delta\tilde{\rho}(\mathbf{x}) = \sum_r \int_0^T dt \ C^i(\mathbf{x}_r,t,\mathbf{x}_s \mid \mathbf{x})_n \delta\tilde{u}^i(\mathbf{x}_r,t;\mathbf{x}_s) \ . \tag{87}$$

This gives

$$\delta\tilde{\rho}(\mathbf{x}) = \sum_r \int_0^T dt \left[ -\Gamma^{ij}(\mathbf{x}_r,t;\mathbf{x},0)_n * \ddot{u}^j(\mathbf{x},t;\mathbf{x}_s)_n + (\alpha^2(\mathbf{x})_n - 2\beta^2(\mathbf{x})_n) \frac{\partial\Gamma^{ij}}{\partial x^j}(\mathbf{x}_r,t;\mathbf{x},0)_n * u^{mm}(\mathbf{x},t;\mathbf{x}_s)_n \right.$$

$$\left. + 2\beta^2(\mathbf{x})_n \frac{\partial\Gamma^{ij}}{\partial x^m}(\mathbf{x}_r,t;\mathbf{x},0)_n * u^{jm}(\mathbf{x},t;\mathbf{x}_s)_n \right] \delta\tilde{u}^i(\mathbf{x}_r,t;\mathbf{x}_s) \tag{88}$$

Using identities 1, 2, and 3 of appendix 1, this gives

$$\delta\tilde{\rho}(\mathbf{x}) = \int_0^T dt \left[ \dot{u}^i(\mathbf{x},t;\mathbf{x}_s)_n \delta\dot{\Psi}^i(\mathbf{x}) + (\alpha^2(\mathbf{x})_n - 2\beta^2(\mathbf{x})_n) u^{mm}(\mathbf{x},t;\mathbf{x}_s)_n \delta\Psi^{kk}(\mathbf{x},t;\mathbf{x}_s) \right.$$

$$\left. + 2\beta^2(\mathbf{x})_n u^{km}(\mathbf{x},t;\mathbf{x}_s)_n \delta\Psi^{km}(\mathbf{x},t;\mathbf{x}_s) \right] \ , \tag{89}$$

where the field $\delta\Psi$ has been defined by equations (78).

*j ) Methods of resolution.* My personal experience in the present problem suggests the following strategy. First, as for all nonlinear problems, it is important to start iterating at a point as close as possible to the final solution, for the nonlinearity of the problem to be as small as possible. In the present context, this means to start from a model for which the data residuals can be explained as well as possible using Born's approximation, i.e., using a model for which the low spatial frequencies of the P-wave and S-wave velocities are as good as possible.

The three parameters $IP(\mathbf{x})$ , $IS(\mathbf{z})$ , and $\rho(\mathbf{x})$ have been chosen to be as independent as possible. Furthermore, the importance of these parameters is very different. Most of the data features can be explained with P-waves alone. This suggests to start iterating using a gradient method for the P-wave impedance alone (i.e., maintaining fixed S-wave impedance and density ). This requires a reasonably good model of the low spatial frequency part of the P velocity.

Once a good model $IP(\mathbf{x})$ has been obtained, the remaining data residuals will in particular contain S waves. If a reasonably good model for the long spatial wavelengths of the S velocity can then be obtained, some gradient iterations should be performed to obtain a good model of S-wave impedance. The remaining residuals will contain, if any, some information on the density. Some gradient iterations for the density will end the process.

As the total problem is nonlinear, the entire process should in principle be iterated until convergence. As the chosen parameters are acceptably independent, I hope the first model obtained after a single loop will be good enough (if the long spatial wavelengths of the P-wave and S-wave velocity in the starting model are right ).

The starting models of the low spatial frequency part of the P velocity and S velocity have to be obtained using an independent method (not discussed here ). It is not clear at present into which extent the gradient iterations for $IP(\mathbf{x})$ , $IS(\mathbf{x})$ , and $\rho(\mathbf{x})$ will be able to correct for the imperfections of these velocity models. I am not very optimistic on that, because preliminary results on nonlinear inversion for one dimensional models suggest that the number of iterations need to modify the low spatial frequencies of the model using gradient methods may be enormous.

In what follows, it is assumed that an Earth model ($\rho(\mathbf{x})_n$ , $IP(\mathbf{x})_n$ , $IS(\mathbf{x})_n$) that contains the long spatial wavelength component of the P-wave and S-wave velocities is given, and I will discuss how to ameliorate this model, i.e., how to obtain a model with lower value of the functional (35). From the previous discussion, it follows that I can separately discuss the problem of ameliorating the P-wave impedance model, the S-wave impedance model, and the density model.

*k ) Optimization of the P-wave impedance.* Denote by $\rho(\mathbf{x})_n$ , $IP(\mathbf{x})_n$ , and $IS(\mathbf{x})_n$ the model already obtained, which we wish to further optimize for the impedance $IP(\mathbf{x})$ . I will use a gradient iterative method which will give models $IP(\mathbf{x})_{n+1}$ , $IP(\mathbf{x})_{n+2}$ , ...

Using the steepest descent algorithm (9)-(10) and the results already obtained for the transpose operators, we obtain the equations corresponding to an iteration of the steepest descent method for the P-wave impedance:

$$\delta \tilde{u}^i(\mathbf{x}_r t;\mathbf{x}_s)_n = \frac{u^i(\mathbf{x}_r,t;\mathbf{x}_s)_n - u^i(\mathbf{x}_r,t;\mathbf{x}_s)_{obs}}{\sigma^2(\mathbf{x}_r t;\mathbf{x}_s)} \tag{90a}$$

$$\delta \Psi^i(\mathbf{x},t;\mathbf{x}_s)_n = \sum_r \Gamma^{ij}(\mathbf{x},0;\mathbf{x}_r,s)_n * \delta \tilde{u}^j(\mathbf{x}_r,t;\mathbf{x}_s)_n \tag{90b}$$

$$\delta \tilde{IP}(\mathbf{x})_n = -2\alpha(\mathbf{x})_n \sum_s \int_0^T dt\, u^{ii}(\mathbf{x},t;\mathbf{x}_s)_n \delta \Psi^{jj}(\mathbf{x},t;\mathbf{x}_s)_n \tag{90c}$$

$$\delta IP(x,y,z)_n = K_p \int_0^{Z_{max}} dz' Min(z,z') \delta \tilde{IP}(x,y,z')_n \tag{90d}$$

$$IP(\mathbf{x})_{n+1} = IP(\mathbf{x})_n - \mu_n [\delta IP(\mathbf{x})_n + IP(\mathbf{x})_n - IP(\mathbf{x})_{prior}] \ , \tag{90e}$$

where $\mu_n$ is the real constant which makes $S(\mathbf{IP}_{n+1}, \mathbf{IS}_n, \rho_n)$ minimum.

Let me now turn to the physical interpretation of this result.

- *Equation (90a):* $u^i(\mathbf{x}_r,t;\mathbf{x}_s)_n$ are the data predicted for the model $\mathbf{IP}_n$ , $\mathbf{IS}_n$ , $\rho_n$ . Its effective computation requires a numerical resolution of the system (44). $\sigma^2(\mathbf{x}_r,t;\mathbf{x}_s)$ represents the (squared ) estimated error at time t in the i-th component of the dispacement measured at $\mathbf{x}_r$ , for the source at $\mathbf{x}_s$ . Then $\delta \tilde{u}^i(\mathbf{x}_r,t;\mathbf{x}_s)$ clearly represent the weighted residuals.

- *Equation (90b):* $\Gamma^{ij}(\mathbf{x},t;\mathbf{x}_r,0)_n$ is the Green's function for the model $\mathbf{IP}_n$ , $\mathbf{IS}_n$ , $\rho_n$ . We have already seen that equation (90a) has not to be used as it stands. Instead, the field $\delta\Psi_n$ has to be obtained solving numerically the system (79) (backwards in time ). As $\delta\Psi^i(\mathbf{x},t;\mathbf{x}_s)_n$ is obtained when propagating *backwards in time* the wheighted residuals $\delta\tilde{u}^i(\mathbf{x}_r,t;\mathbf{x}_s)$ it can be intuitively interpreted as a "current missing field".

- *Equation (90c):* This is the most important of the equations, because inversion is being performed here. After some corrections (90d-e-f ), $\delta I\tilde{P}(\mathbf{x})_n$ will essentially be the correction to be applied to $\mathbf{IP}_n$ for obtaining $\mathbf{IP}_{n+1}$ (as shown by (90f )). Equation (90c) shows that this correction at a given point $\mathbf{x}$ , for given shot $\mathbf{x}_s$ , equals the time correlation of the dilatation $u^{ii}(\mathbf{x},t;\mathbf{x}_s)_n$ of the current predicted field with the dilatation $\delta\Psi^{jj}(\mathbf{x},t;\mathbf{x}_s)_n$ of the current missing field. The physical interpretation is as follows: if for a given source point $\mathbf{x}_s$ , and at a given point $\mathbf{x}$ , the dilatation of the current predicted field is time correlated with the dilatation of the missing field, we should create this missing field by adding a P impedance diffractor at point $\mathbf{x}$ . This interpretation is strikingly similar to the imaging principle of Claerbout (1971), but is here in an elastic context and results from a very general optimization criterion.

- *Equation (90d):* The "migrated" field $\delta I\tilde{P}(\mathbf{x})_n$ is here operated with the covariance operator incorporating a priori information. Here I have chosen the kernel Min(z,z') corresponding to the hypothesis that real impedance sequences look like random walks. The sum here essentially corresponds to taking twice the primitive of $\delta I\tilde{P}(x,y,z)_n$ with respect to z . The parameter $K_p$ will control the trade off between the importance of the a priori information and the information obtained from the data set.

- *Equation (90e):* The new model $IP(\mathbf{x})_n$ is obtained here. An optimum value of $\mu_n$ (for which the costs function is minimum ) is obtained by trial and error.

Each iteration corresponds to a sort of generalized elastic "prestack" migration. Hopefully a few iterations will suffice (if we do not wish to improve the long spatial wavelengths ). Readers not interested in inversion, but only in prestack migration, may consider these equations as a serious candidate for replacing acoustic migration equations.

*l ) Optimization of the S-wave impedance.* Turning now to S-wave impedance, denote by $\rho(\mathbf{x})_n$ , $IP(\mathbf{x})_n$ , and $IS(\mathbf{x})_n$ the model already obtained, which we wish to further optimize for the impedance $IS(\mathbf{x})$ . The gradient iterative method will give models $IS(\mathbf{x})_{n+1}$ , $IS(\mathbf{x})_{n+2}$ , ...

Using the steepest descent algorithm (9)-(10) and the results already obtained for the transpose operators, we obtain the equations corresponding to an iteration of the steepest descent method for the S-wave impedance:

$$\delta\tilde{u}^i(\mathbf{x}_r t;\mathbf{x}_s)_n = \frac{u^i(\mathbf{x}_r,t;\mathbf{x}_s)_n - u^i(\mathbf{x}_r,t;\mathbf{x}_s)_{obs}}{\sigma^2(\mathbf{x}_r t;\mathbf{x}_s)} \tag{91a}$$

$$\Psi^i(\mathbf{x},t;\mathbf{x}_s)_n = \sum_r \Gamma^{ij}(\mathbf{x},0;\mathbf{x}_r,t;\mathbf{x}_s)_n * \delta\tilde{u}^j(\mathbf{x}_r,t;\mathbf{x}_s)_n \tag{91b}$$

$$\delta I\tilde{S}(\mathbf{x})_n = -4\beta(\mathbf{x})_n \sum_s \int_0^T dt \, [u^{km}(\mathbf{x},t;\mathbf{x}_s)_n \Psi^{km}(\mathbf{x},t;\mathbf{x}_s)_n - u^{ii}(\mathbf{x},t;\mathbf{x}_s)_n \Psi^{jj}(\mathbf{x},t;\mathbf{x}_s)_n] \quad (91c)$$

$$\delta IS(x,y,z)_n = K_s \int_0^{Z_{max}} dz' \, Min(z,z') \, \delta I\tilde{S}(x,y,z')_n \quad (91d)$$

$$IS(\mathbf{x})_{n+1} = IS(\mathbf{x})_n - \mu_n (\delta IS(\mathbf{x})_n + IS(\mathbf{x})_n - IS(\mathbf{x})_{prior}) \ , \quad (91e)$$

where $\mu_n$ is the real constant which makes $S(\mathbf{IP}_n, \mathbf{IS}_{n+1}, \rho_n)$ minimum. The physical interpretation is as for the P-wave impedance.

*m ) Optimization of the density.* Finally, turning to the density, denote by $\rho(\mathbf{x})_n$ , $IP(\mathbf{x})_n$ , and $IS(\mathbf{x})_n$ the model already obtained, which we wish to further optimize for the density $\rho(\mathbf{x})$ . The gradient iterative method will give models $\rho(\mathbf{x})_{n+1}$ , $\rho(\mathbf{x})_{n+2}$ , ...

Using the steepest descent algorithm (9)-(10) and the results already obtained for the transpose operators, we obtain the equations corresponding to an iteration of the steepest descent method for the density:

$$\delta\tilde{u}^i(\mathbf{x}_r t;\mathbf{x}_s)_n = \frac{u^i(\mathbf{x}_r,t;\mathbf{x}_s)_n - u^i(\mathbf{x}_r,t;\mathbf{x}_s)_{obs}}{\sigma^2(\mathbf{x}_r t;\mathbf{x}_s)} \quad (92a)$$

$$\Psi^i(\mathbf{x},t;\mathbf{x}_s)_n = \sum_r \Gamma^{ij}(\mathbf{x},0;\mathbf{x}_r,t)_n * \delta\tilde{u}^j(\mathbf{x}_r,t;\mathbf{x}_s)_n \quad (92b)$$

$$\delta\tilde{\rho}(\mathbf{x})_n = \sum_s \int_0^T dt \Big[ \dot{u}^i(\mathbf{x},t;\mathbf{x}_s)_n \dot{\Psi}^i(\mathbf{x},t;\mathbf{x}_s)_n + (\alpha^2(\mathbf{x})_n - 2\beta^2(\mathbf{x})_n) u^{ii}(\mathbf{x},t;\mathbf{x}_s)_n \Psi^{jj}(\mathbf{x},t;\mathbf{x}_s)_n$$
$$+ 2\beta^2(\mathbf{x})_n u^{km}(\mathbf{x},t;\mathbf{x}_s)_n \Psi^{km}(\mathbf{x},t;\mathbf{x}_s)_n \Big] \quad (92c)$$

$$\delta\rho(x,y,z)_n = K_p \int_0^{Z_{max}} dz' \, Min(z,z') \, \delta\tilde{\rho}(x,y,z')_n \quad (92d)$$

$$\rho(\mathbf{x})_{n+1} = \rho(\mathbf{x})_n - \mu_n (\delta\rho(\mathbf{x})_n + \rho(\mathbf{x})_n - \rho(\mathbf{x})_{prior}) \ , \quad (92e)$$

where $\mu_n$ is the real constant which makes $S(\mathbf{IP}_n, \mathbf{IS}_n, \rho_{n+1})$ minimum. The physical interpretation is as for the P-wave impedance.

**Acknowledgements**

I thank our sponsors CGG, IFP, SCHLUMBERGER, SNEA, and TOTAL. This work has also been supported by CNRS (ATP Tomographie Géophysique INSU)

## Appendix

I demonstrate here the following identities:

$$\sum_r \int_0^T dt \, \frac{\partial \Gamma_0^{ij}}{\partial x^j}(\mathbf{x}_r, t; \mathbf{x}, 0) * u_0^{mm}(\mathbf{x}, t) \delta \bar{u}^i(\mathbf{x}_r, t) = \int_0^T dt \, u_0^{mm}(\mathbf{x}, t) \Psi_0^{kk}(\mathbf{x}) \ , \tag{A1}$$

$$\sum_r \int_0^T dt \, \frac{\partial \Gamma_0^{il}}{\partial x^m}(\mathbf{x}_r, t; \mathbf{x}, 0) * u_0^{lm}(\mathbf{x}, t) \delta \bar{u}^i(\mathbf{x}_r, t) = \int_0^T dt \, u_0^{lm}(\mathbf{x}, t) \Psi_0^{lm}(\mathbf{x}, t) \ , \tag{A2}$$

and

$$\sum_r \int_0^T dt \, \Gamma_0^{ij}(\mathbf{x}_r, t; \mathbf{x}, 0) * \ddot{u}_0^j(\mathbf{x}, t) \delta \bar{u}^i(\mathbf{x}_r, t) = -\int_0^T dt \, \dot{u}_0^i(\mathbf{x}, t) \dot{\Psi}_0^i(\mathbf{x}, t) \ , \tag{A3}$$

where $\Psi_0^i(\mathbf{x},t)$ is defined by

$$\Psi_0^i(\mathbf{x}, t) = \sum_r \Gamma_0^{ij}(\mathbf{x}, 0; \mathbf{x}_r, t) * \delta \bar{u}^j(\mathbf{x}_r, t) \ , \tag{A4}$$

and where $\Psi_0^{ij}(\mathbf{x},t)$ is the associated strain :

$$\Psi_0^{ij}(\mathbf{x}, t) = \frac{1}{2} \left\{ \frac{\partial \Psi_0^i}{\partial x^j}(\mathbf{x}, t) + \frac{\partial \Psi_0^j}{\partial x^i}(\mathbf{x}, t) \right\} \ . \tag{A5}$$

*Demonstration:* From equations (A4)-(A5) we have

$$\Psi_0^{ik}(\mathbf{x}, t) = \sum_r \frac{1}{2} \left\{ \frac{\partial \Gamma_0^{ij}}{\partial x^k}(\mathbf{x}, 0; \mathbf{x}_r, t) + \frac{\partial \Gamma_0^{kj}}{\partial x^i}(\mathbf{x}, 0; \mathbf{x}_r, t) \right\} * \delta \bar{u}^j(\mathbf{x}_r, t)] \ ,$$

and, using the reciprocity property of the Green's function,

$$\Psi_0^{ik}(\mathbf{x}, t) = \sum_r \frac{1}{2} \left\{ \frac{\partial \Gamma_0^{ji}}{\partial x^k}(\mathbf{x}_r, 0; \mathbf{x}, t) + \frac{\partial \Gamma_0^{jk}}{\partial x^i}(\mathbf{x}_r, 0; \mathbf{x}, t) \right\} * \delta \bar{u}^j(\mathbf{x}_r, t) \ . \tag{A6}$$

For the trace $\Psi_0^{kk}(\mathbf{x},t)$ this gives

$$\Psi_0^{kk}(\mathbf{x}, t) = \sum_r \frac{\partial \Gamma_0^{jk}}{\partial x^k}(\mathbf{x}_r, 0; \mathbf{x}, t) * \delta \bar{u}^j(\mathbf{x}_r, t) \ . \tag{A7}$$

For the time derivative $\dot{\Psi}_0^i(\mathbf{x},t)$ we successively have

$$\dot{\Psi}_0^i(\mathbf{x}, t) = \frac{\partial}{\partial t} \sum_r \Gamma_0^{ij}(\mathbf{x}, 0; \mathbf{x}_r, t) * \delta \bar{u}^j(\mathbf{x}_r, t)$$

$$= \frac{\partial}{\partial t} \sum_r \int_0^T dt' \, \Gamma_0^{ij}(\mathbf{x}, 0; \mathbf{x}_r, t - t') * \delta \bar{u}^j(\mathbf{x}_r, t')$$

$$= \frac{\partial}{\partial t} \sum_r \int_0^T dt' \, \Gamma_0^{ij}(\mathbf{x}, t' - t; \mathbf{x}_r, 0) * \delta \bar{u}^j(\mathbf{x}_r, t')$$

$$= - \sum_r \int_0^T dt' \, \dot{\Gamma}_0^{ij}(\mathbf{x}, t' - t; \mathbf{x}_r, 0) * \delta \bar{u}^j(\mathbf{x}_r, t')$$

$$= - \sum_r \int_0^T dt' \; \dot{\Gamma}_0^{ij}(\mathbf{x},0;\mathbf{x}_r,t-t') * \delta\bar{u}^j(\mathbf{x}_r,t')$$

$$= - \sum_r \dot{\Gamma}_0^{ij}(\mathbf{x},0;\mathbf{x}_r,t) * \delta\bar{u}^j(\mathbf{x}_r,t) \ , \tag{A8}$$

and, using the reciprocity property,

$$\dot{\Psi}_0^i(\mathbf{x},t) = - \sum_r \dot{\Gamma}_0^{ji}(\mathbf{x}_r,0;\mathbf{x},t) * \delta\bar{u}^j(\mathbf{x}_r,t) \ . \tag{A9}$$

We successively have

$$\sum_r \int_0^T dt \; \frac{\partial \Gamma_0^{ij}}{\partial x^j}(\mathbf{x}_r,t;\mathbf{x},0) * u_0^{mm}(\mathbf{x},t) \delta\bar{u}^i(\mathbf{x}_r,t) =$$

$$= \sum_r \int_0^T dt \int_0^T dt' \; \frac{\partial \Gamma_0^{ij}}{\partial x^j}(\mathbf{x}_r,t-t';\mathbf{x},0) u_0^{mm}(\mathbf{x},t') \delta\bar{u}^i(\mathbf{x}_r,t)$$

$$= \sum_r \int_0^T dt \int_0^T dt' \; \frac{\partial \Gamma_0^{ij}}{\partial x^j}(\mathbf{x}_r,0;\mathbf{x},t'-t) u_0^{mm}(\mathbf{x},t') \delta\bar{u}^i(\mathbf{x}_r,t)$$

$$= \sum_r \int_0^T dt' \; \frac{\partial \Gamma_0^{ij}}{\partial x^j}(\mathbf{x}_r,0;\mathbf{x},t') * \delta\bar{u}^i(\mathbf{x}_r,t') u_0^{mm}(\mathbf{x},t') \tag{A10}$$

from where, using (A7), Identity (A1) follows.

We successively have

$$\sum_r \int_0^T dt \; \frac{\partial \Gamma_0^{il}}{\partial x^m}(\mathbf{x}_r,t;\mathbf{x},0) * u_0^{lm}(\mathbf{x},t) \delta\bar{u}^i(\mathbf{x}_r,t) =$$

$$= \sum_r \int_0^T dt \int_0^T dt' \; \frac{\partial \Gamma_0^{il}}{\partial x^m}(\mathbf{x}_r,t-t';\mathbf{x},0) u_0^{lm}(\mathbf{x},t') \delta\bar{u}^i(\mathbf{x}_r,t)$$

$$= \sum_r \int_0^T dt \int_0^T dt' \; \frac{\partial \Gamma_0^{il}}{\partial x^m}(\mathbf{x}_r,0;\mathbf{x},t'-t) u_0^{lm}(\mathbf{x},t') \delta\bar{u}^i(\mathbf{x}_r,t)$$

$$= \sum_r \int_0^T dt' \; \frac{\partial \Gamma_0^{il}}{\partial x^m}(\mathbf{x}_r,0;\mathbf{x},t') * \delta\bar{u}^i(\mathbf{x}_r,t) u_0^{lm}(\mathbf{x},t')$$

$$= \sum_r \int_0^T dt' \; \frac{1}{2} \left\{ \frac{\partial \Gamma_0^{il}}{\partial x^m}(\mathbf{x}_r,0;\mathbf{x},t') + \frac{\partial \Gamma_0^{im}}{\partial x^l}(\mathbf{x}_r,0;\mathbf{x},t') \right\} * \delta\bar{u}^i(\mathbf{x}_r,t) u_0^{lm}(\mathbf{x},t') \ , \tag{A11}$$

the last equality being due to the symmetry of the strain tensor ($u_0^{lm}(\mathbf{x},t) = u_0^{ml}(\mathbf{x},t)$). Using (A6), Identity (A2) follows.

For Identity (A3), we need first to obtain an intermediate result. We have

$$\Gamma_0^{ij}(\mathbf{x}_r t,\mathbf{x},0) * \ddot{u}_0^j(\mathbf{x},t) = \int_0^T dt' \; \Gamma_0^{ij}(\mathbf{x}_r t-t',\mathbf{x},0) \ddot{u}_0^j(\mathbf{x},t') \tag{A12}$$

and, integrating by parts,

$$\Gamma_0^{ij}(\mathbf{x}_r,t,\mathbf{x},0)*\ddot{u}^j(\mathbf{x},t)=\Gamma_0^{ij}(\mathbf{x}_r,t-T;\mathbf{x},0)\dot{u}_0^j(\mathbf{x},T)-\Gamma_0^{ij}(\mathbf{x}_r,t;\mathbf{x},0)\dot{u}_0^j(\mathbf{x},0)$$
$$+\int_0^T dt'\,\dot{\Gamma}_0^{ij}(\mathbf{x}_r,t-t',\mathbf{x},0)\dot{u}_0^j(\mathbf{x},t)\ . \qquad \text{(A13)}$$

As the initial conditions (equations 12c-d of the previous problem ) impose $\Gamma_0^{ij}(\mathbf{x}_r,t;\mathbf{x},t')=0$ for $t<t'$, and $\dot{u}_0^j(x,0)=0$, this gives

$$\Gamma_0^{ij}(\mathbf{x}_r,t,\mathbf{x},0)*\ddot{u}_0^j(\mathbf{x},t)=\dot{\Gamma}_0^{ij}(\mathbf{x}_r,t,\mathbf{x},0)*u_0^j(\mathbf{x},t)\ . \qquad \text{(A14)}$$

Now, using this last equality we succesively obtain

$$\sum_r\int_0^T dt\ \Gamma_0^{ij}(\mathbf{x}_r,t;\mathbf{x},0)*\ddot{u}_0^j(\mathbf{x},t)\delta\bar{u}^i(\mathbf{x},t)=$$
$$=\sum_r\int_0^T dt\ \dot{\Gamma}_0^{ij}(\mathbf{x}_r,t;\mathbf{x},0)*\dot{u}_0^j(\mathbf{x},t)\delta\bar{u}^i(\mathbf{x},t)$$
$$=\sum_r\int_0^T dt\int_0^T dt'\ \dot{\Gamma}_0^{ij}(\mathbf{x}_r,t-t';\mathbf{x},0)\dot{u}_0^j(\mathbf{x},t')\delta\bar{u}^i(\mathbf{x}_r,t)$$
$$=\sum_r\int_0^T dt\int_0^T dt'\ \dot{\Gamma}_0^{ij}(\mathbf{x}_r,0;\mathbf{x},t'-t)\dot{u}_0^j(\mathbf{x},t')\delta\bar{u}^i(\mathbf{x}_r,t)$$
$$=\sum_r\int_0^T dt'\ \dot{\Gamma}_0^{ij}(\mathbf{x}_r,0;\mathbf{x},t')*\delta\bar{u}^i(\mathbf{x}_r,t')\dot{u}_0^j(\mathbf{x},t')\ , \qquad \text{(A15)}$$

from where, using (A9), identity (A3) follows.

*Chapter 7*

# Crosshole transmission tomography

*S. Ivansson*

## 1. Introduction

### 1.1 Crosshole seismics in general

The seismic crosshole technique offers a means to investigate the rock mass between two or more boreholes. Already in 1917 Reginald Fessenden proposed a method to locate ore bodies by crosshole measurements. Fig. 1 is redrawn from his original paper (Fessenden 1917). It is a plan view showing four vertical boreholes and two ore bodies in between. Using a source of seismic energy in one of the boreholes and a detective device in one of the others, the traveltime of waves that have been reflected at or transmitted through the ore bodies can be determined. Combining traveltimes from a number of source/receiver locations at different depths in different boreholes, it should be possible to roughly locate the ore bodies by hand interpretation using elementary geometry. This is the basic idea as outlined by Fessenden. Although the procedure requires a fairly simple medium (only one or two ore bodies, homogeneous surroundings), it points forward to the present use of crosshole tomography.

Among recent applications of crosshole seismics we may note Fehler and Pearson (1984). These authors made use of amplitude measurements to estimate the quality factor Q and locate fractures in crystalline rock at a hot dry rock geothermal reservoir. During heat extraction a decrease in average Q was noted due to extensive fracturing. A water-filled fracture will affect waveform character, frequency, and also amplitude. Shear wave amplitudes, in particular, will be greatly reduced. These effects constitute the basis of fracture location techniques.



*G. Nolet (ed.), Seismic Tomography, 159–188.*

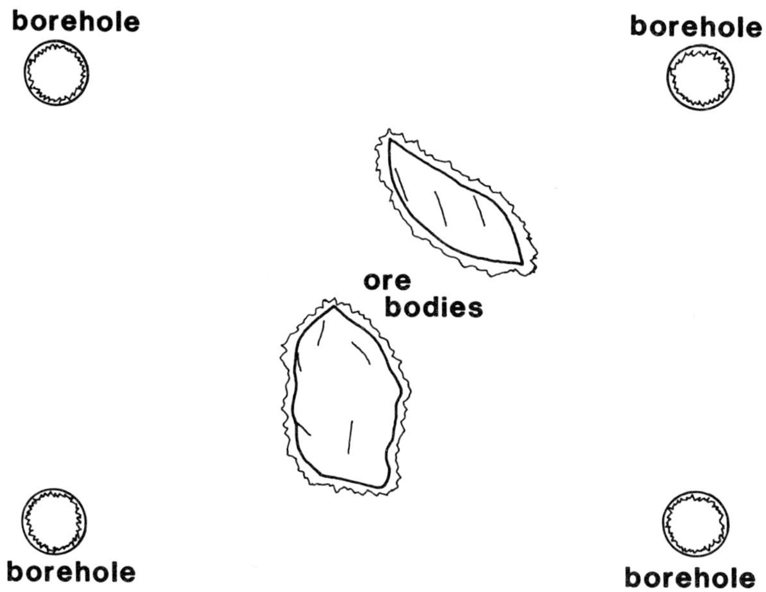


*Figure 1.* Plan view showing four boreholes, and two ore bodies.

Paulsson et al. (1985) reported successful monitoring of rock parameters in a small-scale crosshole experiment, the distance between the holes being a few m only. Electric heaters were used to simulate the thermal effect of nuclear waste. P-wave velocities were found to increase linearly with temperature. Changes in attenuation of the seismic waves were shown to be indicative of fracture closure and pore pressure changes during the heating process. Thermal damage to the rock mass resulted in permanently reduced P-wave velocities.

Ultrasonic crosshole measurements in a Swedish iron mine were performed by Nordqvist (1986). Rock mass classification was carried out by using the P-wave velocity, attenuation, and signal duration measures. Attenuation and signal duration proved to be more sensitive to joint frequency than P-wave velocity. Signal duration, a new parameter, was defined as the quotient between two root-mean-square (rms) values of the signal, taken over intervals some time after and immediately after the first-arrival, respectively. Joints and other discontinuities will give rise to a large signal duration because of later arrivals caused by scattering, reflections, and wave conversion. Nordqvist also showed that structural changes in the rock, caused by blasts or a changing stress field, could be effectively monitored by crosshole measurements. Applications to civil engineering were described by McCann et al. (1986). In one of their field examples, a railway tunnel between two boreholes was shown to have a significant effect on the velocity of propagation and amplitude of the transmitted seismic signals. Fig. 2 shows the tunnel and

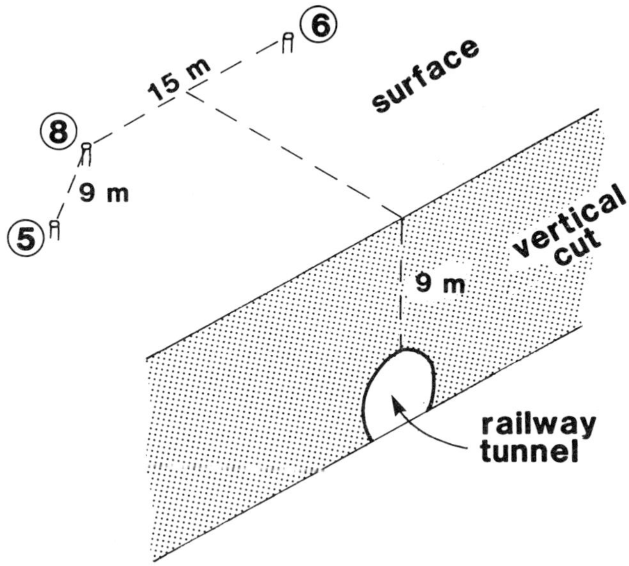


*Figure 2.* A test site for cavity location. The known cavity is an abandoned railway tunnel. The three numbers indicate boreholes. (From McCann et al. 1986.)

three boreholes (Nos. 8, 6, 5) which were used for crosshole measurements with source and receiver at the same depth. Apparent P-wave velocity as a function of depth for two crosshole scans is given in Fig. 3, note the delay introduced by the tunnel. The decrease in amplitude due to the tunnel is similarly shown in Fig. 4.

In another example, McCann et al. showed how crosshole measurements of the same type, performed between boreholes surrounding a building, could be used in conjunction with geologic information to delineate the boundaries of a fracture zone in the rock mass underlying the building.

### 1.2 Incorporating the tomographic formalism

We have seen that crosshole measurements are useful for studying properties of a rock volume situated between boreholes. The papers referred to above do not, however, make full use of the possibility to exactly locate zones of weak rock etc. in the crosshole region. For this, a more complete scan, using a vast number of source/receiver combinations, is needed. It is also necessary to make use of geophysical inverse theory, geotomography in particular.

Let us describe the tomographic formalism for the problem of determining the seismic P-wave slowness (reciprocal velocity) distribution within a crosshole region from first-arrival traveltimes. In Fig. 5 two typical crosshole geometries are shown. Case (a) concerns two vertical boreholes driven into the ground from the surface. The seismic source can be located at various positions in one of the boreholes and receivers are

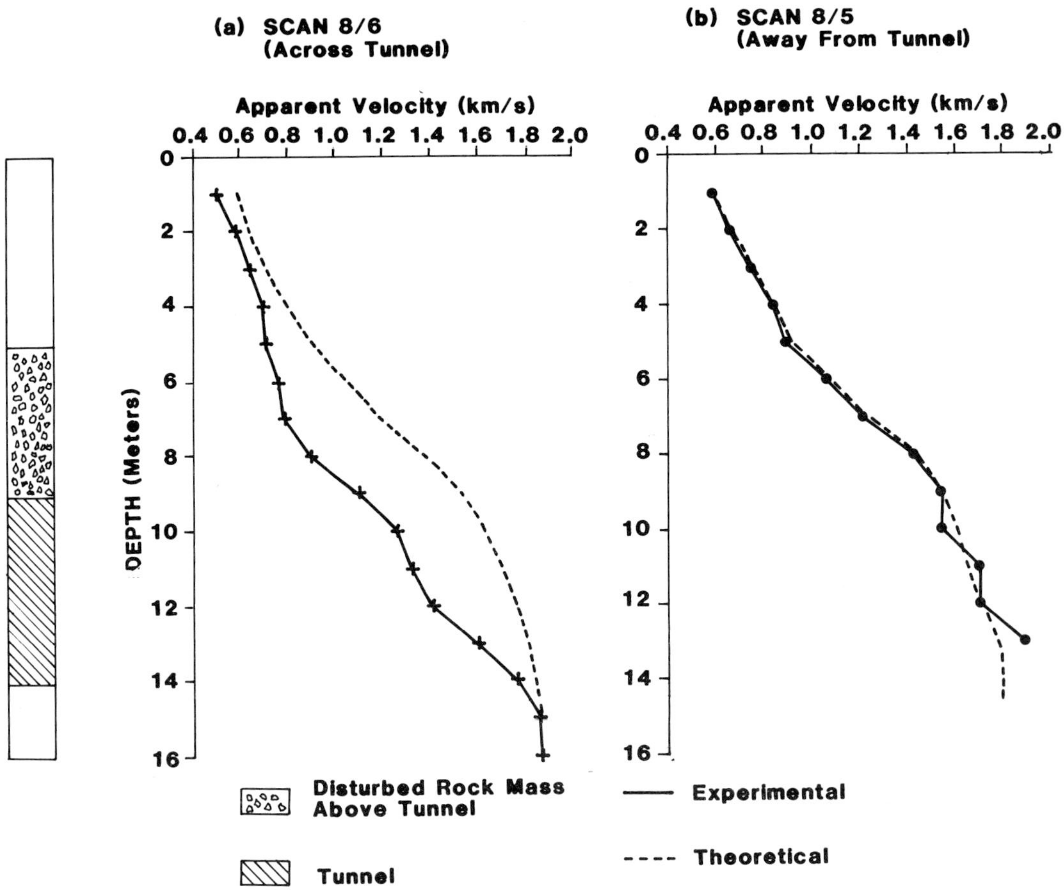


*Figure 3.* Crosshole P-wave velocity data showing the delay introduced by the tunnel. The theoretic curves were constructed using ray tracing on a velocity structure derived from seismic refraction surveys made parallel to and off the tunnel axis. (From McCann et al. 1986.)

deployed in the opposite borehole and on the surface. In case (b) a fan geometry is shown. It can be used in a mine, for example. Two boreholes are drilled, starting from approximately the same point in a gallery, into the rock in front. The source is placed in one of the boreholes and a movable receiver chain in the other.

Using an explosive source of reasonable strength, the energy can be recorded at a distance of several hundred meter in granite rock. Even with a less powerful source, like a hammer, a sparker, or a piezoelectric device, a distance of one or two hundred meter can be covered if signal stacking techniques are employed.

Now, let $\mathbf{x} = (x_1, x_2)$ denote coordinates within the crosshole region and let $m(\mathbf{x})$ be the unknown slowness function (m for model). Suppose there are N data values, i.e., first-

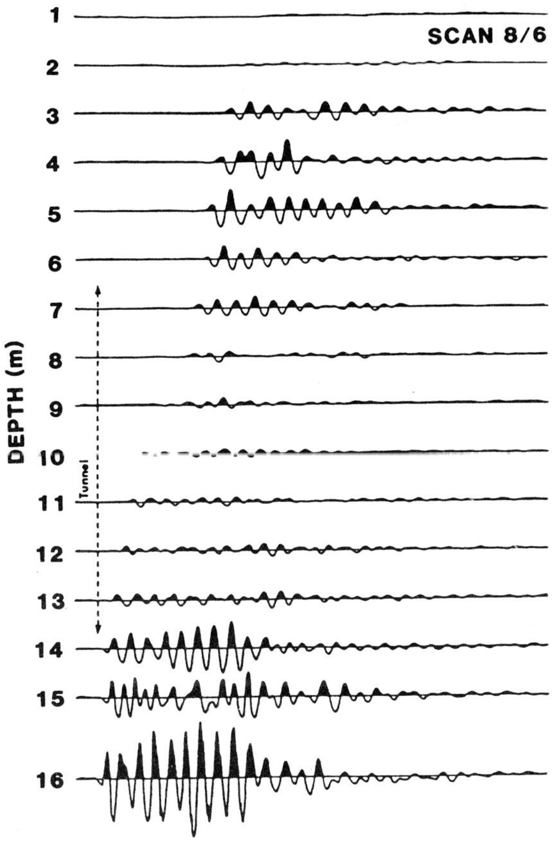


*Figure 4*. Crosshole P-wave signals. Note the decrease in amplitude introduced by the tunnel. (From McCann et al. 1986.)

arrival traveltimes, which we denote by $d_i$, $i = 1,2,..,N$ . They are obtained for different source/receiver locations. Each $d_i$ may be written (high-frequency approximation)

$$d_i = g_i(m) \equiv \int_{T_i(m)} m(\mathbf{x})\, ds \tag{1}$$

where $g_i$ is a nonlinear functional, ds is arc length, and $T_i(m)$ denotes the curve, connecting source and receiver, which yields the least possible traveltime value (cf. Fermat's principle). The nonlinearity is due to the complex dependence of $T_i(m)$ upon $m$. With the obvious vector notation we will also write $\mathbf{d} = \mathbf{g}(m)$ . Characteristic of geophysical tomography is that the data are connected to the model by curve integrals.

With a suitable choice of space for slowness models m, the functionals $g_i$ will become Frechet differentiable. This follows, using a physical argument, from Fermat's principle

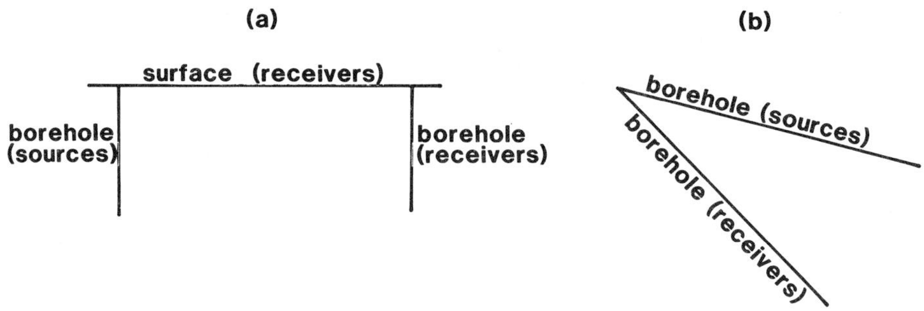


*Figure 5.* Two typical crosshole geometries.

(see chapter 1). We obtain

$$g_i(m+\delta m) - g_i(m) = \int_{T_i(m)} \delta m(\mathbf{x})\, ds \tag{2}$$

when $\delta m$ is small (in the appropriate norm). It follows that when $m$ is approximately proportional to $\bar{m}$ where $\bar{m}$ is an a priori model, the system $\mathbf{d} = \mathbf{g}(m)$ may be linearized by

$$d_i = \int_{T_i(\bar{m})} m(\mathbf{x})\, ds \tag{3}$$

For a homogeneous $\bar{m}$ the paths $T_i = T_i(\bar{m})$ will be straight. In fact, the straight-ray approximation has been very common in the applications. Solution schemes for the equation system (3) will be discussed later on.

The formalism introduced by equations (1)-(3) is also applicable to amplitude tomography for determination of the variation within the crosshole region of the quality factor, Q. Let $Q(\mathbf{x})$ and $c(\mathbf{x})$ denote the Q and velocity functions, respectively. If $\omega$ is angular frequency and $A(\omega)$ is the obtained amplitude spectrum at a certain receiver, corrected for source spectrum, geometric spreading, and instrumental response, we have

$$A(\omega) = \exp\left[-\frac{\omega}{2}\int_T \frac{ds}{Q(\mathbf{x})c(\mathbf{x})}\right] \tag{4}$$

Here T denotes the ray path. Taking logarithms one obtains

$$-\frac{2}{\omega}\ln A(\omega) = \int_T \left[\frac{1}{Q(\mathbf{x})c(\mathbf{x})}\right] ds \tag{5}$$

With $m(\mathbf{x})$ interpreted as $1/Qc$ the formalism of equations (1)-(3) is still usable. Incorporating amplitude data for a number of source/receiver combinations (and frequencies), the Q structure of the rock can thus be determined by tomography if the velocity structure is known.

It should be noted that amplitude tomography is much more difficult in practice than is traveltime tomography. Complications do certainly arise because of frequency dependence and because energy is sucked out of the wave train by reflection and mode conversion at interfaces. Such losses are not included in a simple line integral of reciprocal Q, nor are the effects of multi-pathing that may increase the signal amplitude at finite frequencies. Furthermore, the corrections for source spectrum, source- and receiver- directivity and geometric spreading may be difficult to make with sufficient accuracy. One might also suspect that with receivers in a borehole the effect of the borehole could be serious. Fortunately, however, Blair (1984) has shown that borehole effects are negligible for wavelengths greater than 10 times the borehole circumference. Even for a borehole of 165 mm diameter in crystalline rock (P-wave velocity about 5000 m/s), this corresponds to frequencies as high as about 1000 Hz.

A few examples dealing with reconstruction of absorbing anomalies were published by Neumann-Denzau and Behrens (1984). The authors concluded that inversion of amplitude decays is indeed more critical than inversion of traveltimes. An attempt to carry out amplitude tomography in a mine was reported by New (1985a). His sources and receivers were located in galleries in the mine. Unfortunately, the signal spectra were severely disturbed by the presence of the mine openings, they were dominantly controlled by local rock discontinuity effects at the source and receiver positions. Thus, the measured amplitudes were not found sensitive to the variations in natural rock condition and were not usable for tomographic inversion. A seismic shadow caused by the presence of a major void was observed, however, and the author concluded that crosshole measurements may not be disadvantaged to such an extent because a borehole will cause much less disturbance to the rock mass than the mine openings. In fact, we will later show examples of reasonably successful amplitude tomography based on the simplistic formalism given above.

Of course, a seismic signal carries much more information than the instant of its first onset and the amplitude spectrum of its initial portion. In principle any measurable quantity that can be connected to some rock parameter by an equation of type (1) could be used for tomographic inversion. It may also be possible to stabilize the usual traveltime or amplitude tomography by incorporating additional information obtained from the registrations. Del Pino and Nur (1985) proposed a method to make use of wave polarization as obtained from three-component geophones.

### 1.3 Areas of application

Seismic crosshole tomography has been used for a number of applications of different kinds. Among these we may mention mineral exploration in mines (Gustavsson et al. 1986), fault detection in coal seams (Mason 1981; Bodoky et al. 1985), stress monitoring in coal mines (Kormendi et al. 1986), cave detection (Vazquez et al. 1985), delineation of a salt dome flank (Peterson et al. 1985), dam investigation (Cottin et al. 1986), and rock investigation in connection with disposal of nuclear fuel waste (Wong et al. 1983, 1984, 1985; Ivansson 1985; Peterson et al. 1985; Gustavsson et al. 1986; Cosma 1986; Hammarstrom et al. 1986). The distance between the boreholes has generally been between 100 and 400 m. Particular examples of crosshole tomographic reconstructions will be given in a later section. Concerning the coal-mine applications we might remark that "cross-gallery" rather than "cross-hole" would be the appropriate word since the sources and receivers are used in galleries. However, the basic principles remain unaffected.

Of particular importance in recent years is the research that has been generated by the search for crystalline rock of high quality, suitable for radioactive waste disposal. It might be of interest to give a brief account of the geologic background and of two such research programmes, one Swedish and one Canadian, where seismic crosshole tomography is included.

Granite rock is considered to be suitable for long-term storage of heat generating radioactive waste from nuclear power plants. Granites are typically very hard, massive, crystalline, igneous rocks. Their great stability means that they are resistant to erosion and weathering, and therefore a repository placed deep (300-700 m) is very unlikely to be disturbed by climatic or geological events. The crystallinity of granite means that the rock itself has a low permeability with groundwater flow being restricted to joints and fissures. Since ground water flow is the most probable mechanism by which radioactivity could be transported back to man, it becomes highly important to develop techniques by which fracture zones in granite can be detected and groundwater flow be studied. It is also necessary to carry out in situ experiments under conditions which closely resemble those of an actual repository.

Following a few years of initial experiments, a major research programme, the International Stripa Project, began in May 1980. The name is derived from the Stripa Mine, an abandoned iron ore mine in central Sweden where most of the experiments have taken place. Apart from Sweden, eight other countries take part in the project. Research is carried out under four headings: hydrogeological investigations of the Stripa granite and migration within fracture systems, hydrogeochemistry of groundwater at the Stripa Mine, behaviour of bentonite clay as a backfilling and sealing material, and detection and characterization of fracture zones in granite by crosshole measurements. Concerning the last heading, seismic, radar, as well as hydraulic testing methodologies are being examined. The tomographic approach has been used with success for both seismic and radar data. In conjunction with radar reflection measurements it has proved helpful for determining the location, extent, thickness, and dominant physical properties of fracture zones.

Apart from participating in the International Stripa Project, Canada has maintained an own research and development programme concerning nuclear fuel waste disposal since the mid-seventies. Geophysics plays a particularly significant role within this program. An overview was given by Soonawala (1984). Seismic crosshole tomography has been successfully exploited at the Canadian Underground Research Laboratory in Manitoba.

## 2. Some basic theoretical considerations

The discussion here will be given in terms of traveltime tomography under the straight-ray approximation.

## 2.1 Uniqueness properties

The great success of medical tomography is largely due to two facts: the propagation paths of the X-ray energy are straight and the object (the human body) can be scanned from all directions. In geophysics the situation is much less satisfactory in both these respects, therefore we cannot expect as detailed images as in medicine. Nevertheless, theoretically some of the usual crosshole measurement geometries are sufficient for determining the seismic velocity distribution uniquely, at least if the straight-ray approximation is valid. This will now be shown using an elegant argument due to Strichartz (1982).

As in section 1.2 $\mathbf{x} = (x_1, x_2)$ denote coordinates within the crosshole region but here we prefer to denote the unknown slowness function by $f(\mathbf{x})$ instead of $m(\mathbf{x})$. Also, it will be advantageous to define $f(\mathbf{x})$ as being zero outside the crosshole region. We are then dealing with a function that vanishes outside a bounded set and the data integrals (cf. equation (3)) do not change if the straight crosshole integration paths are (artificially) extended to infinity.

Consider now a set L of straight lines fulfilling the following three hypotheses:

a. for each line $l$ in L, there is a cone with $l$ at one end such that all lines in this cone are also in L (see Fig. 6a)

b. each line $l$ in L may be parallel-translated to a region where f vanishes in such a way that all the passed parallel-translates of $l$ are also in L (see Fig. 6b)

c. each point outside the region where f is known to vanish is traversed by at least one line in L.

Such a set of lines L will be called a *Strichartz set for f.*

*Proposition 1:* Let L be a Strichartz set for f and suppose that

$$\int_l f(\mathbf{x})\, ds = 0 \tag{6}$$

for all straight lines $l$ in L. Then $f(\mathbf{x})$ vanishes identically. (We assume that f is reasonably smooth although it can be shown by standard approximation arguments that such restrictions are not essential.)

*Proof:* Choose any line $l$ in L. For simplicity we consider first the case when l is the $x_1$-axis and suppose that the translation of (b) can be performed to the line $x_2 = b$ lying in a region where f vanishes. For each $x_2$ between 0 and $b$, choose $h(x_2)$ as the $x_1$-coordinate of the vertex of the cone referred to in (a) for the line at $x_2$ parallel to the $x_1$-axis. Define the function $F(x_2, k)$ by

$$F(x_2, k) = \int f(x_1, x_2 + k(x_1 - h(x_2)))\, dx_1 \tag{7}$$

From the hypothesis (6) we know that F vanishes identically for $x_2$ between 0 and $b$ and some $k$ close to zero. Using (a) we obtain

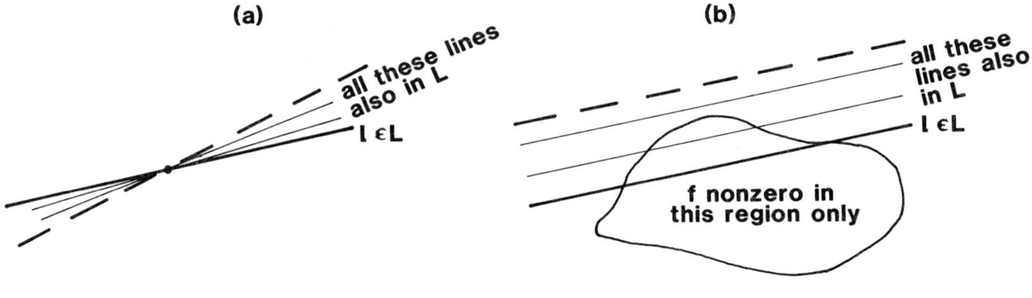


*Figure 6.* Illustration for the definition of a Strichartz set L of straight lines for f.

$$
\begin{aligned}
0 &= \frac{\partial F}{\partial k}(x_2, 0) = \int (x_1 - h(x_2)) \frac{\partial f}{\partial x_2}(x_1, x_2)\, dx_1 \\
&= \frac{\partial}{\partial x_2}\left[\int x_1 f(x_1, x_2)\, dx_1\right] - h(x_2) \frac{\partial}{\partial x_2}\left[\int f(x_1, x_2)\, dx_1\right] \\
&= \frac{\partial}{\partial x_2}\left[\int x_1 f(x_1, x_2)\, dx_1\right]
\end{aligned} \tag{8}
$$

Integrating (8) and using (b) it follows that

$$
0 = -\int_0^b \frac{\partial F}{\partial k}(x_2, 0)\, dx_2 = \int x_1 f(x_1, 0)\, dx_1 \tag{9}
$$

It is obvious that also

$$
0 = \int x_2 f(x_1, 0)\, dx_1 \tag{10}
$$

Realizing that the restriction of $l$ to be the $x_1$-axis is not essential, we conclude that the hypothesis (6) is fulfilled with $f(\mathbf{x})$ replaced by $p(\mathbf{x})f(\mathbf{x})$ where $p(\mathbf{x})$ is any first-order polynomial in the two variables $x_1$ and $x_2$. Iterating the argument we note that $p(\mathbf{x})$ may actually be any polynomial. Considering any line $l$ in L and recalling that the one-dimensional polynomials form a complete set of functions for a closed and bounded interval (Weierstrass' approximation theorem), it follows that f must vanish identically on l. By (c) the proof is complete.

*Proposition 2:* Let L be a Strichartz set for f. Then $f(\mathbf{x})$ is determined uniquely if all the straight-line integrals

$$
\int_l f(\mathbf{x})\, ds \quad , \quad l \in L \tag{11}
$$

are known.

*Proof:* Consider two possible functions $f_1$ and $f_2$ . Apply Proposition 1 to their difference.

Proposition 2 immediately yields results of interest for some typical seismic crosshole geometries:

*Proposition 3:* - Consider Fig. 5(a). If traveltime data are available for all source/receiver combinations (idealized straight paths) when the source is located in the left borehole and the receiver is located higher up in the right borehole or on the shown part of the surface, then the crosshole slowness distribution is uniquely determined.
- Likewise, consider Fig. 5(b). Traveltime data for all source/receiver combinations with the source in one borehole and the receiver in the other suffice in this case for a unique reconstruction.

*Proof:* All that is needed is to verify that the sets of integration paths form Strichartz sets, and this is immediate.

On the other hand, pure crosshole geometries, i.e., as Fig. 5(a) but without receivers on the surface, are not sufficient for a unique reconstruction. A low-velocity strip parallel to the boreholes, for example, will not be discernible since all traveltimes will be increased in the same proportion and the crosshole region will appear as being homogeneous with a low average velocity. Nevertheless, pure crosshole geometries can be quite successful for mapping local velocity anomalies interior to the crosshole region. Again we may refer to Proposition 2, if the slowness function is known a priori close to the surface, for example, we may form a Strichartz set and infer the possibility of a unique reconstruction.

For a few remarks concerning uniqueness when non-straight paths are used, see Ivansson (1986).

### 2.2 Inversion schemes

The uniqueness results presented in the previous section are important but one must remember that in practice it is not enough that there exists a unique solution for an inverse problem to be manageable. The solution must also be a smooth function of the data so that the impact of measurement errors is not too large. In this respect, the crosshole geometries discussed above are not particularly favourable. It would be much better if the whole border of the region could be made accessible which would allow a more complete coverage by ray paths in different directions. Furthermore, in practice the number of measured data values is always finite which implies that uniqueness as proved above will only "appear in the limit" when more and more measurements are made.

Most applications of crosshole tomography so far have been based on the series-expansion method for solving the basic linearized equation system (3). Introduce M linearly independent basis functions $\phi_j(\mathbf{x}), j=1,2,..,M$ .The idea is to seek a slowness model of the form

$$m(\mathbf{x}) = \sum_{j=1}^{M} b_j \ \phi_j(\mathbf{x}) = \mathbf{b}^T \phi(\mathbf{x}) \tag{12}$$

which is compatible with the data $\mathbf{d} = \mathbf{g}(m)$. Here $b_j$ are coefficients to be determined and the obvious vector notation has been introduced. Note that the selected model is restricted to the linear span of the $\phi_j$'s. It is essential that the basis functions are chosen flexible enough to allow an accurate representation of the actual slowness function.

Using the expansion of m in terms of the $\phi_j$'s, we may rewrite the linearized data functionals

$$\begin{aligned} d_i = g_i(m) &\approx \int_{T_i} m(\mathbf{x})\, ds \\ &= \sum_{j=1}^{M} b_j \int_{T_i} \phi_j(\mathbf{x})\, ds = \sum_{j=1}^{M} G_{ij} b_j \end{aligned} \tag{13}$$

if the matrix $\mathbf{G}$ is defined by

$$\mathbf{G}_{ij} = \int_{T_i} \phi_j(\mathbf{x})\, ds \tag{14}$$

Thus our data simply become N linear equations, $\mathbf{d} = \mathbf{Gb}$, for M unknown parameters $b_j$.

In general this system of equations is both overdetermined and underdetermined. To pick a stable preferred $\mathbf{b}$ some criterion is needed, e.g., the least squares criterion. Furthermore, additional constraints will in general have to be imposed. It is natural to take the preferred $\mathbf{b}$, $\hat{\mathbf{b}}$, as the vector which minimizes $|\ \mathbf{d} - \mathbf{Gb}\ |^2 + |\ \mathbf{Cb}\ |^2$ where $\mathbf{C}$ is a suitable matrix taking a priori "information" into account (see also chapters 1,2 and 6). This implies that

$$\begin{aligned} \hat{\mathbf{b}} &= (\mathbf{G}^T \mathbf{G} + \mathbf{C}^T \mathbf{C})^{-1} \mathbf{G}^T \mathbf{d} \\ &= \bar{\mathbf{b}} + (\mathbf{G}^T \mathbf{G} + \mathbf{C}^T \mathbf{C})^{-1} \mathbf{G}^T (\mathbf{d} - \mathbf{G}\bar{\mathbf{b}}) \end{aligned} \tag{15}$$

for any $\bar{\mathbf{b}}$ which satisfies $\mathbf{C}\hat{\mathbf{b}} = 0$. Thus it often makes no difference whether one works with absolute or residual traveltimes. The strength of "damping" in different parts of the rock volume can be regulated by changing $\mathbf{C}$.

The most common basis functions are the box-wise constant ones. The rock volume is divided into a number of rectangles (cubes in the 3D case) and the basis functions are chosen as the characteristic functions of these rectangles. Another possibility is to use bilinear elements. The division into rectangles is kept, but only functions obtained by bilinear interpolation in the grid formed by the midpoints of the rectangles are considered. When bilinear elements are used in the following, the coefficients $\hat{b}_j$ will be the estimates of the slowness function at the midpoints of the rectangles.

When the number of basis functions is not too large, $\hat{\mathbf{b}}$ can be computed using (15) and Gaussian elimination. On a VAX 11/750 computer, for example, $M = 700$ can be handled reasonably fast (a few hours of computing time). In order to keep M small it is important to choose a suitable type of basis functions. For box-wise constant basis functions it is necessary to use a comparatively large M in order to avoid serious artifacts in the tomographic images (cf. Ivansson 1985). Smooth basis functions, like bilinear elements, are much more favourable in this respect.

A few examples of tomographic reconstructions, based on synthetic data, will now be given. They give some appreciation of the potential of the method for mapping fracture zones. The following procedure was followed:

- Given a number of shot/receiver positions and a specified velocity distribution for the crosshole area, all the minimum traveltimes were computed by raytracing.
- Then the straight-line tomographic inversion procedure was used to invert the traveltime data back to a velocity structure that can be compared to the originally specified one. The damping matrix **C** introduced above was chosen as a suitable matrix with row sums zero thus favouring smooth slowness models.

Our source/receiver geometry is shown in Fig. 7, each of the 493 source/ receiver pairs is connected by a straight line. Two examples of reconstructions using synthetic data are given in Fig. 8. The background velocity of the rock was assumed to be 6.0 km/s.

Different low-velocity strips, simulating fracture zones and having a velocity of 4.615 km/s, were introduced. In the figure these strips are labeled. Also included was a low-velocity circle simulating a drift. The only difference between the two examples is the zone D.

There are two reasons why the reconstructions in Fig. 8 are not perfect. First, ray-bending was not accounted for in the inversion and second, the ray-path coverage was not complete which necessitates a certain loss of resolution.Note that a few areas with higher velocity than 6.0 km/s are obtained, they are typical ray-bending artifacts. The loss of resolution is apparent from the broadening of the fracture zones and the irregular shape of the drift. Also, fracture zone D is not mapped very sharply. Nevertheless, the essential features are successfully recovered.

The reconstructions of Fig. 8 were based on bilinear elements, using a division into 5 m squares leading to about 600 unknown slowness parameters, and $\hat{\mathbf{b}}$ was computed from (15) using Gaussian elimination. It is the opinion of the author that this is a most satisfying solution procedure for 2D applications where M can be kept small enough. An important advantage is that statistical errors can be assessed and resolution matrices be computed, as we will see shortly. Nevertheless, most workers on crosshole tomography have used iterative methods to solve (13) or similar equation systems. The best known types of iterative methods are SIRT, ART, and CG (several variants are possible). They have been presented in detail by Van der Sluis and Van der Vorst (chapter 3), so we only make a few remarks.

SIRT and ART were discussed in connection with borehole geophysics by Lager and Lytle (1976) and Dines and Lytle (1979). Applications of CG seem to be more recent, see Nolet (1985) (who used CG in the shape of Paige-Saunders' algorithm) and Ivansson (1986). The start solution for the iterative methods is usually computed by some kind of

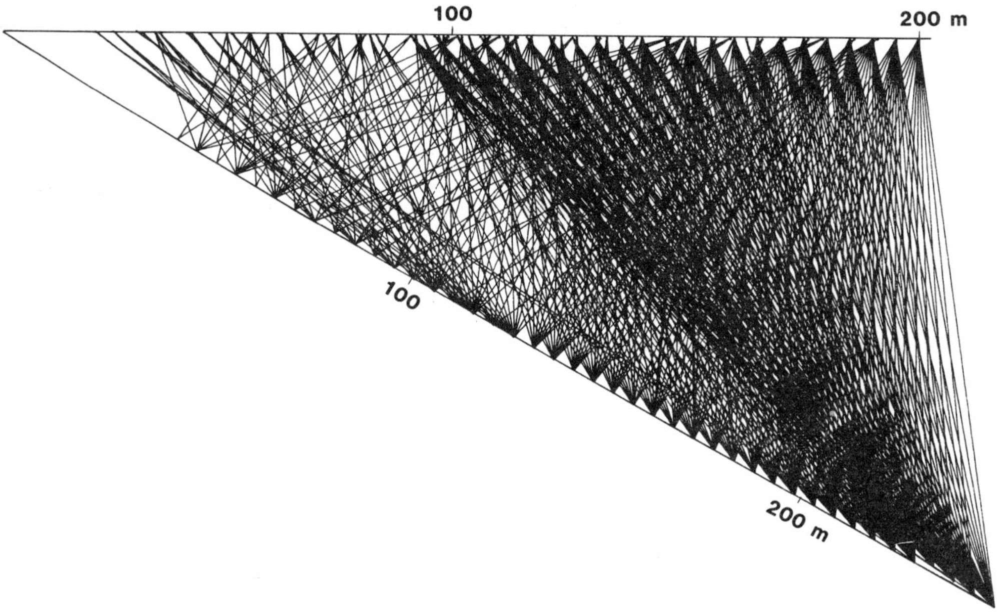


*Figure 7.* Source/receiver geometry with two boreholes.

back projection, for a few variants see Ivansson (1985), Mason (1981), and New (1985b). For a stopping criterion for the iterative procedure, the root-mean-square (rms) value of the residual traveltimes has often been used. It may, however, be rather insensitive and it may be better to use a distance measure applied directly to the successive slowness solutions and their predecessors. One should also be aware that often the main structure of the true solution is visible after comparatively few iterations whereas a considerable number of iterations may be needed to reach the correct magnitude of the velocity fluctuations.

## 2.3 Statistical aspects and resolution

Our preferred coefficient vector $\hat{\mathbf{b}}$, see (15), can be written $\hat{\mathbf{b}} = \mathbf{Hd}$ with

$$\mathbf{H} = (\mathbf{G}^T \mathbf{G} + \mathbf{C}^T \mathbf{C})^{-1} \mathbf{G}^T \tag{16}$$

Supposing the measurements are contaminated with uncorrelated random noise so that $cov(\mathbf{d}) = \sigma^2 \mathbf{I}$ (where $\mathbf{I}$ is the identity matrix), we see that $\hat{\mathbf{b}}$ is a random vector with $cov(\hat{\mathbf{b}}) = \sigma^2 \mathbf{HH}^T$. The impact of measurement errors on the estimate $\hat{m}(\mathbf{x}) = \hat{\mathbf{b}}^T \phi(\mathbf{x})$, cf.(12), can thus be assessed. An example, based on the geometry of Fig. 7 and using the same inversion operator as for Fig. 8, is given in Fig. 9. The figure shows the expected errors in $\hat{\mathbf{b}}$ (i.e., $\hat{m}(\mathbf{x})$ at the centers of the squares) in percent for $\sigma = 0.2$ ms at the typical velocity of 5.0 km/s. They are reasonably small throughout the area.

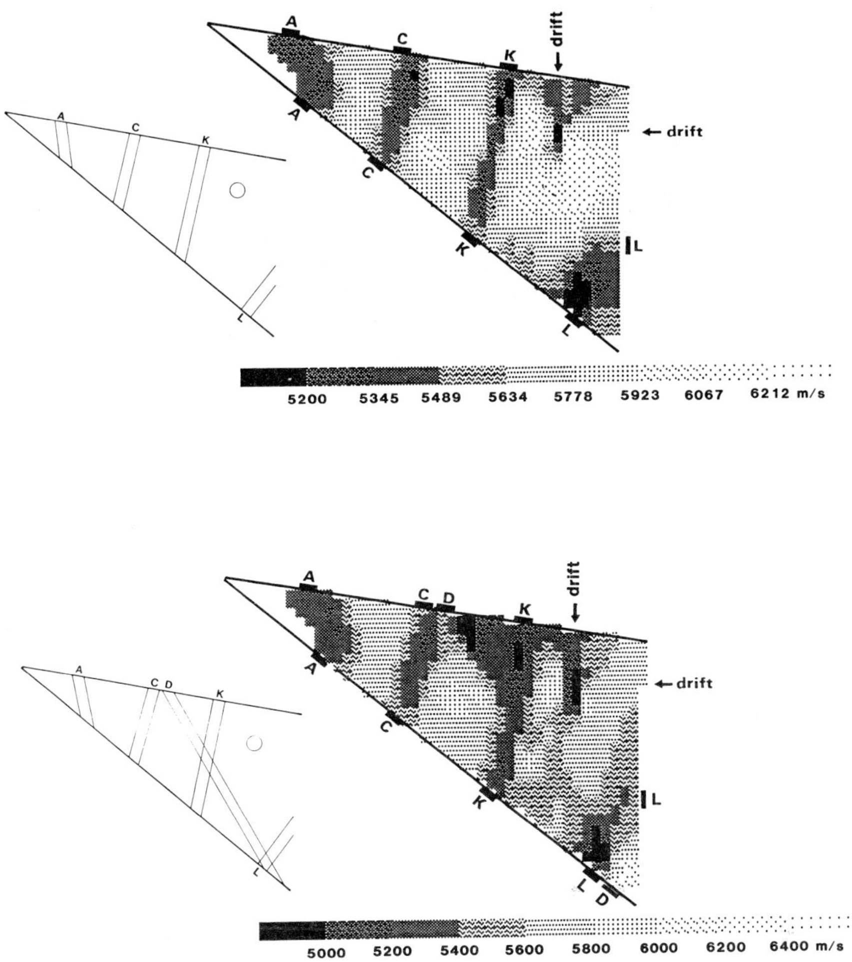


*Figure 8.* Two examples of tomographic reconstruction of fracture zones using synthetic data. In each case the original velocity model with the low-velocity strips, and a drift, is shown on the left and the reconstruction on the right.

Supposing the true slowness model $m(\mathbf{x})$ may be written $m(\mathbf{x}) = \mathbf{b}^T \cdot \phi(\mathbf{x})$ and that $\mathbf{d} = \mathbf{G}\mathbf{b}$, we obtain $\hat{\mathbf{b}} = (\mathbf{HG}) \cdot \mathbf{b}$. The matrix $\mathbf{R} = \mathbf{HG}$ is called the resolution matrix since it indicates the possibilities of resolving models of type $\mathbf{b}^T \phi(\mathbf{x})$. Ideally, we would prefer to have $\mathbf{R}=\mathbf{I}$ but in general $\hat{\mathbf{b}}$ will be a blurred version of $\mathbf{b}$. Our matrix $\mathbf{R}$ may also be written

$$\mathbf{R} = \mathbf{I} - (\mathbf{G}^T \mathbf{G} + \mathbf{C}^T \mathbf{C})^{-1} \mathbf{C}^T \mathbf{C} \tag{17}$$

If the row sums of the damping matrix $\mathbf{C}$ equal zero it follows that all row sums of $\mathbf{R}$ are equal to 1. This is a desirable property since it means that each $\hat{b}_j$ will be an average of the true $b_\nu$, but the weight coefficients will not always be positive. Fig. 10 complements Fig. 9 with a few resolution results. Part (a) shows the diagonal of the resolution matrix whereas

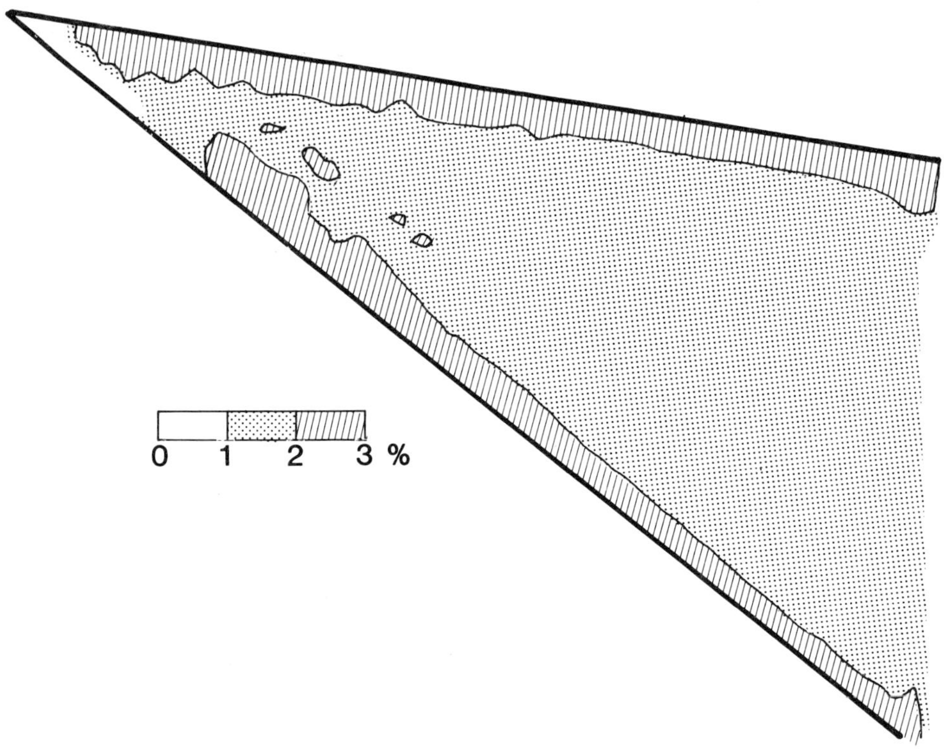


*Figure 9.* Expected magnitude of noise-induced errors.

parts (b) and (c) show the resolving power at two different locations in the crosshole area. Concerning part (a) we note that the best resolution is obtained where the ray-path pattern is dense with crossing rays in many directions (cf. Fig. 7). Shown in parts (b) and (c) are actually the values, put out at the appropriate places in the crosshole region, in two rows of the resolution matrix **R**.

In each case a normalization to the maximum value 1.0 was done. Note the comparatively sharp resolution that is obtained in (b). In (c) there is a smearing in the vertical direction, the main direction of the rays there.

Actually there is a tradeoff between confidence (small statistical errors) and resolution. Fig. 9 and Fig. 10 must thus be viewed in conjunction. The tradeoff is controlled by the choice of damping matrix (denoted **C** above).

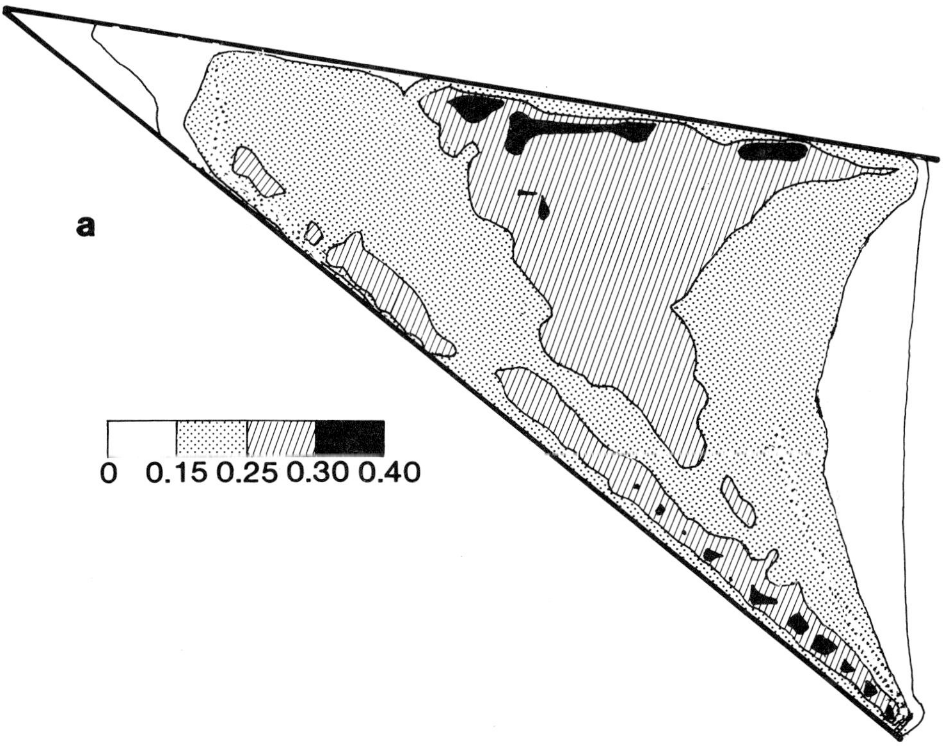


*Figure 10a.* Obtainable resolution. (a) The diagonal of the resolution matrix.

## 3. Complications: Ray-bending and anisotropy

### 3.1 Ray-bending

So far the data functionals $\mathbf{d} = \mathbf{g}(m)$ used for the inversion procedure have been linearized by using fixed (straight) integration curves $T_i$ (see equations (1)-(3)). This may result in certain artifacts when the velocity structure is not homogeneous. Iterative techniques to take the nonlinearity into account have been suggested. The traditional approach for a nonlinear geophysical inverse problem is as follows. At first a parameterization of some suitable space of Earth models is done. Each data value will be connected to the finite number of model parameters by a certain functional (cf.,(1)). A preliminary Earth model is chosen and improved iteratively. Each iteration step involves linearization of the data functionals around the current Earth model (cf.,(3)) and solution, in the least-squares sense, of a suitable linear equation system for the residual model parameters. In our case, extensive two-point ray-tracing must be performed. This will increase the computing time most significantly.

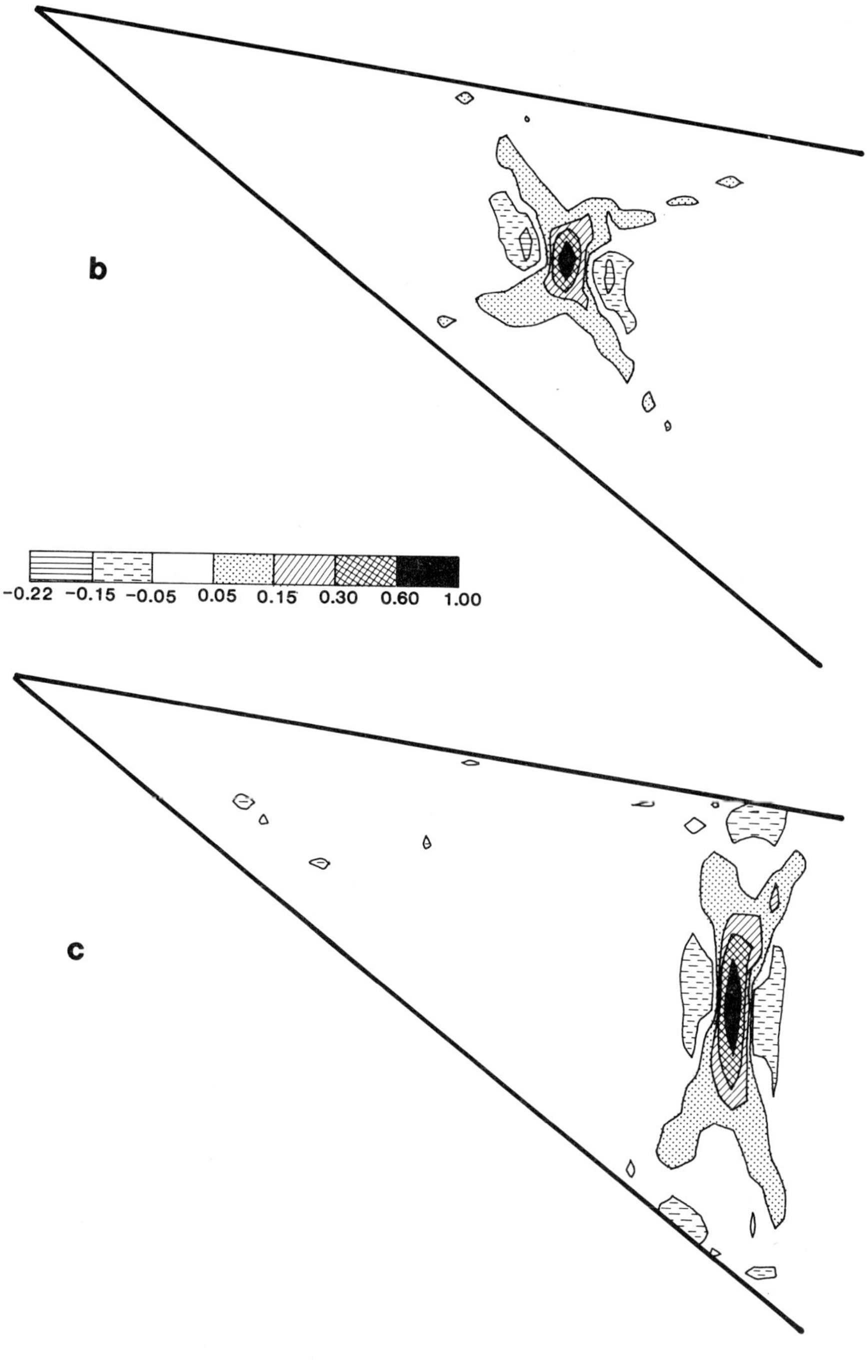


*Figure 10b and c.* Obtainable resolution. (b) Resolving power at a point in the middle. c) Resolving power at a point in the middle to the right.

Crosshole applications of this technique have been described by Bois et al. (1972), Lytle and Dines (1980), Hermann et al. (1982), Cottin et al. (1986), and Bregman et al. (1986). By using a suitable parameterization of the crosshole velocity structure it is possible to achieve reasonably fast ray-tracing computations. Bregman et al. (1986) divided the crosshole region into small triangles in each of which the velocity gradient was kept constant. This resulted in ray paths composed of circular segments which could be calculated analytically. A successful application to real data was reported.

If the initial solution is sufficiently good, it is often possible to obtain improvements by the method described. Unfortunately, however, the difficulties arise when an initial solution (usually computed by the straight-ray linearization) is bad and the corrections are really needed. The rays may get trapped by the artifacts, the procedure may diverge, or it may converge to a wrong solution. Noting these difficulties, Radcliff et al. (1984) suggested "path elimination" (and smoothing between the iterations) to stabilize the procedure. Basically, the idea is simply to temporary neglect data that do not match the current velocity solution well enough. For synthetic test cases, Radcliff et al. succeeded to obtain significant improvements in this way.

An iterative procedure taking account of ray-bending but still using the straight paths in the inversion was tried by Bates and McKinnon (1979) and Ivansson (1985). They estimated the straight-line integrals of the true slowness model by adding to the measured data the appropriate time differences computed for the current model candidate. This made it possible to use the same inversion operator throughout the iterations, but the reconstruction quality was similar to that obtained with the traditional procedure discussed above.

A less ambitious way to "correct" for ray-bending errors is the following (Ivansson 1985). Choose a number of suitable velocity models and compute the corresponding first-arrival traveltimes by ray-tracing. Invert these traveltimes by some straight-line tomographic method and consider the images. By learning how different types of structures are imaged, where artifacts are present, and so forth, it may then be easier to interpret results from real data. The interpretation process will be one of trial-and-error comparisons with forward modeling as an essential ingredient. Recall Fig. 8 where we learned how certain low-velocity strips were mapped in a certain situation. By studying synthetic examples of this kind, one also gets an appreciation of the impact of incomplete ray coverage in certain regions, which is very important. We must expect that the images will never be perfect and that ambiguities will often remain. Additional information and ideas about what is geologically plausible may be needed in order to resolve these ambiguities.

Actually it is the experience of several workers that straight-line tomography will do reasonably well when the velocity never departs from the average by more than 10% (say) (e.g., Dines and Lytle 1979). This is fortunate since explicit ray-bending corrections imply highly increased complexity and computational work. One should also bear in mind that ray-bending corrections for velocity variations traverse to the planar crosshole region will hardly be possible without a 3D source/receiver coverage.

An entirely different way to incorporate ray-bending effects was introduced to geophysics by Devaney (1984). As data he used Fourier spectra of the registered signals.

His technique, first applied to ultrasound, has been called "diffraction tomography" and is theoretically capable of reconstructing the velocity distribution of a weakly inhomogeneous rock formation.

### 3.2 Anisotropy

Consider Fig. 11. It is a comparison of vertical and horizontal velocities obtained in a crosshole experiment in crystalline rock at Krakemala in southern Sweden. The distance between the two vertical boreholes, 450 m deep and leading down from the surface, was 625 m. Explosions were detonated at different depths in one of the boreholes. The vertical velocities were obtained from traveltimes registered by a surface geophone very close to this borehole, whereas the horizontal velocities were similarly obtained by a movable geophone array in the opposite borehole. The most interesting feature concerning Fig. 11 is the significant gap between the vertical and horizontal velocities. To a certain extent this gap can be explained by the horizontal paths actually being refracted. Suppose for a moment that the true velocity distribution is actually isotropic and laterally homogeneous with depth dependence given by the thick curve (the one for the vertical velocities). By computing synthetic traveltimes for this structure and inverting them in the same way as the experimental ones, a small artificial gap between reconstructed vertical and horizontal velocities was definitely obtained, but it was nowhere near as large as that in Fig. 11. The fact that the gap of Fig. 11 is so much wider indicates that refraction alone cannot explain the experimental result. A natural explanation is effective anisotropy, probably caused by a region of horizontal fractures at depths between 180 and 350 m. Support to this idea was also obtained by the borehole logs, analysis of tube waves, and a study of artifacts in the tomographic image (Ivansson 1985).

When anisotropy is present it may cause severe problems concerning the tomographic inversion. In fact, instead of the uniqueness results in section 2.1 we now have the following nonuniqueness result.

*Proposition 4:* Let $s_1(\mathbf{x}, \theta)$ and $s_2(\mathbf{x}, \theta)$ be two functions, they are meant to represent the slowness at $\mathbf{x}$ in direction $\theta$ . The line integrals of $s_1(\mathbf{x}, \theta)$ and $s_2(\mathbf{x}, \theta)$ along any line of direction $\theta$ are equal, for all $\theta$ , provided only that

$$s(\mathbf{x}, \theta) = s_1(\mathbf{x}, \theta) - s_2(\mathbf{x}, \theta)$$

$$= f(\mathbf{x}) \cos^2(\theta) + g(\mathbf{x}) \sin^2(\theta) \tag{18}$$

where f and g are any two functions such that

$$f = \frac{\partial^2 h}{\partial x_2^2} \quad \text{and} \quad g = -\frac{\partial^2 h}{\partial x_1^2} \tag{19}$$

for some twice continuously differentiable function h which vanishes outside a bounded set.

*Proof:* We need to show that

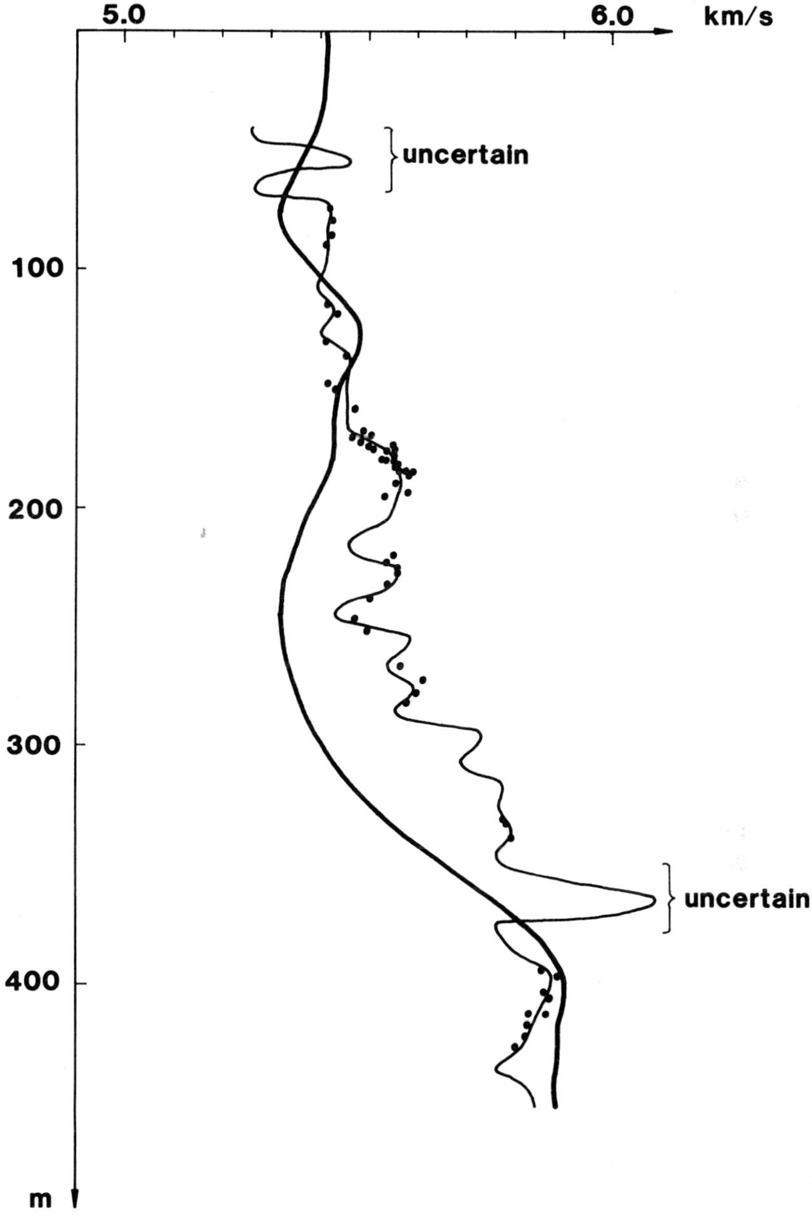


*Figure 11.* Experimental Krakemala results. Bold line: vertical velocities , thin line: horizontal velocities, dots: mean velocities of certain individual horizontal rays.

$$\cos^2(\theta) \int f\left[t\cos(\theta) - s\sin(\theta), t\sin(\theta) + s\cos(\theta)\right] ds$$

$$+ \sin^2(\theta) \int g\left[t\cos(\theta) - s\sin(\theta), t\sin(\theta) + s\cos(\theta)\right] ds = 0 \qquad (20)$$

for all t and $\theta$ . Introduce the function $F_{\theta,t}(s)$ by

$$F_{\theta,t}(s) = \cos(\theta) \frac{\partial h}{\partial x_2}\left[t\cos(\theta) - s\sin(\theta)\ , t\sin(\theta) + s\cos(\theta)\right]$$

$$+ \sin(\theta) \frac{\partial h}{\partial x_1}\left[t\cos(\theta) - s\sin(\theta)\ , t\sin(\theta) + s\cos(\theta)\right] \qquad (21)$$

We realize that the left-hand side of (20) equals $\int F_{\theta,t}(s)\, ds$ , and this integral must vanish since h was assumed to vanish outside a bounded set.

Actually the angular dependence is much more complicated than the simple elliptical type assumed here (e.g., Leary and Henyey 1985). This makes it problematic to allow for anisotropy explicitly in the inversion process, artifacts may arise because of failure to model the slowness/direction curve accurately by few parameters.

One should not be too discouraged because of problems with anisotropy, however. First, if the anisotropy is regular throughout the crosshole region, it may be possible to "correct" the measured traveltimes in some simple way and then apply the usual "isotropic" tomographic inversion. Second, it is often the case that an isotropic model is sufficiently accurate. A simple first test to see whether anisotropy is present is to plot the average velocities of all the rays as a function of direction.

## 4. Field results

A few field results based on seismic crosshole tomography will now be presented. They are taken from some of the references already given in section 1.3.

### 4.1 Mapping an ore body in a mine (after Gustavsson et al. 1986)

The seismic crosshole method was tested in the Kiruna Research Mine in northern Sweden. The experiment was performed between two vertical holes at a level of 320 m below the ground (Fig. 12). The holes, which were core drilled, were 165 m apart and 50 m deep. They were placed on opposite sides of a magnetite body, taking the shape of a thin (20-25 m) slab slanting $60^o$ from the horizontal. In addition, there were galleries along a line between the two holes and perpendicular to the magnetite sheet. These galleries ran at the 320 and 370 m levels. As observed in the galleries, the character of the boundaries of the magnetite and surrounding waste rock was somewhat different on each side of the sheet. The quartz-porphyry on the right side of the magnetite body showed a very sharp contact which was in general covered by a few centimeter of chlorite. On the left side, an intermediate zone of breccia changed to syenite porphyry in some ten meter. Microexplosions (10 g) were detonated in the holes at various depths to generate seismic signals. Fourteen horizontal geophones (digital grade wide band, natural frequency 10 Hz,

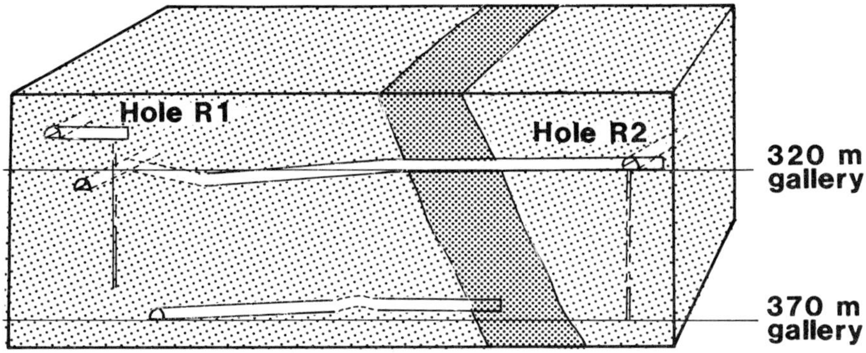


*Figure 12.* The test area in the Kiruna Research Mine.

Sensor-Geosource) were located along the line in the gallery between the two holes according to the ray-path diagram shown in Fig. 13. The signals from microexplosions in the R1 hole (Fig. 12.) were also recorded by a borehole geophone array with three units separated by 10 m. The array was locked at various depths in hole R2.

Recordings were obtained from 50 explosions, 30 and 19 were set off in holes R1 and R2, respectively. An additional microexplosion was set off in the gallery at the 370 m level (Fig. 12). The signals were recorded at a sampling rate of 2 kHz on each of 25 channels. The result from a tomographic inversion of the traveltimes is shown in Fig. 14 . It is based on 400 rays (assumed straight) according to Fig. 13. A decomposition into rectangles of size 3.25 m $x$ 4 m was used, leading to 640 unknown parameters which could be handled by direct inversion of the resulting linear equation system (see section 2.2). The boundaries of the magnetite and breccia as observed in the galleries at the 320 and 370 m levels are marked in the figure and agree with velocity contrasts in the tomogram. The boundaries are smoothed in the image as a result of damping in the solution. There is a large low-velocity area within the syenite-porphyry which is statistically significant. Unfortunately, no other geological or geophysical investigations have been done in this part of the mine.

The in situ velocities shown in Fig. 14 were compared with values computed from E-modules and densities of core samples. A qualitative agreement was obtained. In both cases the high and low velocities for quartz-porphyry and magnetite, respectively, were very pronounced. No values from the syenite-porphyry are available.

### 4.2 Stress monitoring and fault detection in coal mines (after Kormendi et al. 1986 and Bodoky et al. 1985)

Monitoring of the rock stresses and their changes is of great importance for the safety in coal mines. It has been found that under certain circumstances the seismic velocities are monotonic functions of the stresses. Thus traveltime tomography becomes a feasible method in this context.

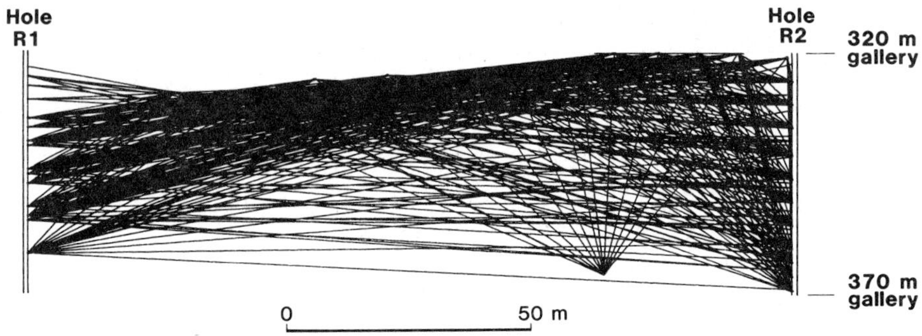


*Figure 13.* Kiruna ray-path diagram.

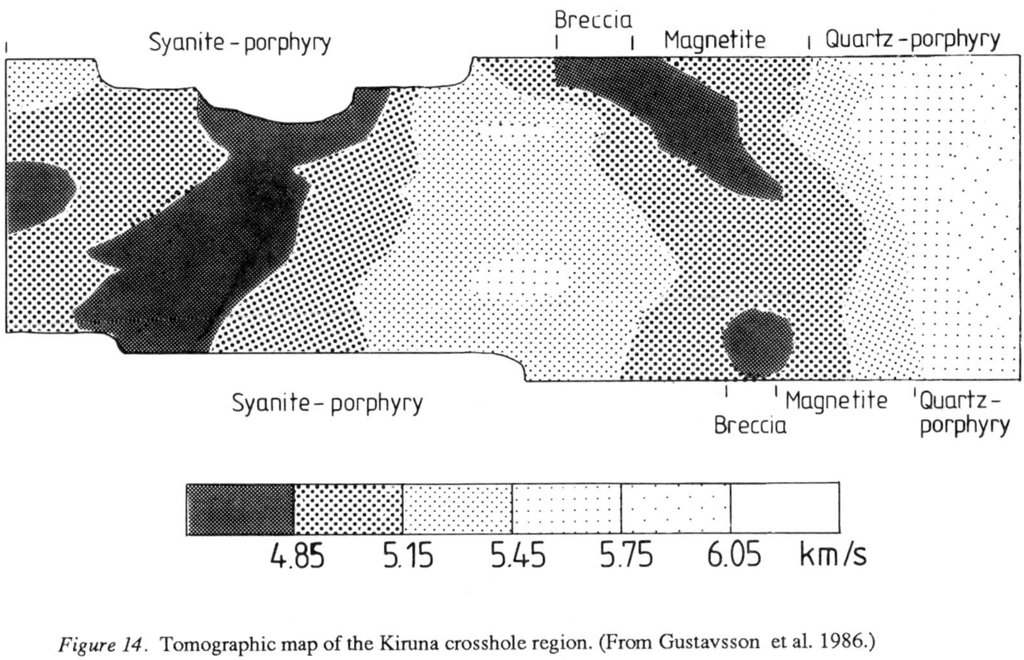


*Figure 14.* Tomographic map of the Kiruna crosshole region. (From Gustavsson et al. 1986.)

A field test was performed in a Hungarian coal mine. The coal seam, about 6 m thick, was situated at a depth of 180 m, between two roads, the top road and the tail road. The distance between the roads was 80 m. Seismic energy was produced by hammer blows striking on rock-bolts fixed horizontally into the seam in the top road with a spacing of 4 m. The signals were picked up by piezoelectric transducers placed in the tail road, also with a spacing of 4 m. Data from twelve receivers were recorded digitally with a sampling interval of 0.125 ms.

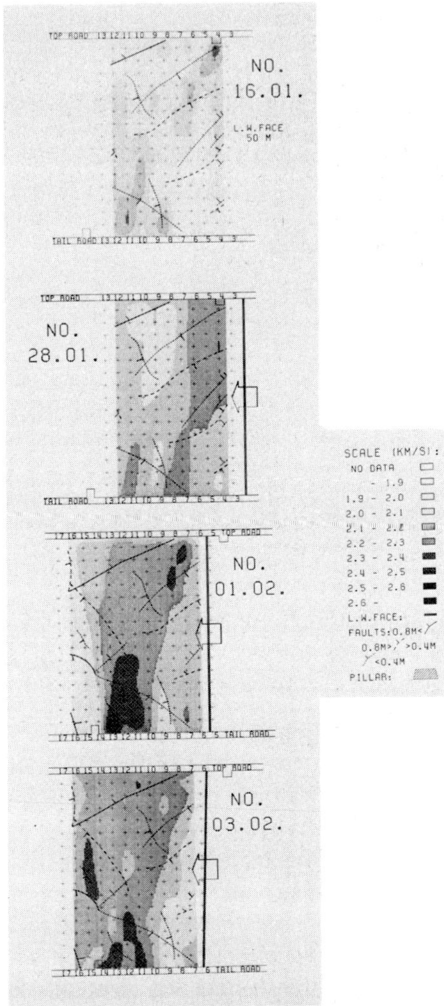


*Figure 15.* Sequence of velocity maps showing how increased stresses appear in the forefield of the workings and are shifted ahead by the advancing longwall face (which is marked by an arrow). (From Kormendi et al. 1986.)

In theory the first onsets should be the P-waves refracted in the host rock. In practice, however, these waves were found to be too weak. Instead the fist high amplitude onsets corresponded to direct-run P-waves propagating in the coal seam.

A sequence of tomographic velocity maps, obtained from direct P-arrival traveltimes in the mine, are shown in Fig. 15. They refer to different dates, given in the figure, during a mining operation where the long-wall face was driven into the seam. The first map, No. 16.01, shows no special character. The velocity minimum belongs to a broken zone with many micro-faults. In the three following maps a zone of increased velocity appears as a consequence of the approaching face at a distance of 15 to 25 m from it. The shape of the

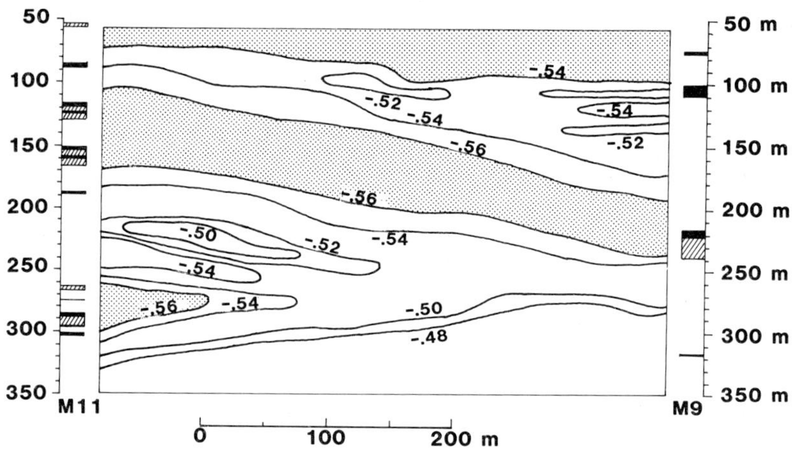


*Figure 16.* Seismic transparency map for the region between boreholes M11 and M9. (Logarithmic units, low transparency values represent mechanically poor rock). Results from fracture counts in the boreholes are shown at the edges, with a dark shading indicating many fractures per m. (From Wong et al. 1985.)

high-velocity anomaly is distorted by microtectonics. A decrease in velocity can be observed in the neighbourhood of the face. This is in good agreement with mining experience. Several other nice examples were given by Kormendi et al. (1986).

Bodoky et al. (1985) used transmission tomography for detecting faults and dikes within a coal seam. The low-frequency part of the SH component of the seam wave, the Evison wave, has a significant part of its energy in the form of inhomogeneous plane waves propagating outside the seam while the high-frequency part propagates almost entirely within the seam. Thus, if the seam is interrupted by some discontinuity, the high-frequency waves will loose much more of their energy than will the low-frequency waves. Bodoky et al. introduced the quotient of the energy within a high-frequency gate and a low-frequency gate of the amplitude spectrum of the Evison wave. Using this measure and a simple backprojection algorithm they succeeded to obtain transmittance maps of the seam that were of help in locating faults and dikes.

### 4.3 Fracture location in Canadian crystalline rock (after Wong et al. 1985)

Wong et el. (1985) have performed seismic amplitude tomography at the Canadian Underground Research Laboratory in Manitoba. As transmitter and receivers they used piezoelectric transducers, the transmitter being designed to apply a unidirectional force perpendicular to the borehole wall. The peak power delivered to the transmitter was a pseudo-random binary sequence with predominant frequencies in the range from 1 to 5 kHz. By stacking repeatable signals and cross-correlating with the reference waveform it was possible to greatly enhance the signal-to-noise ratio and register signals over paths up to 500 m. The sampling frequency was 50 kHz.

Fig. 16 shows the result obtained from measurements between two boreholes, called M11 and M9, with a separation of 450 m. For every seismogram, the amplitude spectrum of the direct arrival was estimated by Fourier transforming a short time segment

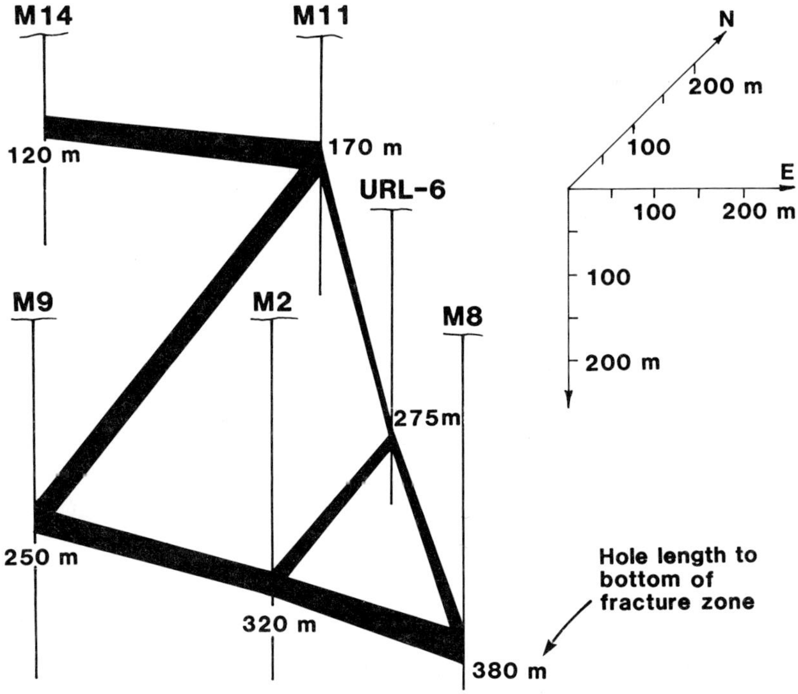


*Figure 17.* Perspective representation of a fracture zone as obtained from seismic transparency tomography. M14, M11, M9, M2, M8 and URL-6 denote the six vertical boreholes used for the five crosshole scans (M14/M11, M11/M9, M9/M8, URL-6/M2). (From Wong et al. 1985.)

immediately following the arrival time. The logarithms of amplitudes were then used in a simple backprojection algorithm to form tomographic images of seismic transparency. The shaded areas indicate highly lossy rock that is possible to correlate with fracture zones. The fracture zone intersections indicated at the edges of Fig. 16 were obtained from standard geophysical logs. The image indicates that the fracture intersections between 120 and 170 m in borehole M11 are connected structurally to the fracture intersections at 200 to 250 m in borehole M9. This does not necessarily imply hydrogeological connection, however.

Crosshole transparency maps were also obtained for four other pairs of boreholes. Taken together they form a set that reveals the three-dimensional nature of a certain fracture zone. This is shown in the perspective plot of Fig. 17. The fracture zone has a dip of $8^o$ to the south and about $14^o$ to the east, confirming previous interpretations.

The tomographic images also reveal distinct changes in rock quality. The upper 100 to 150 m of granite are heavily weathered or have a high density of microfractures and joints. The granite from about 150 m down to the top of the fracture zone is moderately transparent. Below the fracture zone the rock quality is in general quite good.

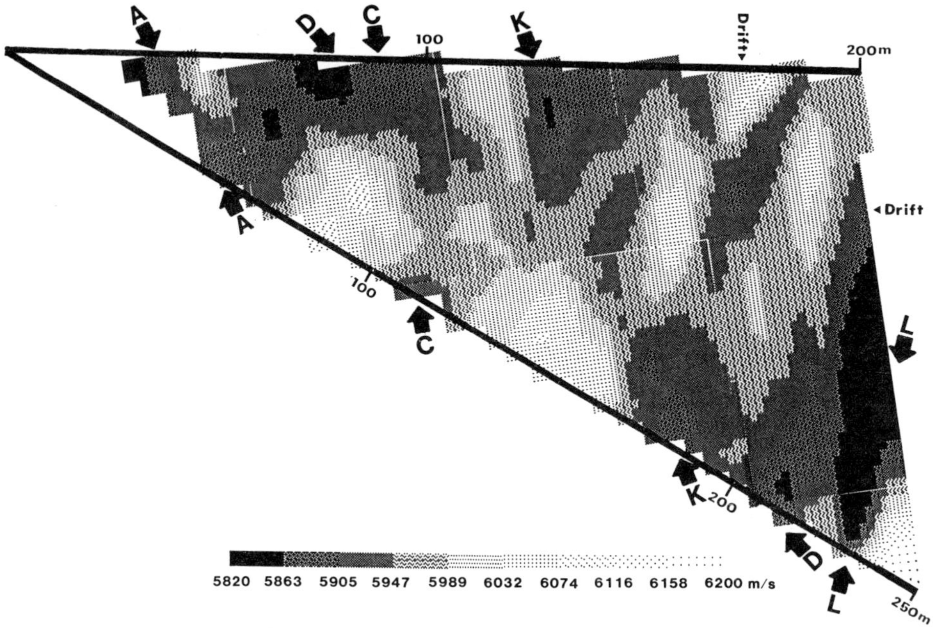


*Figure 18.* Tomographic P-velocity map obtained from crosshole measurements at the Stripa mine. (From Hammarstrom et al. 1986.)

## 4.4 Fracture location in Stripa crystalline rock (after Hammarstrom et al. 1986, and Cosma 1986)

Fig. 18 shows the tomographic P-velocity map obtained by Hammarstrom et al. (1986) from a crosshole experiment at the Stripa mine. Essentially the same type of equipment was used as that described in section 4.1. Some 400 traveltimes were used for the inversion. Combining the map of Fig. 18 with previous information, that obtained from the borehole logs in particular, one has arrived at an interpretation in terms of five low-velocity zones. Synthetic model calculations similar to those in Fig. 8 were also helpful in this process. The zones are marked A, D, C, K, and L in the figure. The main feature is the fractured zone L. Also note the low-velocity region due to a drift (3 m in diameter) crossing the section. The comparatively low velocity contrasts indicate that the Stripa granite is fairly homogeneous.

Cosma (1986) uses a steel hammer to produce mechanical shocks in boreholes. The hammer can slide along the borehole inside a waterproof cylindrical enclosure. The driving force is provided by a stiff spring which, when compressed and released, pushes the hammer against an anvil wedged in the borehole. Loading, firing, and fastening to the borehole wall are achieved hydraulically, the whole operation being controlled through a composite cable from a control panel. The hammer blows can be repeated at 5-10 second

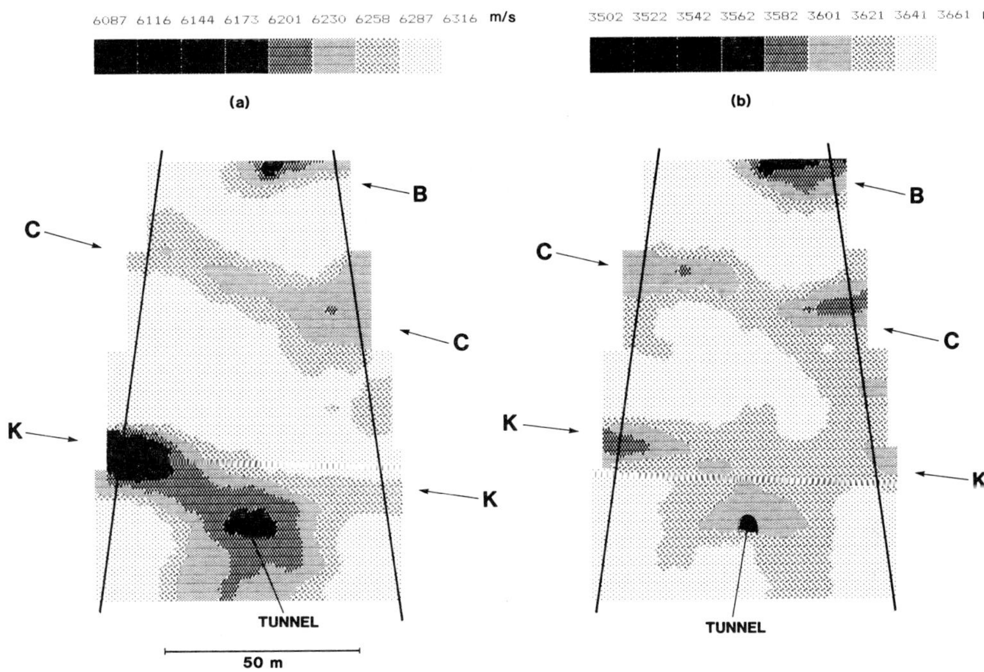


*Figure 19.* Tomographic velocity maps obtained from crosshole measurements at the Stripa mine. Part (a) for P-velocities .Part (b) for S-velocities. (From Cosma 1986.)

intervals. In general several hammer pulses have to be added before a sufficiently clear signal can be recorded. The zero-time signal is picked up by a pulse sensor placed on the anvil.

This source can be used in vertical, inclined, horizontal, or upgoing boreholes. Both P- and S-waves are generated, the maximum S-wave output is obtained in the transverse direction. The unit used at Stripa was equipped with a 200 m cable and a hammer with mass 2.5 kg delivering 50 J per blow.

As detectors Cosma uses accelerometers (100 mV/g). Good coupling to the borehole wall is achieved by a motor-driven clamping mechanism. The dominating frequencies of the signals used for traveltime picking are generally between 3 and 5 kHz. Fig. 19 shows tomographic velocity maps obtained from measurements between two boreholes in the Stripa mine. Part (a) is for the P-velocities whereas part (b) is for the S-velocities. Some 200 traveltimes were used in each case. Accurate determination of the S-arrivals was occasionally hindered by interference with P-wave reflections but the quality of the S-velocity map is nevertheless reasonably good. The features displayed in parts (a) and (b) are concordant with each other and they are also concordant with the information available from other types of surveys performed at the same site. The arrows marked B, C, and K show the location of fracture zones as determined by previous measurements. In the lower part of the sections there is a tunnel crossing. The inversion algorithm used belongs to the SIRT family. Because of the small velocity variation through the whole section ray-tracing procedures did not bring about any noticeable improvement.

*Acknowledgments* I am grateful to Nina Bregman, Calin Cosma, Laszlo Dianiska, Monica Hammarstrom, Per Moren, and Nash Soonawala for providing recent material that was of use when preparing this text. Thanks also to Jorgen Pihl and Magnus Hagwall for critically reading the manuscript.

*Chapter 8*

# Tomography from seismic profiles

*P. Firbas*

The solution of the tomographical problem for seismic profiles is based on the *linearization principle* and the *generalized inverse* . It is independent of measurement geometry (e.g. surface profile, cross-hole, etc.). It may be applied to surveys of various scales (local, regional, DSS, etc.). It allows simultaneous usage of various types of waves (refracted, reflected, transformed). An outline of the structure of the program package is offered and both model and field data examples are presented.

## 1. Introduction

The recent improvement of seismic data quality calls for more sophisticated interpretation procedures, by means of inhomogeneous layered models. The fast development of computers in the last decade makes the application of such procedures possible. One such procedure, already tested both on model and field data, is presented. The discussion in this chapter concentrates more on various pitfalls and limitations than on lists of lengthy formulae. Emphasis is on the basic ideas concerning the selection of the type of model, input data errors, and various assumptions underlying the digital solution as all these may influence the resulting solution even more than some details of computations done deeply in the computer program. The approach presented is based on a generally proved "linearization principle" and "generalized inverse" which will be discused in separate sections. The solution presented is generally applicable to all types of waves, various measurement geometries, and various a priori information.



*G. Nolet (ed.), Seismic Tomography, 189–202.*

## 2. Type of model

It is well known from experience that the geological medium is generally: a) inhomogeneous (physical properties vary in space) b) layered (there are discontinuities with physical parameters abruptly changing) c) anisotropic (physical properties depend on the direction of measurement) d) composed of more components (solid, fluid, and gaseous) e) in an inhomogeneous temperature field f) in an inhomogeneous stress field, etc. A model which would allow for all these features simultaneously would be so complicated that it would hardly be of any use even for direct problem solution. Experimental physics removes this complexity by the preparation of well defined "samples" for measurements. Geophysics can only take the Earth as it is and at most has a limited possibility of selecting an appropriate geometry of measurement.

So in any case geophysics faces the problem of carrying out the measurements on a complex medium in spite of the fact that only a limited amount of information is required (and in fact a limited amount can be obtained). In consequence the model simplification, which puts emphasis on the properties measured and neglects other features of the medium, is a very important step. Insensitive approaches may easily distort even a very sophisticated interpretation. This very important point, to a great extent conditioning the eventual success, is almost always solved only intuitively. The approach used in this chapter is quite general, although only 2D models (isotropic, laterally inhomogeneous with a few number of layers) will be demonstrated by the computed examples. The possible extensions to more complicated models are discussed in section 11.

## 3. Linearization principle

The tomographical problem in seismology is based on an inversion of a non-linear equation (see also chapter 1):

$$t(S,R) = \int_{L(S,R)} N(x,y,z)\, ds \tag{1}$$

where $N(x,y,z)$ represents the "slowness" of the medium (i.e. the inverse value of the velocity distribution $V(x,y,z)$ ), $L(S,R)$ is the ray connecting a source S and a receiver R, and $t(S,R)$ is the measured "time-of-flight". This equation remains non-linear in spite of the fact that we substituted slowness $N$ for velocity $V$ , since the integration path $L(S,R)$ depends on the unknown slowness distribution $N$ according to Fermat's principle:

$$\delta t(S,R) = \delta \int_{L(S,R)} N(x,y,z)\, ds = 0 \tag{2}$$

The general solution of non-linear problems such as (1) is not available and therefore the so called "linearization principle" is of great value, because it allows us to find the solution by a sequence of iterations. The linearization approach has been intuitively used by many authors (at most verbally recalling the ray path stationarity due to the Fermat principle). Romanov (1972) gave perhaps the first mathematical proof for the case of an isotropic non-layered medium (with a velocity distribution having continuous second derivatives) based on the wave equation. Červený (1982) derived a linearized formula for a non-layered anisotropic medium. Firbas (1984b) analyzed a general case of 3D anisotropic inhomogeneous layered medium. This proof took as its starting point the explicit computation for the case of two isotropic homogeneous half-planes separated by a fixed linear interface. This case was successively generalized for an arbitrarily high number of interfaces (smoothly varying medium), curved fixed interfaces, anisotropic media between interfaces, and the 3D medium. Only such a full proof allows us to formulate the linearized algorithm for the tomographical inverse problem for complex media with curved interfaces. Nevertheless, caution has to be exercised at spikes on interfaces, where both the simple ray theory and the linearization approach can hardly be applied. To formulate the linearization principle, let us have a model $M_0$ characterized by the slowness distribution $N_0$ for which the rays $L_0(S,R)$ and times-of-flight $t_0(S,R)$ are known. Let us further suppose that the function $N_0$ does not differ too much from the real slowness $N$ in the real medium $M$ . Let us call the function $N_0$ "the initial model". We can write

$$N(x,y,z) = N_0(x,y,z) + N_1(x,y,z) \tag{3}$$

$$|N_1(x,y,z)| \ll N_0(x,y,z) \tag{4}$$

Let us call the function $N_1$ "the perturbation function". It can be proved that

$$t(S,R) = t_0(S,R) + t_1(S,R) + e(S,R) \tag{5}$$

where

$$t_1(S,R) = \int_{L_0(S,R)} N_1(x,y,z)\, ds_0 \tag{6}$$

and $e(R,S)$ stands for a small term which approaches zero like $O(t_1(S,R))$ and so can be neglected in regard to the small term $t_1(S,R)$ and we can write approximatively

$$t(S,R) = t_0(S,R) + t_1(S,R) \tag{7}$$

The equation (7) is now linear in the unknown function $N_1$ since (what is most important) the term $t_1(S,R)$ is computed by an integration along a known old ray $L_0(S,R)$ in the medium $M_0$ . It has to be stressed at this point that after linearizing (1) through neglecting the term $e(S,R)$ we cannot expect to arrive at the true solution $N(x,y,z)$ after one iteration, but only at some estimate of $N$. Nevertheless, the linearization principle outlined makes it possible repeatedly to substitute the obtained estimate of $N$ for the initial slowness $N_0$, compute appropriate rays $L_0(S,R)$ and repeat the inversion resulting in a new estimate of $N$ . The convergence of such a process to a true solution is not straightforward and depends also on the particular approach used for solving the linear inverse problem in every iteration and has much in common with the global/local minima problem.

## 4. Input data and their errors

Input data used for the tomographical inverse problem are the onset times of various types of waves. The most important between them are the times-of-flight read in the first onsets as these can be determined with the highest precision. In the case of surface profile measurements these onsets are mostly formed by refracted waves. When reading the onsets on seismogram sections we can commit at least two principal errors which may distort the solution. First, it may happen that instead of the proper onset some later phase is read (e.g. if the true onset is hidden in the noise or in the coda of other onsets). Second, it may happen that especially for wide angle surveys, the type of wave may be wrongly assigned. These facts entail the requirement that the interpretation process should be at least partly interactive in order to have the possibility to remove these types of errors even between iterations. There are, of course, many other sources of errors originating both in the field and in the reading of input data. For regional survey profile data these errors may be in the range of 10-100ms, when earthquake data are used the errors may be in the range of 50-500ms. These errors can be estimated by a variance (diagonal elements in the data covariance matrix) and so used in calculation. The fact that the travel-time curves are smooth and continuous maybe in principle covered by non-zero off-diagonal elements in the data covariance matrix.

## 5. Two-point ray tracing

The linearized approach to the tomography inverse problem presupposes the knowledge of rays $L_0(S,R)$, i.e. the solution of the two-point ray tracing problem. In this investigation the well developed ray tracing method of Červený and Pšenčík (1981 - see also chapter 5) has been used. It allows to compute rays in fairly complex 2D inhomogeneous layered models with curved interfaces for various types of waves (refracted, reflected, transformed). The flexibility of this ray tracing system may be fully utilized both for the complexity of the model in the tomography inverse problem and for utilization of onsets of various types of waves as input data. There are numerous problems connected with the wave field complexity. For example, many triplications of travel-time curves computed by ray tracing cannot be distinguished in the field data. In such a situation one can always align real data only with the earliest travel-time curve branch of the triplication. There are also other points worth mentioning. It is hardly ever true that in each iteration all input data correspond to arrivals in the model used for ray tracing; nevertheless, it should generally be true that the higher the number of the iteration, the higher the number of fitted input data should be. Pathological effects like serious decrease in the number of fitted rays from one iteration to another is a signal for the interpreter to check his interpretation of wave types or check the model for unwanted low velocity zones which may produce shadow zones. It should be stressed that the simple fit of a restricted subset of input data for which computed rays exist cannot be considered a final result. There are at least three other conditions: first, all input data must have their counterparts in computed rays; second, the input data should cover the whole range of existence (detectability) of a particular wave type; third, there should be no rays found in areas where shadow zones are found in seismograms. Some of these criteria do not rely only on kinematic but also on dynamic information and may sometimes be hard to prove.

## 6. Initial model, perturbation function and their approximation

It can be easily seen that the closer the initial model $N_0$ to the real distribution $N$ , the easier it is to find the solution. Nevertheless, it has been experimentally found that even for initial models that are far from the true solution, the algorithm suggested works well in many particular cases (Firbas, 1984d). Generally, the initial model is taken from an overall knowledge of the surveyed region, from generalized 1D inverse problem solutions or (which we consider most profitable) by making a raw fit solving interactively the 2D direct problem and simultaneously changing the model. For the last mentioned approach an interactive program package for a desk-top computer Hewlett-Packard 9845 has been written (Firbas and Skorkovská, 1986). Essential for the solution are a suitable approximation of the initial model and a suitable parametrization of the perturbation function in each iteration. Various types of parametrization have been tested, the condition of a smooth slowness distribution between interfaces being observed throughout. A simple approach utilizing the Legendre polynomial expansion in the area covered by rays in one iteration was discussed in Firbas, 1981. It has been found that for more complicated models and for computing higher iterations it is more suitable to perform the expansion over the fixed rectangular area $(x_{min},x_{max})\times(z_{min},z_{max})$ and to use B-spline functions (Giloi, 1978) modified at the margins of the area. In such a case both the initial model and the unknown perturbation function are approximated consistently by the same type of function, i.e. spline polynomials. The continuous inverse problem concerning an unknown perturbation function is so transformed into a discrete one with a rather limited number of unknown parameters (spline coefficients). Moreover, this approximation is fairly general and so does not impose any serious limitation on the shape of the perturbation function even when the number of spline coefficients (4 to 10 ) in each axis for one layer is kept rather low. It has to be noted that such smooth approximation with a low number of parameters fits quite easily even a rather complex distribution while a "uniform box" approximation would need a tremendously high number of boxes and moreover will still contain artificial discontinuities on the box boundaries. The spline approximation can be used in the form

$$N_1 = \sum_{k=1}^{K} \sum_{l=1}^{L} \sum_{m=1}^{M} C_{klm} F_k\left(\frac{x-x_k}{\Delta x}\right) F_l\left(\frac{y-y_l}{\Delta y}\right) F_m\left(\frac{z-z_m}{\Delta z}\right) \tag{8}$$

where $C_{klm}$ are the unknown spline coefficients and $F_i$ are the modified B-spline functions (depicted as an example on Figure 1)

## 7. Linear system and the inverse operator

The linearization approach and the linear parametrization of the perturbation function enable us to formulate an iterative algorithm which at each iteration requires the solution of a system of linear equations derived from (6) and (7)

$$t_1(S,R) = t_{exp}(S,R) - t_0(S,R) \tag{9}$$

After utilizing the known raypath $L_0(S,R)$ and the expected shape of the perturbation function (8), the left hand side can be written for every ray in the form

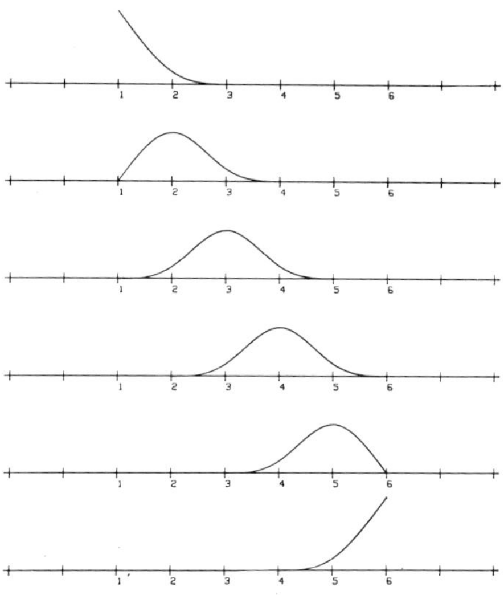

*Figure 1* Example of modified B-spline function (for one axis) used for approximation of the perturbation function.

$$t_1(S,R) = C_{klm}\, G_{klm}(S,R) \tag{10}$$

In each iteration we have to solve a linear system which consists in a typical 2D case of hundreds of equations with the first tens of unknowns. There are generally more equations than unknowns, which helps in suppressing the noise contained in the data. This does not imply that the system is purely overdetermined. There may be no rays in the lower corners of the model or, due to ray geometry, areas may form in the model which cannot be fully resolved. Such a system is sometimes called "mixed-determined" and cannot be directly solved by either the simple least-squares or by the minimum length method. The solution of such systems is relatively well understood (e.g. Menke, 1984; Jackson 1972,1979; Franklin, 1970; etc.).

Van der Sluis and Van der Vorst (chapter 3) show that there are two basic ways to create a generalized solution of the linearized problem:
a) damped least squares utilizing both a weight (covariance) matrix for the input data space (taking into account different data precision and mutual data correlation) and a weight (covariance) matrix for the model space (taking into account the a priori information).
b) singular value decomposition (again complicated by both a covariance (weight) matrix for input data and a covariance (weight) matrix reflecting the a priori information.) When applied these two methods give nearly the same results. Neither leads to a perfect resolution both in the data and in the model spaces and both methods involve one (more-or-less) trial-and-error parameter.

The first method (a) minimizes the sum of the weighted norm of data misfit and that of the solution using a trial-and-error weight parameter, which emphasizes either the data misfit norm or the model norm (a priori information). This solution is rather fast to compute but one does not have much insight into the solution structure and even parameters which would be perfectly resolved by the data only may be influenced by implemented a priori information. The second method (b) allows to distinguish those parameters (or their linear combinations) which are resolved by the data (fitting them in the least square sense) from those for determination of which the a priori information is crucial. The procedure is more time consuming but gives much more insight into the structure of the inverse operator, because both the spectrum of singular values and the set of the base vectors in the model space are known. The theoretical advantage of forming two distinct groups of singular values (non-zero and zero) is plagued by the limited accuracy of computers and our information about the coefficients of the linear system ( in our case in ray tracing, calculating the elements $G_{klm}$ and the singular value decomposition). In such a situation, no sharp distinction can be made between non-zero and zero singular values, but one more-or-less "continuous" spectrum ranging from large singular values to small or even zero ones can be obtained. So it is necessary to select a threshold (which is a more-or-less trial-and-error parameter) below which the small singular values are substituted for by zeroes and not taken into account in the generalized inverse operator. The knowledge of the spectrum of singular values affords the possibility of relativily easy control of the variance of the resulting solution by various settings of the threshold. Thus in any approach one has to make some compromise between a good data and model resolution and the variance of the solution. Various tests can be performed in order to establish whether the chosen number of unknown parameters is not too low or too high, whether there is a statistically good fit to all the data (i.e. whether the misfit errors have an appropriate distribution, whether the data obey the Gaussian statistics well enough or whether there are some "outliers"), etc. In any case one has to keep in mind that it is only a linearized and not a liner inverse problem which is solved. In solving the seismic tomographical problem both mentioned methods were used with the result that the one based on a singular value decomposition was found more instructive with the computer time consumption tollerable up to round 100 of unknown parameters. Various types of the a priori information were used but mainly the norm of difference of computed and a priori model was minimized and simultaneously the smoothness of slowness distribution in individual layers maximized. The general formulae for such solutions, for the data and model resolution matrices, and for the model covariance matrix were derived (Firbas, 1984d). It is worth mentioning here that usage of the polynomial approximation of type (8) allows the partitioning of rather complicated integrals expressing the smoothness condition which was used in the form

$$\iiint \left\{ \frac{\partial^6 (N_0 + N_1)}{\partial^2 x \, \partial^2 y \, \partial^2 z} \right\} dxdydz \rightarrow MIN \tag{11}$$

## 8. Program package

Utilizing the theory of the "linearized approach" and "generalized inverse", a program package was developed aiming at a routine processing simultaneously treating the refraction and the reflection data. The package is relatively large comprising approximately

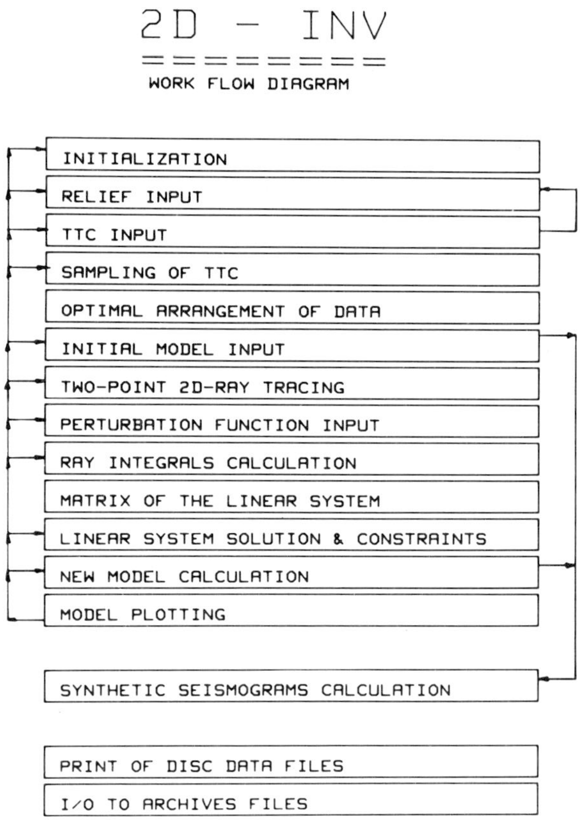


*Figure 2* Structure of the developed program package showing individual program modules (labelled by their tasks) and possible sequences of using individual modules.

15000 FORTRAN lines. The overall structure is outlined in Figure 2 showing basic system modules corresponding to the basic tasks performed. The whole package is written to be so modular as to allow nearly each step to be repeated if necessary; in this way it gives the interpreter a good chance to work interactively. The common data are stored in shared disk files. It is possible simultaneously to use travel-time curves of various types of waves, those originating from different shot points and those acquired through various measurement geometries (profile, cross-hole, etc.). They all may be added and deleted between iterations. The input model is a general 2D model with all features afforded by the direct problem solution (Červený, Pšenčík, 1981). The perturbation function may be formed by a sum of terms of type (8). A term may cover either one layer independently or a set of layers (or even the whole model) in its entirety. Consequently the velocity contrast at interfaces is either kept constant or allowed to vary. It is possible to fix the slowness distribution in individual layers in individual iterations and in this way solve the tomography problem, for instance, from the uppermost layers stepwise down. Both approaches of creating the

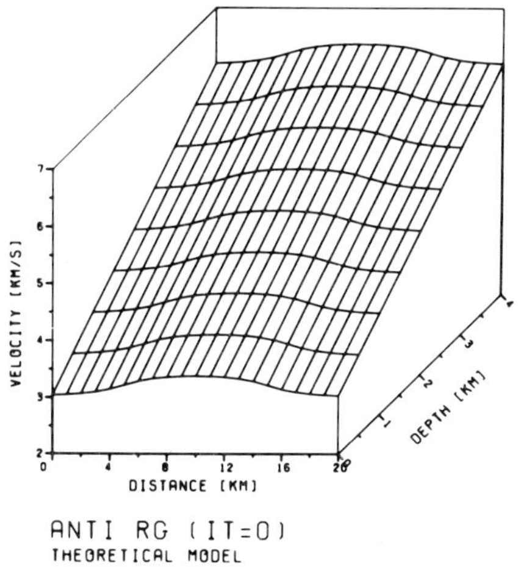


*Figure 3a* Theoretical 2D model of an anticline.

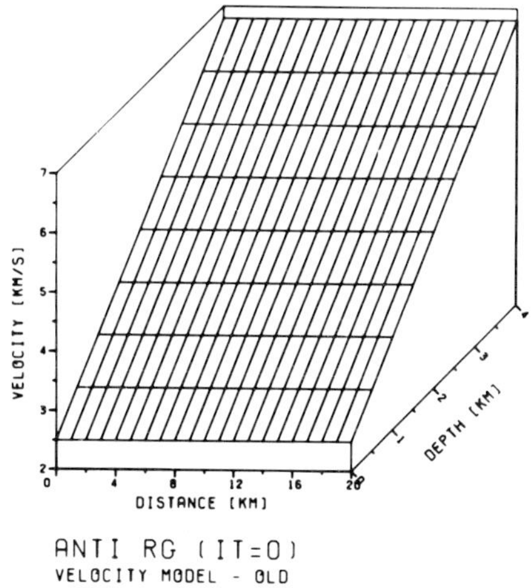


*Figure 3b* 1D starting model.

"generalized inverse" mentioned in part 7 are possible. The whole system is completed with a variety of print, plot, archive and diagnostic program modules.

## 9. Model examples

A large number of test model examples were computed (Firbas, 1984d).Figure 3a shows a theoretical 2D model of an anticline. Figure 3b shows a 1D starting model. The surface profile geometry had 5 shot-points and as input data only refracted waves onsets were used. The computed model can be seen in Figure 3c as a variant to which the model smoothness condition was not applied. Figure 3d shows the computed model after application of the smoothness condition. In comparison with Figure 3c, Figure 3d shows an improvement which can be seen especially at maximum depth where nearly no rays penetrate. The results shown were obtained using 20 unknowns. After the first iteration the absolute deviations from the theoretical model were in the range of 0.1 km/s. Figure 4a shows a "secret model" which was used to compute synthetic sections for the CCSS IASPEI workshop 1984 at Einsiedeln, Switzerland. These sections were distributed to the participants wishing to demonstrate how their algorithms for the inverse problem worked. The resulting model computed by the algorithm presented is shown in Figure 4b. Both refracted and reflected waves in the first layer were used to compute the velocity distribution in the first layer including the low velocity zone on the left side. This remarkable success in resolving this low velocity layer in the proper shape and position may have been achieved also due to the simple shape of the first interface (Firbas, 1984c). The starting model used was a simplified 2D model without the low velocity zone which was obtained by the mentioned interactive program for 2D direct problem solution.

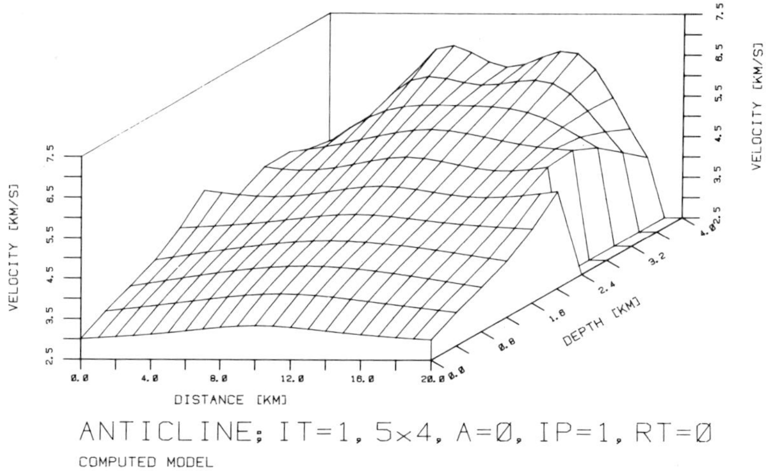


*Figure 3c* Result of generalized inversion after 1 iteration, without the smoothness condition.

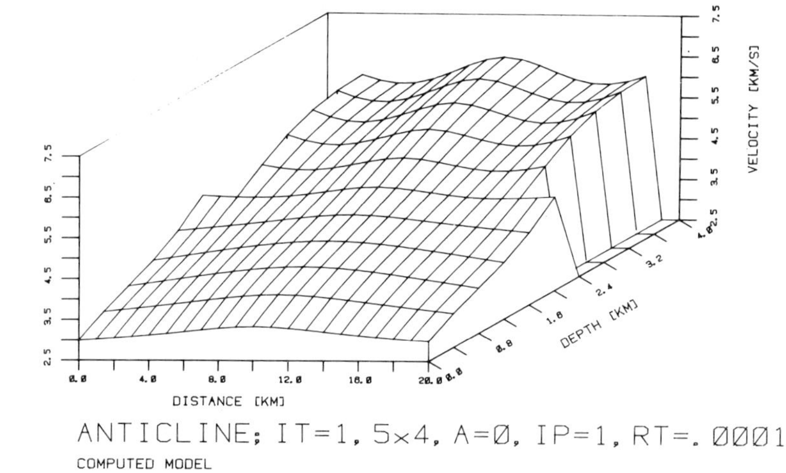


*Figure 3d* Result of generalized inversion after 1 iteration, with the smoothness condition applied.

## 10. Field data tomography examples

This part presents some of the results arrived at in the processing of the field data, and adds a brief commentary.

Figure 5 shows an interpretation for the profile R4, located in the German/Czechoslovakian region. The result was obtained after one iteration. Refracted waves from 8 shot-points were used.

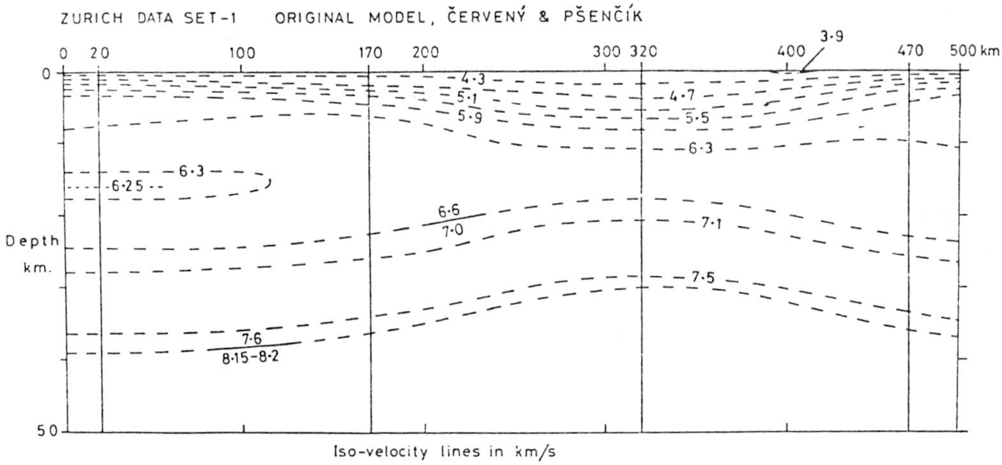


*Figure 4a* The theoretical 2D model.

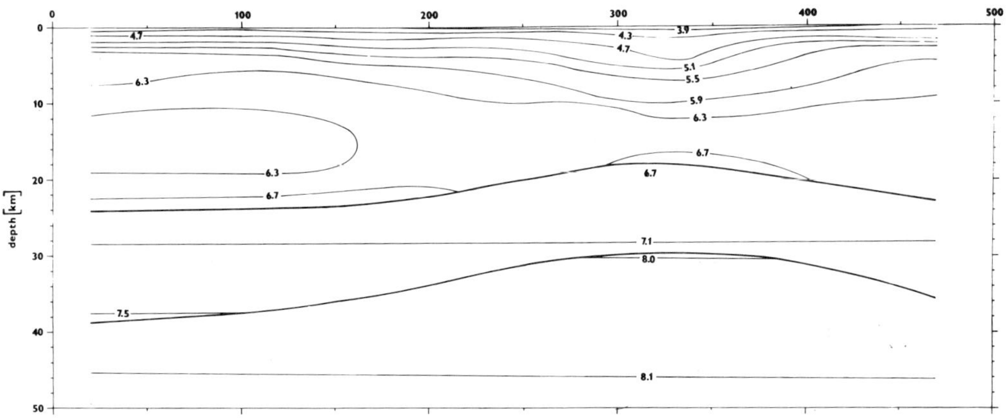


*Figure 4b* The resulting 2D model.

Figure 6, taken from Firbas (1984a), depicts the result for a DSS profile in Saudi Arabia. The picture clearly shows the transition zone between the oceanic and the continental types of crust at the margin of the Arabian Shield and the Red Sea sediments (km 0 - 200). The M-discontinuity is situated somewhere at the contour line 8.0 km/s. The result was obtained after one iteration. The starting model used was only 1D.

Figure 7, finally, shows a velocity profile of comparatively high lateral heterogeneity measured across the Carpathians with an area where high velocity bedrock reaches the

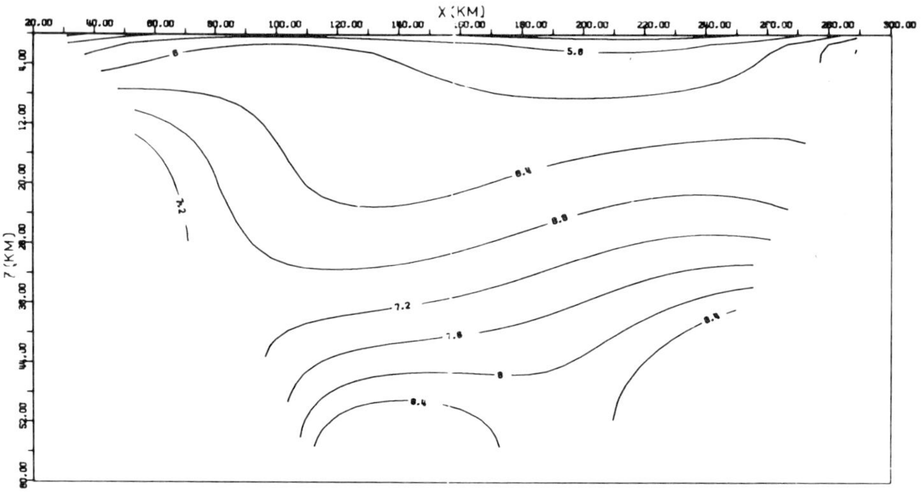

*Figure 5* Velocity model of the international profile R4 (CSSR & FRG), running in the SW-NE direction between Prague and Regensburg.

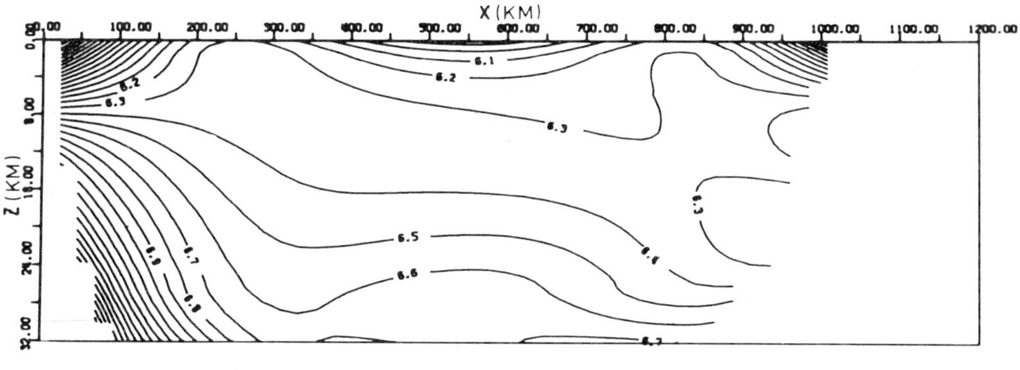

*Figure 6* Velocity model for the DSS profile across Saudi Arabia (Firbas, 1984a). Profile starts in the Red Sea at the SW corner of the Arabian Peninsula, crosses diagonally Saudi Arabia and ends near the Iraq border.

surface and is surrounded on both sides by low velocity sediments. This result was obtained after one iteration. The starting model used was only 1D. There was almost a perfect agreement with a model obtained by many tedious trial-and-error forward modelling steps.

## 11. Conclusions and generalizations of the tomographical approach

The preceding sections offered a brief outline of the theory of the "linearized approach", thoughts on the application of "generalized inversion", a short description of the program package, and both model and field data inversion examples. The theory is not limited by the extent of the survey. It is applicable to surveys ranging from local microsurveys to global

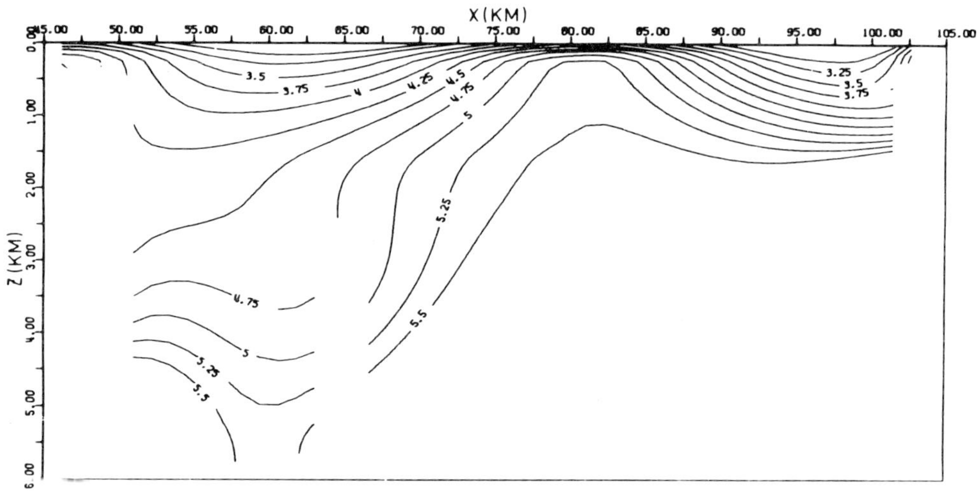

*Figure 7* Velocity model for the Carpathians. Profile crosses the Western Carpathians from N to S on the territory of Czechoslovakia

macrosurveys. A good result (confirmed by direct site examination) was obtained even with a shallow survey at the limits of validity of the ray theory with a model only some meters deep. Even the geometry of measurement does not represent any strong limitation as it mainly influences the two-point ray tracing stage. The examples presented were computed for the most common geometry (surface profile) which is one of the less suitable for the stability of the inverse problem. The program was also tested on model examples for cross-hole measurement with great success. The problem of implementing the 3D medium concerns chiefly the step of model approximation and requires an efficient two-point ray tracing in complex structures, because the outlined spline approximation of the perturbation function as well as the linearization approach and the generalized inversion are directly applicable. Perhaps, generally speaking, possible, but not fully solved up now is the direct implementation of the automatic interface shape determination. Up to now this problem has been solved by successive trial-and-error changes of interfaces between iterations only. The application of such inverse theory to anisotropic media was discussed elsewhere (Červený and Firbas, 1984). There are perhaps limits of rather practical character due to lack of data and due to the complex structure of anisotropic materials. A simple theoretical extension is represented by the inclusion of teleseismic data or data from local earthquakes. Such inversion procedures were published in the literature (Aki,Christofferson,Husebye, 1977; Crosson, 1976; Nolet,1981; Pavlis and Booker, 1980; Spencer and Gubbins,1980; etc.). In the first case only the direction of the teleseismic ray at the border of the model must be known additionally. In the second case some additional parameters related to hypocenter origins must be taken into account additionally.

*Acknowledgments.* This project was carried out under the State Development Plan P-10-347-451 in Geofyzika Brno. The author thanks Prof. Červený (Faculty of Mathematics and Physics, Charles University, Prague) and Dr. Pšenčík (Geophysical Institute of

Czechoslovak Academy of Sciences,Prague) for their valuable advice and for making possible to use their two-point ray tracing program to start with. The author wishes to thank the Management of Geofyzika Brno for permission to publish this paper.

# Chapter 9

# Seismic rock properties for reservoir descriptions and monitoring

A. Nur

## 1. Introduction

It has long been recognized that seismic wave characteristics measured at the earth's surface can provide information not only about the attitude and distribution of interfaces between rock types within the earth, but also about the mineralogy, and the state of the subsurface rocks. In fact much of our knowledge about the internal *constitution* of the earth has been derived from seismic wave characteristics such as velocities and amplitudes.

However seismic methods, notably reflection methods in exploration geophysics have been used extensively in the mostly past to delineate rock interfaces in the earth's shallow crust, to evaluate structures which might bear hydrocarbons. Relatively little use has been made of seismic waves for the determination of the rock properties of direct interest to hydrocarbon recovery (e.g. porosity, permeability), or the direct detection of hydrocarbons. Even in acoustic logging, only the estimation of porosity from velocities has been developed as a regular service. The estimates of permeability or saturation are based on other, nonseismic, methods.

Because of the increasing importance of oil recovery, the growing complexity of recently discovered oil fields, and the growing realization that reservoirs and recovery are more heterogeneous than assumed in the past, a major shift in the use of seismic methods has taken place during the past one or two decades. One of the central aspects of this shift involves the need to establish and understand the relation between the seismic properties of reservoir and reservoir related rocks, and their production properties (porosity,



*G. Nolet (ed.), Seismic Tomography, 203–237.*

permeability) and state (mineralogy, saturation, pore pressure, etc.). Some obvious applications are the evaluation of stratigraphic traps, fracture detection, and the spatial distribution of porosity and permeability.

Seismic methods are almost never used in hydrocarbon recovery assessment, in spite of the growing need to better understand recovery. A major problem which has emerged in the area of reservoir evaluation and production is the realization of the complexity of most reservoirs, leading to large uncertainties in estimated total recovery, recovery rates, and recovery method. Reservoir complexity is typically related to the significant spatial heterogeneity in porosity, permeability, clay content, fracture density, etc. These spatial variabilities cannot be inferred at any level of detail from well teting data, logs, or cores. They may only be obtained, hopefully, from remote geophysical measurement, especially seismic measurement.

A direct consequence of the heterogeneous nature of reservoirs is the complexity of their recovery processes, ranging from problems like the migration of the gas cap in reservoirs with discontinuous shales, overpressure zones, and the tracking of steam or temperature in thermal recovery in reservoirs with large spatial variations of permeability.

There is little doubt that seismic methods will play, in the future, a major role in helping to solve production and recovery problems. But we first need a better understanding of what it is that seismic waves can tell us about reservoir rocks, and how to extract the desired information. This chapter provides a review of what we know at present about the relation between velocities in rocks and (1) porosity, (2) clay content, (3) overburden pressure and stress, (4) saturation, (5) pore pressure, (6) fluid phase behavior, and (7) hydrocarbons and temperature.

## 2. Effects of porosity and clay content

Shaly sandstones and shales are a major component of sedimentary basins and are of foremost relevance to hydrocarbon reservoirs. The acoustic properties of these rocks are thus of great interest in seismic and well log interpretation. For years, the time average equation of Wyllie et al. (1956, 1958) has been used to obtain porosities from acoustic velocity logs. The equation for P wave velocity $V_p$ in water saturated rock is

$$\frac{1}{V_p} = \frac{(1-\phi)}{V_m} + \frac{\phi}{V_f} \tag{1}$$

where $V_m$ is the P wave velocity of the rock matrix, and $V_f$ is the velocity of the pore fluid. When both $V_m$ and $V_f$ are fixed, the only variable in the equation is porosity. To a first order this simple equation appears adequate for clean sandstones in the middle range of porosity ($10\% < \phi < 25\%$). It is well known however that acoustic velocities of sandstones are also related to mineralogy, pore geometry, degree of consolidation, cementation, confining pressure, pore fluid, pore pressure, and temperature. Consequently, the shortcomings of the time average equation have been extensively discussed. A newer empirical equation based on well log data was obtained by Raymer et al. (1980)

$$V_p = (1-\phi)^2 \times V_m + \phi \times V_f \tag{2}$$

as an alternative to the time average equation. Because only the porosity as a parameter is involved in this equation, it and equation (2) cannot be directly applied to shaly sandstones. As indicated by earlier work of De Martini et al. (1976), Tosaya and Nur (1982a) and Kowallis et al. (1984), in shaly sandstones and shales the time average equation significantly overestimates velocities, as does Raymer's model. The question then is how can the effect of clay best be represented in the velocity equation for shaly sandstones?

Because shear wave velocities are now available from well logs and seismic reflection measurements, it is of interest to better understand also the relation between shear velocity and porosity. An empirical relation between shear velocity and porosity has been proposed in the past by modifying the time average equation (e.g. Domenico, 1984). However, as shown in a later section, this equation cannot be used very well to interpret shear velocity values in shaly sandstones.

Han et al., 1986 have carried out an experimental study in which they measured compressional velocity $V_p$ and shear velocity $V_s$ as functions of pressure in 80 sandstone samples with varying clay content and porosity. In addition, the relations among the velocity ratio $V_p/V_s$, water saturation, elastic moduli, porosity and clay content are also examined.

Compressional velocity $V_p$ and shear velocity $V_s$ versus porosity $\phi$ for all samples are shown in figures 1a and 1b. Despite the significant scatter, clear trends indicate that both $V_p$ and $V_s$ decrease with increasing porosity. As a first trial, the modified time average equation $V_p = V_o - A_1 \times \phi$ was fitted to the data by least square regression. The fitted results are presented in the form of relative deviations versus porosity as shown in figure 2. The matrix compressional velocity computed from the fit is $V_o$ = 5.15 km/s, which is much lower than the expected value $V_p$ = 6.05 km/s for nonporous quartz aggregates (Robert, 1982). Furthermore the relative deviations of the data from the values predicted by the equation versus porosity are quite large. However, these deviations clearly depend on the clay content as seen in fig. 2.

The results above (from Han et al., 1986) imply that any model used to fit both $V_p$ and $V_s$ data in shaly sandstones must include a clay content term. Two simple equations including such terms were found to best describe the data by least square regression at confining pressure of 40 MPa and pore pressure of 1.0 MPa

$$V_p\ (km/s) = 5.59 - 6.93 \times \phi - 2.18 \times C\ (km/s) \tag{4}$$

$$V_s\ (km/s) = 3.52 - 4.91 \times \phi - 1.89 \times C\ (km/s) \tag{5}$$

In the very clean sandstones the measured velocities are higher than those predicted by equations (4) and (5) by about 7 percent for $V_p$ and 11 percent for $V_s$ (figs. 3, 4).

The good fit of shaly sandstone velocities, represented by equations (4) and (5) suggest that velocities in sandstones are nearly independent of the type of clays or the location of clay particles within the rock matrix. The coefficients of the linear fits in equations (4) and (5) for both $V_p$ and $V_s$ are fairly constant with differential pressure over 10 MPa.

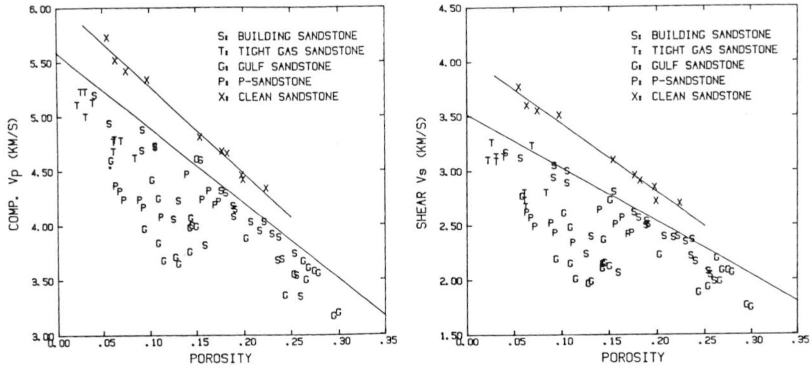


*Figure 1* Measured (a) compressional ($V_p$) and (b) shear ($V_s$) velocities in 80 saturated sandstone samples at confining pressure of 40 MPa and pore pressure of 1 MPa. Straight lines are best linear fits to clay-free sandstones (10 samples) and clay-bearing ones (70 samples) (from Han et al., 1986).

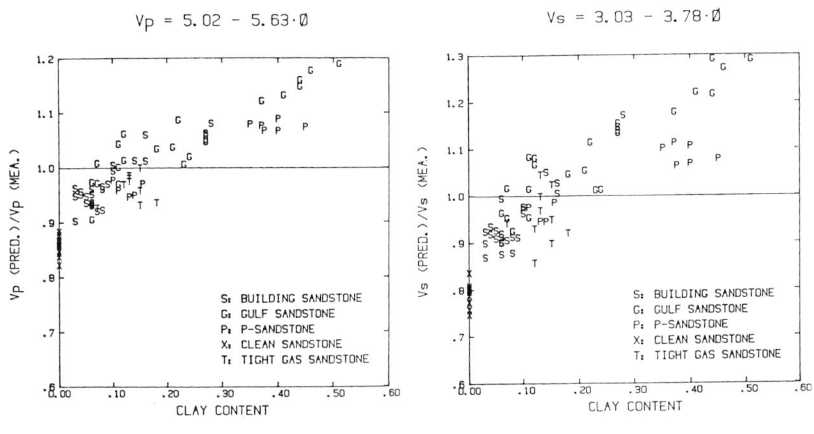


*Figure 2* Deviations of measured (a) $V_p$; and (b) $V_s$ in 80 sandstones from the best linear fit in porosity alone $V = a_0 - a_1\phi$, where $\phi$ is the porosity, plotted against clay content. Note the systematic increase in the difference with increasing clay content in both $V_p$ and $V_s$ (from Han et al., 1986).

Finally, the coefficients in equations (5) and (6) indicate that the influence of clay content (by volume) is about 1/3.2 that of porosity for $V_p$ and 1/2.6 for $V_s$. These ratios were found to be also fairly independent of pressure.

As shear velocity data are becoming more available in seismic exploration and well logging, the velocity ratio $V_p/V_s$ is becoming a useful parameter in the determination of rock properties (e.g. Domenico, 1984; Rafavich et al., 1984; Costagna et al., 1985). Our data show that the velocity ratio for water-saturated shaly sandstones depends on both

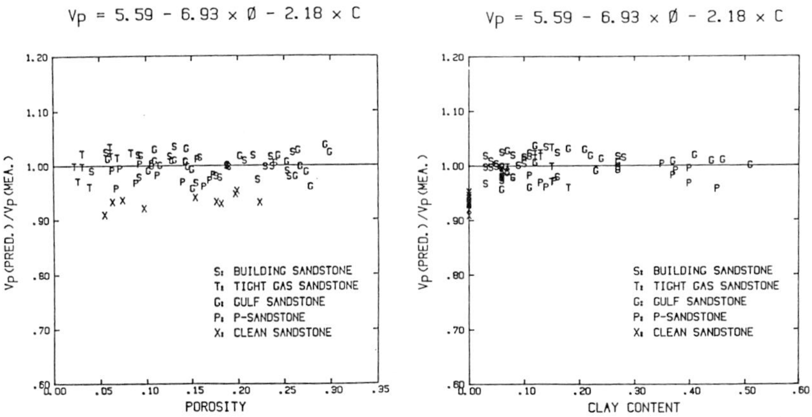


*Figure 3* Deviation of measured $V_p$ values in 80 sandstones from the best linear fit to the data $V_p = a_0 - a_1\phi - a_2 c$, where $\phi$ is porosity and c is volume clay content. Deviations are shown vs. (a) porosity, and (b) clay content. Note that the measured clean sandstones (clay-free) are systematically higher than predicted (from Han et al., 1986).

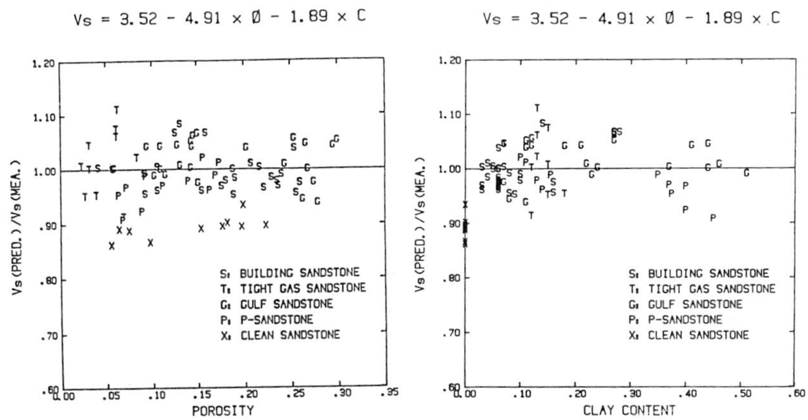


*Figure 4* Deviations of measured $V_s$ values in 80 sandstones from the best linear fit to the data $V_s = b_0 - b_1\phi - b_2 c$ where $\phi$ is the porosity and c is volume clay content. Deviations are shown vs. (a) porosity; and (b) clay content (from Han et al., 1986).

porosity and clay content. By least square regression, this dependence is found to be

$$V_p/V_s = 1.55 + 0.56 \times \phi + 0.43 \times C \qquad (6)$$

The results show that increasing porosity or clay content increases $V_p/V_s$ and that the velocity ratio is more sensitive to porosity changes, in agreement with the results of Costagna et al. (1985). Sandstones with high clay content have velocity ratios and Poisson's ratios similar to carbonate rocks. The resulting ambiguity in the interpretation of velocity data may be resolved by the combined use of the velocity as well as the velocity ratio, providing a useful tool for reliable lithology discrimination.

Finally, Costagna et al. (1985) found that shear velocity is nearly linearly related to compressional velocity for water saturated clastic silicate sedimentary rocks by the equation

$$V_p\ (km/s) = 1.16 \times V_s + 1.36\ (km/s) \tag{7}$$

Our data also show $V_s$ to be nearly linearly related to $V_p$ with somewhat different coefficients than in equation (7). For 75 samples, the best linear least square fit yields

$$V_p\ (km/s) = 1.26 \times V_s + 1.07\ (km/s) \tag{8}$$

## 3. Stress and crack induced velocity anisotropy

Velocities in rocks are often sensitive or even very sensitive to overburden pressure or stress. Figure 5 shows a typical dependence of $V_p$ and $V_s$ on pressure in a sandstone sample. The large increases in the velocities are caused by the closure of cracks and thin gaps at grain contacts under pressure, which induces increasing overall stiffness of the rock. When the cracks in a rock are randomly distributed, and the rock is subject to equal stress in all directions, velocity increases are isotropic. However, rocks with a *nonrandom* distribution of cracks exhibit elastic wave anisotropy (Nur and Simmons, 1969; Nur, 1971). The preferred orientation distribution of cracks may be either intrinsic, such as in shales or metamorphic rocks, or induced by nonhydrostatic stress. An example of the effects of nonhydrostatic stresses on the elastic properties of rocks is shown in figure 7, obtained for a granite cylinder, which was loaded uniaxially in a simple press (fig. 6). Four sets of measurements were made: (1) compressional waves normal to the cylindrical axis, (2) shear waves propagating normal to and polarized normal to the axis, (3) shear waves propagating normal to the axis and polarized parallel to axis, and (4) shear waves propagating parallel to the axis. Each set consists of velocity as a function of uniaxial stress and as a function of the angle $\theta$ (fig. 7).

The results show that velocities increase with stress in all directions, but that the magnitudes of these increases depend on the angle between the direction of the applied stress and the propagation direction of the waves. For P waves the largest effect on velocity is observed when the wave propagates in the direction of the applied stress, the smallest when the wave propagates in a direction perpendicular to the stress (fig. 7). For shear waves, the increases depend also on the direction of wave polarization. The velocity of the shear wave polarized parallel to the axis of the cylinder exhibits large dependence on direction whereas that of the wave polarized normal to the axis are fairly independent of direction.

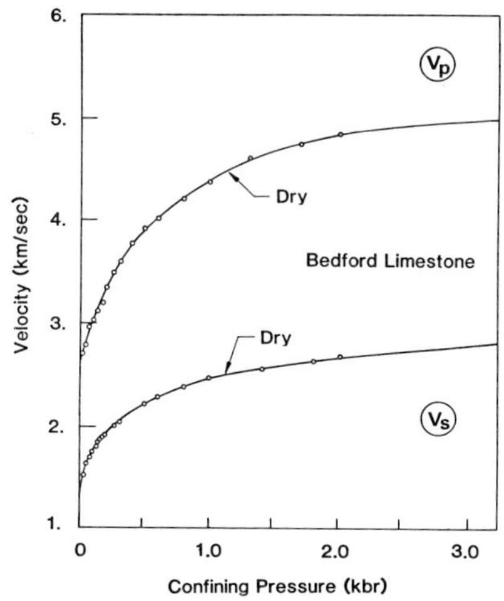


*Figure 5* Typical dependence of $V_p$ and $V_s$ in a dry rock on confining (or overburden) pressure. The large increase of the velocities is due to the closure of the most compliant portions of the rock's pore space under external pressure (from Nur and Simmons, 1969a).

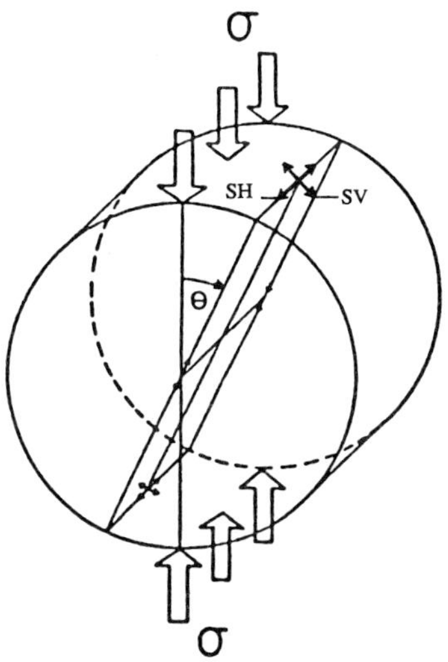


*Figure 6* Geometrical relations between the direction of applied stress α, the direction of wave propagation θ, and the direction of particle motion for quasi P, SV, and SH waves. SV stands for polarization in the plane which includes the direction of applied stress, whereas SH stands for the polarization which is orthogonal to SV (from Nur and Simmons, 1969b).

The results show that rock becomes acoustically anisotropic under uniaxial stress conditions. As it turns out the shear and compressional velocities change with direction, at a given stress level, in a manner expected from the elasticity of crystals. Distinct shear waves with different velocities exist in any direction of propagation when uniaxial stress is

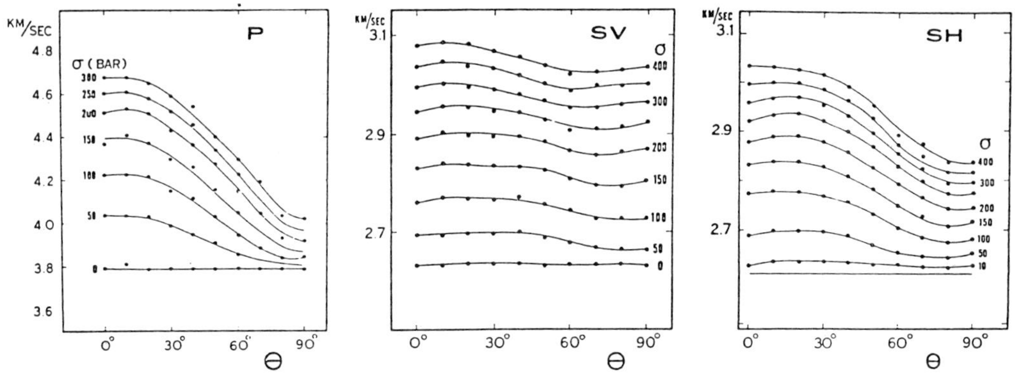


*Figure 7* The dependence of the compressional and the two shear velocities on direction of propagation θ, relative to the direction of applied stress (θ = 0°). The directional variations imply that the stress has induced velocity anisotropy, and the differences between SV and SH imply the velocity birefringence is also induced (from Nur and Simmons, 1969b).

applied. Therefore, the influence of stress on velocity can be described in terms of the anisotropy elements of an elastic crystal. Using this theory, it can be shown (Nur, 1971) that the anisotropy due to *uniaxial stress* corresponds to hexagonal symmetry with the approximate expressions

$$P^2 \approx P_{11}^2 \sin^2\theta + P_{33}^2 \cos^2\theta$$

$$S_1^2 \approx S_{12}^2 \sin^2\theta + S_{13}^2 \cos^2\theta$$

$$S_2^2 \approx S_{13}$$

where subscript 3 refers to the direction of the applied stress. Similar results are obtained when a rock is subject to *pure shear stress*, with orthotropic symmetry of the wave field. We find that the measured values of $S^2$ for the polarized wave in the plane of the applied stress are almost independent of direction, again in excellent agreement with equation. Finally, the theory predicts the occurrence of acoustic birefringence that becomes more pronounced with increasing stress as shown in figure 8.

The stressed rocks were assumed to possess an initial microcrack distribution with spherical symmetry. Most rocks with cracks, however, exhibit significant initial anisotropy. We can describe the correspondence between the symmetries of a general stress field and the stress-induced velocity anisotropy. Various combinations are presented in table 1, indicating several symmetry types (Paterson and Weiss, 1961). Particularly important is the direction of stress application with respect to the principal directions of initial anisotropy. The resulting induced velocity anisotropy depends on this relative direction. If, for example, the material possesses an initial axial symmetry of elastic properties and the applied stress is uniaxial, the resulting anisotropy may be axial, orthorhombic, or monoclinic, depending on the direction of the principal stress.

Interesting compressional wave velocity measurements were published also by Thill et al. (1969). They measured P velocity in various directions in a spherical sample of Salisbury granite and clearly showed that the orthorhombic pattern of velocity must be

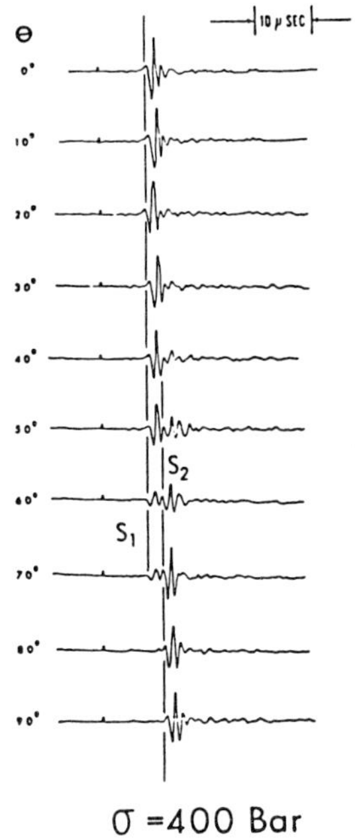


*Figure 8* Observed arrival of shear wave forms traveling along the axis of the cylinder at 400 bars at various angles θ between the direction of applied stress and direction of polarization of the transducers. Acoustic birefringence is most apparent at $\theta = 70°$, where the SV amplitude is about one-third the SH amplitude. At $\theta \ll 70°$, SV amplitude tends to mask the later SH arrival. At $\theta > 70°$, SV arrival is so weak relative to SH that it is hard to discern (Nur and Simmons, 1969b).

related to the distribution of orientations of small cracks in quartz grains (fig. 9).

Another rock type in which velocity anisotropy is the rule are shales. Figure 10 shows measured compressional and shear velocities in a shale vs. overburden pressure, with values given for waves travelling parallel, perpendicularly and at 45° to the bedding, or shale parting planes. The magnitudes of the velocity anisotropy $M = (V_{max} - V_{min})/V_{average}$ are 9 percent and 15 percent for $V_p$ and $V_s$ respectively at room pressure, but *increase* with pressure, or equivalently with depth to 12 percent and 15 percent respectively at 1000 kb (15,000 psi).

## 4. Velocity, saturation and pore pressure

Some of the main factors which control compressional $V_p$ and shear velocities $V_s$ in porous rocks with fluids are (a) confining pressure, (b) pore pressure, and (c) saturation (e.g. Nur and Simmons, 1969). Figure 11 shows these effects for Indiana Limestone, with a pattern which is typical of many sedimentary rocks. As shown in figure 11, the compressional and shear wave velocities in dry and saturated conditions depend very differently on the saturation of samples, on confining pressure, and on pore pressures.

**Table 1. Dependence of symmetry of induced velocity anisotropy on initial crack distribution, applied stress, and its orientation**

| Symmetry of initial crack distribution | Applied stress | Orientation of apllied stress | Symmetry of induced velocity anisotropy | Elastic constants |
|---|---|---|---|---|
| Random | Hydrostatic | | Isotropic | 2 |
| | Uniaxial | | Axial | 5 |
| | Triaxial | | Orthorombic | 9 |
| Axial | Hydrostatic | | Axial | 5 |
| | Uniaxial | Parallel to axis of symmetry | Axial | 5 |
| | Uniaxial | Normal to axis of symmetry | Orthorombic | 9 |
| | Uniaxial | Inclined | Monoclinic | 13 |
| | Triaxial | Parallel to axis of symmetry | Orthorombic | 9 |
| | Triaxial | Inclined | Monoclinic | 13 |
| Orthorombic | Hydrostatic | | Orthorombic | 9 |
| | Uniaxial | Parallel to axis of symmetry | Orthorombic | 9 |
| | Uniaxial | Inclined in plane of symmetry | Monoclinic | 13 |
| | Uniaxial | Inclined | Triclinic | 21 |
| | Triaxial | Parallel to axis of symmetry | Orthorombic | 9 |
| | Triaxial | Inclined in plane of symmetry | Monoclinic | 13 |
| | Triaxial | Inclined | Triclinic | 21 |

The compressional and shear velocities in dry rock increase markedly with overburden pressure, as was shown and discussed already in connection with figure 5. When the same sample is fully saturated at room pressure, a large increase in $V_p$ is obtained, whereas a small or no change is observed in $V_s$.

The effect of pore pressure is to counteract that of overburden pressure. Consequently, in $V_s$ and particularly $V_p$ in dry (air or gas in the pore space) rock, are very much lower when the gas pressure is equal to lithostatic ($P_p = P_c$) than when it is equal to atmospheric pressure ($P_p = 0$). In the brine-saturated rock, $V_p$ and $V_s$ are also lower when pore pressure is equal to overburden pressure $P_p = P_c$ than when it is equal to atmospheric pressure $P_p = 0$. However, because $V_p$ in saturated rock is relatively high, the relative change in $V_p$ due to $P_p$ is smaller than in $V_s$.

The strong dependence of velocity on pressure and saturation is confined to low overburden pressures. At pressure above 1 or 2 kb and without pore pressure all velocities show only a small increase with increasing stress. The velocities of the saturated sample were measured also when pore pressure was equal to the external pressure. The velocities changed little with external pressure and the value of dV/dp, a constant over our range of

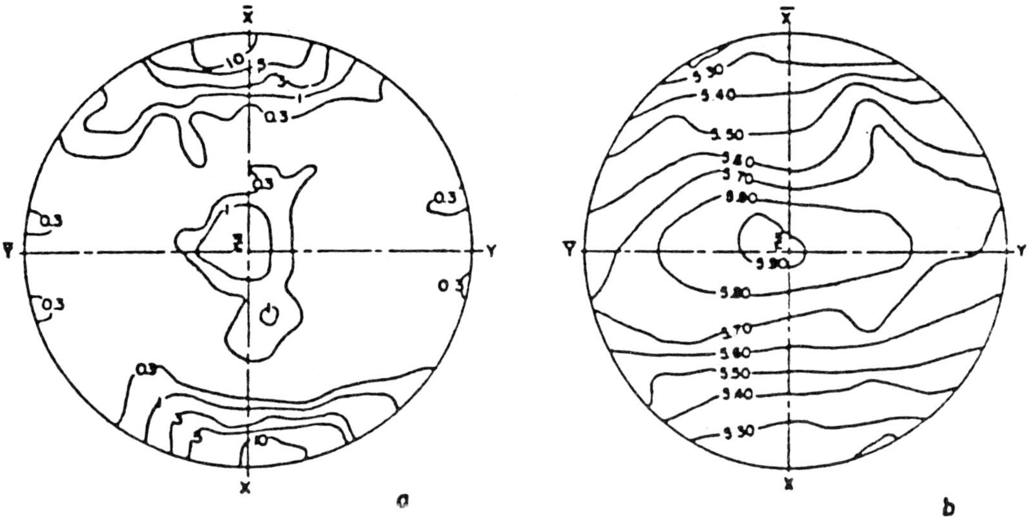

*Figure 9* Comparisons between microcrack and velocity anisotropy. Poles of microfracture planes in quartz grains and longitudinal wave velocities in Salisbury granite. Contours indicate (a) concentration of poles and (b) velocity in km/s. Data are present on equal-area projection (from Thill et al., 1969).

pressure, is approximately the same as for (1) saturated, and unconfined rock and (2) dry rock at high pressure. Although $V_p$ of the confined sample of Indiana Limestone (fig. 11) is lower by 10 percent, and $V_s$ by 35 percent, from the corresponding unconfined velocities the slopes are identical within experimental error.

From the measured velocities we can obtain values of the effective elastic constants of the dry and saturated samples. We assume that the effective elastic constants are related to the velocities in the same way that these quantities are related in a linear elastic material. Thus the effective dynamic bulk modulus

$$K = \rho[V_p^2 - \frac{4}{3} V_s^2]$$

and the effective shear modulus

$$G = \rho V_s^2$$

where ρ is the density of the sample. Shown also are the effective Young's modulus E and Poisson's ratio ν, for both dry and saturated cases. The resulting values (fig. 12) reveal why fluid saturation so greatly influences $V_p$ and not $V_s$. As seen in figure 12 it is the effective bulk modulus of the rock which is responsible for the entire change whereas the shear modulus is almost independent of saturation. The Poisson's ratio ν is of some interest, too. Dry rocks exhibit very small, sometimes even negative values of Poisson's ratio while saturated rocks exhibit abnormally high values. From the expression for Poisson's ratio

$$\nu = (3K - 2G)/(6K + 2G)$$

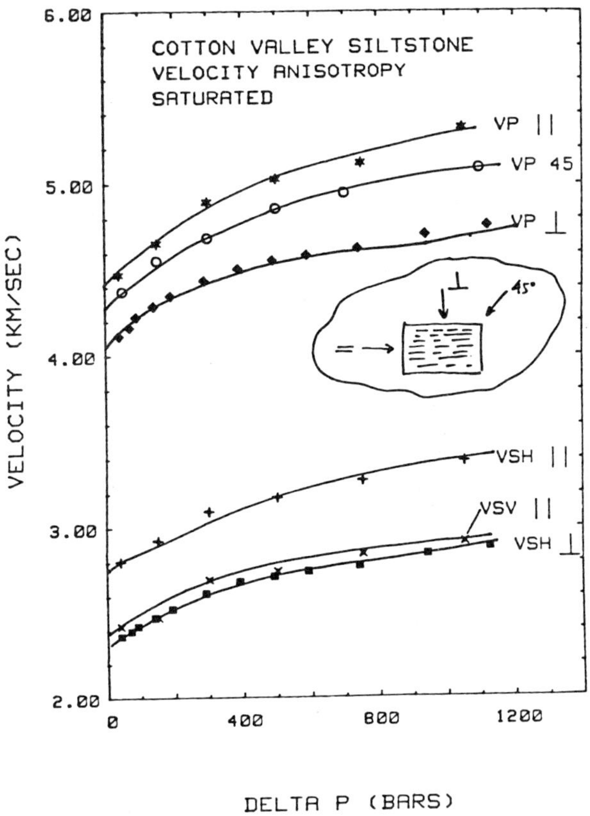


*Figure 10* Directional velocity data for fully saturated anisotropic Cotton Valley shale. Note that the magnitude of the anisotropy increases with increasing differential pressure or equivalently depth in the earth (from Tosaya and Nur, 1982).

it is apparent that a negative Poisson value indicates that $K < \frac{2}{3}\mu$ and $V_p < \sqrt{2}\cdot V_s$. Such low values in dry rocks are observed at very low pressures only. The effective value at higher pressures is near the intrinsic value.

The effect of confining pressure on rock is to deform the most compliant part of the pore space (e.g. microcracks and loose grain contacts) and thus increase the stiffness of the rock, i.e. the effective bulk and shear moduli. The effect of high pore pressure is to mechanically oppose the closing of cracks and grain contacts by the confining pressure, thus leading to low effective moduli and velocities.

The influence of the pore fluid, as separate from its pressure, is related to its compressibility. When pore fluid is relatively incompressible (brine), the effective bulk modulus of the rock is high. In contrast, the shear modulus is barely changed, because the

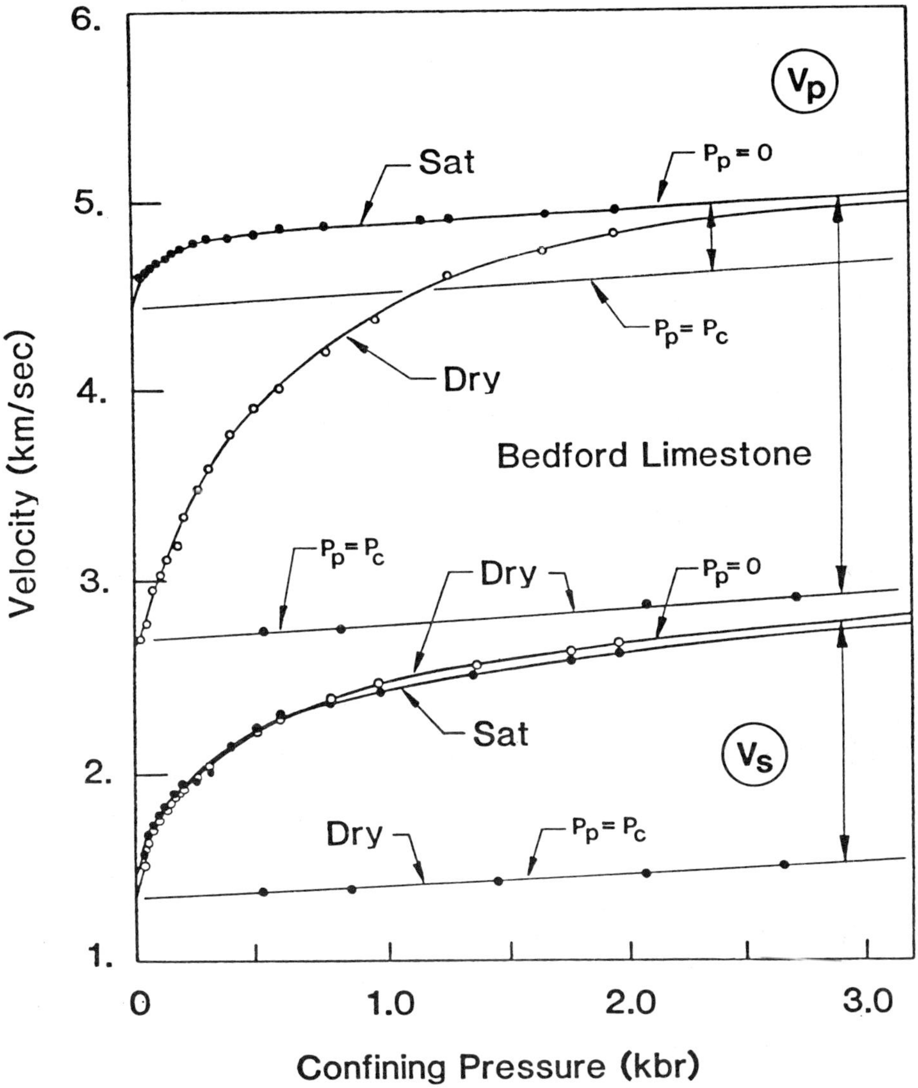


*Figure 11* Velocities in dry, and saturated Bedford limestone as a function of confining pressure ($P_c$). Results are shown for atmospheric pore pressure ($P_p = 0$) and lithostatic pressure ($P_p = P_c$). Note the large differences between dry and saturated $V_p$, and in contrast the small difference between dry and saturated $V_s$. Note also the large effect of high pore pressure in decreasing both $V_p$ and $V_s$.

viscosity of the pore fluid is low, so that the stiffness in shear does change when the pore fluid is changed from air to brine.

Two questions arise: first, how do velocities vary between the two extreme pore pressure cases of $P_p = 0$ and $P_p = P_c$; and second, how does $V_p$ vary between the low

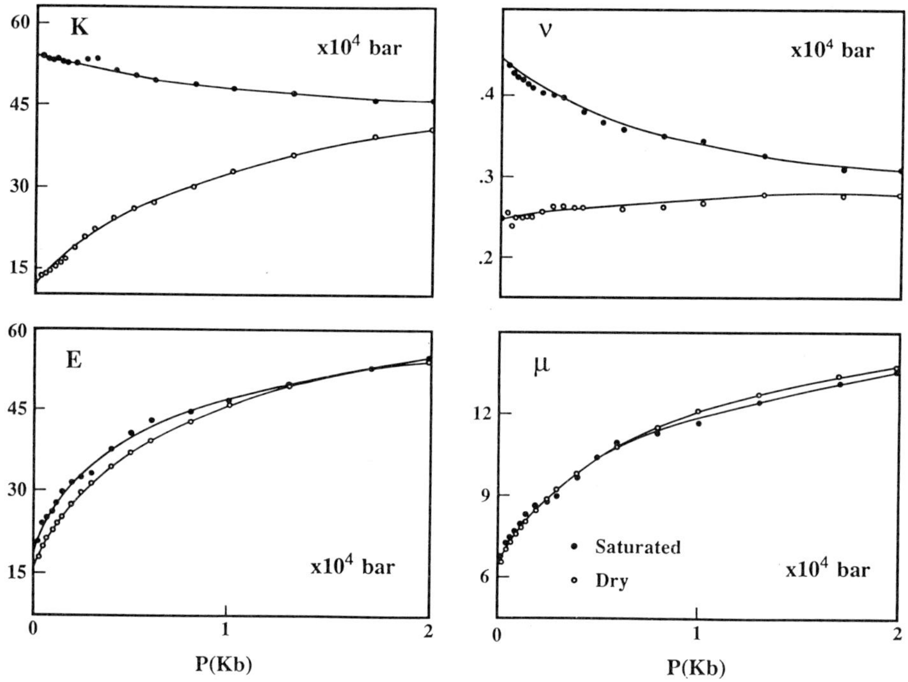


*Figure 12* The effective dynamic bulk (k) and shear moduli (μ), Poisson's ratio (ν), and Young's modulus E derived from the compressional and shear velocities of figure 4.1 for Bedford limestone (from Nur and Simmons, 1969a).

value when the rock is dry, and the high value when it is saturated at given confining and pore pressure. Extensive data (e.g. Christensen and Wang, 1986) show that to a first approximation, velocities are governed by the effective pressure

$$P_{eff} = P_c - \alpha \cdot P_p$$

where $\alpha$ is a constant. For many rocks the value of $\alpha$ is found close to 1 for both $V_p$ and $V_s$, although significantly smaller values are common in low porosity rocks.

Figure 13 shows the dependence of $V_p$ and $V_s$ on the degree of saturation S (Murphy, 1982). It is remarkable that neither $V_p$ nor $V_s$ shows any sensitivity to saturation, except when S becomes close to 1.0, when $V_p$ markedly increases from the low, "dry" $V_p$ value to the high fully saturated $V_p$ value. These results indicate that velocity measurements *cannot* yield information on the degree of saturation in reservoir rocks. However, amplitude or attenuation data, as shown in figure 13, is somewhat sensitive to saturation (Winkler and Nur, 1982). When attenuation, or $Q^{-1}$ is low for both P and S waves, and the P and S velocities are low, too, the rock has low ($S_w < 50\%$) water saturation. When both velocities are low and $Q_s^{-1}$ is low but $Q_p^{-1}$ is high the rock has low gas saturation ($95\% > S_w > 50\%$), and when $V_p$ is high, $Q_p^{-1}$ low, $V_s$ low, the rock is fully saturated ($S_w = 100\%$).

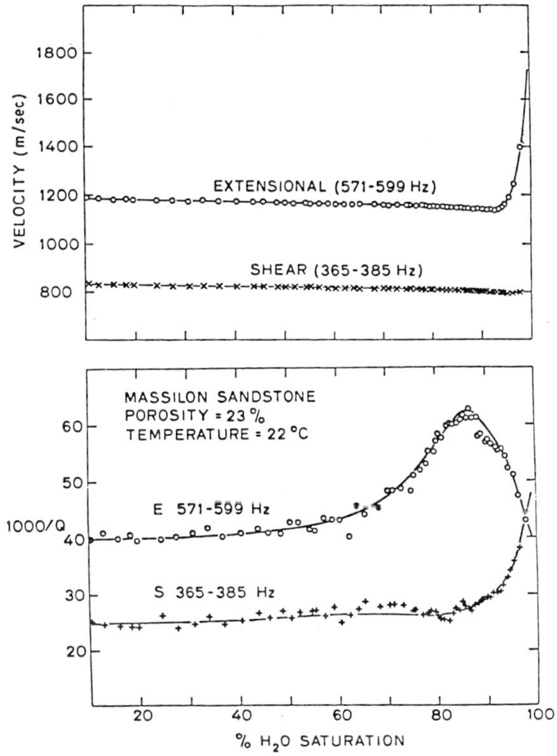


*Figure 13* The dependence of compressional and shear velocities and their specific attenuation on partial saturation in sedimentary rock. Note the absence of velocity changes with saturation, except for $V_p$ when saturation is close to 100 percent. In contrast, $Q^{-1}$ data suggest that it might be possible to distinguish between low water saturation (low Poisson's ratio, modest $Q_p^{-1}$ and $Q_s^{-1}$), high water saturation (low Poisson's ratio, high $Q_p^{-1}$ and modest $Q_s^{-1}$) and very high water saturation (high Poisson's ratio, low $Q_p^{-1}$ and high $Q_s^{-1}$) (from Murphy, 1982).

Figure 14 provides a schematic summary of these relations represented by the ratios of $V_p/V_s$ and $Q_p^{-1}/Q_s^{-1}$. These relations might someday be used for in situ estimation of the degree of saturation.

The results above immediately suggest a number of useful characteristics of velocity in rocks in situ: (a) The dependence of velocities - both $V_p$ and $V_s$ - on confining pressure implies that velocities should generally increase with overburden, or depth, in the crust; (b) The large effect of saturation on $V_p$, and relatively small effect on $V_s$ is the reason for two important exploration concepts: the use of brightspots to detect gas pockets, which show up as low compressional velocity zones, and the use of shear waves in exploration, because $V_p$ carries different information about the reservoir rock than $V_s$; and (c) The effect of pore pressure suggests that seismic velocities may be used to infer in situ pore pressure.

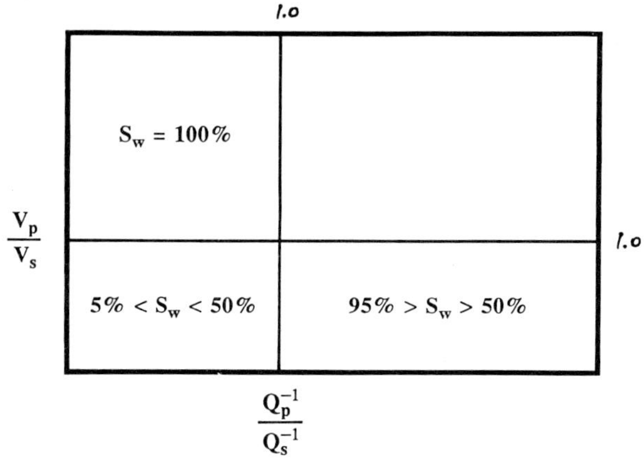


*Figure 14* The ratios of $V_p/V_s$ and $Q_p^{-1}/Q_s^{-1}$ and their relation to the degree of water saturation ($S_w$) in porous rocks (after Winkler and Nur, 1982).

## 5. Velocity-phase transitions

A variety of interactions between passing seismic or acoustic waves, and rock with pore fluid systems undergoing phase transitions or chemical reactions are possible, and a few have been identified and investigated. The effects of such phase transitions on wave propagation can be divided into several types: simple changes in velocities due to changes in the moduli or densities of the material involved, for example upon the melting of solid hydrocarbons in rocks with increasing temperature. Often these simple velocity changes are associated with high attenuation peaks, because wave energy is used up to help drive the transformation. Changes involving minima in velocities as well as Q occur when the wave interacts strongly with the transformation process - e.g. when the transformation rate is equal to the wave period and when compressibility becomes very high - e.g. the beginning of the separation of gas out of solution in the pore fluid.

Geophysically there are a number of transformations and reaction which are of particular interest: the transition between hot water and steam transition for geothermal reservoir exploration, delineation and monitoring; the melting of heavy hydrocarbons in situ during thermal recovery; the freezing and thawing of water-bearing rocks in permafrost conditions; and chemical reactions such as cracking of hydrocarbons and their oxidation.

### 5.1 Water-Steam Transition

Ultrasonic compressional and shear velocity measurements were made on rock samples (Ito et al., 1979; DeVilbiss, 1980), sealed in an impervious jacket. Each sample was subjected to a constant confining pressure with pore pressure of 15 bars in an externally heated pressure vessel with silicon fluid as the pressure medium. After raising the temperature to 150° C, the pore pressure was decreased stepwise while pulse-transmission velocity and first arrival wave amplitude were measured. At a pore pressure of approximately 4.7 bars, the saturation pressure of water at this temperature, most of the water in the pores evaporates or is displaced by steam (Keenan et al., 1969). Upon reaching atmospheric

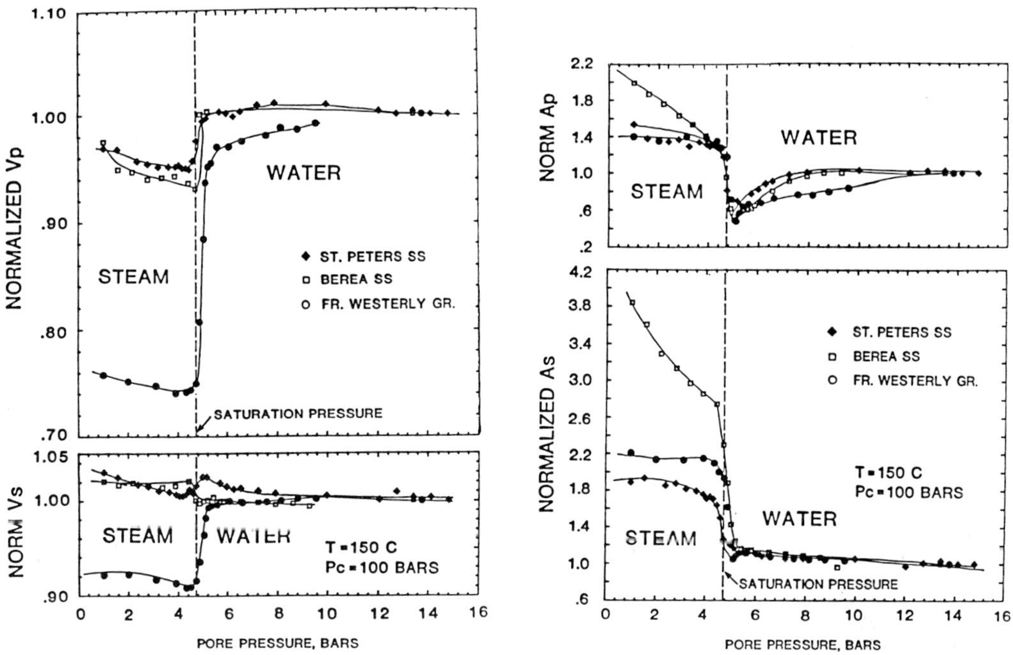


*Figure 15* Compressional and shear wave (a) velocities and (b) amplitudes in rocks with steam and hot water in their pore space. The transition between hot water and steam and 150°C occurs at 4.7 bars. In general, velocities increase sharply upon the transition from steam to hot water in the pores, and attenuation shows a sharp decrease in shear, and a sharp minimum in compressional amplitudes.

pressure the procedure was reversed.

Figure 15 shows the measured velocities as a function of pore pressure. The values have been normalized with respect to the velocity at the highest pore pressure corresponding to the transition rock. The dotted line indicates the water-steam transition (Keenan et al., 1969) at the temperature of the experiment.

The P velocity in the fractured granite shows a large change at the water-steam transition, due to the change in bulk modulus of the pore fluid. The data indicate that the transition in the granite is smeared over 2 bars around the transition pressure of water. Furthermore the variation in velocity with transition in the fracture granite is much greater than in the sandstone. Part of this difference may be due to density changes in the pore fluid upon the transition, which counteracts the effect of the decreases of the bulk modulus of the fluid (Ito et al., 1979).

In Berea Sandstone the abrupt increase in shear velocity at the water-steam transition in only 50 percent of the change expected from the density change due to the total expulsion of water from the pores, meaning that the effective shear modulus of the rock decreases as steam replaces water in the pores. The data also show a 9 percent change in shear velocity in the fractured granite when the pore pressure is near the water-steam transition, corresponding to shear modulus change of 18 percent. Again this change is clearly due to a change in effective shear modulus. Although the magnitude of the change in effective shear modulus for the St. Peter Sandstone is much smaller, it appears to be present as well.

Figure 15b shows values of peak amplitudes of the first arrival reflecting the attenuation of waves in the sample. The values are again normalized with respect to amplitude at the highest applied pore pressure. All three samples show a sharp minimum in P wave amplitude $A_p$ just at the saturation pressure of water, with an increase to 0.5 at pore pressure below the saturation.

Shear wave amplitude $A_s$ for the three samples is distinctly different from $A_p$. Unlike $A_p$, shear attenuation increases monotonically through the water-steam transition in the pores. $A_s$ is almost constant with pressure above the saturation pressure where the P wave attenuation is decreasing, and it changes with fluid pressure much more than $A_p$ at low pressure.

The results may be explained as due to the mixture of steam vapor and hot water in the pores at the phase transition conditions. With a few percent steam the density of the pore mixture is relatively high and similar to water, but the bulk modulus is low and similar to steam. The shear velocity, which is insensitive to the bulk modulus of the fluid inclusions, is barely influenced, whereas the compressional velocity is sensitive to the bulk modulus and undergoes a measurable change. Furthermore, the increased attenuation of the compressional waves near the liquid-vapor transition is probably due to localized fluid flow, and thus similar to the peak in $Q_p^{-1}$ in figure 13.

The experimental results may be relevant for geothermal exploration, where P wave attenuation may be much greater than S waves, and the ratio of the amplitudes $A_p/A_s$ may be a particularly sensitive indicator of the presence of steam underground. For example, Mahood and McEvilly (personal communication) observed a decrease of the P to S wave amplitude ratio in the La Primavera caldera, Jalisco, Mexico. Anomalously low values of the ratio were observed at the center of the caldera where surface steam manifestations are evident.

### 5.2 Water and Ice

In figure 16 are shown the velocity and attenuation of compressional waves in $H_2O$ as a function of temperature, across its freezing point. As might be anticipated the velocity is relatively high in solid ice (below $-40^o$ C) and lower in water (above $0^o$ C). The large increase in velocity between $-40^o$ C to about $-2^o$ C must be related to the change in ice's crystal structure close to melting. However the most significant change in velocity is the large drop upon melting, around $0^o$ C. As shown in figure 16 this drop in velocity is associated with an equally large peak in attenuation, which clearly spans the range of temperature of the phase transition.

The reason for the drop in velocity upon melting is mostly due to the disappearance of the shear modulus, as solid ice becomes liquid water, and the associated changes in the compressibility. The large loss of wave energy during the transition is most likely caused by the interaction of the wave with the phase transformation itself - with some of the wave energy being used to promote the transformation.

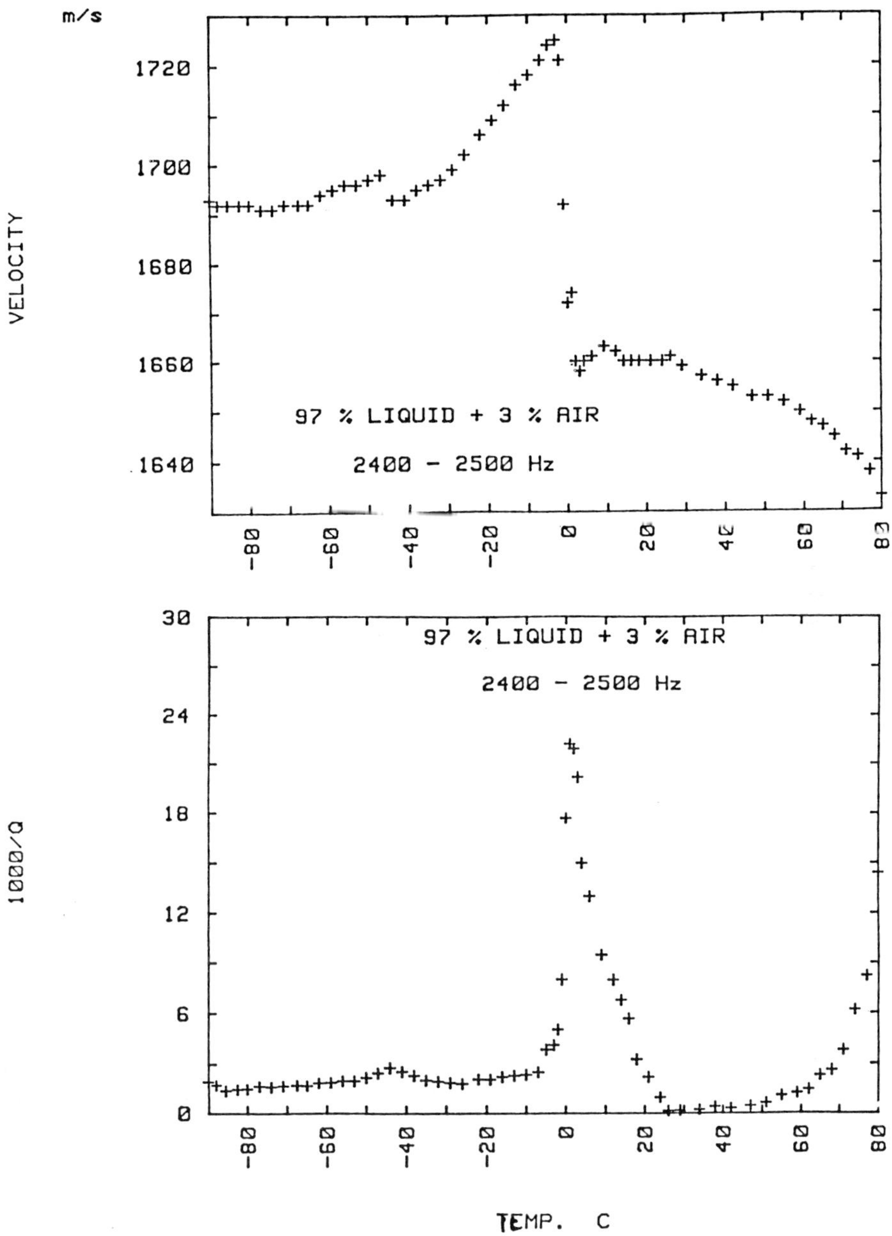


*Figure 16* Effect of freezing on compressional (a) velocity, and (b) attenuation in $H_2O$. Similar changes are found in rocks saturated with $H_2O$ upon freezing (from Ito et al., 1979).

Figure 17 shows wave velocities in porous rock containing $H_2O$, vs. temperature. Much like the ice-water system itself, the rock shows also a significant decrease of velocity

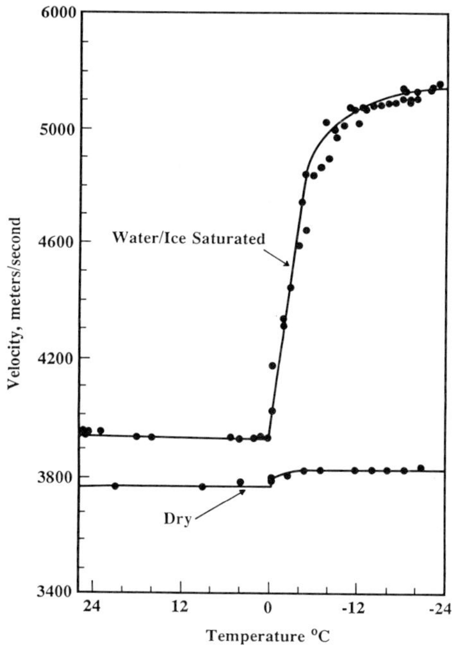


*Figure 17* The effect of temperature on velocity in rocks with $H_2O$ , as it undergoes melting of the ice in the pore space (after Timur, 1968).

upon melting (Timur, 1968; King, 1977).

These results suggest the kind of seismic characteristics which might be indicative of the bottom of permafrost zones, which are often highly irregular and consequently need mapping. The results suggest also that monitoring of the thawing front of permafrost, e.g. around deep production wells through which warm oil is being pumped up, may be quite feasible.

## 6. Velocity-hydrocarbons-temperature relation

In 1984, Tosaya and Nur (1984) first discovered surprisingly large decreases with temperature of velocities in core of heavy oil or tar sands from Venezuela, California, and Canada. In some rocks decreases of velocity reached 40 percent and more with temperature increasing from $20^o$ to $120^o$C. Because these changes were not anticipated, a systematic study was undertaken to identify their causes and understand what controls their magnitude.

Measurements of velocities were first done in (1) purified hydrocarbons Eicosene and wax; and (2) natural heavy oil and tar from several fields around the world. Secondly, a

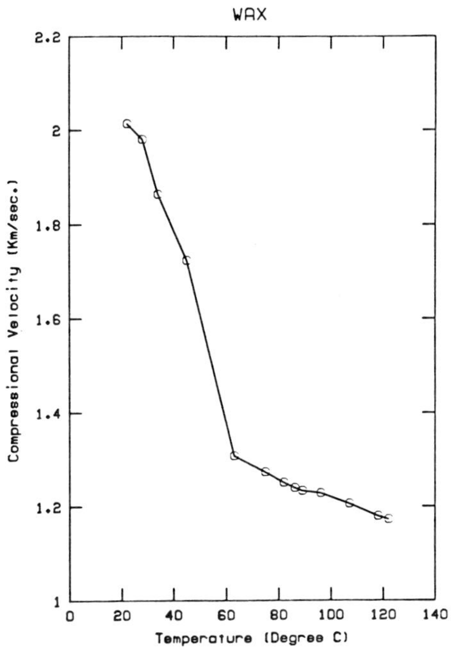


*Figure 18* The effect of temperature on the compressional velocity in paraffin wax. Note the large change associated with melting.

laboratory study was made of the effects of temperature on wave velocities in rocks (well-cemented Massillon sandstone and unconsolidated Ottawa sand) saturated with the above heavy hydrocarbons, as well as natural heavy oil sands and tar sands from Venezuela, California, and Canada.

## 6.1 Wave Velocity in Hydrocarbons

One of the most striking results of the study (Wang and Nur, 1986) is that compressional velocities $V_p$ in hydrocarbons decrease very much with temperature. For example in the wax (fig. 18) a large decrease of the velocity is found in the temperature interval of $20^o$ to $65^o$C. At higher temperatures further decreases in $V_p$ are also observed. The compressional velocity in pure Eicosene similarly decreases in its melting interval of $27^o$ to $29^o$C (fig. 18).

Temperature clearly has the largest effect on the compressional velocities in wax, and similarly in Eicosene (Wang and Nur, 1986) near their melting temperatures. As these solid hydrocarbons soften and melt their shear moduli decrease rapidly with temperature leading in turn to large decreases in the compressional wave velocities. Once the solid hydrocarbons are liquid their velocities depend only on their bulk moduli which are less sensitive to temperature, so that further changes in velocity with increasing temperature are smaller.

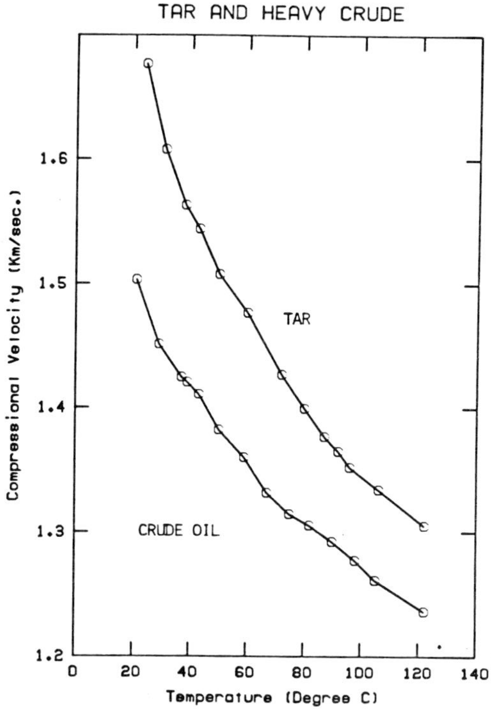


*Figure 19* Compressional velocities in heavy crude oil and in tar vs. temperature. Note the large temperature sensitivity of the velocity, which in this case is clearly not due to melting (from Wang and Nur, 1986).

The compressional wave velocities in samples of natural recovered heavy crude and the tar (fig. 19) decrease with temperature almost linearly to 80$^o$ or 90$^o$ C. These decreases are most likely caused by the increasing compressibility of the crude and the tar with temperature. Figure 19, indicates also that the rate of decrease with temperature of the compressional wave velocity in the crude and the tar is smaller beyond 70$^o$ to 90$^o$ C.

Figure 20 shows measured velocity values vs. temperature in pure alkanes and alkenes. The results show clearly a systematic relation between $V_p$, temperature, and the inverse of the molecular weight of the hydrocarbons. The results suggest that it might be possible eventually to obtain information from seismic velocities about the type of hydrocarbon present in rock.

## 6.2 Velocities in Rocks with Hydrocarbons

Compressional and shear velocities in Massillon Sandstone were measured after saturation with wax For comparison, velocities in Massillon saturated with water and with air were also measured, as shown in figure 21.

Compressional and shear velocities in both Massillon Sandstone and clean Ottawa sand saturated with wax are significantly higher than the water-saturated values in the temperature interval of 20$^o$ to 45$^o$ C. This difference is due to the higher effective elastic

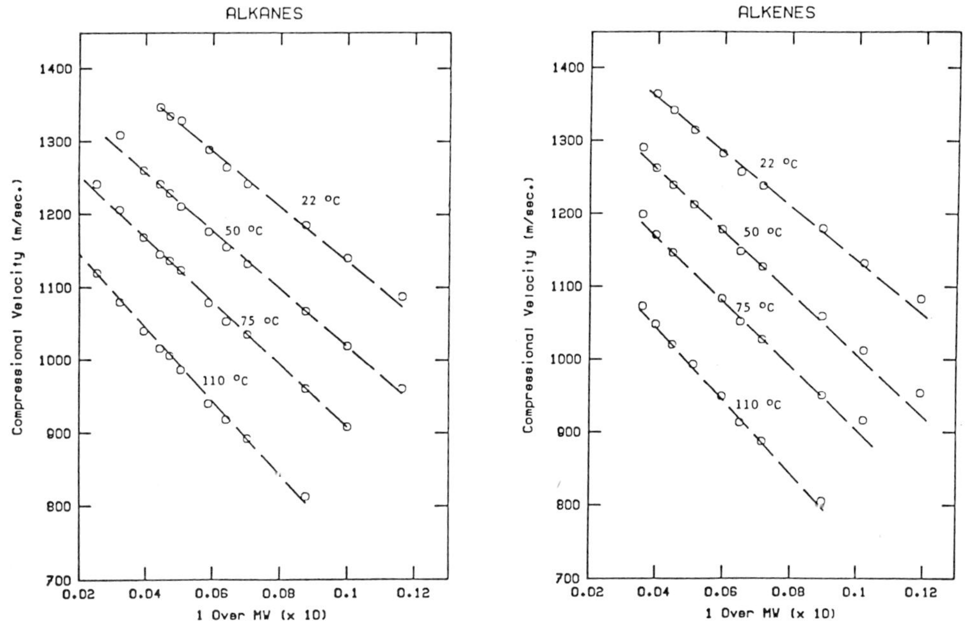


*Figure 20* Compressional velocities in selected alkanes and alkenes vs. (*molecular weight*)$^{-1}$ and temperature at room pressure (from Wang, 1986).

moduli of the rocks when water in the pores is replaced by the solid wax. The effect is particularly large for the shear velocity, because shear waves do not propagate through liquids.

As the wax in the pores completely melts at temperatures above 65° C, both compressional and shear velocities in both samples significantly decrease. The decrease in $V_p$ is mainly due to the lowered $V_p$ in the wax itself. As the wax in the pores turns to liquid at temperatures above 65° C, $V_s$ in the wax-saturated rock becomes close to $V_s$ in water-saturated rock, because both liquids cannot support shear stresses, and thus have little effect on the shear velocities (King, 1966; Nur and Simmons, 1969; Murphy, 1982).

The compressional wave velocity in Ottawa sand saturated with heavy oil at confining pressure of 20 MPa and pore pressure of 5 MPa shows a rapid decrease with increasing temperature from 20° to 100° C (fig. 22), whereas $V_p$ in the water-saturated sample decreases much less. The rapid decrease of $V_p$ in rock with crude is most likely due to the rapid decrease of $V_p$ in the crude itself, which exhibits similar dependence on temperature (fig. 19).

The compressional wave velocities versus temperature for the two tar concentrations are very close to each other, suggesting that the amount of the tar in unconsolidated sand does not affect $V_p$ too much.

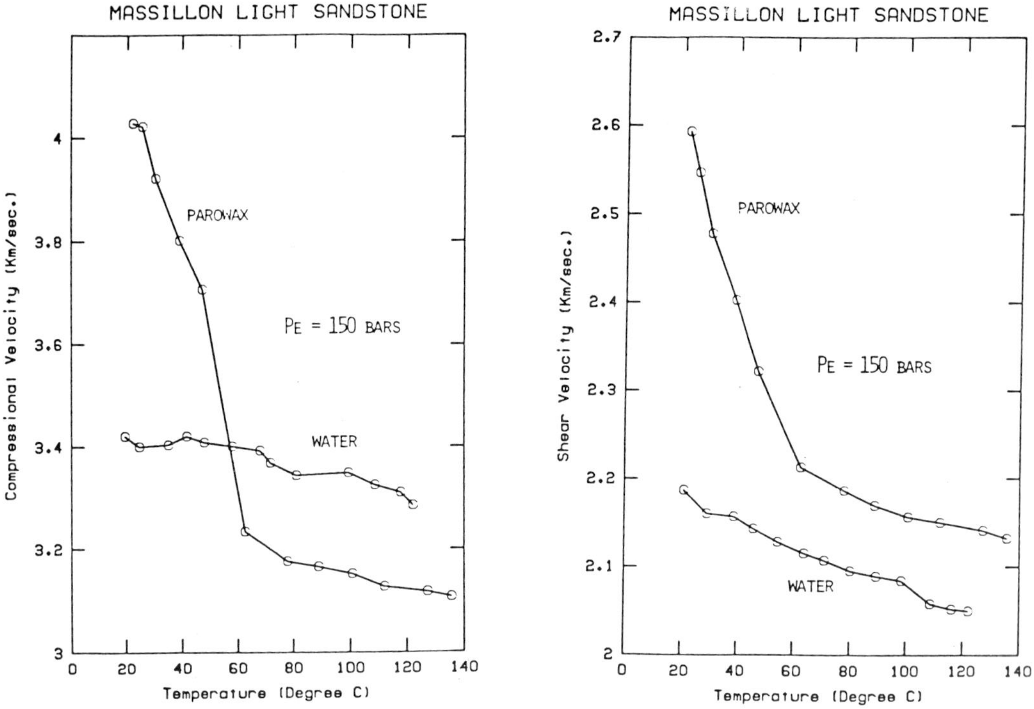


*Figure 21* The effect of temperature on (a) $V_p$, and (b) $V_s$ in sandstone saturated with paraffin wax. Note again the large decrease of $V_p$ and $V_s$ in the melting interval of the paraffin wax (from Wang and Nur, 1986).

The unconsolidated Ottawa sand sample with 10.7 percent tar contains also 16 percent volume air. When this space is filled with water at controlled pore pressures, the compressional wave velocity increases (fig. 22). Below $70^o$ C, $V_p$ in this tar and water sand decreases significantly with increasing temperature. Above $70^o$ C the decrease of $V_p$ with increasing temperature is smaller. This suggests that the decrease in $V_p$ at temperatures below $70^o$ C is mainly caused by the decrease in $V_p$ of the tar itself, and that the decrease above $70^o$ C may be due to high pore pressure.

### 6.3 Velocities in Natural Oil and Tar Rock Sands

Results were also obtained on the effects of elevated overburden pressure, pore pressure, temperature, and oil/brine ratio on velocities in *natural* oil and tar sand samples from: (1) Venezuela sand (2) California and (3) Canada.

A sample from the Lake Maracaibo area was used to study the effect on velocities of oil-to-brine ratio for samples with (1) 100 percent oil, (dead crude of API gravity 12), (2) 50 percent oil, 50 percent brine, and (3) 100 percent brine in the pores.

Measurements were made of the P and S wave travel times and first-arrival amplitudes of transmitted ultrasonic pulses. The velocity values in the samples (fig. 22) show that in

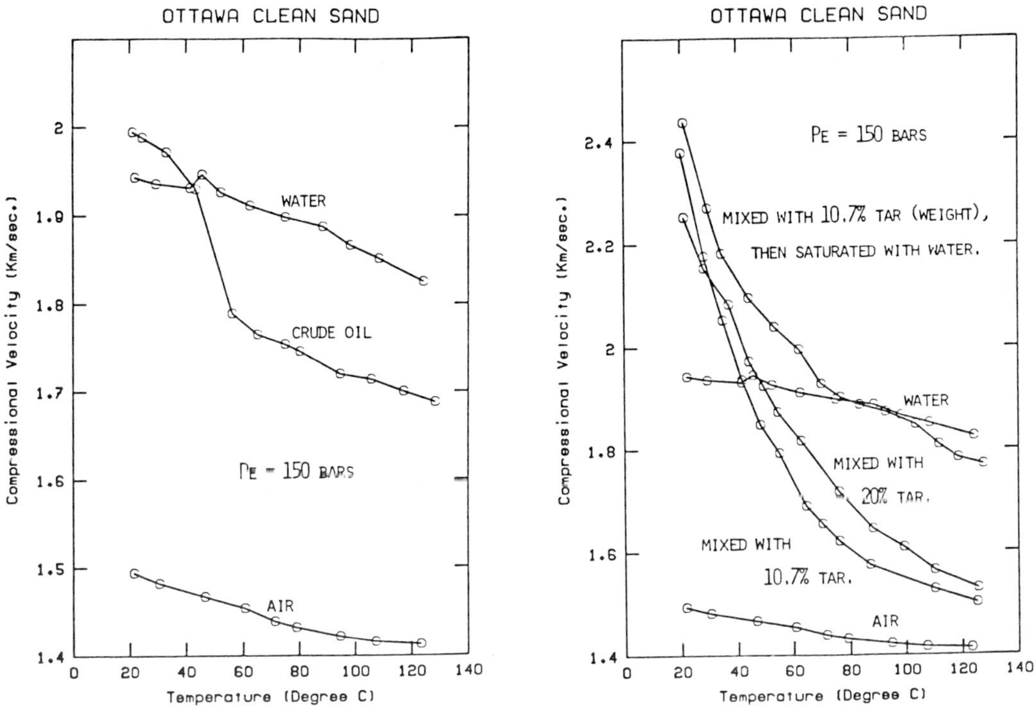


*Figure 22* Compressional velocities in Ottawa sand with (a) crude, and (b) tar in the pore space vs. temperature. Data for samples with brine and air are also shown (from Wang and Nur, 1986).

oil-saturated sands velocities are extremely sensitive to temperature, while only nominally dependent on stress. This large decline in compressional velocity (nearly 40 percent) over the limited temperature interval of 25° to 150° C at constant differential pressure, suggests that velocities may serve as highly accurate thermometers, with potential application in monitoring temperatures in heated oil zones around injector wells in steam floods or fire floods.

The pressure and temperature effects are strikingly reversed when oil is replaced by brine as the pore phase. With hot brine occupying the pore space, velocities are found to be strongly dependent on differential pressure, and fairly independent of temperature. The strong temperature dependence of the oil-saturated core (fig. 23) is thus evidently due solely to the presence of oil. A third Venezuelan sample was tested with 50 percent oil/50 percent brine in the pore space. The results, shown in figures 23, indicate that the pressure and temperature effects are intermediate to the two end cases. Velocities show a moderate dependence on both temperature and differential pressure.

In addition to the velocity data, first-arrival P-wave amplitudes were collected (Tosaya et al., 1984). Large decreases are found in the normalized P-wave amplitude with increasing temperature in the samples that contain oil, due to attenuation. In the sample

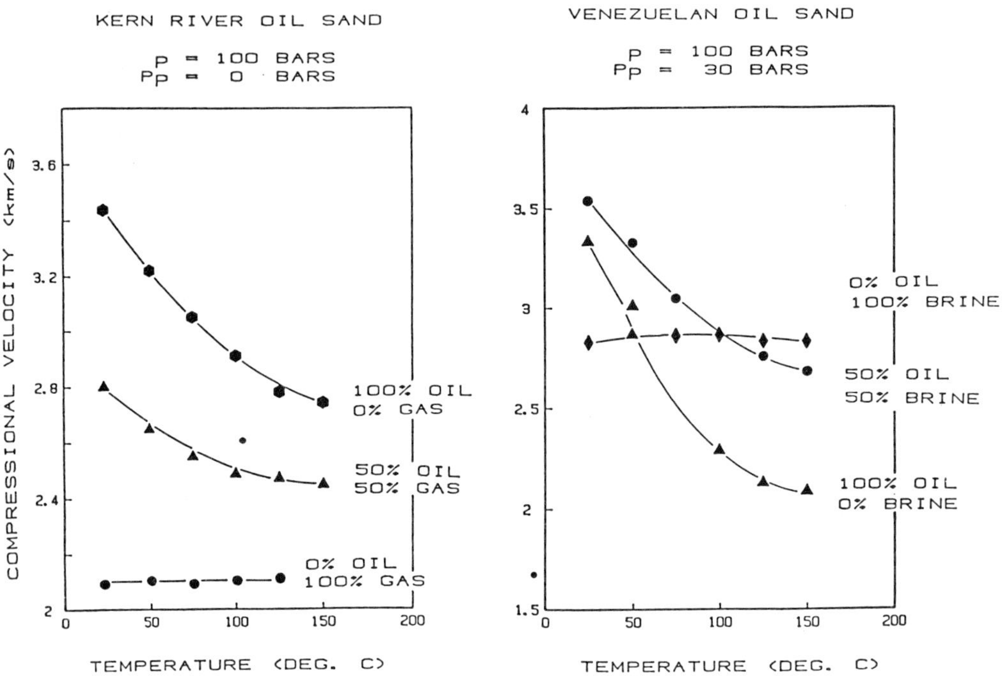


*Figure 23* Compressional velocities in heavy oil sands from California and Venezuela, vs. temperature. The largest decreases in velocity of 20 percent and 43 percent respectively are found in samples saturated with oil. No temperature dependence remains in the clean, brine-saturated samples. Mixtures of 50 percent brine and 50 percent oil yield intermediate dependence on temperature. These results imply that the oil is responsible for the very large temperature dependence in these rocks (from Tosaya and Nur, 1982).

with 100 percent brine pore fluid the amplitude loss over the temperature interval from $25^o$ to $150^o$C is about 15 percent; for the sample with half oil and half brine the effect is about 45 percent; and for the sample that was 100 percent oil saturated the effect is about 60 percent.

Velocity data for three samples of Kern River sand with oil contents of 100 percent, 50 percent, and 0 percent are also presented in figure 23. Although the dependence of velocity on temperature that is characteristic of this reservoir is considerably lower than that of the Venezuelan sand samples, the general behavior is similar, with the largest decrease in compressional velocity with temperature shown by the 100 percent oil-saturated sand, no dependence shown by the 0 percent oil-saturated sand, and an intermediate dependence shown by the 50 percent oil-saturated sand. Similar results were also obtained for saturated Athabasca tar sand sample (Tosaya et al., 1984) showing a 70 percent decrease in compressional velocity over the temperature interval from $25^o$ to $200^o$C - nearly twice the effect in the Lake Maracaibo sample.

In figure 24 we compare the compressional to the shear velocities vs. temperature in Kern River heavy oil sands. The results show that both $V_p$ and $V_s$ are sensitive to temperature, and that consequently it is not only the bulk modulus, or compressibility of the hydrocarbons vs. temperature which is involved, but also the viscosity, as well as the

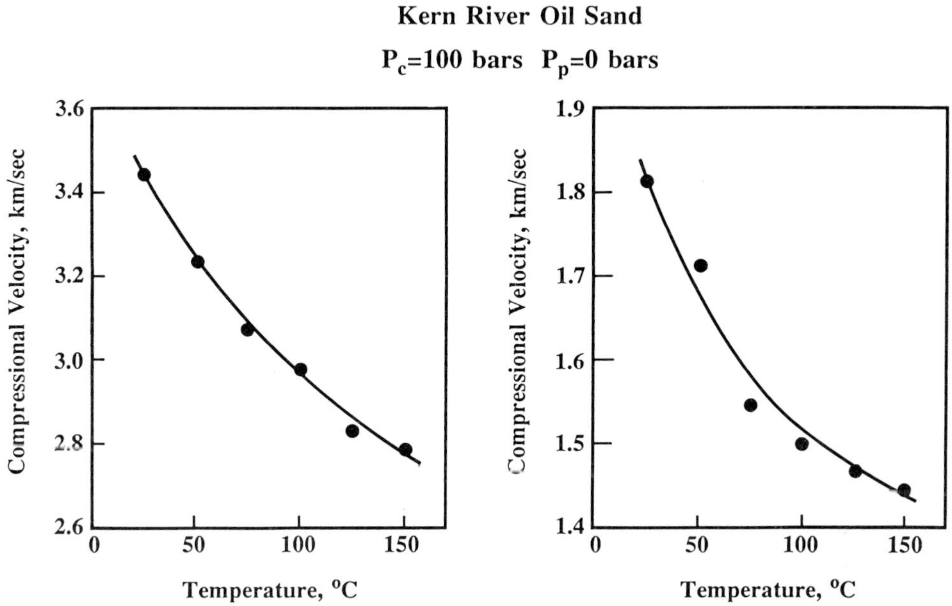


*Figure 24* Compressional and shear wave velocities in Kern River oil sands vs. temperature (from Tosaya, 1982).

interactions of these hydrocarbons with the rock. The details of these interactions are at present not well understood, as the results fail to agree with the existing simple models of velocities in rocks with fluids.

Figure 25 shows the dependence of $V_p$ in the Maracaibo heavy oil sands vs. temperature as a function of pore pressure. The dependence on temperature is the same at low and high pore pressure, implying that the large decrease of velocity with temperature is *not* due to the generation of free gas upon heating, as might be suggested by some. Instead, the results further support the conclusion that the velocity-temperature sensitivity is related to the behavior of the crude itself.

The very large magnitudes of the effects of temperature and steam on the measured seismic properties of these laboratory samples strongly suggest that efforts to monitor thermal EOR fronts, including hot water and steam floods, should be highly successful if seismic signal strength and spatial resolution are adequate. In addition, the velocity and amplitude data shown as functions of temperature in heated reservoir sands in this report indicate that seismic properties can be used as a thermometer to map the spatial distribution of heated oil within the reservoir.

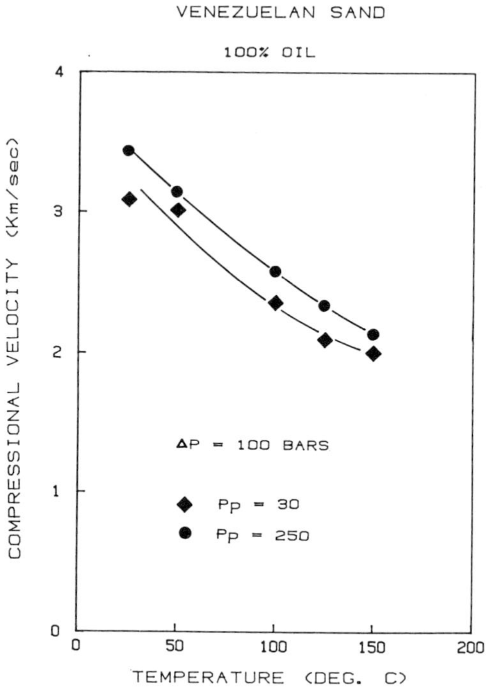


*Figure 25* The effect of pore pressure on the temperature dependence of the compressional velocity in a saturated Venezuelan oil sand. These results imply that the effect is due to intrinsic properties of the oil, and is not due to a free gas phase which might be liberated during heating (from Tosaya et al., 1982).

## 7. Applications

Table 2 lists some of the more obvious applications of seismic velocities and attenuation in the description of reservoirs and the monitoring of their recovery.

### 7.1 Porosity and Permeability Mapping

Knowledge of the spatial distribution of the three first-order reservoir parameters - porosity, permeability, and saturation - is one of the most desirable applications of seismic wave measurements. Such measurements can be analyzed using for example the results of section 2. An interesting study along this line was published by Robertson et al. (1983), indicating that inferring spatial porosity variation from interval velocities is quite feasible. Robertson's study, like some other attempts to use the interval velocities extracted from seismic data to map lateral variations of porosity are based on simple regression formula or semi-empirical relations between porosity and velocity. Because of the complex sensitivity of acoustical velocities to many different rock parameters, the use of an assumed *unique* velocity-porosity relation is unrealistic.

Doyen et al. (1984) have proposed, instead, a geostatistical estimation technique for porosity mapping from seismic data in petroleum reservoirs. In their method interval velocities in the reservoir layer are combined with well porosity measurements to predict the spatial distribution of porosity throughout the entire reservoir even where wells are

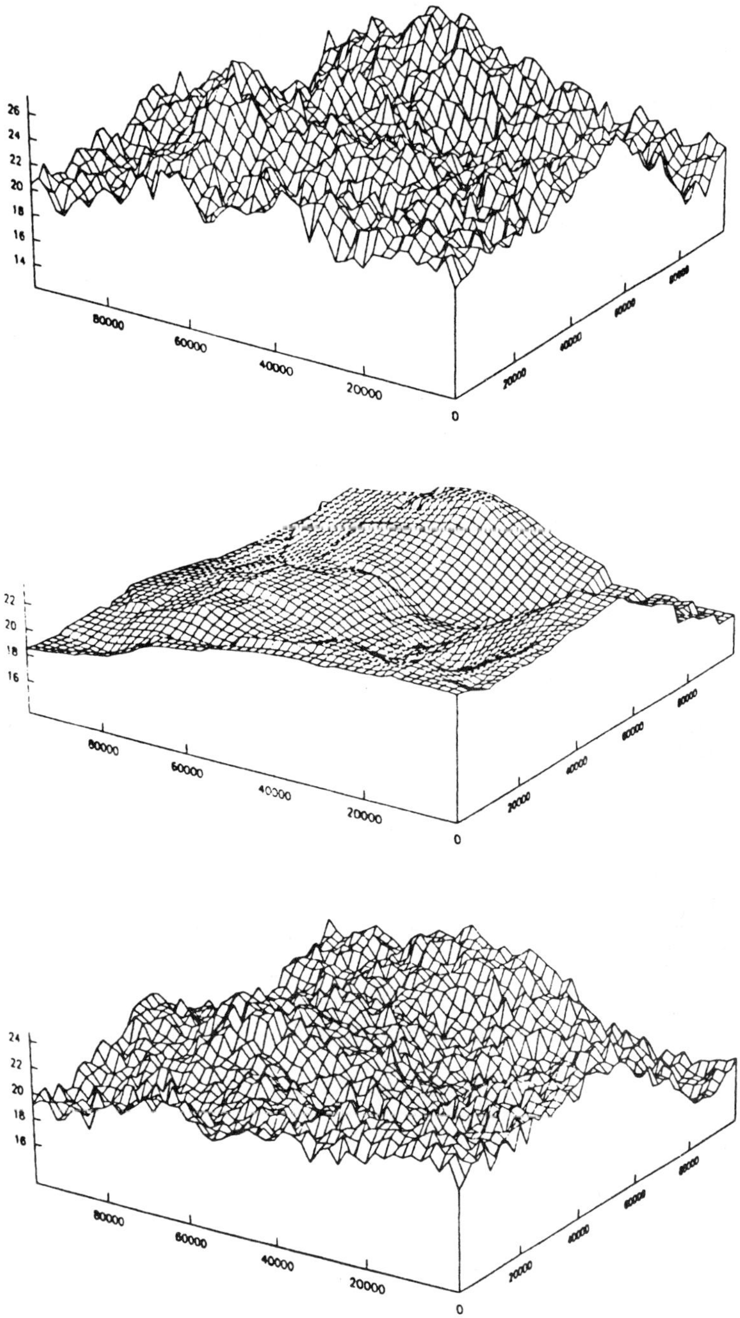


*Figure 26* Mapping porosity using few direct well data for porosity and abundant indirect seismic data (from Doyen, personal communication). The upper figure shows a simulated porosity map. The middle figure shows a porosity map estimated from 44 wells "drilled" in the simulated field. The lower figure shows the estimated porosity distribution based on the 44 wells plus abundant seismic data (Doyen et al., 1984).

sparse or absent and porosity is consequently greatly undersampled. This is done by inferring the pattern of joint spatial variability of porosity and velocity across the reservoir from the accurate but sparse well data and the abundant but inaccurate (as far as porosity is concerned) seismic information. The spatial crosscorrelation between the two parameters is then used to derive linear mean square estimates of porosity. The mapping technique was tested on a numerically simulated reservoir model. Figure 26 shows for comparison, Doyen et al. (1984) estimated porosity distribution based on wells alone, and the estimated porosity distribution using the interval velocity data as well.

It is conceivable that in the future permeability may similarly be estimated from seismic velocity data, although the uncertainties associated with such estimates may be much greater than those for porosity.

### 7.2 Anomalous Pore Pressure Detection

Figure 27 shows a schematic of a low-velocity zone in both $V_p$ and $V_s$ associated with a zone of anomalously high pore pressure, or so-called geopressured zone. This behavior follows from the dependence of velocities on pore pressure and confining pressure, as shown in figure 11. These results imply that *both* compressional and shear velocities should be anomalously low in a zone of high pore pressure. This is in contrast with a gas-bearing zone, in which $V_p$ shows a pronounced low velocity, but $V_s$ does not. Consequently, compressional wave data alone cannot distinguish very well between gas zone or a geopressured zone, but the addition of shear velocity data provides the distinction between the two situations.

The detection of geopressured zones is important in the exploration for such zones, and in early detection ahead of the drill bit, in order to prevent blowouts due to insufficient mud weight. The latter problem is widespread in the Gulf Coast for example, and many parts of California, where geopressured zones tend to be present at depths of a few thousand feet.

High pore pressure may also be responsible for seismic reflection from deeper within the earth's crust, where poorly understood extensive horizontal reflectors are found, sometimes crossing rock type and structural boundaries. It is possible too, that some deep crustal faults which offset basement against basement are reflective, in spite of the petrological similarity of the two sides of these faults, due to the presence of geopressured zones (Jones and Nur, 1984), although such pressures must be close to lithostatic to be easily observable (Walder and Nur, 1984).

### 7.3 Fracture Detection and Stress Determination

The results of section 3 immediately suggest the possibility of detecting natural or induced fractures, or the use of seismic measurements to infer the in situ state of stress, via the stress-induced changes in the configuration of microcracks in rocks. Early applications of this idea include attempts to detect tectonic stress changes near active faults (Tocher, 1957; Nur, 1972; Aki et al., 1970) and the detection of near-surface velocity changes induced by earth tides. Active searches for velocity changes associated with dilatancy - the presumed earthquake precursor - were made by several Russian (Nersesov et al., 1969; Semenov, 1969) and American (Aggarwal et al., 1973; Whitcomb et al., 1973) investigators. In these studies, the opening of microcracks under high shear stress were expected to give rise to

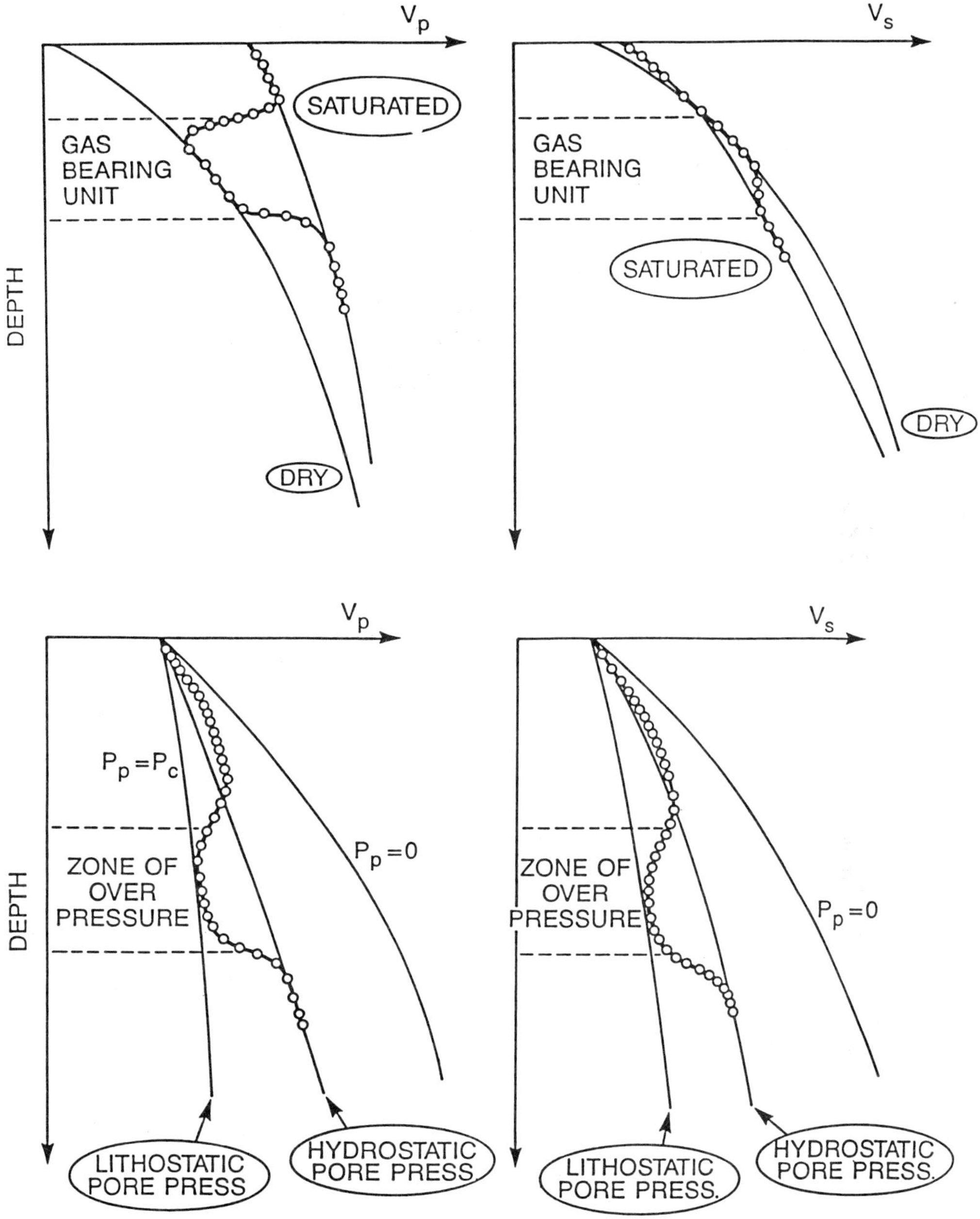


*Figure 27* Schematic velocity profiles associated with (a) a gas-bearing zone; and (b) a geopressured zone. Note the diagnostic difference in shear velocity behavior.

velocity decreases. Seismic birefringence and S wave splitting in dilatant rocks was investigated in situ and in the laboratory (Bonner, 1974). Crampin (1985) has furthered the study of these effects in a series of papers dealing also with ways to infer crack densities and orientation from seismic data. Finally, Lynn and Thomson have very recently (1986) presented results of a study in which S wave polarization was successfully used to determine in situ fracture distribution.

Stress sensitive velocities have been utilized recently also in core and borehole studies. Velocity anisotropy measured in oriented cores soon after recovery sometimes correlates with strain release, so that the direction of the principal stresses can be estimated. Finally, velocity anisotropy measured around the well bore (Mao and Sweeney, 1986) may have also been suggested as a means for in situ stress determination.

### 7.4 Tracking Thermal Fronts

One of the problems faced by a variety of EOR schemes is the need to determine, with accuracy not yet possible, the spatial distribution of reservoir properties and their changes with time. Of particular interest, and promise, are changes which occur during fire flooding, and steam flooding. In these operations, it is especially important to try to determine the direction of propagation and details of the shape, rate of movement and spatial heterogeneity of the fire or steam fronts. To accomplish such determinations we ideally would like to continuously monitor reservoirs throughout their volume, using remote sensing methods.

The most promising method is seismic waves (Nur, 1982). Because waves travel through the medium, they sample the rock properties only along their path, and can thus provide information on the material along this path. By measuring wave characteristics through a variety of paths it is in principal possible to image the entire volume.

The basis for using seismics as a tool for thermal front monitoring is the great sensitivity of velocities in rocks with hydrocarbons to temperature, as described in section 6 of this chapter. Consider the case of steam flooding: As the steam in the reservoirs heats the oil, changes in time and space of both the velocities and dissipation properties take place. These changes can be detected by seismic sources and receivers, distributed around and in the reservoir volume. The travel time $t_{ij}$ between a source (i) and a receiver (j) is

$$t_{ij} = \int_S V^{-1}(x_n)dS$$

where p is the path of minimum $t_{ij}$, V is the appropriate seismic velocity, and dS is an increment of the path length. For a fixed pair of source/receiver, the observed travel time of a signal can change with time directly due to changes of the velocity distribution $V(x_i)$, and due to the bending of the path S, also caused by changes in the velocity. The rays between a fixed source at depth and an array of fixed surface receivers change shape as the steam front crosses them. The associated travel times also increase.

The amplitude of the wave traveling along S is determined by the geometrical spreading - which is approximately a function of path length, and by the intrinsic attenuation of the rocks along S. For plane waves (or equivalently, ignoring geometrical

**Table 2. Some applications of velocity data**

| |
|---|
| Porosity and permeability mapping |
| Anomalous pore pressure detection |
| Fracture detection |
| Tracking thermal fronts |
| Gas cap movement |
| Water flooding |
| Steam reservoir boundaries |
| Distribution of permafrost |

spreading for the moment), and assuming that most of the amplitude variation is due to $Q^{-1}$ variations, we have approximately

$$ln(A/A_o) = -\frac{\pi f}{V} \int_S Q^{-1}(S)dS$$

As the $Q^{-1}$ distribution throughout the reservoir changes, so does the amplitude distribution of the waves arriving at the array. This suggests that waves might be used to monitor reservoirs during some enhanced recovery operations.

To obtain the velocities and amplitudes above it is necessary to invert the measured travel time $t_{ij}$ and amplitude $A_{ij}$ to obtain $V(x_i)$ and $Q(x_i)$ throughout the reservoir, from which we may then infer the distribution of steam, for example. Such schemes have been developed extensively in medicine, and to some degree in geophysics (e.g. Dines and Lytle, 1979).

The accuracy to which the changes in the reservoir can be determined, depends to some extent on the method used for inversion. Because the shape of the rays as well as travel times depend on the velocity distribution, the inversion is nonlinear, posing computational difficulties which must still be dealt with. The resolution is also dependent on the amount of data available: clearly a sufficient number of criss-crossing rays are required to adequately sample the reservoir. This implies that a large number of sources and receivers may be needed. Furthermore, their location might have a profound effect on the resolution. Because steam flooded zones tend to be flat and horizontal, a cross-hole imaging system, with its pervasive horizontal wave rays will provide lower resolution of the extent of flooding than the bottom-to-surface configuration, in which rays are perpendicular to the zone.

The frequency of the waves determines the possible spatial resolution. If the position of the point needs to be determined to a few meters, frequencies at the 100 Hz to 1000 Hz may be needed. Although waves at such high frequencies can be generated, the distance to which they travel depends on the attenuation properties of the reservoir system. When attenuation is very high, the received signal may become very weak.

### 7.5 Gas Cap Movement

The results in section 4, in which the presence of gas causes low compressional velocities, has been utilized extensively to discover gas pockets using the now well-established "bright spot" method. The same effect may be utilized in the future in recovery, when gas is involved. One recovery process which may be particularly amenable is when a reservoir with a gas cap is involved, particularly when the gas cap grows, or "moves" as recovery progresses. The problem which is of interest to the production manager is when and where to expect gas to break through oil-producing wells. It is of great advantage to delay this breakthrough as much as possible. However, no method is available at present to accurately determine where the gas cap is at a given time. Although numerical simulations may be capable of estimating the migration of the gas, such simulations are highly uncertain in heterogeneous reservoirs, where early gas breakthrough is often associated with the presence of low-permeability shale streaks.

The idea of seismic detecting and monitoring of the gas cap is based on the effect of gas on $V_p$, as shown in figure 11. It should work quite well when the oil has no free gas in it, because the velocity contrast is usually large. However when some free gas is present in the oil below the gas cap it may be quite difficult to detect the boundary between the two and its movement with time.

### 7.6 Water Flooding

The most common EOR process is water flooding - where water injected in some wells enhances oil flow through producing wells. One of the most common difficulties encountered in water flooding is channeling, or the non-uniform flow, e.g. along more permeable streaks in producing zones. The geophysical tracking of the water front could therefore be of great economic value.

Two effects make this kind of tracking potentially possible: First, the water is injected at higher pressure than the ambient formation pressure. This increase in pore pressure should lead to a *decrease* in seismic velocities (fig. 11). However because most water floods are fairly deep, this effect on velocities is generally not very large. The second effect is associated with the change of temperature, as discussed in section 6. Because the injected water is often much colder than the ambient reservoir temperature, some cooling of the oil in the flooded zone may take place, with an associated increase in velocities as suggested by figure 20. Whether this effect is large enough to be detected seismically depends on the magnitude of the temperature change due to water flooding, on the sensitivity of the velocity in the specific reservoir rock to temperature.

## 8. Conclusion

Table 3 lists the main factors which significantly influence wave velocities in crustal or reservoir rocks with fluids. As reviewed in this chapter, these factors include porosity, clay content, stress (overburden pressure and non-hydrostatic), pore pressure, pore fluid type, phase behavior of the pore fluid, and temperature when hydrocarbons are present in the pores.

Table 2 lists some of the most obvious applications of velocity measurements in situ. These include the determination of spatial heterogeneities, e.g. spatial variation of porosity, gas content, or pore pressure, or temporal changes, such as occur during steam or

**Table 3. Parameters which can be related to seismic velocities and attenuation in porous rock**

| |
|---|
| Porosity |
| Clay content |
| Saturation |
| Partial saturation |
| Stress and overburden pressure |
| Hydrocarbon type |
| Temperature |
| Phase transformation |

describe reservoirs in more detail, and monitoring their recovery processes, using high resolution seismic methods. Much of the methodology required remains to be developed. Although 3-D and VSP surveys already contribute significantly to reservoir description, cross-hole tomography and inverted VSP, using downhole sources and a very large number of surface receives, are just beginning to emerge. With data densities which are much greater than those needed for exploration through rock volumes (reservoirs, production zones, etc.) which are quite small, it should thus become very practical to use seismic probing routinely in development and production. The velocities and amplitude data obtained can then be converted to desired reservoir parameters, using the effects described in this chapter.

*Acknowledgements*. It is a pleasure to acknowledge the many contributions by the students and other researchers in the Stanford Rock Physics laboratory: I am particularly indebted here to the work by Hisao Ito, John DeVilbiss, William Murphy, Kenneth Winkler, Terry Jones, Carol Tosaya, Vo-Than, Te-Hua Han, Zhijing Wang, and Philippe Doyen. The research which led to these results was funded by the office of basic research of the U.S. Department of Energy, and by the Stanford Rock Physics (SRP) consortium of oil and oil field service companies.

*Chapter 10*

# Seismic data collection platforms for satellite transmission

*G. Poupinet*

## 1. Introduction

The application of tomographic techniques requires that a large amount of data is collected. The structure has to be criss-crossed by as many seismic waves as possible, in order to extract an evenly distributed information. For global seismological studies, we are concerned with wavelengths larger than one kilometer. We will not consider the specific problems related to the oil industry. The industry has already performed tomographic experiments with several hundred sensors.

For mantle tomography or for crustal studies using microearthquakes, natural events are continuously recorded and the geophones are spread on surfaces of several thousands square kilometers. Many different recording techniques are available and are chosen according to the duration of the experiment. Paper recorders and magnetic tape recorders are handy for short duration field experiments. For long duration observations, telemetry, either by phone or by radio, is more convenient. This standard field equipment is proposed by manufacturers and developed by academic institutions (see Lee and Stewart,1981 or Prothero,1984). A remarkable effort is presently undertaken within the framework of the US PASSCAL project (IRIS,1985; Meyer and Mereu,1983) to build a general purpose portable seismic station. The difficulty for a multipurpose equipment is that it should store a very large amount of data in the field (of the order of 100 Megabytes); magnetic tapes, cassettes or streamers, are the only available solutions. Magnetic tapes remain a fragile information support for extreme weather conditions and they require more maintenance than telemetry. They also suppose more data handling and tedious work for data interpretation.



*G. Nolet (ed.), Seismic Tomography, 239–250.*

The point of view adopted here is that short period seismological equipment should be adapted to specific scientific experiments and that a general purpose equipment may be too heavy to maintain for the academic teams in charge of teleseismic studies with a large number of stations.

## 2. Data for lithospheric tomography: the need for portable teleseismic arrays

First, let us examine the basic data needed for the study of the 3-D structure of the lithosphere. Most inversion techniques use first arrival times (Iyer,1975;Aki et al,1977) and seldom amplitudes (Haddon and Husebye,1978).

Large arrays have been designed to detect nuclear explosions shot anywhere on the globe and they record many distant earthquakes. The travel times residuals from LASA and NORSAR were inverted to compute a block structure of the lithosphere beneath Montana and Southern Norway. Teleseismic data are also collected in other regions on permanent or temporary microseismic networks. Figure 1 shows the example of a Hindu Kush earthquake of magnitude 6.1, recorded on the 42 stations of NORSAR. A very good correlation is observed for P-waves recorded by the same sub-array and relative arrival times are precisely measured by correlating the first wave trains. Correlation gets poorer for later arrivals or when different sub-arrays are compared. The hypothesis for the inversion, is that the impinging wave front can be approximated by a plane wave and this is more exact for first arrivals,like P-waves. Very big variations in amplitude are observed from close-by stations and are difficult to interpret univocally.

On a network like NORSAR, essentially composed of short period vertical seismometers, the reading of S-arrival times is nearly impossible. Although in theory the relative arrival times of secondary phases can be measured, no study of the S-structure has been attempted with vertical sensors. Broad band 3-components arrays will be necessary to acquire data for S-tomography studies of the upper mantle. For the present time, P-tomography is a more achievable objective.

A station devoted to P-teleseismic tomography could be a recorder of the first part of seismograms, in order to correlate P and PKP from one site to the other.

Seismic arrays like LASA or NORSAR have been exceptional tools and lead to important results to decipher the earth structure. The present challenge is to deploy equivalent arrays in any place where the structure of the mantle should be described. Any tectonic problem, like a mountain range, a sedimentary basin, a rift or a shield, cannot be understood when the structure of the lithosphere is unknown. The controversy on the continental tectosphere (Jordan,1975) and the real depth of the lithosphere beneath cratons shows how little we know on the deep structure of continents. Surface waves sample wavelengths of several hundred or thousand kilometers, and most paleogeographic units have scales of a few ten kilometers. Variations in seismic properties are observed everywhere and they reflect the past history of continents (Poupinet, 1979). Therefore, the study of the deep structure of geological units requires that large temporary short period arrays be moved from region to region (IRIS,1985). Spatial sampling will be adapted to the size and depth of the region studied.

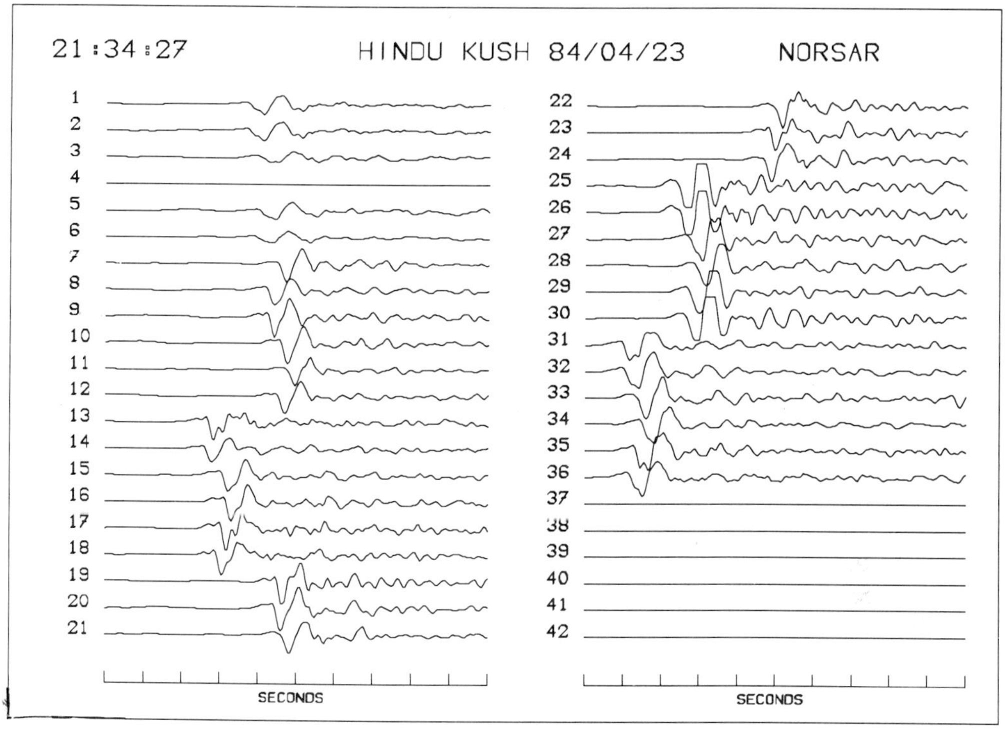


*Figure 1* Example of an earthquake from the Hindu Kush recorded by the NORSAR seismic array. Distance is 44.5 degrees and magnitude 6.0.

Such a network will record teleseisms and should detect small signal to noise ratios, particularly in regions distant from the most active seismic belts. The equipment should be easy to maintain and be run by the small teams involved in mantle studies. The ideal seismological network would be composed of low cost stations transmitting all data to a central computer. The network should be installed anywhere on earth, in any climatic condition, during several months and monitored from the investigator's laboratory. On a small scale, radio or phone telemetered networks, either digital or analog, connected to a central mini-computer are efficient to survey a volcano or a seismogenetic zone. Ground telemetry is nearly impossible in mountainous regions and on very long distances; phone lines are not available everywhere. Tape recorders or any kind of local storage require a large manpower for maintenance: they are seldom used for long term studies, the exception being the USGS one week autonomy tape recorders. There is a need for telemetry on the scale of tomographic mantle studies.

In the framework of the French LITHOSCOPE project, we have explored a solution using the satellite data collection systems developed for environmental monitoring, like meteorology and hydrology. Several equipments specialized in microseismic and teleseismic studies in regions of difficult access have been built and are presently tested. Few satellite based data collection systems are available and they impose severe constraints on the flow of data.

## 3. Environmental satellite data collection systems

A data collection system is composed of

- Data Collecting Platforms (DCP) spread on the ground.
- A radio relay on board of a satellite
- One or several data receiving and storage centres.

### 3.1 INTELSAT

The INTELSAT system provides a worldwide phone coverage; it is not an environmental data collection system but could centralize data as any ground phone network. Until recently, the access to INTELSAT or other phone satellite relays supposed very complex emitters and antennas with a diameter of several meters. Seismological data from RSTN stations or from the NORESS array in Norway, and also oil exploration data are currently transmitted by phone satellites. INTELSAT would become interesting for mantle tomography,if it could be accessed by small terminals. A technique called Wide Spread Spectrum , derived from RADAR applications, is proposed by commercial firms (Melrose et Vernucci, 1983); data are transferred at 1200 bauds from small terminals with a 0.8 m diameter antenna. Several permanent stations upgrading the World Wide Seismological Standard Network will transmit data in real time to the United States with such a system (Dziewonski, 1984). In a near future, the fast development of mobile phone systems, may open new perspectives for relaying geophysical data. Actually, the only operational systems are meteorological services.

### 3.2 Geostationary satellites

Five geostationary satellites are positioned above the equator at an altitude of 36000 km. They transmit pictures of the cloud cover of the earth and relay data transmitted by environmental sensors (WMO, 1982). A publication from the World Meteorological Organization (1985) describes the technical features of the available data collecting systems. The five geostationary satellites are compatible and provide so-called regional and international data links. 2 GOES are covering North America and the Pacific, METEOSAT the Atlantic, Africa and Europe and GMS the Far East. Each satellite sees about one third of the surface of the earth. METEOSAT has 66 channels on which messages of at most 5192 bits are transmitted every 3 hours. The European Space Agency allows to transmit more messages, so that a capacity of 1,8 Megabytes per year per DCP is a minimum. An alert channel is reserved for short messages of less than 64 bytes, when a threshold is overpassed. Each self timed DCP is attributed a frequency between 402-402.2 Mhz and a time window. Data are collected by the European Space Operating Centre in Darmstadt (FRG) and dispatched in nearly real time on the WEFAX channel of the satellite,which disseminates images. Data are received on a Secondary Data User Station (SDUS), built with a 1.2 meter antenna, a low cost receiver, a bit synchronizer and a microcomputer (figure 2). This system will even be more interesting for seismology when the precise date of reception of messages will be transmitted; a DCP will then be time synchronized by comparing the DCP time at emission with the satellite time at reception.

GOES has already transmitted P-arrival times for the large earthquakes that may generate tsunamis in the Pacific Ocean (Clark and Medina,1976). Webster and al (1981) presented the design of a GOES seismological station. Geostationary satellites are extremely interesting for seismology and particularly for mantle tomography and regional earthquakes monitoring: their visibility zone is large and they handle several thousands DCPs. The portability of a station depends on the size of the antenna: a 10W emitter should be used for portable station. The SDUS gives fast access to distant users for a small investment.

*Figure 2* A METEOSAT direct read-out station which receives all METEOSAT messages multiplexed in WEFAX images.The microcomputer decodes seismological messages and plot the transmitted seismograms.

### 3.3 ARGOS

ARGOS is a location and a data collection system. It is installed on two NOAA meteorological spacecrafts which are orbiting on polar sun synchronous orbits at an altitude of about 800 km. ARGOS DCPs are simple compared to METEOSAT: they emit randomly on a single frequency (401.65 Mhz) and they do not need a precise clock. The transmission capacity of the system is very small. Messages of 32 bytes are transmitted every 150 seconds and the DCP does not know if they have been received. For normal visibility conditions, an average of 300-400 bytes are transmitted at latitudes close to 45 degrees. The transmission capacity decreases on the equator and increases at the poles, where

ARGOS is the only data transmission system as geostationary satellites do not see the poles. An average of 100 Kilobytes is transmitted per year and per DCP. The world wide coverage of the system may be important for some applications. The system can handle more than 1000 DCPs. The time of reception of messages in the satellite is clocked with a 20 milliseconds precision and is transmitted with the messages. Data received by the satellite are relayed on a VHF down link. A direct read out station receives about 70 % of the messages within a circle of 1000 kilometers. All messages are stored on board the satellite and then transferred to the ARGOS processing Centre in Toulouse (F) and in the United States; they are archived on tapes and distributed on telex and phone lines. A delay of about 3 hours is required to access data from any point on the globe. Monthly tapes are handy for experiments that do not require real time observation. In a pioneering experiment, Endo et al (1974) collected tilt and seismic activity (number of earthquakes) measurements on 15 volcanoes through the ERTS1 satellite.

## 4. A seismological portable array transmitting via ARGOS or METEOSAT

### 4.1 A seismological DCP

*Hardware:*

A seismological DCP is an event detector interfaced with a satellite transmitter. Most functions are the same as those of a field recorder and a station with internal memories can be the basis for a seismological DCP. A specific hardware has been designed, in order to achieve the following goals:

- very low power to achieve an autonomy of several months with car batteries or of years with a tiny solar panel and an internal battery.
- robustness. Stations are often transported in remote sites.
- aptitude to run in very difficult and variable weather conditions without any maintenance.
- easiness in installation.
- possibility to increase the number of stations at will.

Figure 3 is a block diagram of a METEOSAT DCP. Three electronic boards are enclosed in a waterproof metal box:

- the analog board
- the microcomputer board
- the satellite emitter. Each station has its one emitter, either ARGOS or METEOSAT.

The analog board is protected against field emission by a steel casing, fixed inside the main box. Perturbations of the low voltage seismometer signal have been a very serious problem at the beginning of our tests; we wanted the entire equipment to be packed in one single box and to run on one single power supply. First the seismometer output is preamplified ; a gain is preset at values of 1, 10 or 100 by an internal switch. The signal is filtered to avoid aliasing. A feature specific to teleseismic studies has been included in the design. 1 Hz seismometers are sometimes fragile and 2Hz geophones would be preferable: they are smaller and less expensive. A band pass rectifier enhances the low frequencies relative to the high frequencies. This feature needs more field testing, and can be switched out when using 1 Hz geophones. The filtered signal is amplified by a gain ranging amplifier. The gain is set by the microcomputer board.

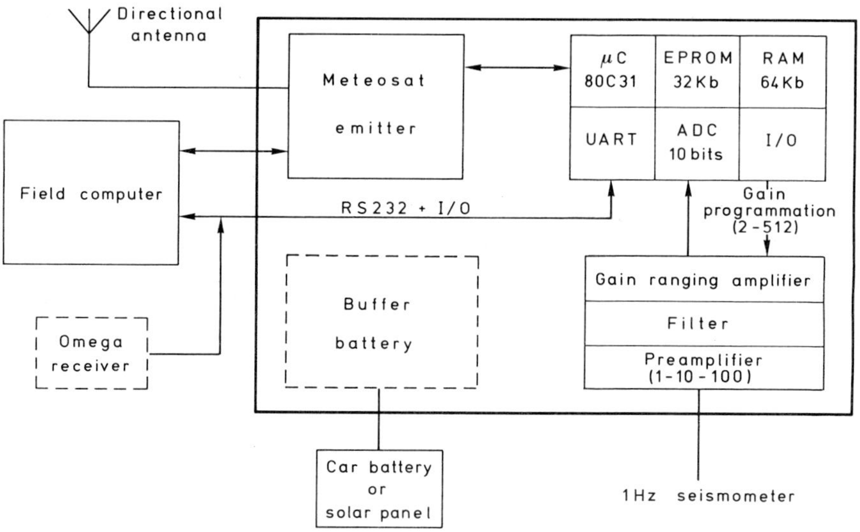


*Figure 3* Functional diagram of a METEOSAT seismological Data Collecting Platform. Box in dashed line are optional. For instance, the OMEGA receiver will not be necessary when the European Space Agency will date the time of reception of messages with a better precision.

The digital board includes
- a microprocessor
- a 10 bits analog to digital converter
- a calendar-clock chip
- an EPROM (4 to 32 Kbytes)
- a static RAM (24 to 64 Kbytes)
- a UART
- a watchdog

The architecture is similar to an ocean bottom seismometer (Prothero,1984). Two CMOS microcontrollers, a Motorola MC6805 and an Intel 80C31 have been experimented. The last board is built with the 80C31 which is more powerful than the 6805, and which includes the multiplication and has several interrupts. The 80C31 can be programmed with an advanced language whereas the 6805 was programmed with machine code. External communication through a UART allows the connection of a field computer to visualize the noise, set program parameters and retrieve the events stored in the DCP. These two boards consume about 15 mA on 12 V.

*Software:*

The software has several functions:
- it runs the clock
- it digitizes the ground motion and adjusts the gain of the amplifier

- it detects seismic transients
- it stores events in the RAM memory
- it computes a magnitude
- it sorts events according to magnitude
- it prepares events for transmission to ARGOS or METEOSAT
- it communicates with the radio transmitter
- it communicates with a field terminal
- it stores the largest events in a large RAM
- in the actual METEOSAT DCP, it synchronizes the clock on OMEGA.

Seismic transients are detected when the Short Term Average (STA) on Long Term Average (LTA) ratio is larger than a given threshold (Allen,1982; Evans and Allen,1985 for teleseismic applications). Each event is given a DCP time and a raw "magnitude": to enhance strong arrivals a "magnitude" is computed as the sum of the absolute values on a certain length of time. Only the largest magnitude events are stored and transmitted.

With ARGOS, each event is coded on 64 bytes and transmitted on two successive messages. The noise is kept for 16 points before detection; 8 points at the beginning of the event are stored with a digitization rate of 0.01s in the microseismic version and 0.02s in the teleseismic version. The event is coded on 24 bytes after detection, with a digitization rate adapted to the wave form; we try to transmit the first 2 or 3 archs of the signal.The message transmitted to ARGOS contains the DCP time of the event on 3 bytes, 1 byte for the gain and digitization rate, 24 bytes of data, 3 bytes with the time at the beginning of emission. ARGOS has a capacity for time synchronization of individual DCPs. Each message received by the satellite is given a date of reception, with a precision of 0.02s. By comparing the drift of a DCP clock with the satellite clock, on several satellite passages, the drift of DCP clocks is computed and a precision better than 0.01s on the time is achieved, although DCP quartz are not calibrated. The investment for the satellite emitter replaces that for a precise radio synchronized clock. This time synchronization function of ARGOS has been used in our first analog DCPs installed on ETNA (Poupinet and Glot,1982; Glot et al,1986). On Etna, we recorded synchronous magnetic transients on DCPs distant by 20 kilometers and got a verification of the precision of the decoding of DCP clocks; it was better than 0.01 second.

For METEOSAT, the length of a message can be up to 649 bytes. We actually transmit one event on 307 bytes and as we test the DCP for local and distant earthquakes, we digitize at 0.01 Hz. Each event is given a time and magnitude and the two largest events since the last transmission are telemetered. If there was no detection, a noise sequence is transmitted. The time synchronization procedure is not possible on METEOSAT and we do not know if it is feasible on GOES and GMS. The European Space Agency dispatches a time of reception coded in second on the SDUS. There is no technical problem to code this time more precisely and ESA intends to do it in 1987. The planning of satellite takes years, and the problem of time synchronization, using the transmission link, should be addressed to Space Agencies during the early design of data collection systems. On our METEOSAT prototype, the DCP clock is synchronized by an OMEGA receiver. In the next version, the seismic board clock will be driven by the transmitter clock, which is very stable: the OMEGA receiver will not be needed. On this prototype, two events coded on 307 bytes

*Figure 4* An ARGOS seismological Data Collecting Platform installed in Ubaye, French Alps. 4 DCPs are complementing a microearthquake network. Two are at an altitude of 2400 and 2700 meters and work during winter.

are transmitted every hour. Changes in the format of the data are easy to implement. A Walsh transform algorithm will improve the detection of teleseisms in noisy sites (Goforth and Herrin,1981); software has been written for the 80C31.

### 4.2 Installation in the field

Figure 4 shows an ARGOS DCP installed in Ubaye (French Alps) for a microearthquake study. In this version, the DCP is powered by a lithium cadmium cell inside the electronic box recharged by a small solar panel. The panel and the round shape antenna are fixed on top of a mast. In other installations, the DCP is powered by a 60 Ah car battery with an autonomy of 5 months. The battery and the electronic box are put inside a plastic garbage container and the antenna is fixed on the cover. After installing the seismometer, the ARGOS DCP is started by plugging the power: there is no need to set the gain (except to decrease the gain of the preamplifier in noisy sites). The clock does not need to be set. The noise can be visualized by connecting a field terminal. This terminal is also used to retrieve the 300 64 bytes largest events, stored in the RAM. This procedure is rather slow and should be improved in the future, with the new board which has a larger internal storage memory.

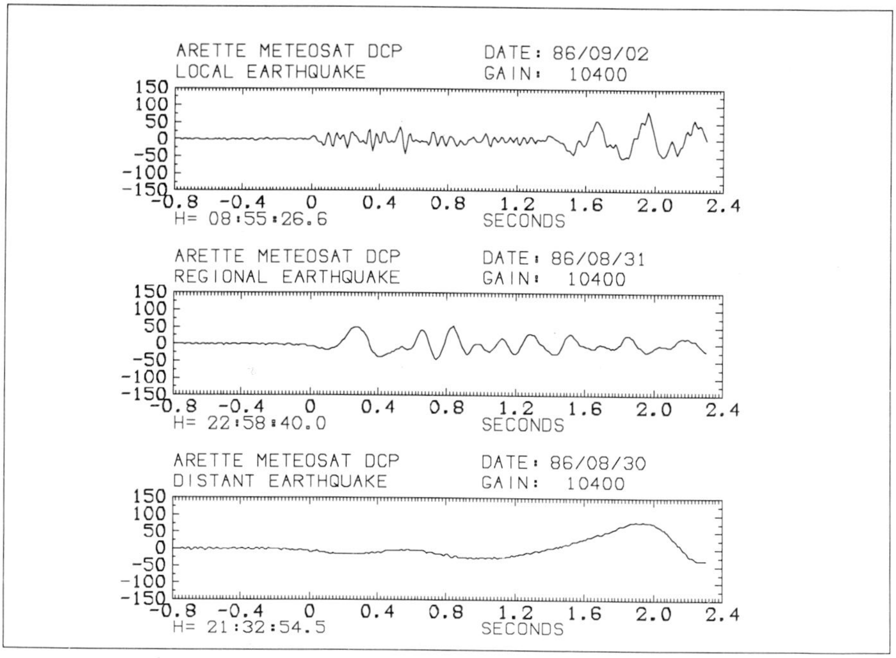


*Figure 5a* Example of seismograms transmitted by METEOSAT. METEOSAT messages were received on a direct read-out station.

The installation of a METEOSAT DCP is slightly more complex. The antenna has to point towards the satellite and the clock has to be set precisely, to prevent interferences with other DCPs.

### 4.3 Decoding of messages

Messages are received through three different channels:

- a direct read-out station for real or nearly real time reception
- phone interrogation of the data center in charge of operation of the data collecting system
- monthly magnetic tapes archived by the data center and sent by mail.

A fourth possibility is to receive in direct view on a radio receiver, like in normal ground telemetry.

The direct read-out station provides fast access to the data transiting through the satellite and is a necessary tool to monitor the correct functioning of DCPs . The main advantage is that it works independently of the Space Centre: there is no need for good quality phone transmission to retrieve data. With METEOSAT, SDUS receives all data and can assume the function of the Space Centre archiving computer. Such is not the case for

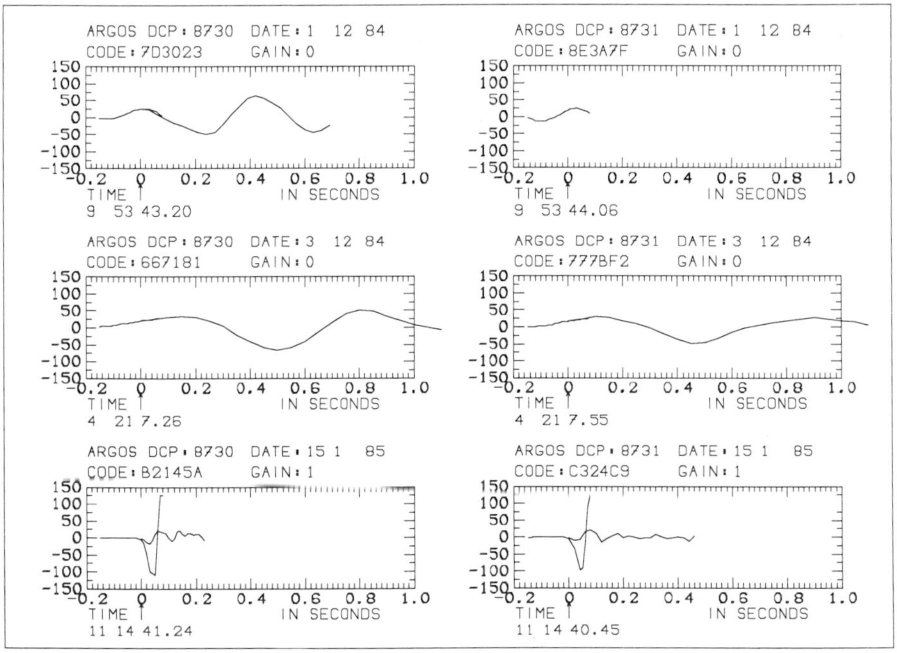


*Figure 5b* Example of seismograms transmitted by ARGOS. ARGOS messages were decoded from the ARGOS Centre monthly tapes.

ARGOS as the performances of the ARGOS centre are always better than those of a direct read-out station. When the DCPs and the reception station are too close of high mountain ranges, like in Grenoble, less than 30 % of messages are received. The METEOSAT SDUS microcomputer decodes messages and is the core of an automatic seismic network.

ARGOS DCPs more than 1000 km from the laboratory are accessible in nearly real time through the ARGOS centre, which dispatches data with a delay of a few hours. Transmission on packet switching lines is now a current procedure and automatic interrogation of the ARGOS centre is feasible with a micro or a minicomputers.

For teleseismic studies, real time operation is not necessary and the Space Centre archives data on a monthly basis. The tapes are then decoded on a mini-computer and the processing of data are similar to those of a centralized telemetered network.

Figure 5 shows examples of seismograms transmitted by METEOSAT and ARGOS. For METEOSAT, the prototype DCP digitizes 100 points per second to detect local earthquakes as well as distant shocks. The third low frequency event in figure 5a is the P-wave from the Romania earthquake of 86/08/30 recorded in Arette (French Pyrenees). For a teleseismic version of the METOSAT DCP, the digitization rate will be of the order of 0.05 s and longer time windows will be transmitted. A METEOSAT DCP transmit 48 such messages per day. In figure 5b, we selected some typical ARGOS records from DCPs installed in Ubaye (French Alps). In other versions of the ARGOS DCP, we only transmit arrival times, polarity, a quality of detection and a first zero crossing, coded on 4 bytes per event (Glot et al,1986): such short coding allows an automatic location of up to 100 earthquakes per day. This system is tested on Etna. Other types of messages can be adapted for instance for the study of the noise in one site.

## 5. Conclusion

Satellite telemetry has seldom been used in seismology. For local earthquakes studies and for seismic crisis, the limited amount of data transmitted requires that some records be selected before transmission. For mantle tomography, a data collection system like METEOSAT may already transmit too large data flows and the capacity of ARGOS may be sufficient with efficient sorting algorithms. The comparison of individual DCP clocks to the satellite clock is an easy solution for time synchronization. Satellite telemetry solves most of the problems related to the handling of data coming from many different geographical sites. The centralization of data by the satellite, and the possibility to access them from many small direct read-out stations from different laboratories, make satellite data collection systems a competing tool for international seismological projects.

*Chapter 11*

# The harmonic expansion approach to the retrieval of deep Earth structure

*A. Morelli and A.M. Dziewonski*

## 1. Introduction

In the study of the structure of the earth, the most effective approximation is that of spherical symmetry. One dimensional models are employed in seismology with success: the variation with depth of the variables describing the state of the earth is much more important for predicting its functionals than the changes which take place over the horizontal scale. There are, however, important reasons for departing from this simple model and considering lateral variations in the parameters describing the planet. A one-dimensional, spherically-symmetric model of the medium should represent a correct global average. A problem arises in the definition of the average: data distribution is highly non-uniform. Thus, if the variations seen in the data are not the result of a random process, but, on the contrary, contain some systematic effect, then the average could be geographically biased by the areas that are more densely sampled. This is in fact the case for the earth: we have evidence, coming from different sources, that there are significant, although small, systematic geographical variations. Consequently, spherical models are weighted averages of the structure of the earth, with a weight which varies geographically. The search for three-dimensional models is then of interest also for the sake of a better knowledge of the average earth.

Geographical variations of the value of functionals of the earth are by no means new to seismology. As far as travel times are concerned, for example, it was observed long ago that there are stations which tend to record arrivals of body wave phases systematically early or late (Gutenberg, 1953; Herrin and Taggart, 1962). This is in great part attributed to



*G. Nolet (ed.), Seismic Tomography, 251–274.*

crustal or lithospheric heterogeneity, which is characterized by relatively short wavelength and large amplitude. An analysis of travel times made on a global scale also shows the existence of patterns characterized by spatial coherence on a large scale, uncorrelated with the known lithospheric structure (Julian and Sengupta, 1973; Dziewonski, 1984). The origin of this effect is to be searched at greater depths. In fact, the presence of heterogeneity in the interior of the planet is certainly to be expected: a dynamic earth must be heterogeneous. Seismology represents the only way of probing the interior of the earth with resolution sufficient to provide useful information in three dimensions, necessary for a better understanding of the dynamics of the planet.

Different portions of the spectrum of the seismic signal are used to retrieve the earth structure at different depths. The approaches followed make use of surface waves (Nakanishi and Anderson, 1982, 1983, 1984), body-wave travel times (Dziewonski et al., 1977; Clayton and Comer, 1983; Dziewonski, 1984), normal-mode spectra (Masters et al., 1982; Ritzwoller et al., 1986; Giardini et al., 1987), and complete mantle and body-wave waveforms (Woodhouse and Dziewonski, 1984, 1986). The different types of data, containing information travelling with frequencies which range from 1 Hz to 0.5 mHz, present very different characteristics in terms of resolving power, manageability, and global availability of observations.

The existence of heterogeneities is expected at all depths in the earth. In the mantle they must be associated with the generation of convective motions. Density anomalies in the mantle will cause the core-mantle boundary (CMB) to depart from the figure of hydrostatic equilibrium. Assuming density and viscosity models for the lower mantle, Hager et al. (1985) inferred the presence of topography at CMB with a few kilometers of peak-to-peak amplitude. The core is the subject of recent interest. Two different kinds of observation suggest that aspherical variations in its structure are present: Masters and Gilbert (1981) noticed anomalous behaviour of free-oscillation modes that interact with the core; similarly, considering the travel times of *PKIKP*, Poupinet et al. (1983) found significant lateral variation of the values reported over the globe. However, there was no precise indication about the site of the heterogeneity.

In this study we shall confine ourselves to the analysis of travel times of compressional waves corresponding to *PcP*, BC branch of *PKP*, and *PKIKP* phases (see Figure 1). Assuming the large-scale structure of the lower mantle already known from previous studies (we will refer to Dziewonski, 1984), these phases allow us to study deeper regions. Their diverse sensitivity allows us to resolve different structural elements. Considering *PcP* and $PKP_{BC}$ travel times, Morelli and Dziewonski (1987) derived a model for the topography of CMB, and inferred lateral homogeneity for the outer core. Using *PKIKP* travel times, Morelli et al. (1986) showed evidence of anisotropic behavior of the inner core. We shall follow the approach adopted in the latter studies.

Perhaps the most important choice to make in a study like ours regards the approximate representation of the earth that we intend to adopt. We have to find *i)* which is the physical mechanism responsible for the effect observed, and *ii)* the most appropriate way to represent (i.e. parameterize) it. The first point is crucial: a wrong choice can lead to a meaningless answer. We start out with a working hypothesis, usually derived from

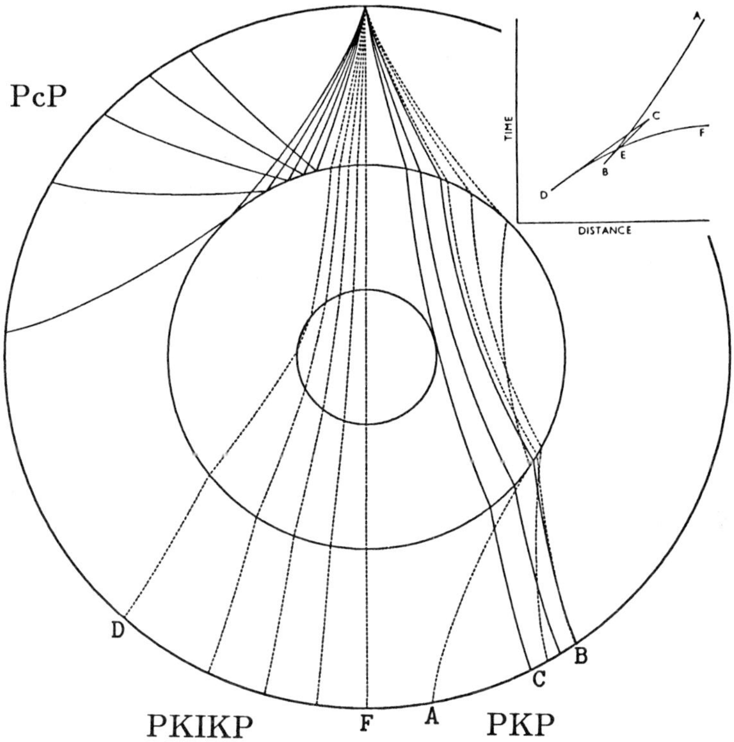


*Figure 1.* Ray paths of phases interacting with the earth's core. The letters correspond to the cusps of the travel time curve (see inset). Branch *DF* corresponds to the phase *PKIKP*.

previous studies. Following the inversion, the working hypothesis must be checked against the data and the statistical significance of the model must be addressed. The standard error analysis can fail in such a situation, and we ought to resort to empirical verifications suggested by the geometry of the problem and by the availability of data.

Travel times of body waves with periods of a few seconds sample the interior of the planet with high resolution. However, they are also strongly affected by small-scale local heterogeneities and by errors in the hypocentral parameters. These limitations can only be overcome resorting to very large datasets, with the assumption that random variations can be statistically filtered out. Body-wave travel times represent a common choice for studies of the lower mantle, since the work of Julian and Sengupta (1973). Two possible representations of the structure can be adopted. We can divide the lower mantle into a number of blocks, which in published studies ranges from 150 (Dziewonski et al., 1977) to about 75,000 (Comer and Clayton, 1984). The values of $P$-wave velocity perturbation inside each block are the unknowns of the inversion. Alternatively, we can make use of the spherical harmonic expansion (Dziewonski, 1984; Morelli and Dziewonski, 1985). It easily allows correlations and comparisons with other geophysical fields and, if the number of unknowns is reduced to a level allowing for an exact matrix inversion, it permits evaluation of the reliability of the results, as will be seen in the following sections.

The methods of solving a discretized inverse problem is the subject of many books and papers. A summary introductory discussion can be found in Aki and Richards (1980). There, references are listed to the original works and to more detailed discussions. In particular, Wiggins (1972) gives a comprehensive exposition of the problem cast for a realistic case.

## 2. The data

### 2.1 Phase selection

We shall consider *P*-wave travel times. The low-degree structure of the lower mantle is assumed to be known with enough accuracy: model L02.56 (Dziewonski, 1984), specified in terms of an expansion up to degree and order 6, will be our reference. Our aim is to apply a similar approach to the regions below the lower mantle. For this, the first step is to find data suitable for such an inquiry. The *Bulletins of the International Seismological Centre* (1964-1982) represent the only readily accessible source for a global study. The Centre collects travel times reported by practically all the operational seismic stations of the world; the Bulletins are available on magnetic tape since 1964. From the information contained there we select phases that are sensitive to the structure to be studied.

The *PcP* phase corresponds to waves reflected from the CMB interface. If we assume known the lower mantle structure, and correct for its effect, then we should be able to derive a model of CMB topography using these data alone. This is true only to some extent, however. The amplitude of *PcP* on a seismogram is not large, and it comes after a strong direct *P*. For a large earthquake it is often concealed by *P* coda and is difficult to read. Furthermore, its travel-time curve intersects the *PP* branch at about 45°, and *PPP* at 37° ; before 23° it crosses over with *S* , and at large distances it merges with *P* diffracted at the core-mantle boundary (Jeffreys and Bullen, 1958). Because of these complications *PcP* is not frequently reported (sixty times less frequently than *P*) - and its readings after a strong *P* are often not reliable. Figure 2 shows the reflection points on CMB corresponding to the set built with the selection procedure that will be explained later. The geographical coverage is less than perfect - the southern hemisphere is particularly poorly sampled.

The travel-time curve of the core phases is shown in Figure 3. The most reliable branch is *DF* , corresponding to *PKIKP*: it is the first arrival phase over a broad range of distances. Each arrival carries with it a coda which often makes the recognition of later phases problematic. The *BC* branch coming after *DF* is an example. The arrivals associated with $PKP_{BC}$ have, however, large amplitudes which makes it relatively easy to identify them in the seismogram. Figure 4 shows histograms of *PKP* reports in the catalog with a given travel-time residual, defined with respect to the theoretical *DF* arrival time. Each histogram is for a one degree wide distance range centered on 145.5°, 150.5°, and 153.5° . The *BC* and *DF* arrivals are roughly coincident at 145° and it is not possible to discriminate between them. As the distance increases, however, the two

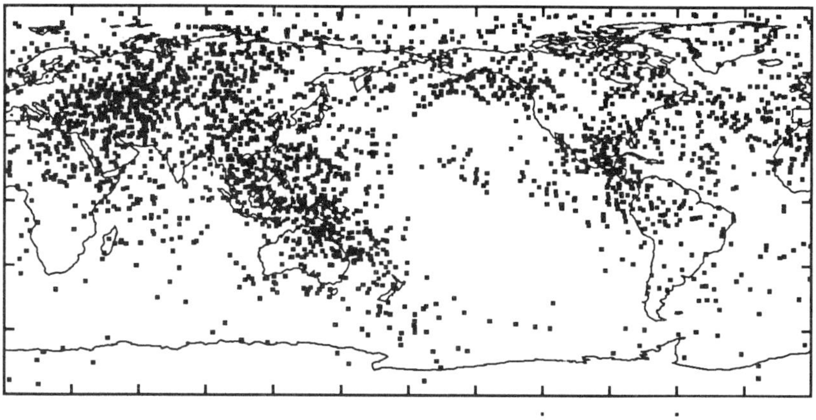

*Figure 2.* Map of the reflection points, at the core-mantle boundary, of the 'summary' paths belonging to the *PcP* phase. Each summary path derives from averaging the bundle of rays connecting the same pair of bins of the subdivision described in Section 2.2. The difference of sampling between the northern and the southern hemispheres is large.

peaks split. Away from the caustic $B$, the amplitude of $BC$ decreases, and so does the number of reports on the Bulletins. On the other hand, the contribution of $DF$ is rather uniform, and the splitting allows us to evaluate its amplitude. Figure 4 confirms the observation of Anderssen and Cleary (1980), that most of the observations interpreted as $DF$ at distances close to the intersection of the two branches correspond to $BC$ instead. We will consider all arrivals within ±4 s from the theoretical $PKP_{BC}$ time in the distance range between 145° and 155°, assuming from the preceding considerations that the contamination from $DF$ is negligible.

## 2.2 Quantization of coordinates

The global distribution of receivers and sources is highly non-uniform. Subduction zones present a strong concentration of seismic activity and seismic stations are mostly confined, with a strong cultural bias, to continental areas. This can severely affect the resulting model, and precautions must be taken. First, we consider only shallow earthquakes, which present a more uniform distribution. Then, we discretize geographical coordinates so that sources and receivers will be located inside the meshes of a grid. Only the coordinates of the cell, along with its averaged residual, will be considered during the inversion.

A more uniform sampling is only one advantage of the discretization. A second purpose for introducing it is computational efficiency. The number of paths for which to compute the differential kernels needed for the inversion is dramatically reduced. Yet another advantage of this summary-path approach is the filtering of small-scale local variations that occur among earthquakes or stations corresponding to the same mesh. These high-frequency variations are seen as noise in a study aimed at the retrieval of very-large scale features.

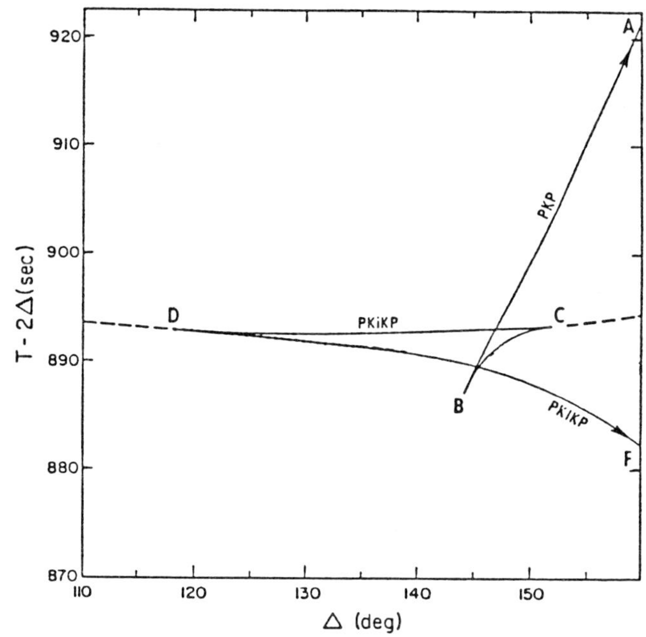


*Figure 3.* Travel time curve for *PKP* phases (after Choy and Cormier, 1983).

We use different grids for sources and receivers. First, we define the grid of 1656 roughly equi-areal cells of Figure 5; this is, in practice, identical to the grid adopted by Clayton and Dziewonski (1984). Then, for each of these we construct two more grids, shown in Figure 6, one for $P$ and $PcP$ (from 25° to 104°), and one for the core phases (from 110° to 180°). For each phase, we stack all the observations for earthquakes located in the same source-cell onto these two receiver-grids. This pattern of residuals - resulting from an average made for the bundle of rays connecting the same pair of cells - will enter the inversion as a *summary earthquake* (see Figure 1 in Dziewonski, 1984). At the end of the process, 650 of such *summary earthquakes* with their $P$, $PcP$, $PKP_{BC}$, $PKIKP$ residual patterns constitute the whole dataset.

## 2.3 Relocation

A wrong location of events can bias the resulting model. The hypocentral parameters (geographical coordinates $\lambda$ and $\phi$, depth $d$, and origin time $t_0$) are, in general, not precisely known. A location procedure finds the point in the space $(\lambda,\phi,h,t_0)$ which minimizes some function (usually the sum of the squares) of the travel-time residuals computed with respect to some model. I.S.C. uses Jeffreys-Bullen travel-time tables. Differences between the real earth and the model can translate into a mislocation. Such errors are particularly important in a study like this, because a shift in the location can induce artificial lateral heterogeneities, or, conversely, can mask the existing ones. It is important that the location be determined using the best time tables available. As will be seen in the next section, our determination of lateral variations will be expressed in terms of

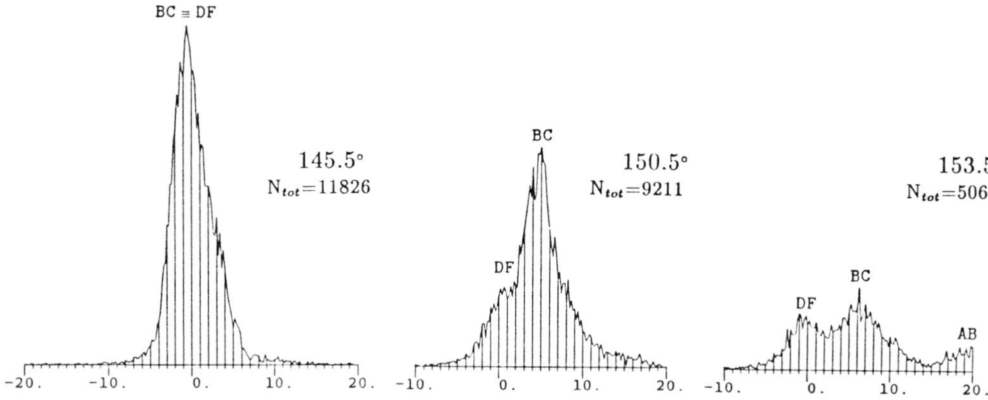


*Figure 4.* Histograms of *PKP* arrival times in 1° intervals. The preponderance of *BC* in the reports on the ISC Bulletin is apparent.

perturbations to a spherically-symmetric starting model. A spherical model is needed for the computation of path integrals, and, for consistency, earthquake locations are found using the same reference earth structure.

Events should be relocated again after a model is found, using corrections for heterogeneities, and the inversion repeated again, until a satisfactory convergence is achieved. This is the scheme followed by Dziewonski (1984) for the lower mantle model L02.56. His final model, however, did not undergo major changes with respect to the one obtained at the first step. In addition, relocation is mainly controlled by *P* data, sensitive only to lower mantle structure, which we hold fixed. We will therefore assume that the effect of relocation on models for the core and core-mantle boundary is negligible.

The data selection proceeds as follows.

- Only events with focal depth less than 50 km and with at least 30 *P* arrivals spread over three azimuthal quadrants are considered. The location of events with a poor azimuthal coverage is not well controlled.
- Travel times are computed using a spherically-symmetric starting model derived from the surface focus data in Table 4 of Dziewonski (1984) as a perturbation to PREM (Dziewonski and Anderson, 1981). Travel times are then corrected for ellipticity (Dziewonski and Gilbert, 1976), and azimuth independent terms of station residuals of Dziewonski and Anderson (1983).
- For all the earthquakes which satisfy the requirements we apply a standard iterative least-squares relocation technique using *P* and *PKIKP* travel times. These phases are not able to break the trade-off between depth and origin time, so that we keep the depth fixed to the value listed on the catalog. Events which fail to converge are discarded.
- Finally, the corrected residuals for each phase (*PcP*, $PKP_{BC}$, PKIKP) are projected onto the receiver grid. To each bin is attributed the average resulting from all the

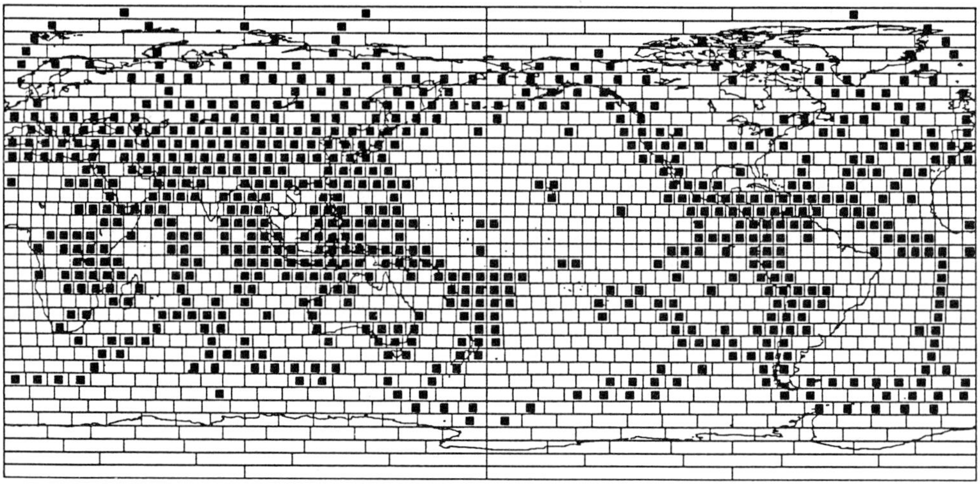

*Figure 5.* Grid used to discretize epicentral coordinates. All earthquakes within the same cell form a *summary earthquake.*

earthquakes located in the same source cell (Figure 5). All the non-empty bins constitute the residual pattern of the *summary earthquake* , to be written on a file used as input for the inversion.

This procedure selects 32,000 events which result in 650 *summary earthquakes* . The number of readings and summary rays for each phase is reported in Table 1.

## 3. The inversion scheme

### 3.1 Discretization

The determination of the velocity structure of the earth or of the topography of its boundaries is a continuous inverse problem. We want to determine a scalar function given a set of measurements of a functional of it of the form:

$$\int_{\gamma_i} G_i(\mathbf{r})\cdot g(\mathbf{r})\, d\zeta = \delta t_i \qquad i = 1,2,\ldots,n \tag{1}$$

The observable is the travel time residual $\delta t_i$ corresponding to the $i$-th path $\gamma_i$; the function $g(\mathbf{r})$ represents the structure ($P$-velocity field, elevation of CMB, ... ) at the point $\mathbf{r} = = (r,\theta,\phi)$, and $G_i(\mathbf{r})$ is the kernel describing the dependence of travel time on the structure. The variable $\zeta$ is a coordinate along the ray path. The adoption of first-order perturbation theory allows us to refer the quantities appearing in equation (1) to a reference initial state. Thus, $g(\mathbf{r})$ will be the perturbation (assumed small) to a reference model in

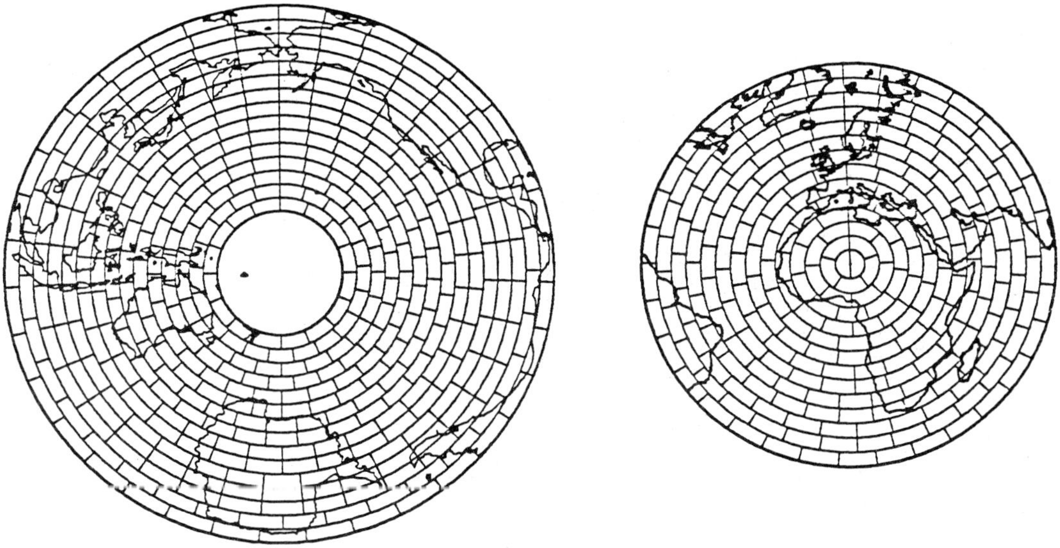

*Figure 6.* Grids used to discretize receiver coordinates for a source located at 17°S, 173°W. Left: grid used for $P$ and $PcP$ (circles are at 25° and 104° from the center). Right : grid used for the $PKP$ phases (outer circle corresponds to 110°, the center is at 180°). Each summary path is the average of the bundle of rays which connect the same source and receiver cell.

which the value of the kernel and the path $\gamma_i$ are computed, and $\delta t_i$ the corresponding travel-time residual.

To determine the continuous function $g(\mathbf{r})$ in (1) from a finite set of measurements $\delta t_i$ we must discretize the problem (see, for example, Aki and Richards, 1980, p.676). We choose to expand $g(\mathbf{r})$ in a series of orthogonal basis functions truncated at some maximum degree. For horizontal variations the obvious choice is spherical harmonics. Thus, for the $P$-wave velocity perturbation field in a spherical shell (like, say, the lower mantle) we have:

$$\delta v_P(\rho,\theta,\phi) = \sum_{k=0}^{K} \sum_{l=0}^{L} \sum_{m=0}^{l} ({}_k c_l^m \cos m\phi + {}_k s_l^m \sin m\phi) \cdot f_k(\rho) \cdot p_l^m(\cos\theta) \tag{2}$$

and for the topography of a boundary:

$$\delta r(\theta,\phi) = \sum_{l=0}^{L} \sum_{m=0}^{l} (c_l^m \cos m\phi + s_l^m \sin m\phi) \cdot p_l^m(\cos\theta) \tag{3}$$

In these expressions $\theta$ is colatitude, $\phi$ is longitude, $\rho$ is radius normalized in $(-1,1)$, $f_k(\rho)$ is the $k$-th Legendre polynomial with unit norm, and:

$$p_l^m(\cos\theta) = \left[ (2-\delta_{m,0})(2l+1)\frac{(l-m)!}{(l+m)!} \right]^{\frac{1}{2}} P_l^m(\cos\theta) \tag{4}$$

where $P_l^m$ is the associated Legendre function of degree $l$ and order $m$ (Abramowitz and

**Table 1: Number of readings**

| | $PcP$ | $PKP_{BC}$ | $PKIKP$ |
|---|---|---|---|
| Readings | 26125 | 82403 | 93863 |
| Summary rays† | 5182 | 4610 | 11487 |
| rms (sec) | 1.61 | 1.38 | 1.28 |

†*Summary rays* derive from the averaging scheme described in section 2.2

Stegun, 1964). The normalization is such that the squared norm on the unit sphere $\Sigma$ is $4\pi$:

$$\int_{\Sigma}\left[\cos(m\phi)\cdot p_l^m(\cos\theta)\right]^2 d\Omega = 4\pi \tag{5}$$

for $m = 0,1,2,...,l$, and similarly with the term $\sin(m\phi)$ for $m = 1,2,...,l$. The coefficients ${}_kc_l^m$, ${}_ks_l^m$ (or $c_l^m$,$s_l^m$) parametrize the unknown perturbation. The angular and radial degrees of the expansion (respectively $L$ and $K$) have to be selected considering the coverage provided by the data.

As the degree increases, the basis functions are characterized by higher spatial frequency. By extending the expansion (2) or (3) to a high degree we could approximate the perturbation function $\delta v_P$ arbitrarily well. But then we could not determine it: the resolving power of the dataset available is in fact limited, and we must find a threshold beyond which the expansion looses physical significance. This is equivalent to limiting our search to the low wavenumbers - or large wavelengths - of the perturbation. The parameters of the solution represent the discrete spectrum of the structure, and, given the data at hand, we only attempt to retrieve the low-frequency part of it. The harmonic expansion approach, similar for instance to the modeling framework used in studies of the gravitational and magnetic fields of the earth, presents several advantages with respect to an alternative representation of the field in which each parameter corresponds to the value of the unknown function $g(\mathbf{r})$ in a small region (block, patch) of its domain. We call this other approach *local parametrization*.

The local parametrization applied to the global scale usually requires a very large number of unknowns (Comer and Clayton, 1984, used some 75,000 blocks to model the lower mantle; see also Hager et al., 1985). Using a low order expansion in analytical functions we implicitly resign any claim to retrieve abrupt variations but we deal with a number of parameters (of the order of a few hundred) small enough to allow the computation of the inverse of the inner product matrix of the linearized problem deriving from (1). This gives us the possibility to find formal estimates of covariance and resolution for the model, and an inverse with the optimal combination of the two, through methods like the maximum likelihood or the stochastic inverse.

Furthermore, when dealing with studies extended to the global scale, the harmonic expansion approach offers another class of favorable features which derive from the fact that the solution corresponds to the discrete spectrum of the structure. Any low-pass filtered version is immediately available by considering the corresponding harmonic components. Higher details can be obtained by gradually increasing the order of the expansion. In this way the stability of the results can be tested. Finally, it is worth noticing that very often the comparison with other geophysical fields, for which spherical harmonics represent the natural choice (such as the gravity and the magnetic fields) is needed. With the velocity structure expanded in a similar way, such a comparison is immediately available. If, conversely, a local parametrization is adopted, we must perform the harmonic expansion of the block model. In this way we do not get an expansion which minimizes the misfit to the original data, but an expansion instead which tends to follow closely the block-like pattern. The presence of artificial discontinuities between blocks introduces high frequencies and therefore possible aliasing. The effect is accentuated if the value corresponding to blocks that are not sampled is assumed to be zero. Since we do not know it, the most reasonable estimate is the value interpolated between adjacent areas by means of the lowest degree harmonic possible, which is accomplished by the maximum likelihood inversion for the coefficients of the harmonic expansion.

### 3.2 Inversion

The substitution of (2) into (1) yields the linearized inverse problem:

$$\mathbf{A}\cdot\mathbf{x} = \mathbf{d} \tag{6}$$

Here $\mathbf{x}$ is the vector of parameters representing the model, $\mathbf{d}$ is the data vector ($d_i$ is the travel-time residual of the $i$-th ray with respect to the starting model), and $\mathbf{A}$ is the kernel matrix - $A_{ij}$ is the partial derivative of the $i$-th travel time with respect to the $j$-th parameter $x_j$). If $p$ is the number of parameters and $n$ the number of observations, $\mathbf{A}$ has dimension $(n \times p)$.

In general, we may want to solve for parameters with different physical meaning, such as *P*-wave velocity and boundary shape perturbations, using different kinds of data, such as *PcP* and *PKIKP*. There are two complications:

- *the data* ($\mathbf{d}$) are not homogeneous: even though we can assume that all the observations of the same dataset share the same variance, this is certainly not acceptable for different phases. It is however easy to take this into account by weighting equation (6) according to the error estimates available, in the way shown, for example, by Wiggins (1972, eq.5). Our problem is more general. As will be seen considering core-mantle boundary structure, we will have to add together sets of different size. A systematic mislocation of our sources would reflect onto CMB radius, through *PcP* and $PKP_{BC}$ data, in two opposite ways. The opportunity to eliminate this effect by averaging the *PcP* and $PKP_{BC}$ data would be lost if one of the subsets were much more numerous than the other. To prevent this, we must consider their relative size to balance the model.[1] This can be included in the covariance matrix for

---

1. This is equivalent to a situation in which two operators make measurements of the same quantity, $q$, and operator $A$ takes $N_A$ measures whereas operator $B$ takes $N_B$. If, say, $N_A \gg N_B$, to get rid of possible systematic subjective errors we would probably find the best estimate of $q$ as $\frac{1}{2}[(\sum_i q_i^A)/N_A + (\sum_i q_i^B)/N_B]$ rather than $(\sum_i q_i^A + \sum_i q_i^B)/(N_A + N_B)$.

the observations (**S** in the notation of Wiggins). It is however important to notice that this procedure is not justified if the parametrization adopted is flawed and the models resulting from separate inversions of the sets yield incompatible answers (see Section 5.2).

- *the parameters* (**x**) are not physically homogeneous : in general, we may want to invert for parameters with different physical meaning, such as $P$-velocity and lateral variations of the radius of a discontinuity. We must rescale the parameters. The way to do that is to normalize the unknowns so that the data sensitivity per unit change in the numerical value of each coefficient is of the same order. The elements of the matrix **W** of Wiggins (1972) are such that the problem would be perfectly balanced, in the case of travel times, for a vertical ray. We, however, deal with phases with different sensitivity to the parameters, and must be content with a choice which makes the numbers of the same order of magnitude. This can be regarded as putting some *a priori* information in the model, assuming that if the $P$-wave velocity is of the order of a few percent then the boundary perturbation will be of the order of a few kilometers. Such scaling is important if an approximate inverse is to be computed.

In addition, when judging the quality of the inversion we have to consider several heuristic criteria. These include, for instance, the compatibility between different datasets, and physical prejudices which we would like to see satisfied (as, for example, the lateral homogeneity of the outer core). They are difficult to quantify and must be included in a somewhat subjective way.

The system of equations of condition (6) will be solved in a least-squares sense. This is very practical when dealing with a large number of data because it entails the construction and inversion of the inner-product matrix $\mathbf{A}^{\mathrm{T}}\mathbf{A}$, of dimension $(p \times p)$ regardless of the number of observations. The building of the matrix $\mathbf{A}^{\mathrm{T}}\mathbf{A}$ is time-consuming, but it can be stored for each data set separately. The joint inversion of two, or more, sets is then done by inverting the sum, with the appropriate weights, of the corresponding inner-product matrices, which do not need to be computed again. In this framework, inversions with different characteristics can be performed in a very economical way.

The least-squares solution of equation (6) can be written in terms of the inverse of the inner-product matrix. Rather than computing the exact inverse, we decompose it in terms of its eigenvalues and eigenvectors. The generalized inverse is then constructed setting to zero the least significant eigenvalues. This is equivalent to the maximum likelihood inversion of (6) (see Aki and Richards, 1981). Notice that storing eigenvectors and eigenvalues allows us to compute inverses with different cut-offs in a very fast way.

## 4. Structure of the core-mantle boundary

### 4.1 Method

The travel-time perturbation $\delta t$ for a change in the radius $r$ of a discontinuity is

$$\delta t = -\frac{2\delta r}{r}(\eta_+^2 - p^2)^{1/2} \tag{7}$$

for a ray reflected from the top of a discontinuity, and

$$\delta t = -\frac{\delta r}{r}\left[(\eta_+^2 - p^2)^{1/2} - (\eta_-^2 - p^2)^{1/2}\right] \tag{8}$$

for a refracted ray (Dziewonski and Gilbert, 1976). In these expressions $\eta = = r/v$, $p$ is the ray parameter, and $v$ is the wave speed. Subscripts + and – denote the limiting values above and below the discontinuity, whose unperturbed radius is $r$. As $P$-wave velocity is higher at the bottom of the lower mantle than at the top of the outer core, $\eta_+ < \eta_-$, and the quantity in square brackets in (8) is negative. This means that for a positive change of the radius of CMB the travel-time of a reflected phase, such as *PcP*, becomes shorter, whereas the travel-time perturbation of a transmitted phase (such as $PKP_{BC}$ ) is positive.

If a systematic pattern due to a cause other than CMB topography affects the data, the reflected-wave and the refracted-wave sets would map it onto CMB with opposite sign. In this way we can address two hypotheses that are a common source of uncertainty:
*i)* some structure exists in the real earth which is not modeled, and which is the cause of some detectable effect on the observations; and
*ii)* this structure is artificially mapped by the inversion procedure on the resulting model by means of parameters with a different physical meaning.

In the present example, we choose *PcP* and $PKP_{BC}$ phases as the most suitable for the experiment. At first, we use the two sets independently in two separate inversions. If the resulting models will show significant similarity we should rule out both *(i)* and *(ii)*.

### 4.2 Inversion of PcP

We insert the expansion (3) for $\delta r$ in (7), and consider (7) at the colatitude $\theta_i$ and longitude $\phi_i$ of the reflection point on CMB of the $i$-th ray ($i = 1,2,\ldots,n$). This gives a linear system of $n$ simultaneous equations in $p$ unknowns (if $p$ is the total number of parameters of the expansion). We can solve it by means of the least-squares (LS) technique, and compute the covariance matrix, $\mathbf{C} = \sigma\cdot(\mathbf{A}^T\mathbf{A})^{-1}$. The matrix $\mathbf{C}$ is a $(p \times p)$ square symmetric matrix. In the point of coordinates $(\hat\theta,\hat\phi)$ the value of the model is given by

$$\delta r\,(\hat\theta,\hat\phi) = \mathbf{x}^T\cdot\mathbf{f}(\hat\theta,\hat\phi) \tag{9}$$

and the covariance is

$$\sigma_r^2(\hat\theta,\hat\phi) = \left[\mathbf{f}(\hat\theta,\hat\phi)\right]^T\cdot\mathbf{C}\cdot\mathbf{f}(\hat\theta,\hat\phi) \tag{10}$$

Here $\mathbf{f}$ is the $p$-vector of the basis functions and $\mathbf{x}$ is the $p$-vector of model coefficients.

We can plot the map of the error estimate (10), as in Figure 7. The comparison with *PcP* data distribution (Figure 2) shows high correlation, that was expected. Larger error is found in areas sparsely sampled.

Even though such a solution is not completely unrealistic, we may want to choose a more conservative one, which does not include the most poorly determined features of the

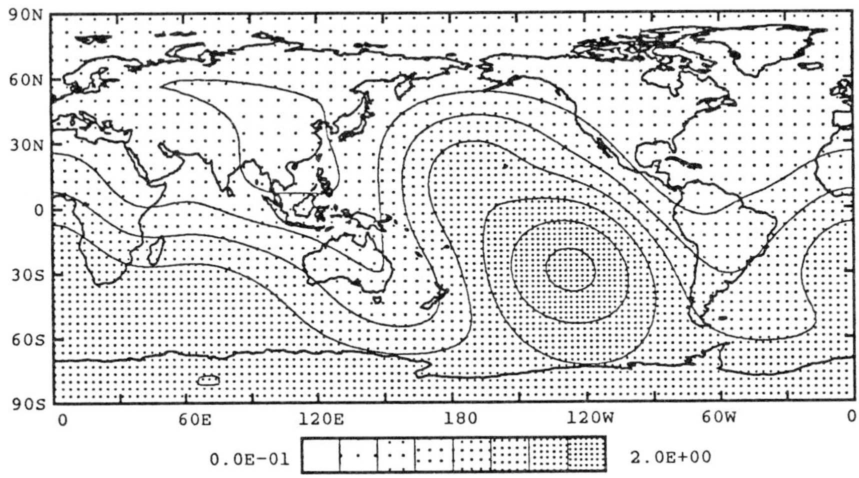


*Figure 7.* Error map for the least-squares inversion of the *PcP* dataset. Large error is present where the data coverage is poor (compare with Figure 2).

model. We construct the generalized inverse to $\mathbf{A}^T\mathbf{A}$ by means of an eigenvalue/eigenvector decomposition, of which we artificially set to zero the smallest eigenvalues (see section 12.3 of Aki and Richards, 1980). The effect is to regularize the solution, and as a consequence the error map is more uniform. Discarding 6 eigenvalues (out of 25) we constrain the solution on the 19 best determined degrees of freedom. The resulting model is shown in Figure 8. Core-mantle boundary shows perturbations of the order of (plus or minus) five kilometers.

### 4.3 Inversion of PKP(BC)

In analogy, $PKP_{BC}$ data alone can be inverted to obtain an independent determination of CMB. In this case, formula (8) has to be used, keeping in mind that for every ray we must consider the travel-time perturbation originated at both entry and exit points. We solve as in the previous case (our best solution is now obtained for 23 eigenvalues) and the result is shown in Figure 9.

The comparison with Figure 8 shows significant correlation in terms of location of the main features. The resolution that can be achieved in an expansion to order and degree 4 is limited, and we cannot aim at reproducing fine details. The cross-correlation coefficient is 0.7. The main difference is in the amplitude of the two maps, with $PKP_{BC}$ leading to larger topography than $PcP$.

Because the sensitivity of $PcP$ and $PKP_{BC}$ to a radius change of CMB is opposite in sign, the result cannot be explained by a source of systematic error (like, for instance, a systematic mislocation of the epicenters or mantle structure unaccounted for). If this were the case, in fact, the systematic error would have been mapped as CMB anomaly with opposite sign by the two phases. Therefore we conclude that, in their major features, the

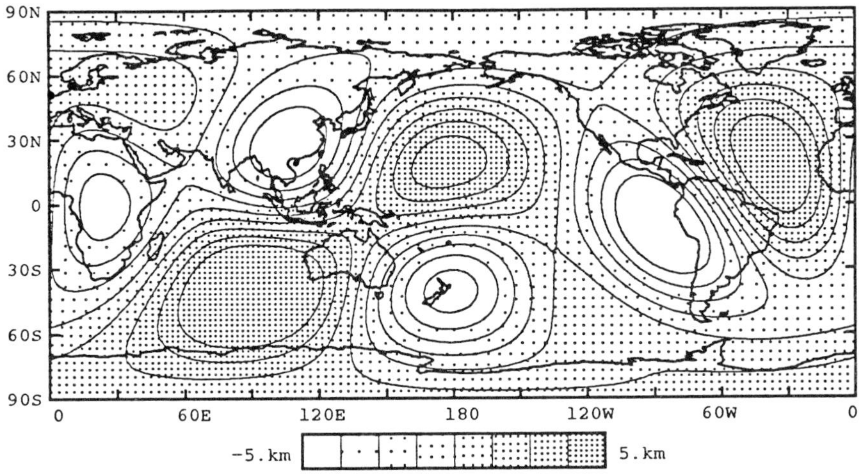


*Figure 8.* Map of topography of the core-mantle boundary from the maximum likelihood inversion of the *PcP* set.

maps of Figures 8 and 9 effectively correspond to lateral variations of similar size of the shape of the core-mantle boundary (Morelli and Dziewonski, 1987).

The other important conclusion that can be drawn following this approach is that the outer core is, within the resolution allowed by seismology, laterally homogeneous. In fact, if $PKP_{BC}$ sampled detectable aspherical anomalies in the outer core, we would see them erroneously mapped onto CMB. The lack of significant and systematic discrepancy between Figures 8 and 9 shows that this is not the case. It is in fact in accordance with the information that we derive from different studies, such as travel-times of multiple internal reflections in the core (*PmKP*, Buchbinder, 1972), and fluidodynamics considerations (Stevenson, 1986).

### 4.4 Joint inversion

Once their consistency is established, and we are assured that the parametrization has physical meaning, we can combine the two datasets together and invert them simultaneously. The most intuitive way of doing this is to join the two systems of equations of condition considered so far. This would give $(n_{PcP}+n_{BC})$ equations in the same $p$ unknowns. The inner-product matrix $\mathbf{A}^T\mathbf{A}$ is obviously still $(p \times p)$ and no change is needed in the inversion procedure.

We do this considering the weighting scheme of section 3.2. The result of the joint inversion with 23 nonzero eigenvalues is illustrated by the map in Figure 10. The spherical harmonic coefficients are listed in Table 2.

Combining the two data sets improves the resolution. These results were, however, obtained under several assumptions, not all of them necessarily valid. The estimate of the variance of each single datum is impossible in the practice. The travel-time residual for the

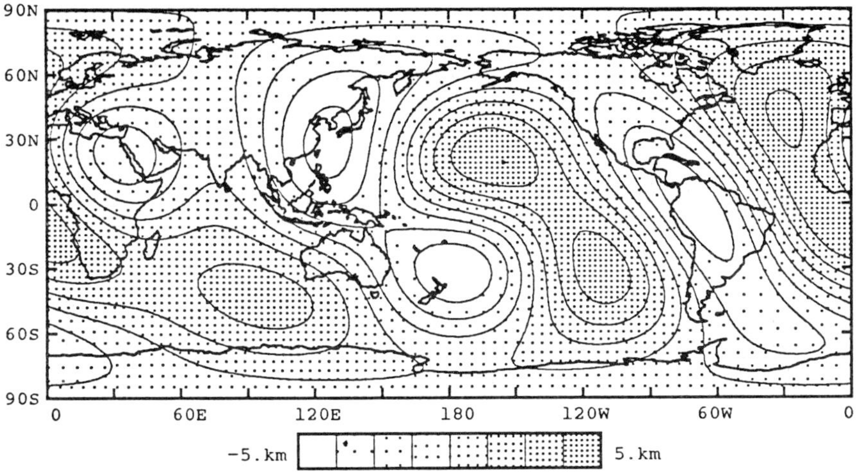


*Figure 9.* Topography of CMB deriving from the inversion of the $PKP_{BC}$ dataset. The comparison with figure 8 shows a good agreement where the structure is sampled by *PcP* (see also Figure 2).

$i$ -th path can be ideally seen as deriving from three terms:

$$\delta t_i = h_i + k_i + \varepsilon \tag{11}$$

The first term $h_i$ is the contribution of the anomalous structure we are seeking. The last element $\varepsilon$ in (11) is the random reading error, with zero average and some variance $\sigma_\varepsilon^2$; source complexity, operator reading errors, instrumental noise all contribute to it. In general, it does not follow a gaussian distribution, as is evident from Figure 4. The first histogram is clearly skewed - the explanation is that an operator is more likely to read an arrival late rather than early, missing the first small impulse hidden in the earth noise. The second term $k_i$ in (11) is another systematic error which originates from the path $\gamma_i$. This is caused by model inadequacy: our parametrization of the earth is incomplete.

We only aim at fitting the term $h_i$. The other two terms can vary considerably for different stations and different earthquakes. Their quantification is not possible, and we in practice resort to the approximation in which each residual is seen as:

$$\delta t_i = h_i + e \tag{12}$$

where now $e$ has zero average and a variance $\sigma_e^2$ assumed to be the same for all the observations pertaining to the same phase. The variance of the phase is then estimated by the variance of the misfit after the inversion. This is the estimate of variance of the data considered in derivation of the error map in Figure 7.

The complicated statistics of the problem forces us to give to the term $e$ properties which are equivalent to $\varepsilon$, although the model inadequacy error $k_i$ can be important. Note that the relative magnitude of $h_i$ and $k_i$ depend on the parametrization chosen, but their sum does not. Also, not necessarily $h_i$ is physically correct. It in fact represents the fit to

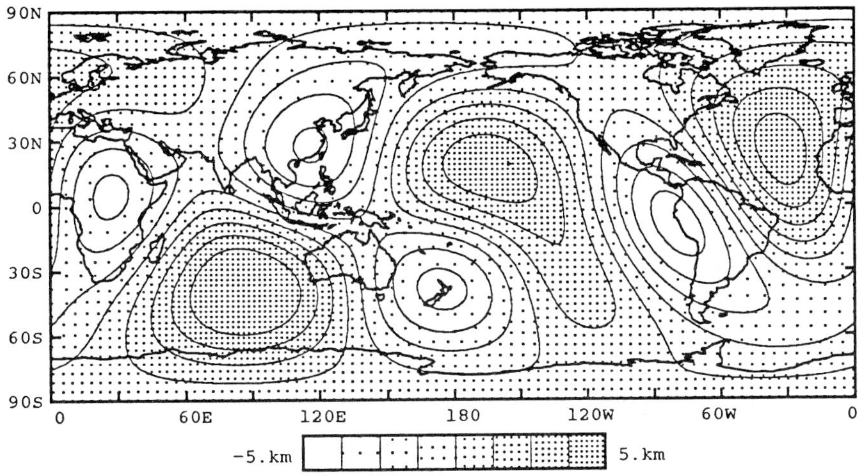


*Figure 10.* Undulations of the core-mantle boundary resulting from the joint inversion of the two datasets.

the original data that can be obtained with such a parametrization. The choice of the parameters is a strong *a priori* information that we force into the model, and may not be correct. The objective of the experiment described in this section was to test the correctness of the representation.

## 5. Anisotropy of the inner core

### 5.1 Method

The travel-time correction for a velocity perturbation $\delta v$ on the infinitesimal ray segment of length $dl$ is $(-\delta v/v^{2})dl$. We assume first-order perturbation theory: the path in (1) is computed in the (spherically-symmetric) unperturbed model, and the integral kernel depends only on radius. We can exploit the radial dependence and rewrite (1) as

$$\delta t_i = \int_{\bar{r}_i}^{R} \frac{-2\eta}{v_0^2(\eta^2 - p_i^2)} \delta v(\mathbf{r})dr \qquad i = 1,2,\ldots,n \tag{13}$$

(cf. Bullen and Bolt, 1985, p.170). The integral is evaluated in the interval between the bottoming radius $\bar{r}_i$ of the ray and the earth radius $R$. The velocity $v_0$ is a function of radius only, but the perturbation $\delta v$ varies laterally. The polar coordinates $\theta$, $\phi$ are expressed as a function of $r$ to give the path in parametric form. They are found for the initial spherical model by means of simple trigonometric formulae.

We represent velocity heterogeneities in the inner core by means of the spherical harmonic expansion (2). The range corresponding to the interval $(-1,1)$ for the variable $\rho$ is from 0 to the inner-core boundary (ICB) radius $R_{ICB}$. Also, the integration in (13) will be extended to $R_{ICB}$ rather than to $R$. To avoid singularities at the centre of the earth we

**Table 2: Coefficients for CMB (cf. figure 10)**

| | | $c_l^m$ | $s_l^m$ |
|---|---|---|---|
| $l = 0$ | $m = 0$ | -2.50 | |
| $l = 1$ | $m = 0$ | 0.34 | |
| | $m = 1$ | 0.07 | -0.03 |
| $l = 2$ | $m = 0$ | 0.14 | |
| | $m = 1$ | 0.55 | -1.02 |
| | $m = 2$ | 0.29 | 0.00 |
| $l = 3$ | $m = 0$ | -0.26 | |
| | $m = 1$ | 0.30 | -0.04 |
| | $m = 2$ | 1.15 | -0.31 |
| | $m = 3$ | -0.11 | -1.25 |
| $l = 4$ | $m = 0$ | -0.11 | |
| | $m = 1$ | 0.22 | 0.36 |
| | $m = 2$ | -0.44 | 0.12 |
| | $m = 3$ | -0.66 | -0.04 |
| | $m = 4$ | -0.23 | -0.68 |

demand the vanishing of $\delta v$ and its first radial derivative at $r = 0$. Furthermore, we limit ourselves to the maximum radial order $K = 2$. Thus, the pattern of heterogeneity will be modulated with depth by a function proportional to $r^2$.

With the expansion (2), (13) becomes a linear equation in the $p$ structural parameters ${}_kc_l^m$, ${}_ks_l^m$. Considering it for all the $n$ available observations gives the linear system of equations of condition of the form $\mathbf{Ax} = \mathbf{d}$. This will be solved as in the previous example.

The harmonic expansion allows to evaluate the integrals resulting from (13) in an economic way (Dziewonski, 1984). We have to integrate numerically expressions which are linear in the spherical harmonic functions. By means of suitable rotations (see Backus, 1964), we can reduce the problem to integrations along the equator, which depend only on the angular distance between source and receiver. These have to be calculated only once, and then transformed by rotation to give the value for the actual path.

Travel times of the *PKIKP* phase are the data we intend to use. We must of course remove the contribution of the structure that we already know: lower-mantle heterogeneities and core-mantle boundary topography (from the model of Figure 10). The ISC Bulletin usually reports core phases for distances from 110° to 180° . For convenience, we divide this dataset into the 5 subsets defined in Table 3.

To avoid complications, like *PKiKP* reflected below the critical angle and contamination by precursors scattered at CMB (cf. Bullen and Bolt, 1985, p.260; Anderssen and Cleary, 1980), we will only consider sets $\mathbf{R}_2$, $\mathbf{R}_4$ and $\mathbf{R}_5$.

**Table 3: Subdivision of PKIKP data†**

| | Range | Turning radius | $N_{conn}$ |
|---|---|---|---|
| $\mathbf{R}_1$ | 110° - 120° | $R_{IC}$ | 2306 |
| $\mathbf{R}_2$ | 120° - 135° | 1200–$R_{IC}$ | 4997 |
| $\mathbf{R}_3$ | 135° - 155° | 880–1200 | – |
| $\mathbf{R}_4$ | 155° - 170° | 370–880 | 2131 |
| $\mathbf{R}_5$ | 170° - 180° | 0–370 | 428 |

†The radius of the inner core $R_{IC}$ is 1221.5 km. The number of connections $N_{conn}$ is not specified for $\mathbf{R}_3$, which includes the intersection of *DF* with *BC*.

## 5.2 Separate inversions

In Table 3 the number of available summary data are reported for each of the distance intervals that we defined. In range $\mathbf{R}_3$, the travel-time curve is complicated by the presence of precursors and intersection of different branches. The arrivals are dominated by the *BC* branch and it is practically impossible to identify the (many fewer) readings pertaining to *PKIKP*. Arrivals in $\mathbf{R}_1$ do not penetrate into the inner core, and could be used to model the inner-core boundary. The increase in $v_P$ which occur entering ICB, as opposed to the decrease found crossing CMB, makes the kernels for reflected and refracted phases of the same sign. We cannot repeat the experiment made in the previous section for CMB. Besides, transmitted waves like *PKIKP* are relatively insensitive to changes in the radius of the inner core because of the relatively small increase in seismic velocity across ICB. We therefore assume the ICB to be spherical.

Rays belonging to $\mathbf{R}_5, \mathbf{R}_4, \mathbf{R}_2$ sampled the inner core at different depths. As listed in Table 3, set $\mathbf{R}_2$ only contains information about the outermost 20 km, whereas $\mathbf{R}_5$ is the only source of information for the innermost region of the earth. Also, the number of summary paths varies considerably for each set $\mathbf{R}$. This is to be ascribed in great part to a geometrical effect, as shown by Anderssen and Cleary (1980). The surface of an annulus one degree wide decreases as the distance increases, reducing the area where large-distance arrivals can be read.

The parametrization that we choose does not allow any degree of freedom with depth: the heterogeneity pattern is required to taper as $r^2$ as $r$ vanishes. Should this parametrization be inadequate and we were to invert simultaneously all three $\mathbf{R}$ sets without relative weighting, then the small $\mathbf{R}_5$ set would be almost completely ignored. It is, thus, important to analyze the three sets independently.

The three heterogeneous models at the top of the inner core are plotted in Figure 11 for, respectively, $\mathbf{R}_5, \mathbf{R}_4, \mathbf{R}_2$; since heterogeneities are modulated by only one radial function, their pattern does not change with depth. For ease of comparison, only harmonic degree $l = 2$ is shown. The $\mathbf{R}_5$ model is dominated by a strong zonal pattern which reaches a perturbation of 0.5 km/sec at the poles. This component is much smaller in $\mathbf{R}_4$ data and, though small, is of opposite sign for $\mathbf{R}_2$.

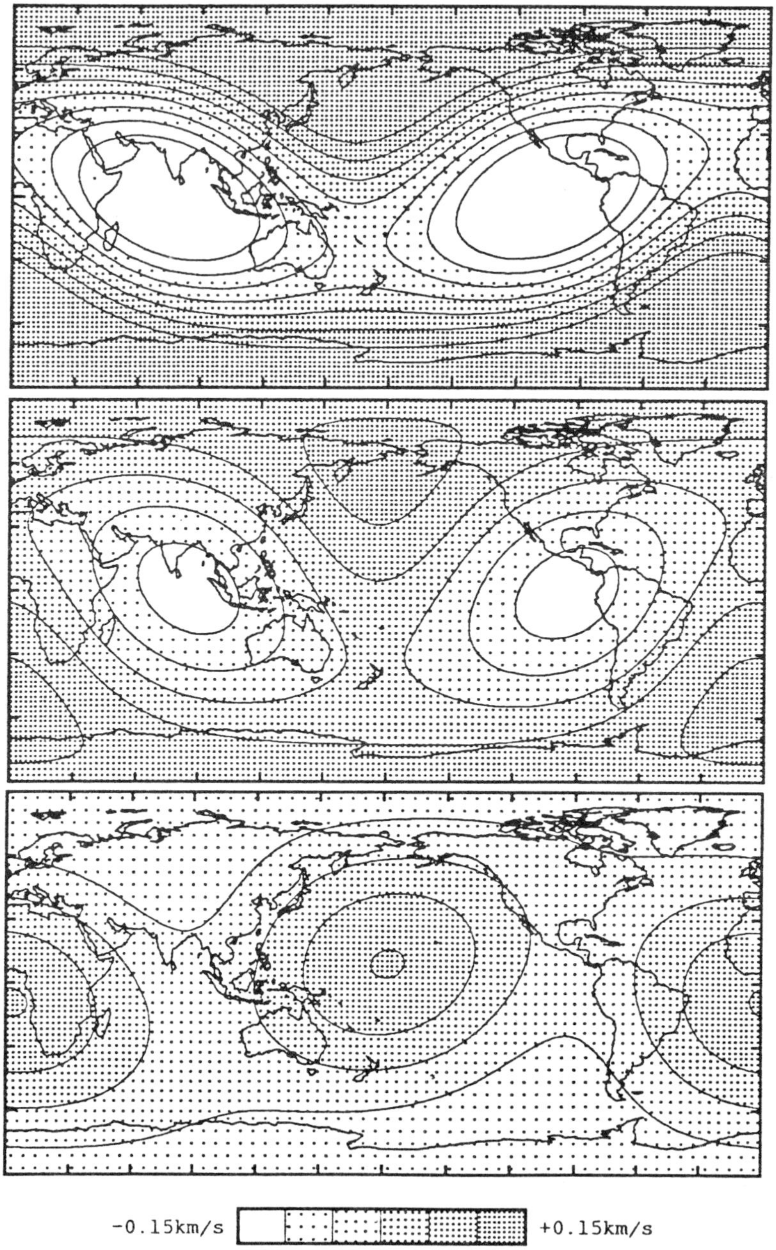


*Figure 11.* Separate inversions for heterogeneity in the inner core under the assumption of isotropy. The pattern decreases as $r^2$, the maps are for the top of the inner core. Only degree 2 is plotted. Top panel: model from the $\mathbf{R}_5$ dataset; middle: model from $\mathbf{R}_4$; bottom: model from $\mathbf{R}_2$. The three results are inconsistent.

The model with the largest variation is obtained from the smallest dataset (only 428 summary rays). The suspicion could arise that we are looking at an artifact of an insufficiently constrained inversion, but two different considerations confirm the validity of the model. First, the separate inversions, made using two subsets into which we randomly partitioned $\mathbf{R}_5$, give models that correlate with the top map of Figure 11 with cross-correlation coefficients of 0.8 and 0.9. Second, visual inspection of the residuals which constitute $\mathbf{R}_5$ (see Figure 12) shows that such a large effect is in fact required.

The obvious conclusion is that the parametrization is not correct. The problem is not in the simplified depth dependence. The very large anomaly shown by $\mathbf{R}_5$ data would have to be contained in a small sphere of 400 km radius, and still vanish at the center of the earth. This is physically very difficult to envision. From these considerations, and from a comparison with normal mode observations, Morelli et al. (1986), Woodhouse et al. (1986) and Giardini et al. (1987) inferred the presence of anisotropy in the inner core.

### 5.3 Cylindrical anisotropy

The large zonal component of the $\mathbf{R}_5$ model implies that for nearly vertical rays a polar path is almost two seconds faster than a path on the equatorial plane. This could in fact be accomplished by an anisotropic velocity field with rotational symmetry about the earth's axis. Rays travelling to short distances spend less time in the inner core, and present more curved paths, so that the effect of anisotropy on them would be small. The principal (or, more probably, only) constituent of the inner core is iron, in a cristallographic form characterized by cylindrical symmetry (hexagonal close packed phase, or ε-iron, Anderson, 1986). A preferential alignment of single crystals can easily produce the effect postulated, characterized by a magnitude of the order of a few percent in $P$ velocity.

Following the notation of Love (1927), an anisotropic velocity field with a symmetry axis can be characterized by the five constants $A, C, F, L, N$. With such an assumption, $P$-wave velocity along a direction forming an angle $\xi$ with the symmetry axis can be expressed as:

$$v_P^2\rho = A\sin^4\xi + C\cos^4\xi + 2(2L+F)\sin^2\xi\cos^2\xi \tag{14}$$

where $\rho$ is density; in the case of isotropy, $A = C = 2L+F$. The effect we are seeking is small, so we can to consider:

$$C = (1+2\varepsilon)A \qquad 2L+F = (1+\sigma)\sqrt{AC} \tag{15}$$

where the anisotropy coefficients $\varepsilon$, $\sigma$ are assumed to be small enough that the terms of second and higher order can be neglected. This is equivalent to perturbation theory applied to small changes of the value of $\varepsilon$ and $\sigma$ from their initial value of zero. Paths and differential kernels are computed in the same isotropic, spherically symmetric, starting model as before. With (15) and neglecting terms of high order we obtain, from (14)

$$v_P(\xi) = v_{eq}\cdot(1+\varepsilon\cos^2\xi+\sigma\cos^2\xi\sin^2\xi) \tag{16}$$

where the equatorial velocity $v_{eq}$ represents the velocity for $\xi = \pi/2$, and by definition is $\sqrt{A/\rho}$. In this way, we only add two more parameters to the model.

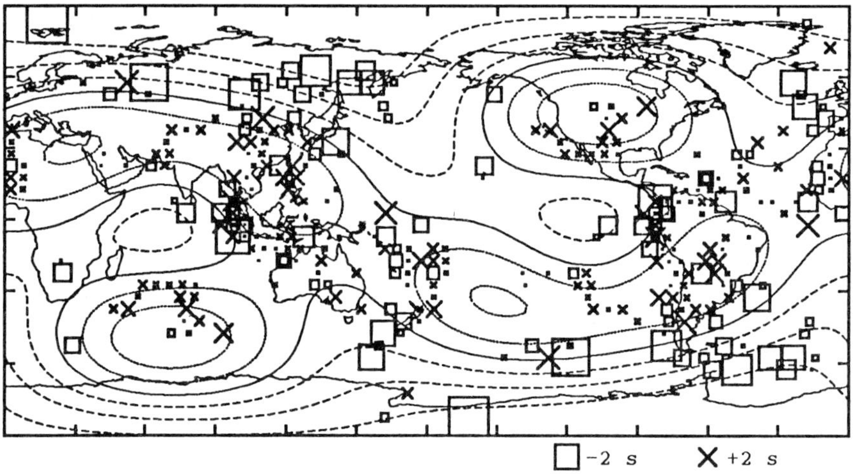


*Figure 12.* Travel time residuals of the $\mathbf{R}_5$ data. The residual for each ray is plotted at both source and receiver points. Contour lines show the even order harmonic expansion truncated at degree 4. A strong axisymmetric component is present.

The values of $\varepsilon = 0.032 \pm 0.005$ and $\sigma = -0.064 \pm 0.015$ result from the inversion. The amplitude of the anisotropic perturbation is required to decrease with depth as $r^2$, in the same way as the heterogeneity perturbation $\delta v$. The polar vertical travel time resulting from this effect alone is about 2.5 seconds faster than the equatorial. With these values, the separate inversions of $\mathbf{R}_5$, $\mathbf{R}_4$, and $\mathbf{R}_2$ are repeated, yielding the models plotted in Figure 13. The effect of anisotropy, in terms of compatibility among the three different models, is significant. As already mentioned, however, the characteristics of the final model, dominated by data relatively insensitive to anisotropy, do not change in a significant way, and because of the small numerical weight of $\mathbf{R}_5$ we would not find the inadequacy of the parametrization without performing the separate inversions for each of the $\mathbf{R}$ sets.

## 6. Conclusions

We have presented two examples of application of terrestrial tomography through spherical harmonic expansion to studies of the deep interior of the Earth. These two examples were chosen to illustrate more than a straightforward use of a particular algorithm to derivation of a laterally varying model.

The topography of the core-mantle boundary inferred from the *PcP* data alone would have to be treated with substantial skepticism because we are not in position to quantify precisely the sources of systematic errors due to mislocation of earthquakes or imperfections of our model of the aspherical structure of the lower mantle. Standard error analysis, which assumes that errors are random and uncorrelated will, in such situations, always underestimate the range of uncertainties and be, therefore, misleading if not meaningless. What makes convincing our estimate of the size and location of the principal

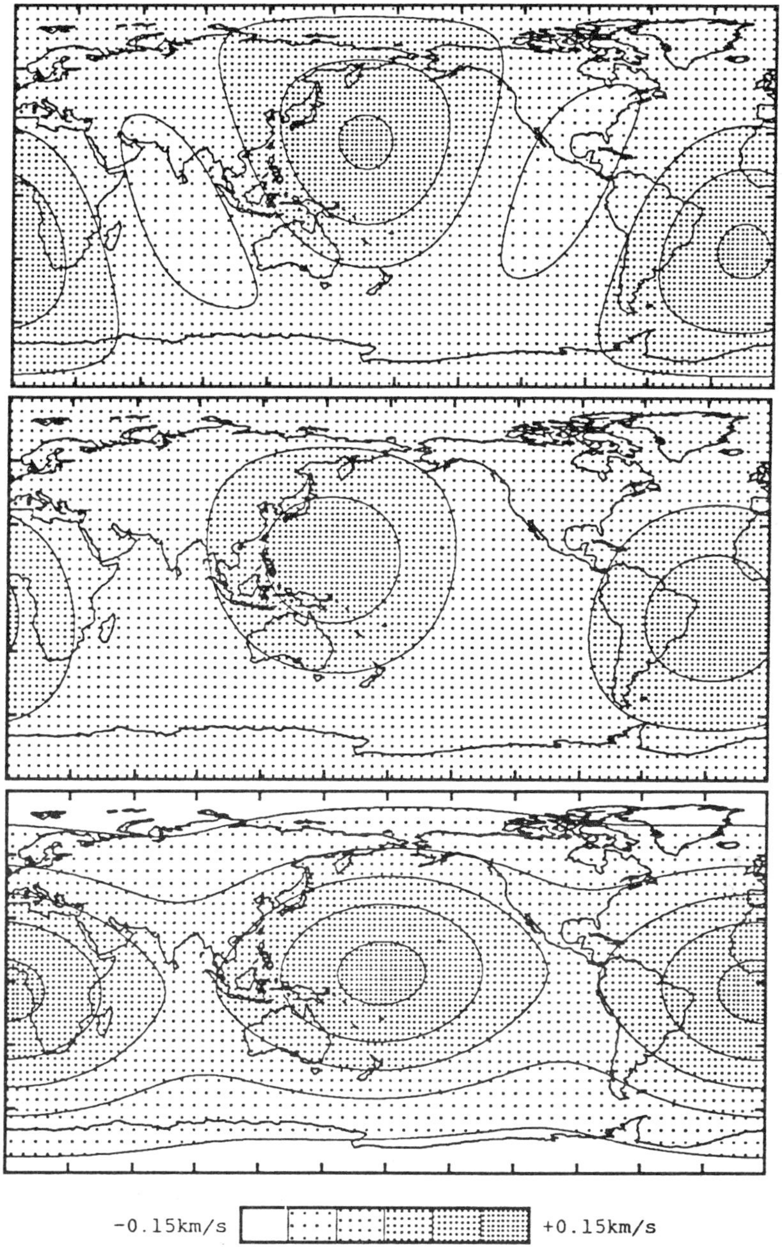


*Figure 13.* Results of inversions for heterogeneity in the inner core when the effect of cylindrical anisotropy, as described in the text, is taken into account. Compare with similar Figure 11. Now the separate inversions yield similar results.

features at the core-mantle boundary is the compatibility of the results obtained using separately *PcP* and $PKP_{BC}$. The situation is particularly advantageous in this case, because the kernels for *PcP* and *PKP* are of opposite sign. Nevertheless, attempts should be always made to test results by analyzing separately different subsets of data.

If one were to suggest a general rule for making a discovery, the foremost advice would be to search for inconsistencies obtained using two (or more) different data sets. It was the incompatibility between the strong zonal splitting of normal modes sensitive to the properties of the inner core (Li et al., 1986) and the lack of such a feature in our initial models of the inner core (Morelli and Dziewonski, 1986), obtained from the entire data set and formulated in terms of a general 'isotropic' heterogeneity, that led to re-examination of the problem. In particular, we began to investigate the properties of *PKIKP* in different distance ranges and found the strong zonal component in the residuals in a 170° - 180° distance range. The suggestion of cylindrical anisotropy is, of course, a non trivial step, but the need to find an unorthodox mechanism was dictated by the results of data analysis.

The ISC data set is very noisy; Morelli and Dziewonski (1987) estimate that for the *P*-waves the variance due to random noise is an order of magnitude greater from that due to lateral heterogeneity. For other subsets of data this ratio is even worse. This means that a direct comparison of model predictions with original observations is not likely to be informative. In such circumstances adoption of proper parameterization is critically important.

*Acknowledgements:* We would like to thank J.H. Woodhouse, D. Giardini, X.-D. Li and J. Bloxham for numerous discussions during the course of study described here. This research was supported by a grant EAR83-17594 from the National Science Foundation.

# Chapter 12

# Ray tracing for surface waves

*N. Jobert and G. Jobert*

## 1. Introduction

Surface waves (at least the fundamental mode, see below) are the latest arrivals on the seismic record at a given epicentral distance. On classical records, they show up as long-period oscillations, predominating in the record for shallow earthquakes and distances larger than a thousand kilometers. Their relative importance increases with the distance, which can be explained by their 2-dimensional propagation in a direction parallel to the surface of the Earth. Indeed they are guided waves with standing wave properties in a direction normal to this surface: along this direction the displacements are in phase at all depths. Thus their geometrical spreading, as guided waves, is less than that of body waves which propagate in 3 dimensions.

Surface waves are classified in two types: the faster Love wave with a displacement transverse (y-component) to the direction of propagation (x- component), and the Rayleigh wave with a displacement in a vertical plane containing the direction of propagation (x and z components). Their theory has been known since 1885 for the Rayleigh wave, 1910 for the Love wave. At the surface of the Earth they behave as waves guided in the crust and upper mantle where the velocities are lower than at depth. Love waves are built with transverse SH waves multiply reflected at the surface of the Earth, Rayleigh waves with multiply-reflected and converted P and SV waves. Thus their local velocities are respectively related to the local velocity structure of these body waves. But it can be shown that except for the short wave-lengths influenced by the very shallow layers, the fundamental mode of Rayleigh waves is sensitive mainly to SV wave velocity.



*G. Nolet (ed.), Seismic Tomography, 275–300.*

For a guided wave, the theory shows the existence of an infinity of modes characterized by an index $m$ . At a given frequency their phase velocities increase with $m$ , and the variation of their displacement with depth shows a number of "nodes" related to $m$ . The fundamental mode corresponds to $m=0$ . For a horizontally layered half-space the frequency range of existence of higher modes ($m>0$) is more and more restricted to high frequencies by a cut-off frequency increasing with $m$. But the fast higher modes are less excited by shallow earthquakes than the fundamental mode which is generally well observed.

As long as the guide which supports them does not present a strong lateral variation with distance, surface waves can be used to retrieve average structural information at depth between two stations (or source and station) which can be widely separated. Hence their efficiency to study oceans or inaccessible regions, as long as the structure remains not far from laterally homogeneous between the two points. The different frequencies present in the records have only to be analyzed and separated. Due to dispersion, which is a property characteristic of surface waves, each monochromatic wave propagates with its own velocity. The fundamental mode behaviour is such that each of them is most sensitive to the structure at a depth which is a fraction of the wave-length, about 1/3 for Rayleigh waves, 1/4 for Love waves (Knopoff, 1972). Consequently, the resolution of the inverse problem for the variation of velocity structure with depth is all the better as the bandwith of data is broader. However, compared to that of body waves, the resolution is limited by the long-period character of the surface waves and informations are averaged over a certain depth interval. The maximum depth down to which information is obtained is independent of the distance between stations, so that large depths can be reached with a few long-period stations. Surface or mantle waves carry also along their path long-period information about the source.

Quantitative information began to be obtained with the development of long-period instruments (Press, 1956, Sato, 1958). A first group velocity tomography of the surface of the Earth was presented by Santo (1966). Since 1960, the instrumental techniques have greatly improved. Now to seismologists are available high-quality long-period data, digitally recorded in the stations of arrays such as GDSN, IDA or GEOSCOPE. The precision of measurements increased, and long-period tomography has developed to bring out large-scale laterally heterogeneous models of the Earth, such as that by Nataf, Nakanishi et Anderson (1986), obtained by inversion of velocity measurements, or M84C (Woodhouse et Dziewonski, 1984) obtained by waveform modelling. These new results show large scale structure anomalies in the mantle that can be related to convection patterns.

In the classical methods of surface wave tomography, the waves are supposed to propagate around the Earth along great circles. Considering Fermat's principle, this is valid in the geometrical optics approximation, and first order perturbation theory, for smooth and weak lateral heterogeneities. In the great circle tomographic technique, a regionalization is performed to retrieve regional "pure-path" velocities in regions considered as laterally homogeneous. The slowness observed along a path is considered to be the sum along the great circle path of the local slownesses weighted by the relative lengths of the corresponding "pure paths". This is an expression for finite lengths of great circle paths of the "phase integral approximation" using the local phase velocities. In the geometrical optics approximation, for body waves, except at first order discontinuities only transmitted waves are considered. The weak reflected and converted waves due to velocity gradients are neglected. Diffraction effects are not taken into account. In the same way, in this approximation for surface waves, reflected and diffracted waves are neglected, as well as waves converted from one mode to another except at discontinuities.

In fact, as has been pointed out by Gjevik (1974) for surface waves on the Earth the "short wavelength approximation" of geometrical optics deals with smooth, weak lateral heterogeneities only at wavelengths long enough to let the waves sample at depth smoother and weaker lateral changes in velocity. Indeed, it is well known that the most surficial layers of the Earth sampled by short-period surface waves are strongly heterogeneous, with small scale heterogeneities such as continent-ocean transition, and recent models of the upper mantle show that lateral heterogeneities decrease with increasing depth.

The classical great circle approximation was questioned after anomalous oceanic surface waves observations by Evernden (1953, 1954), McGarr (1969), Capon (1970, 1971) using LASA and NORSAR arrays, in the period range 20-30 s. At these short periods "multipathing effects" can occur, i.e. interference between waves that arrive at the station after being refracted differently along different paths. An array enables also the measurement of the azimuth of the direction of arrival of the wave, which can be shown in some cases to deviate from the great circle direction. In particular, refractions can occur at continental margins.

The interpretation of these anomalies as due to distorsions of the propagation paths by lateral variations of local velocities brought to light the need for a surface waves equivalent of ray tracing for body waves. Ray tracing equations for a spherical surface were proposed by Julian (1970) and Gjevik (1974). After presenting a perturbation method, this last author performs numerical integrations of the ray tracing equations and obtains time and azimuth deviations in agreement with oceanic observations. Sobel and von Seggern (1978) consider phase, azimuth and amplitude anomalies observed at 20 s. They show by ray tracing, using Julian's equations, that focusing and defocusing effects can partly explain amplitude anomalies responsible for the scatter of magnitude $M_s$ estimations. A numerical problem arises with the representation of the phase velocity geographic distribution, which must be conveniently chosen. Velocities are given at points on a grid, with interpolation schemes for velocity and derivatives. These authors express a need to refine their models, particularly in regions of high velocity gradient. Ray tracing is performed over Eurasia by Patton (1980) at a period of 40 s, to interpret anomalous records of Rayleigh waves. As ray theory amplitude calculations are invalid near caustics, a more refined modelling is performed for the short Rayleigh waves (T<80s) in the Pacific ocean by Yomogida (1985) using the Gaussian beams technique to synthetize monochromatic groups of waves. By means of this technique, Yomogida (1986) inverts oceanic records for a map of phase velocity anomalies in the Pacific.

At longer periods, the effects of lateral variations of the phase velocity appear mainly as amplitude anomalies, since the propagation and travel time along a ray are affected by the lateral velocity gradient while amplitudes depend on second spatial derivatives. For mantle waves in the period range 100-300 s anomalous amplitude observations are attributed to propagation through regions of high velocity gradients by Lay and Kanamori (1985) and Schwarz and Lay (1985) by ray tracing using the techniques of Sobel and von Seggern (1978). They use the recent global models of large-scale lateral heterogeneity of the Earth of Nakanishi and Anderson (1984) and M84C (Woodhouse and Dziewonski, 1984). Still, their geographic distribution of phase velocities have smooth, large-scale weak lateral heterogeneities, being built respectively with 6 and 8 spherical harmonics. But it is enough to induce anomalies of the order of those observable on the high quality long period records now available. Time and spectral anomalies of mantle waves observed on IDA and GEOSCOPE records are reported by Roult and al. (1986). With ray tracing and Gaussian beam synthesis (Jobert, 1986a, 1986b) these time anomalies can be interpreted as due to deviations of the paths from the great circle induced by lateral velocity gradients. Woodhouse and Wong (1986) also show that phase and amplitude anomalies predicted by the above models have a character similar to the observations.

Up to now, in surface wave tomography amplitude informations were scarcely used. First order perturbation schemes are proposed for free oscillations (Romanowicz, 1986; Park, 1986) and ray theory (Woodhouse and Wong, 1986). Since the amplitude of surface waves are sensitive to spatial derivatives of order higher than the first order involved for the phase, their consideration in the inverse problem will increase the resolution.

Ray theory, with phase and amplitude modelling, is an essential tool for the interpretation of surface wave observations and the refinement of the inversion process in surface wave tomography to retrieve the 3-dimensional structure of the Earth. Similarly it can be used to refine the inversion for earthquake source mechanisms.

## 2. Theory of surface rays

The ray method, which proved its efficiency for seismic body waves, may also be applied to study the propagation of waves along the surface of a laterally heterogeneous elastic body. This approach was taken for general surface waves by Karal and Keller (1960) who introduced complex numbers in the ray theory, and by Bretherton (1968) who made a systematic application of the W.K.B. approximation to the propagation of waves in slowly varying waveguides. It was developed for seismic waves on surfaces by Babich and his coworkers: (Babich, 1961a,b; Babich et al., 1976; Nomofilov, 1978); detailed accounts in English may be found in (Babich and Rusakova, 1963; Kirpichnikova, 1976), see also (Červený, Molotkov and Pšenčík, 1977). Woodhouse (1974) made a detailed study of the propagation of surfaces waves over a laterally varying structure with layer boundaries and free surface differing only slightly from a plane.

### 2.1 Ray theory for seismic body waves and surface waves

In the classical seismic ray theory one first seeks for the wave fronts; they are the characteristic surfaces on which the solution of elastodynamics may have a discontinuity (Courant and Hilbert, 1966) and are obtained as solutions of a first order partial differential equation, the characteristic equation, associated to the elastodynamic equation. Every such solution is generated by a family of curves: the bicharacteristics or rays, or more precisely by bicharacteristic strips, along which the fronts have the same tangent planes.

The characteristic equation may be obtained by assuming an asymptotic representation for the displacement as a series in powers of $\omega^{-1}$, the inverse of the frequency, or as a series of terms less and less singular, and introducing it in the elastodynamic equation; the cancellation of the most singular terms leads to a condition of compatibility which is the characteristic equation. This equation - or more precisely the different factors into which it can be decomposed - may be considered as a Hamilton-Jacobi equation for a function called the eikonal and the rays are curves which are solutions of the differential system derived from it. The different terms in the series representing the displacement are propagated along the rays. The conditions of validity of the asymptotic development have been discussed for instance by Červený and al. (1977) and Kravtsov and Orlov (1980). Generally for a signal of wave-length $\Lambda$ the following conditions must be satisfied:

a) The wave-length $\Lambda$ must be much smaller than any characteristic length $l_0$ in the medium:

$$l_0 \gg \Lambda \tag{1.1}$$

$l_0$ may be $f / \| \nabla f \|$ where $f$ is any physical property, $H_i$ the thickness of layers, $R_i$ radii of curvatures of boundaries...

b) The ray method fails in the vicinity of an interface where (1.1) is not valid. It can be used only up to a distance $n \gg \Lambda$ of such a surface.

c) The ray method is not applicable if the distance of propagation $l$ on the ray is too great. One must have:

$$l \ll l_0^2 / \Lambda \tag{1.2}$$

The propagation of discontinuities along the surface of an elastic body may be studied, as did Babich and his coworkers, (for instance Babich and Rusakova, 1963) by using a theorem by Petrovski (1946) who proved that a discontinuity along any smooth surface S is propagated at a point M of S with a velocity given by the Rayleigh equation for a homogeneous half-space with the velocities of the body at M. Complex solutions of the eikonal equation are sought, corresponding to waves attenuated in the direction normal to the surface. In a general way the displacement may be written:

$$\mathbf{U} = \sum_{\alpha} \mathbf{U}_p^{\alpha} f_p^{\alpha}(t-\tau_p(x,z)) + \mathbf{U}_s^{\alpha} f_s^{\alpha}(t-\tau_s(x,z)) \tag{1.3}$$

where $z$ is the distance to S measured along the unit normal $\mathbf{v}$ to S, the functions $f(t)$ are analytic for $-\varepsilon < \mathrm{Arg}\, t < \pi + \varepsilon$ , $\varepsilon$ small positive, and the functions $\tau_i$ $(i=p,s)$ are complex solutions of the eikonal equations:

$$\nabla^2 \tau_i = n_i^2 \tag{1.3a}$$

with the local squared slownesses of the body waves at the point $\mathbf{x}$ on S:

$$n_p^2 = \rho/(\lambda+2\mu) \qquad n_s^2 = \rho/\mu \tag{1.3b}$$

The complex wavefronts with equation: $t=\tau_i$ must cut S on the same real curve $\Gamma$, corresponding to the wave-front on S, so that: $\tau_i(\mathbf{x},0)=t(\mathbf{x})$ and the gradients of the functions $\tau$ may be written:

$$\nabla\tau_i = \overline{\nabla}\tau - i\,\mathbf{v}\gamma_i \quad \mathrm{Re}\gamma_i(\mathbf{x},0)>0 \tag{1.4}$$

where $\overline{\nabla}$ denotes the gradient operator on S. The eikonal equations lead to:

$$\overline{\nabla}^2\tau = n_i^2 + \gamma_i^2 = n^2 \tag{1.4a}$$

where $n$ is the slowness of the Rayleigh wave. This equation is the same as that one would find for a field $F(\mathbf{x})$ satisfying the scalar wave equation where $\overline{\Delta}$ is the Laplacian on S :

$$\overline{\Delta}F = n^2\partial_{tt}^2 F \tag{1.4b}$$

This remark gives a justification to the use of a scalar wave equation in (6.7).

However seismologists are mainly interested in guided surface waves which were not considered in these studies and for which some scaling of the coordinates is necessary as discussed below.

### 2.2 The scaling of coordinates

In ray theory for body waves, in the conditions of validity of the high-frequency approximation (1.1), all space dimensions play the same role. Scale considerations are however different for surface waves. They propagate in 2 dimensions along the surface, according to the local phase velocity dispersion, with conditions equivalent to (1.1) involving surface gradients. At each point of the surface, to the local dispersion are associated eigenfunctions describing the variation of amplitude of the wave in a direction normal to the surface. As we will see, the local dispersion derives from local boundary conditions for the surface wave in the medium under the point considered. The scale of the variations of the physical properties in the direction normal to the surface is completely independent of the horizontal scale. Now the conditions of validity of the development which will be used, similar in form to the conditions (1.1), only express that the variations of the properties in a direction along the surface must be weak and smooth at the scale of a wave-length. In other words, through the dispersion relation, the variation of the phase velocity along the surface must be weak and smooth.

In fact the exact theory of surface waves is developed completely only for very simple structures: the layered media where properties vary only with the distance to a plane or to a point. In these cases, waves of an arbitrary given frequency $\omega$ may be propagated along the surface with a wavevector $\boldsymbol{\kappa}$ only if some relation, the dispersion relation, exists between these quantities. Thus when studying waves propagated on the surface of a medium with laterally slowly varying properties by perturbation methods, the scale- lengths for the displacement will not be taken the same in the direction normal to the surface and in the tangential ones, which we will call horizontal directions.

Another difference with non-dispersive body-waves propagation lies in the existence of a local group velocity governing the propagation of energy along the rays, the geometrical properties of the rays depending only on the phase velocity.

Bretherton (1968) presented a very general and formal method. He builds an asymptotic expansion of the displacement in powers of a small parameter $\varepsilon$ ; its first term provides for $\varepsilon=0$ an exact solution to the propagation on a model without lateral variations, for which the normal modes, progressive along the surface S and with a certain structure in the direction normal to S, are supposed known. He assumes also that in the laterally varying medium the solution may be obtained from a variational principle: the stationarity of the action of a Lagrangian L, which implies a system of Euler equations, equivalent to the elastodynamic equation, and boundary conditions. For $\varepsilon$ small but nonzero, locally the solutions are approximately the same as in a medium laterally uniform with the local properties.

The theory is developed only in cartesian coordinates and therefore for media which differ only slightly from horizontally layered ones. Let $x^i$ denote the horizontal directions, $z$ the vertical. The technique involves a stretching of the horizontal coordinates and time $t$:

$$X^i = \varepsilon x^i \ , \quad T = \varepsilon t \ , \tag{2.1}$$

in order to describe slow changes in the parameters of the wavetrain and of the medium, which correspond to $X=O\,(1)$, whereas a wavelength corresponds to $\mathbf{x}=O\,(1)$. The locally sinusoidal character of the oscillation is given by a factor $\exp(i\,\psi)$, where:

$$\psi(\mathbf{x},t) = \varepsilon^{-1}\Psi(\mathbf{X},T), \tag{2.2}$$

from which we derive a local wavevector $\boldsymbol{\kappa}$ and a local frequency $\omega$ given by:

$$\kappa_j = \partial_{x^j}\psi = \partial_{X^j}\Psi \ , \quad j=1,2 \qquad \omega = -\partial_t\psi = -\partial_T\Psi \tag{2.3}$$

$\omega$ and $\boldsymbol{\kappa}$ verify the relations:

$$\partial\kappa_j/\partial T + \partial\omega/\partial X^j = 0 \qquad \partial_{X^1}\kappa_2 = \partial_{X^2}\kappa_1 \tag{2.4}$$

The displacement will be written as a series in powers of the small quantity $\varepsilon$ with the factor $\exp(i\,\psi)$:

$$\mathbf{U} = \mathrm{Re}[\sum_{r=0}^{\infty} \varepsilon^r \,\mathbf{U}^r(\mathbf{X},z)]e^{i\Psi} \tag{2.5}$$

$\mathbf{U}^0$ will be proportional to a given mode in the equivalent uniform medium corresponding to the point $\mathbf{X}$ and will be determined as if $\boldsymbol{\kappa},\mathbf{X}$ were exactly constant over a wavelength. The following terms $\mathbf{U}^r$ are successive corrections for nonzero $\varepsilon$ due to local nonuniformities in the medium. These irregularities may be changes either in geometry of the medium or in its physical properties. The horizontal coordinates enter (2.5) only through the stretched coordinates $\mathbf{X}$ so that the sinusoidal factor $\exp(i\,\psi)$ describes the slow variation of the wavetrain.

Hudson (1981) presents a different method (parabolic approximation), already used by other authors (Kirpichnikova, 1969). He suggests that for a wave number $\kappa$ on an isotropic medium, one should take scale-lengths differing along the three directions: normal to S, propagation direction and binormal. If $f$ is a large number, of the order of the inverse of the forward-scattering angle, the coordinates would be reduced respectively by $\kappa^{-1}$, $\kappa^{-1}f^2$ and $\kappa^{-1}f$. The displacement is then assumed to be smooth in these reduced variables.

### 2.3 Surface rays

Woodhouse (1974), following Bretherton's general method, considers the case of a medium differing only slightly from a homogeneously layered isotropic half-space, with layers boundaries $S_i$ and free surface S differing only slightly from planes; let P be the plane corresponding to S. The small quantity $\varepsilon$ gives an order of magnitude of the slopes of these surfaces. Woodhouse starts from the first-order differential system for the displacement-stress vector $\mathbf{f}$ with respect to the distance z to P and replaces the displacement and the stress vector $\boldsymbol{\tau}$ on P by expressions as (2.5). He finds that the first term satisfies the equation:

$$\mathbf{W}\mathbf{f}^0 \equiv \partial_z \mathbf{f}^0 - \mathbf{A}(\rho,\lambda,\mu,\boldsymbol{\kappa},\omega)\mathbf{f}^0 = \mathbf{0} \tag{3.1}$$

and continuity conditions for the stress vector:

$$[\boldsymbol{\tau}]\Big|_{S_i} = \mathbf{0} \tag{3.2}$$

The matrix $\mathbf{A}$ is the same as for a laterally uniform half-space but its parameters $\rho,\lambda,\boldsymbol{\kappa},\omega$ depend now on $\mathbf{X}$, and also on $T$ for the last two. The solution of (3.1) must satisfy the homogeneous boundary equations (3.2), and therefore be a particular eigenmode of the system, noted by the index $m$, for which $\omega$ and $\boldsymbol{\kappa}$ are related together by the dispersion relation which depends also on $\mathbf{X}$:

$$\omega = \Omega(\boldsymbol{\kappa},\mathbf{X},m) \tag{3.3}$$

and $\mathbf{f}^0$ takes the form:

$$\mathbf{f}^0 = a(\mathbf{X},T)\hat{\mathbf{f}}^0(\boldsymbol{\kappa},\omega,\mathbf{X},m) \tag{3.4}$$

where $\hat{\mathbf{f}}^0$ is a known solution with fixed phase and amplitude and $a(\mathbf{X},T)$ a complex factor to be determined.

Replacing $\omega$ and $\boldsymbol{\kappa}$ in the dispersion relation (3.3) by their expression in function of $\Psi$ we see that $\Psi$ satisfies the Hamilton-Jacobi equation:

$$\partial_T \Psi + \Omega(\partial_{\mathbf{X}}\Psi,\mathbf{X},m) = 0 \tag{3.5}$$

which may be solved by classical methods of characteristics.

Hamilton's system corresponding to (3.5) for $\mathbf{X}(T)$ and $\boldsymbol{\kappa}(T)$ is:

$$\dot{\mathbf{X}} = \partial_{\boldsymbol{\kappa}}\Omega(\boldsymbol{\kappa},\mathbf{X},m) \qquad \dot{\boldsymbol{\kappa}} = -\partial_{\mathbf{X}}\Omega(\boldsymbol{\kappa},\mathbf{X},m) \tag{3.6}$$

where the point notes a derivative with respect to $T$. As the dispersion relation does not depend explicitly on time, the frequency $\omega$ remains constant on a ray:

$$\dot{\omega} = \partial_{\boldsymbol{\kappa}}\Omega\cdot\dot{\boldsymbol{\kappa}} + \partial_{\mathbf{X}}\Omega\cdot\dot{\mathbf{X}} = 0 \tag{3.7}$$

A solution $\mathbf{X}(T)$ of (3.6) represents a curve on S, which may be called a *surface ray*. We note that the ray tracing equations obtained for the surface waves derive from the dispersion equation whereas for body waves they derive from the elastodynamic equation.

The energy of the surface wave is transported along the ray as the group velocity vector $\mathbf{Y}$ equals $\dot{\mathbf{X}}$:

$$\mathbf{Y} = \partial_\kappa \omega = \partial_\kappa \Omega = \dot{\mathbf{X}} \tag{3.8}$$

As shown by Bretherton for such problems deriving from a variational principle, from the equation corresponding to the following term in the development in $\varepsilon$ one may derive an orthogonality relation. Here the equation for $\mathbf{f}^1$ may be written:

$$\mathbf{W}\mathbf{f}^1 = \boldsymbol{\chi} \tag{3.9}$$

where $\boldsymbol{\chi}$ is a functional of $\mathbf{f}^0$, a member of the nullspace of $\mathbf{W}$. As there exists an operator $\mathbf{K}$ such that $\mathbf{KW}$ is self-adjoint one can write:

$$(\mathbf{KWf}^0,\mathbf{f}^1) = 0 = (\mathbf{f}^0,\mathbf{KWF}^1) = (\mathbf{f}^0,\mathbf{K}\boldsymbol{\chi}) \tag{3.10}$$

Woodhouse finds thus the expression of the complex functional of $\mathbf{f}^0$ which must be zero. The vanishing of its imaginary part leads to:

$$\partial_T(\partial_\omega L) - \partial_{X_\sigma}(\partial_{\kappa_\sigma} L) = 0 \tag{3.11}$$

where $L(a,\omega,\boldsymbol{\kappa},\mathbf{X})$ is a conveniently defined Lagrangian. Equation (3.11) expresses the law of conservation of wave action (Bretherton, 1968) which corresponds to the average Lagrangian principle as formulated by Whitham (1965): $\delta\int L(a,\omega,\boldsymbol{\kappa},\mathbf{X})dX_1 dX_2 dT=0$ with the constraints (2.4).

The variational equation for $a$ is simply:

$$\partial_a L = 0 \tag{3.12}$$

This leads to: $L=0$ for the stationary value, which is obtained if the dispersion relation is satisfied. This expresses the equality of average kinetic and elastic potential energies in a vertical column.

The characteristic curves associated with (3.11) are solutions of:

$$dT/\partial_\omega L = -dX^\sigma/\partial_{\kappa_\sigma} L \quad \text{or:} \quad \dot{\mathbf{X}} = -\partial_\kappa L/\partial_\omega L. \tag{3.13}$$

As L vanishes identically when the dispersion relation is satisfied we have then:

$$\partial_\omega L\, \partial_\kappa \omega + \partial_\kappa L = 0 \tag{3.13a}$$

and therefore (3.13) is equivalent to the first equation (3.6). Using equation (3.13) one may write equation (3.11) in the form:

$$\partial_T(\partial_\omega L) + \partial_{X^\sigma}\left[\partial_\omega L \dot{X}^\sigma\right] = 0 \tag{3.13b}$$

or taking into account the relation: $d/dT = \partial_T + \overline{\nabla}$, where the differential operator on S, $\overline{\nabla}$, is taken with respect to $\mathbf{X}$ ,

$$d(\partial_\omega L)/dT + (\partial_\omega L)\overline{\nabla}\cdot\mathbf{Y} = 0 \tag{3.14}$$

To interpret the last term let us suppose that initial conditions are given for the starting point and the direction of the ray, i.e. for $\mathbf{X}$ and $\boldsymbol{\kappa}$ as functions of two coordinates $s^\alpha$ on

S. One can show that:

$$\overline{\nabla}\cdot\mathbf{Y} = \partial_{X^\sigma}\dot{X}^\sigma = \partial_{s^\alpha}\dot{X}^\sigma\partial_{X^\sigma}s^\alpha = d\ln J/dT, \tag{3.15}$$

where $J=\partial(X^1,X^2)/\partial(s^1,s^2)$ and therefore (3.14) becomes:

$$d(J\partial_\omega L)/dT = 0 \tag{3.16}$$

The Jacobian $J$ is a measure of the geometrical spreading of the rays on S and equation (3.16) means that for an observer moving with the local group velocity the product of the local kinetic energy density with the area of his ray tube remains constant. This result is the equivalent to the classical result for body waves.

Nomofilov (1978) obtained similar results for a general surface S but unfortunately with the same scales for horizontal and normal directions, so that his results refer only to very high frequencies. Assuming as in (1.4) complex values of the normal derivatives of the eikonals on S, he transforms the boundary condition of null stress on S, by expressing the normal unit vector in terms of the vector $\boldsymbol{\kappa}$ . He thus derives a dispersion relation for very high freqency surface waves on the boundary of an anisotropic body, and obtains a Lagrangian from which the same relations as before are deduced. Introducing the Hamiltonian $H$ by Legendre's transformation:

$$H = \omega\partial_\omega L - L \tag{3.17}$$

and noting that from (2.4,3.12):

$$\partial_T L = \partial_\omega L\,\partial_T\omega + \partial_\kappa L\,\partial_T\boldsymbol{\kappa} + \partial_a L\,\partial_T a = \partial_\omega L\,\partial_T\omega - \partial_\kappa L\,\overline{\nabla}\omega \tag{3.17a}$$

one sees that the relation (3.11) is equivalent to:

$$\partial_T H + \overline{\nabla}\cdot(-\omega\partial_\kappa L) = 0 \tag{3.18}$$

$H$ is related to the total energy density of the displacement for the first term of the development.

As L remains null on the ray equations (3.11,18) may also be written:

$$\partial_T(H/\omega) + \overline{\nabla}\cdot(\mathbf{Y}H/\omega) = 0 \tag{3.19}$$

$$\partial_T(H) + \overline{\nabla}\cdot(\mathbf{Y}H) = 0 \tag{3.20}$$

The first equation may be considered as a law of conservation of the number of quanta in the wave packet. $H/\omega$ is well known in classical mechanics as the adiabatic invariant for slow modulations of vibrating linear systems. The second shows, by using Gauss-Ostrogradski theorem, in the first approximation, the conservation of the energy present in a domain all the points of which move with the group velocity.

Finally Woodhouse (1974) and Nomofilov (1978) show that a phase term, due to the imaginary part of the orthogonality relation (3.10), must also be included in the expression of the displacement.

## 2.4 Ray-centered coordinates. The paraxial ray approximation and Gaussian beam solution

Let us consider the elastodynamic equations with the trial solution (2.5) for the local displacement **U** . To find a solution in the vicinity of the ray, following Babich and Buldyrev (1972),the development in (2.5) will be made in powers of $\varepsilon^{1/2}$ instead of $\varepsilon$ , and the stretching of local horizontal coordinates will be made, in a way analogous to that of Hudson, with a scaling factor $\varepsilon^{1/2}$ normally to the ray (local coordinate $n$), and $\varepsilon$ along the ray (local coordinate $s$) and for the time. The ray-centered coordinates system is an orthogonal curvilinear system introduced by Popov and Pšenčík (1978). A detailed presentation, following that for body waves by Červený and Pšenčík (1983) can be found in Yomogida (1985). $s,n,z$ being the new local ray-centered coordinates, the length element $dr$ is given by:

$$dr^2 = h^2 ds^2 + dn^2 + dz^2 \quad \text{with:} \quad h(s,n) = 1 + nC^{-1}\partial_n C \tag{4.0}$$

$C(s,n)$ being the phase velocity. In the following $C$ and its derivatives will be taken on the ray, i.e. for: $n$=0 . A Taylor expansion of the local phase velocity is performed up to $n^2$ (i.e. up to $\varepsilon$). Similarly, in the elastodynamic equations and boundary conditions at the surface a development up to $\varepsilon$ is performed, for the three components $U_s$ , $U_n$ , $U_z$. Love waves are characterized by a "principal component" $U_n^0$ normal to the ray, and Rayleigh waves with "principal components" $U_z^0$ and $U_s^0$. These are the non-vanishing components in the zero-order approximation. They respectively satisfy the local equations (3.1) and boundary conditions (3.2) for Love and Rayleigh waves, with local phase velocities $C_L(s,n)$ and $C_R(s,n)$ .

For Love waves an "additional component" $U_s^1$ , of the order of $\varepsilon^{1/2}$ is found, while $U_z^1$ vanishes. The next approximation of order $\varepsilon$ gives a parabolic partial differential equation with respect to $s$ and $n$ for the principal component $U_n^0$ . For Rayleigh waves the "additional component" $U_n^1$ of the order of $\varepsilon^{1/2}$ is transverse to the ray, and similarly parabolic equations are found for the principal components $U_s^0$ and $U_z^0$ .

As for body waves (Babich and Kirpichnikova, 1974), a "paraxial ray approximation" representing the wave-field in the vicinity of the ray is sought, with the form for the principal component:

$$U_j^0 = A(s,t)\hat{U}_j^0(s,z)\exp(i\omega n^2 M(s)/2) \tag{4.1}$$

Replacing $\mathbf{U}^0$ by this expression in the parabolic equations for Love or Rayleigh waves, multiplying both sides by the eigenvector $\hat{\mathbf{U}}^0$, integrating from the surface to infinity and taking boundary conditions into account, one obtains, after separation of the real and imaginary parts, two relations. The first is a differential equation for $M$ :

$$dM/ds + CM^2 + C^{-2}\partial_{nn}^2 C = 0 \tag{4.2}$$

which has the same form as for body waves. Whereas the variation along the ray of the phase $\psi$ in (2.5) is given by (5.13) below, the phase variation at a distance n away from the ray is expressed through $M$, the curvature of the wave-front being $K = CM$ ( Červený and Hron, 1983).

Introducing the geometrical surface spreading $q(s)$ such that:

$$M(s) \equiv C(s)^{-1} d(\ln q)/ds \tag{4.3}$$

the second relation becomes the transport equations for surface waves:

$$\partial_t \left[ A^2 q I_1 \right] + \partial_s \left[ U_g A^2 q I_1 \right] = 0 \tag{4.4}$$

where $U_g$ is the group velocity, and $I_1$ the integral over depth entering in the average kinetic energy $\omega^2 I_1$ . This equation has been obtained from the second order approximation in $\varepsilon^{1/2}$ in a similar way as for body waves for which the transport equation is given by the second order approximation in $\omega$. It is the same as the expression (3.16) given by Woodhouse (1974), as:

$$I_1 = \partial_\omega L/(2\omega), \quad L = \omega^2 I_1 - I_2, \tag{4.4a}$$

$I_2$ being an integral over depth of the average elastic energy.

The transport equation (4.4) shows for $U_j^0$ an amplitude factor $(U_g\, q I_1)^{-1/2}$ so that (4.1) becomes, with initial conditions at $s{=}s_0$:

$$U_j^0 = \Phi(s_0)(N(s_0)/N(s))^{1/2} \exp[i\,\omega n^2 M(s)/2] \tag{4.5}$$

with: $N(s){=}q(s)U_g(s)I_1(s)$ and $\Phi(s_0)$ is an amplitude factor depending on the radiation at the source.

The differential equation (4.2) for M can be rewritten into a system of two differential linear equations of first order with $M{=}p/q$:

$$p = C^{-1} dq/ds \qquad dp/ds = -C^{-2} \partial_{nn}^2 C q \tag{4.6}$$

We will rapidly recall the theory for 2-D body waves (Červený and al., 1982). Any solution of (4.6) can be expressed in terms of two independent real solutions $q_1, p_1, q_2, p_2$. They are chosen with initial conditions at $s{=}s_0$:

$$q_1(s_0){=}1, \quad q_2(s_0){=}0 \quad p_1(s_0){=}0, \quad p_2(s_0){=}C^{-1}(s_0)$$

so that the solution 1 correspond to plane waves, 2 to a line source. Then the general solution of (4.6) is:

$$\begin{pmatrix} p \\ q \end{pmatrix} = z_1 \begin{pmatrix} p_1 \\ q_1 \end{pmatrix} + z_2 \begin{pmatrix} p_2 \\ q_2 \end{pmatrix} \tag{4.6a}$$

In the paraxial ray approximation, $z_1$ and $z_2$ are real constants. $q$ vanishes at caustics, so that $U_j^0$ given by (4.5) cannot be defined at caustics in this approximation. An interesting improvement results from the choice of complex $z_1$ and $z_2$ . Then one obtains a solution concentrated close to the ray (Babich and Kirpichnikova, 1974), called *Gaussian beam* solution as the amplitude of the wave field has a Gaussian variation in the vicinity of the ray: $\exp(-n^2 D(s))$ where:

$$D(s) = \omega \operatorname{Im}(p/2q) \tag{4.6b}$$

The condition along the ray: $q(s){\neq}0$ guarantees a finite amplitude at caustics. The condition: $\mathrm{Im}(p/q)>0$ guarantees the concentration of the solution close to the ray.

Then the complete wave-field in the vicinity of a station and the displacement at the station can be obtained by summation over a finite number of rays on either side of the station. This shows another advantage of the Gaussian beams method: there is no need to compute rays passing exactly through the station, thus removing the problem of a two-point ray tracing. Recalling the superiority of the Gaussian beam method over the classical ray theory in the vicinity of caustics, we may moreover mention that since it considers the velocity field in the vicinity of the rays it takes into account the lateral averaging properties of surface waves which is done also by Woodhouse and Girnius (1982).

## 2.5 Rays along a curved surface

*2.5.1 General ray-tracing equations:* A way to generalize Woodhouse's results to the case of a generally stratified medium - i.e. a medium the physical properties of which are functions of only one curvilinear coordinate -could be to start from the corresponding differential system for the displacement-stress vector with respect to the normal coordinate as derived by G.Jobert (1976) for the partial derivative or Valette (1986) for the covariant one.

One may assume that the dispersion relation could always be written:

$$\omega = \Omega(\kappa_\sigma, X^\sigma, m) \tag{5.1}$$

where $\kappa_\sigma$ ($\sigma$=1,2) and $X^\sigma$ are respectively the covariant components of the wave-vector on S and the contravariant coordinates, $m$ the index characterizing the considered eigenmode. From (5.1) a Hamilton-Jacobi equation will be derived:

$$\partial_T \Psi + \Omega(\partial_{X^\sigma}\Psi, X^\sigma, m) = 0 \tag{5.2}$$

and the ray tracing equations (R.T.E. below) become:

$$\dot{X}^\sigma = \partial_{\kappa_\sigma}\Omega(\kappa_\sigma, X^\sigma, m) \qquad \dot{\kappa}_\sigma = -\partial_{X^\sigma}\Omega(\kappa_\sigma, X^\sigma, m) \tag{5.3}$$

*2.5.2 R.T.E. for an isotropic dispersion relation:* In the isotropic case the frequency $\omega$ depends only on $k = \|\boldsymbol{\kappa}\|$. In the following we will make use of lower-case $t$ , $\mathbf{x}$ only. We may write (Woodhouse and Wong, 1986), with $g^{\alpha\beta}$ contravariant components of the metric tensor on S:

$$\kappa^2 = \boldsymbol{\kappa}^2 = g^{\alpha\beta}(\mathbf{x})\kappa_\alpha\kappa_\beta \qquad \omega = \omega(\boldsymbol{\kappa},\mathbf{x}) = \bar{\omega}(\kappa,\mathbf{x}) \tag{5.4}$$

From the first equation we deduce:

$$\kappa\partial_{\kappa_\sigma}\kappa = g^{\sigma\nu}\kappa_\nu = \kappa^\sigma \tag{5.5}$$

from the second:

$$\partial_{\kappa_\sigma}\omega = \partial_\kappa\bar{\omega}\,\partial_{\kappa_\sigma}\kappa = \kappa^\sigma\partial_\kappa\bar{\omega}/\kappa \tag{5.6}$$

On the other hand the arc increment along the ray is given by:

$$ds = (g_{\sigma\nu}dx^\sigma dx^\nu)^{1/2} = (g_{\sigma\nu}\dot{x}^\sigma\dot{x}^\nu)^{1/2}dt \tag{5.6a}$$

and taking (5.3) into account:

$$ds = (g_{\sigma\nu}\partial_{\kappa_\sigma}\omega\partial_{\kappa_\nu}\omega)^{1/2}dt = (g_{\sigma\nu}\kappa^\sigma\kappa^\nu)^{1/2}dt\,\partial_\kappa\bar{\omega}/\kappa = \partial_\kappa\bar{\omega}\,dt \tag{5.7}$$

$\partial_\kappa\bar{\omega}$ is therefore the group velocity $U_g$ of the signal at the frequency $\omega$. If $|_f$ indicates that $f$ must be kept constant in the derivation, the R.T.E. become:

$$dx^\sigma = \partial_{\kappa_\sigma}\omega dt = \kappa^\sigma\partial_\kappa\bar{\omega}dt/\kappa = \kappa^\sigma ds/\kappa \tag{5.8}$$

$$d\kappa_\sigma = -\partial_{x^\sigma}\omega\,|_{\kappa_\nu}dt = -[\partial_{x^\sigma}\bar{\omega} + \partial_\kappa\bar{\omega}\partial_{x^\sigma}\kappa]dt \tag{5.9}$$

$$\partial_{x^\sigma}\kappa^2\,|_{\kappa_\nu} = 2\kappa\partial_{x^\sigma}\kappa\,|_{\kappa_\nu} = (\partial_{x^\sigma}g^{\alpha\beta})\kappa_\alpha\kappa_\beta \tag{5.10}$$

and $\omega$ is constant along the ray, so that if $\kappa$ is considered as a function $\kappa(\omega,\mathbf{x})$:

$$\partial_{x^\sigma}\bar{\omega} + \partial_\kappa\bar{\omega}\partial_{x^\sigma}\kappa = 0 \tag{5.11}$$

Finally (5.9,10,11) lead to:

$$d\kappa_\sigma = [\partial_{x^\sigma}\kappa - (\partial_{x^\sigma}g^{\alpha\beta})\kappa_\alpha\kappa_\beta/2\kappa]ds \tag{5.12}$$

which with (5.8):

$$dx^\sigma = \kappa^\sigma ds/\kappa \tag{5.12a}$$

are the purely geometrical ray-tracing equations, the time appearing only in (5.7) related to the arc through the group velocity, as one describes the propagation of a wave packet.

For a description of the propagation of a wave front, the phase velocity is involved. Indeed the variation of the phase along the ray is given by:

$$d\psi/ds = \partial_{\mathbf{x}}\psi\,d\mathbf{x}/ds + \partial_t\psi\,dt/ds = \kappa_\sigma\kappa^\sigma/\kappa - \omega\,dt/ds \tag{5.12b}$$

As: $\kappa = \omega/C$ , where $C$ is the phase velocity, we obtain thus the classical result:

$$\psi = -\omega[t - \int C^{-1}(s)ds] \tag{5.13}$$

It is easy to show that (5.8,12) correspond to Fermat's principle. The time of propagation between two fixed points $M_0$, $M_1$ on S is indeed given by:

$$\int_{M_0}^{M_1} F\,dt$$

where

$$F(x^\sigma, \dot{x}^\sigma) = \kappa U_g = \kappa(g_{\sigma\nu}\dot{x}^\sigma\dot{x}^\nu)^{1/2} \tag{5.13a}$$

The stationarity of F leads to:

$$d(\partial_{\dot{x}^\sigma}F)/dt = \partial_{x^\sigma}F \quad \text{or:} \quad d(\partial_{\dot{x}^\sigma}F)/ds = U_g^{-1}\partial_{x^\sigma}F \tag{5.13b}$$

from which one derives (5.8).

*2.5.3 An example of anisotropic dispersion relation:* In this case the dispersion equation depends explicitly on the components of $\boldsymbol{\kappa}$. At a given point the wave number varies with the incidence and the directions of the slowness vector are different for phase and group velocities. Cases of weak azimuthal anisotropy can be found in some regions of the Earth.

Backus (1962) gives an example of anisotropy due to the slow rotation $\Omega$ of the Earth on short period elastic surface waves. The effect of the Coriolis force is treated as a perturbation and it is found that rotation produces in the angular frequency $\omega$ of a non-degenerate mode a first-order perturbation: $|\omega_1| \ll \Omega$ .

For Rayleigh waves on a layered sphere, with spherical coordinates $(a,\theta,\phi)$:

$$\omega_1 = \kappa_\phi \Omega R(\kappa)/a\kappa = \kappa_\phi W(\kappa) \tag{5.13c}$$

where $R(\kappa)$ is a small factor depending on the eigenfunction of the Rayleigh wave and $\kappa_\phi$ the longitude covariant component of $\boldsymbol{\kappa}$ .

A physical explanation is found by considering a reference frame rotating relatively to the geographical frame with a constant angular velocity $W$ around the axis **z**. The trajectory of the wave packet is a great circle in absence of rotation. Taking the rotation of the Earth into account "a Rayleigh wave packet moves at constant speed around a great circle trajectory whose plane maintains its inclination to the rotation axis **z** but precesses with angular velocity $W\mathbf{z}$".

### 2.6 Particular cases

*2.6.1 A plane free surface:* We take cartesian coordinates $x^1=x$ , $x^2=y$ in the plane, with axes Ox, Oy, and define the incidence angle as the angle of the tangent to the ray with Oy. The components of the wave vector $\boldsymbol{\kappa}$ are:

$$\kappa^1 = \kappa_1 = \kappa \sin i \qquad \kappa^2 = \kappa_2 = \kappa \cos i \tag{6.0}$$

and from the definition of $i$ the R.T.E. (5.8,12) become:

$$\begin{aligned} dx/ds &= \sin i & dy/ds &= \cos i \\ d(\kappa \sin i)/ds &= \partial_x \kappa & d(\kappa \cos i)/ds &= \partial_y \kappa \end{aligned} \tag{6.1}$$

The elimination of $d\kappa/ds$ leads to:

$$\kappa di/ds = \cos i\, \partial_x \kappa - \sin i\, \partial_y \kappa \tag{6.2}$$

which added to (6.1) governs the ray tracing in two dimensions.

*2.6.2 A spherical free surface:* $a$ being the radius of the sphere we will take as coordinates of a point on the surface: $x^1=\theta$, $x^2=\phi$, and define the incidence angle $i$ as the angle between the tangent to the ray and the meridian. The contravariant components of the metric tensor are:

$$g^{11} = a^{-2} \quad g^{12} = g^{21} = 0 \quad g^{22} = (a \sin\theta)^{-2} \tag{6.2a}$$

The equations (5.8) become:

$$a\, d\theta/ds = \cos i \qquad a\, d\phi/ds = \sin i/\sin\theta \tag{6.3}$$

the components of the wave-vector $\boldsymbol{\kappa}$ of modulus $\kappa$ being:

$$\kappa_1 = a\kappa \cos i \quad \kappa^1 = a^{-1}\kappa \cos i \quad \kappa_2 = a\kappa \sin\theta \sin i \quad \kappa^2 = (a \sin\theta)^{-1}\kappa \sin i \tag{6.3a}$$

The equations (5.12) become:

$$d(a\kappa\cos i)/ds = \partial_\theta\kappa + \kappa\cot\theta\sin^2 i \qquad d(a\kappa\sin i)/ds = \partial_\phi\kappa \tag{6.3b}$$

The elimination of $d\kappa/ds$ leads to:

$$a\kappa di/ds = -\sin i\,\partial_\theta\kappa + (\cos i/\sin\theta)\partial_\phi\kappa - \kappa\cot\theta\sin i \tag{6.4}$$

As shown by Aki and Richards (1980) (6.3,4) are the equations of Julian (1970) for 3-dimensional ray-tracing in spherical coordinates, where the radius is set to be constant and equal to the external radius $a$ and the angle of incidence is taken $\pi/2$.

*2.6.3 A transformation leading to the plane case:* Jobert and Jobert (1983) proposed to use the Mercator transformation:

$$y = \ln(\tan(\theta/2)) \tag{6.5}$$

together with the transformations for the phase and the group velocities $C_s$ and $U_g$ respectively into $C_p$ and $U_p$:

$$C_p(y,\phi) = C_s/\sin\theta = C_s\cosh y \qquad U_p(y,\phi) = U_g/\sin\theta = U_g\cosh y \tag{6.6}$$

Then the wave equation for a propagating disturbance $F(\theta,\phi,t)$ in spherical coordinates:

$$\overline{\nabla}^2 F \equiv \partial_\theta(\sin\theta\partial_\theta F)/\sin\theta + \partial^2_{\phi\phi}F/\sin^2\theta = C_s^{-2}\partial_t^2 F \tag{6.6a}$$

becomes after transformation:

$$\partial^2_{yy}F + \partial^2_{\phi\phi}F = C_p^{-2}\partial^2_{tt}F \tag{6.7}$$

which is the wave equation in cartesian coordinates, with a local velocity $C_p(y,\phi)$. The equation for the incidence angle in spherical coordinates:

$$\cot i\sin\theta = d\theta/d\phi \quad \text{becomes:} \quad \cot i = dy/d\phi \tag{6.7a}$$

Likewise, the R.T. system (6.3,4) in spherical coordinates is transformed into the R.T. system (6.1,2) in cartesian coordinates, the time governing equation:

$$ds/dt = U_g(\theta,\phi) \quad \text{becoming:} \quad ds'/dt = U_p(y,\phi) \tag{6.7b}$$

with the same transformation for the phase.

Qualitatively, the success of this transformation can be explained by the fact that the arc elements $ds$ on the sphere and $ds'$ on the plane are related by: $ds=ds'\sin\theta$, so that with taking the inverse of $\sin\theta$ in the transformation (6.6) for velocities the time and phase elements remain unchanged.

This transformation enables the use of 2-dimensional R.T. programs deviced for body-waves (Pšenčík, 1983) for surface wave propagation on a spherical surface (Jobert, 1986a, 1986b).

*2.6.4 A transformation of the spherical R.T.E. in the vicinity of the great circle:* As lateral heterogeneities of the Earth are slight, of the order of a few per cent, it is interesting to bring them out in the R.T.E.. This has been done by Woodhouse and Wong (1986) using the transformation:

$$x^1 = \gamma = \cot\theta \tag{6.8}$$

Then the coefficients of the metric tensor are:

$$g^{11} = a^{-2}(1+\gamma^2)^2 \quad g^{12} = g^{21} = 0 \quad g^{22} = a^{-2}(1+\gamma^2) \tag{6.8a}$$

and the R.T.E. (6.3,4) can be reduced to the second-order approximately linear differential equation for $\gamma(\phi)$:

$$\gamma'' + \gamma = (\nu^2/(1+\gamma^2)+1)(\partial_\theta - \nu\partial_\phi)\ln C(\theta,\phi) \tag{6.9}$$

where : $\nu = -\gamma'$.

The first-order spatial derivatives of the phase velocity $C(\theta,\phi)$, which are small, appear in the right part of the equation as differences with the null value of the homogeneous case. In this last case the solution $\gamma$ is a sine function of $\phi$ , the equation for a great circle being: $\gamma \equiv \cot\theta = \pm\cot\theta_0 \sin\phi$.

The authors use a projection such that the source ($\phi=0$) and epicenter ($\phi=\Delta$) are on the equator $\theta=\pi/2$. Thus the values of $\gamma(\phi)$ are almost exactly the deviations due to the lateral heterogeneities of the ray path off the great circle $\gamma=0$ on the equator.

*2.6.5 An ellipsoidal free surface:* A transformation mapping the surface of an ellipsoid of revolution of eccentricity $e$ onto a plane is proposed by Yomogida and Aki (1985) modifying (6.5):

$$y = \ln[\tan\theta/2((1+e\cos\theta)/(1-e\cos\theta))^{e/2}] \tag{6.10}$$

and replacing $\sin\theta$ by $\sin\theta/(1-e^2\cos^2\theta)^{1/2}$ in (6.6,7).

Jobert (1976) showed that there is a precession of orbits around the Earth due to its ellipsoidal shape.

### 2.7 Amplitudes

*2.7.1 Dynamic ray-tracing equations:* The system (4.6) of two differential equations for the geometrical spreading $q$ and the auxiliary variable $p$ :

$$q = -C^2(\partial^2_{nn}C)^{-1}dp/ds \qquad p = C^{-1}dq/ds \tag{7.1}$$

can be solved together with the R.T.E. (5.8,5.12) with initial conditions for a surface point source: $q(s_0)=0, p(s_0)=C^{-1}(s_0)$.

The different methods of computation in use for body waves (Červený et al., 1977) for 2-D media can be used for surface waves. Let us use the geometrical definition: $q=d\sigma/di_0$ where $d\sigma$ is the width of the ray tube along a wave-front between two rays with initial incidences $i_0$ and $i_0+di_0$. The correspondence between the points on the two rays will have to be made through the time of propagation of the phase, i.e. using: $dt=ds/C$ in the R.T.E. (5.3).

For a *spherical surface* in the R.T.E. (6.3,4) we replace $ds$ by $Cdt$ and consider the coordinates $\theta,\phi$ and incidence angle $i$ as functions of $t$ and $i_0$. By differentiation we get a system of three differential equations for $\partial_{i_0}\theta, \partial_{i_0}\phi$ and $\partial_{i_0}i$ which are the dynamic R.T.E. of

Julian (1970):

$$a\frac{d(\partial_{i_0}\theta)}{dt} = (\partial_\theta C\,\partial_{i_0}\theta + \partial_\phi C\,\partial_{i_0}\phi)\cos i\ - C\sin i\,\partial_{i_0} i$$

$$a\sin\theta\frac{d(\partial_{i_0}\phi)}{dt} = \sin i\,[(\partial_\theta C\ - C\cot\theta)\partial_{i_0}\theta + \partial_\phi C\,\partial_{i_0}\phi] + C\cos i\,\partial_{i_0} i \qquad (7.2)$$

$$a\sin\theta\frac{d(\partial_{i_0} i)}{dt} = T\,\partial_{i_0}\theta + F\,\partial_{i_0}\phi + I\,\partial_{i_0} i$$

with:

$$T = \sin\theta\sin i\,\partial^2_{\theta\theta}C\ - \cos i\,\partial^2_{\phi\theta}C\ + \cos i\cot\theta\partial_\phi C\ - \cos\theta\sin i\,\partial_\theta C\ + C\sin i\,/\sin\theta$$

$$F\ = \sin\theta\sin i\,\partial^2_{\phi\theta}C\ - \cos i\,\partial^2_{\phi\phi}C\ - \cos\theta\sin i\,\partial_\phi C \qquad (7.2a)$$

$$I\ = \sin\theta\cos i\,(\partial_\theta C\ - C\cot\theta) + \sin i\,\partial_\phi C$$

From the solution of this system one can evaluate $q$ :

$$q\ = a\,[(\partial_{i_0}\theta)^2 + (\sin\theta\partial_{i_0}\phi)^2]^{1/2} \qquad (7.2b)$$

With the transformation (6.8) of Woodhouse and Wong (1986), one finds:

$$q\ = -a\,\partial_{\nu_0}\gamma(\Delta,\nu_0)[1+\nu_0^2][1+\nu^2(\Delta)]^{-1/2} \qquad (7.3)$$

which reduces to: $q_h\!=\!a\sin\Delta$ for a homogeneous sphere.

*2.7.2 Incidence of a surface wave on a vertical discontinuity:* Let us consider the transmission of a given mode of surface wave impinging on a vertical plane of discontinuity between two elastic quarter- spaces. On each side of the discontinuity the phase and group velocities are respectively $C$ , $U_g$ and $C'$ , $U'_g$. The angles of reflection and refraction are given by Snell's law : $\sin i/C{=}\sin i'/C'$. $T$ being the transmission coefficient for a ray and $A$ the amplitude of the incident wave, the amplitude of the transmitted wave for the same mode is given by:

$$A'\ = AT\,(\cos i' U_g{}' I_1{}'/\cos i U_g I_1)^{1/2} \qquad (7.4)$$

extending to surface waves the concept of continuity of the geometrical spreading for body waves (Červený et al., 1977). If the surface of discontinuity is no longer a plane, it can be approximated locally by a plane. For the conditions of validity of geometrical optics, (1.1) is a condition on the radii of curvature of the surface, which must be much larger than the wave-length.

For the computation of coefficients such as $T$, an approximate method to solve the problem of propagation of a surface wave through a vertical plane of discontinuity is proposed by Alsop (1966). For a normally incident wave-field containing a single mode, if the two elastic media are strongly different, to satisfy the condition of continuity of displacement and stress across the discontinuity, a conversion of modes is necessary. The transmitted and reflected wave-fields contain several possible modes, the generation of body waves being generally neglected. The coefficients of reflection and transmission are computed using the concept of a " coupling coefficient ". An orthogonality relation using a

Lagrangian form given by Herrera (1964) allows to normalize the eigenfunctions for unit energy transport.

An extension to oblique incidence (Gregersen and Alsop, 1974) uses a representation of the guided waves by interfering multiply-reflected body waves. The results obtained by these approximate methods are in agreement with those given by other methods (Lysmer and Drake, 1971; Knopoff and Hudson, 1964). A new approximate method based on the Green function technique is proposed by Its and Yanovskaya (1985). The concept of mode transformation when the condition (1.1) is no longer fulfilled is generalized by Kennett (1984). The surface wave-field is represented by a multimode superposition, the modal coefficients varying with position along the path. This leads to interconversion between the modes and reflection into backward travelling modes. The effects are described by an evolution equation for the modal coefficients. An application is made to the propagation of high frequency $L_g$ wave (Kennett and Mykkeltveit, 1984). A linearized theory is presented by Snieder (1986a).

**2.8 First-order perturbation approximations**

In the case of weak lateral heterogeneities the perturbation in an observable quantity relatively to the case of a laterally homogeneous medium can be represented with a reasonable accuracy by the linear relationship given by the first order term in the power series of the perturbation in velocity. This is the case for the phase and the amplitude.

*2.8.1 Great circle phase integral approximation:* As we have seen (5.13) the phase variation between two points $A_1$,$A_2$ along the ray R is given by:

$$\delta\phi \Big|_{A_1}^{A_2} = \omega \int_{R(A_1A_2)} \frac{ds}{C(s)} \tag{8.0}$$

where $C$ is the phase velocity.

By use of Fermat's principle, neglecting the variation due to the deformation of the path away from the great circle, the phase perturbation can be related to the velocity perturbation by integration along the great circle :

$$\delta\phi \Big|_{A_1}^{A_2} = -\omega \int_{GC(A_1A_2)} \frac{\delta C}{C^2(s)} ds \approx -\omega C_0^{-2} \int_{GC(A_1A_2)} \delta C ds \tag{8.1}$$

where $C_0$ is an average great circle phase velocity. (8.1) is the basis of the classical great circle tomography.

*2.8.2 Great circle amplitude integral approximation:* To be able to interpret observed amplitudes a formula analogous to (8.1) has been proposed by Woodhouse and Wong (1986) for surface waves, Romanowicz (1986) from free oscillations. Following the first authors, the Green's function associated to the operator on the left side of (6.9) with boundary conditions $\gamma(0)=\gamma(\Delta)=0$ is given by:

$$G(\phi,\phi') = -\operatorname{cosec}\Delta[\sin\phi\sin(\Delta-\phi')H(\phi'-\phi) + \sin\phi'\sin(\Delta-\phi)H(\phi-\phi')] \tag{8.1a}$$

so that, $f$ being the right side of (6.9), this equation is equivalent to:

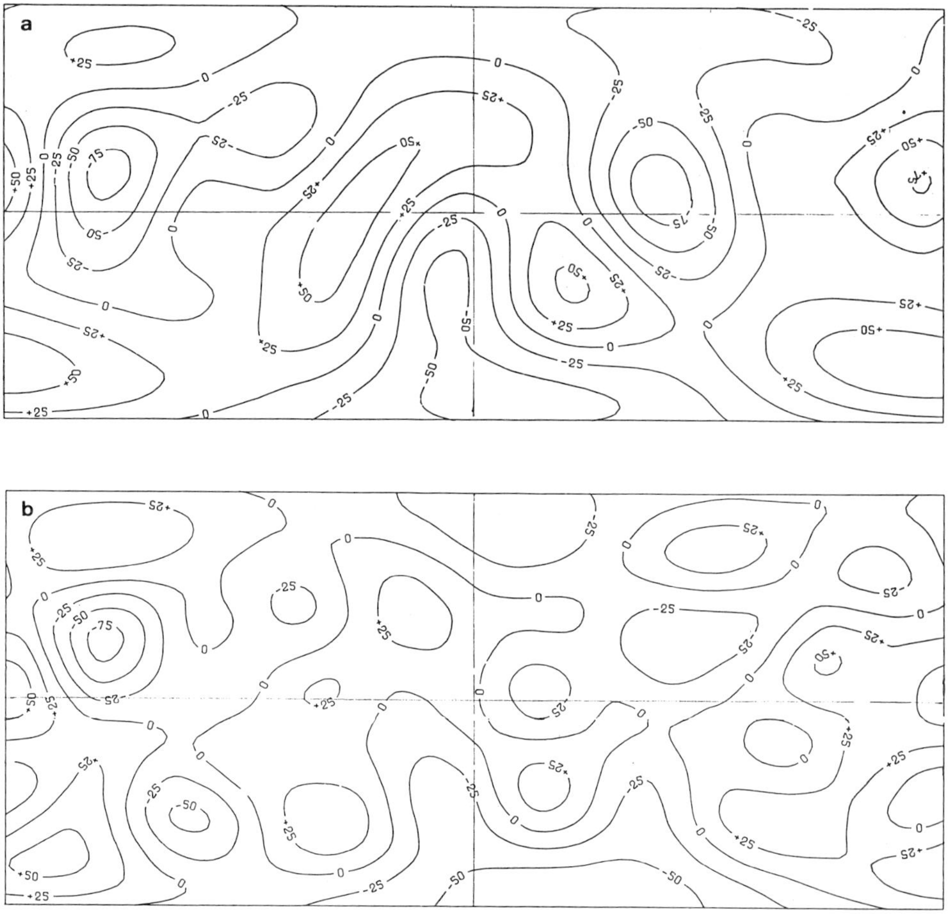


*Figure 1* - a) Rayleigh phase velocity distribution, relative to global average, for the model of Nakanishi and Anderson (1984) at a period of 200 s. The contour lines are in $10^{-3} km\ s^{-1}$
b) Same as a) for the model M84C (Woodhouse and Dziewonski, 1984)

$$\gamma(\phi) = \int_0^\Delta G(\phi,\phi') f\,[\gamma(\phi'),\phi'] d\phi' \tag{8.2}$$

Similarly by differentiation of (6.9) with $g$ on the right side:

$$\gamma'(\phi) = -\sin\phi - \int_0^\phi \sin(\phi-\phi') g\,[\gamma(\phi'),\phi'] d\phi' \tag{8.3}$$

Taking $\delta C$ as a perturbation, the deviations of the path away from the great circle are of the

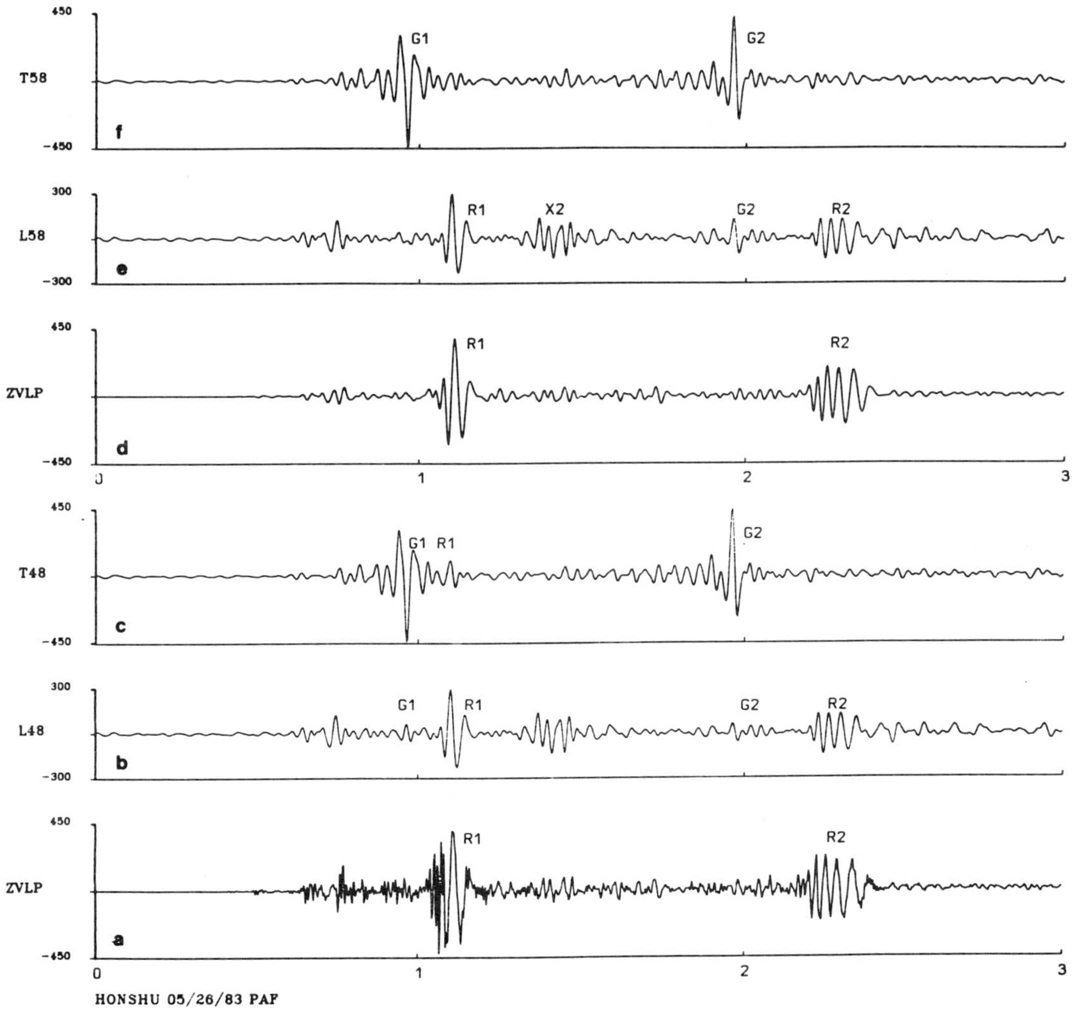


*Figure 2* - Records of the Akita-Oki (Honshu) earthquake of May 26, 1983 at the Geoscope station PAF. a) vertical component b) longitudinal component, c) transverse component, rotated with an azimuth 48°. d) vertical component high- pass filtered at 100 s. e), f) same as b), c) with an azimuth of 58°. All horizontal components are filtered as d). Tick mark interval = 1 h. After Roult et al. (1986).

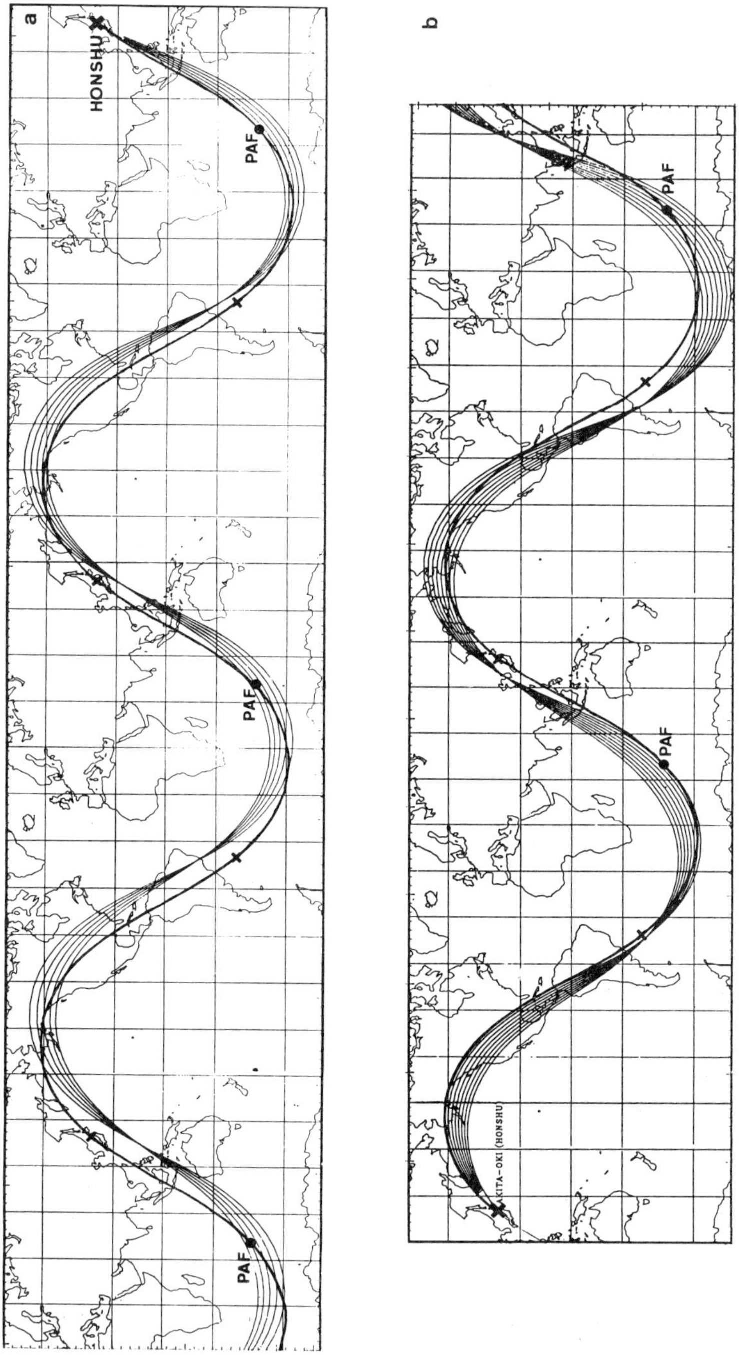


*Figure 3* - a) Ray tracing on Nakanishi and Anderson's model (T=166 s) for the Akita-Oki (Honshu) earthquake of $R_1$, $R_3$, $R_5$ to station PAF. Heavy line is the epicenter-station great circle. Crosses indicate the station and antipodes.
b) same as a) for $R_2$, $R_4$. After N. Jobert (1986b).

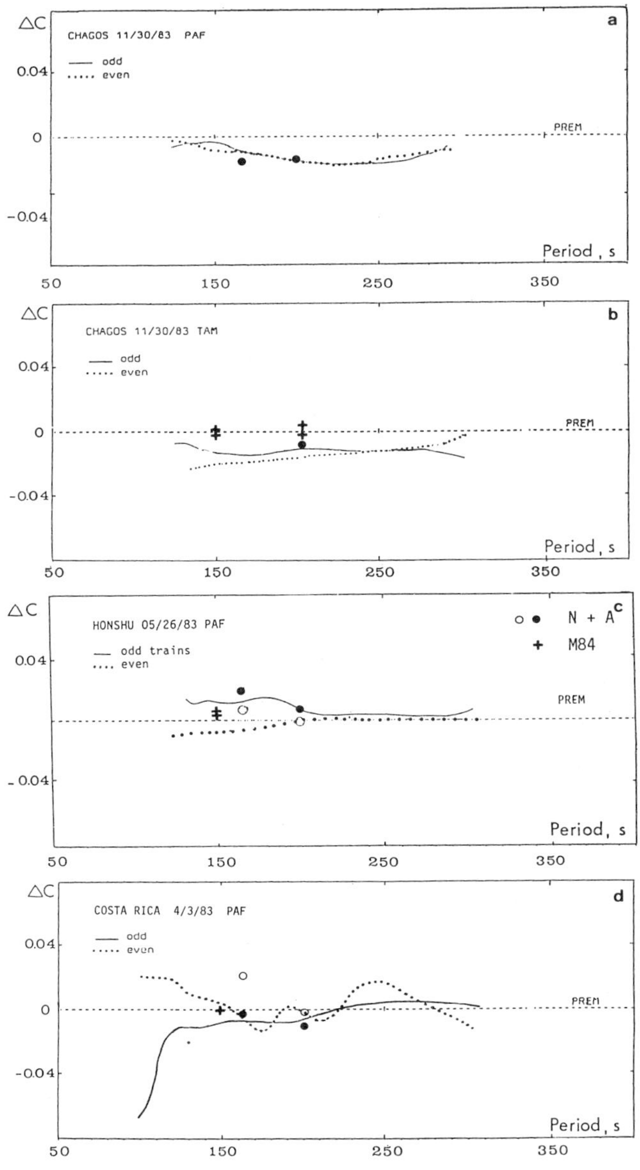


*Figure 4* - Observed average one-orbit phase velocities as a function of period for opposite directions of propagation. Differences with global spherical model PREM (Dziewonski and Anderson, 1981) are shown. a) b) and c) are from Roult and al. (1986). Computations: circles are for model Nakanishi and Anderson (1984), crosses for model M84C (Woodhouse and Dziewonski, 1984).

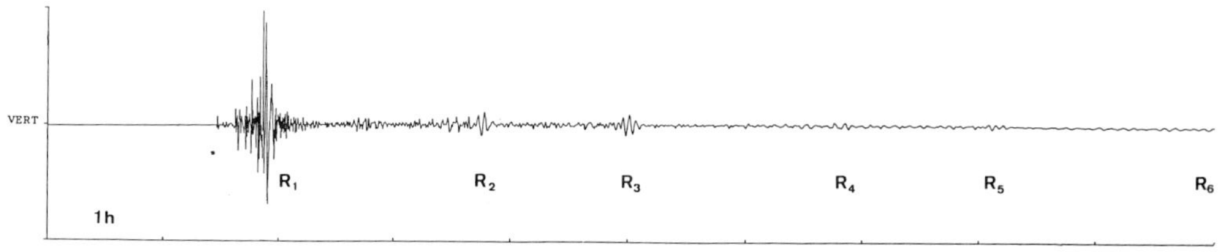


*Figure 5* - Vertical component record of Chagos Islands earthquake Nov.30, 1983 at the Geoscope station TAM. The Rayleigh wave arrivals are indicated up to $R_6$.

same order: $\gamma(\phi)=O(\delta C)$, $\nu(\phi)=O(\delta C)$ and $\gamma'(\phi)=-\sin\phi+O(\delta C)$, $\nu'(\phi)=\cos\phi+O(\delta C)$. Substituting into (8.3) and linearizing we find to first order:

$$\gamma'(\phi) \approx -\sin\phi + \int_0^{\phi} \sin(\phi-\phi') \left[ \sin\phi' \partial^2_{\theta\theta} \frac{\delta C}{C_0} - \cos\phi' \partial_{\phi'} \frac{\delta C}{C_0} \right] d\phi' \quad (8.3a)$$

Taking $\phi=\Delta$ and substituting in the equation (7.3) for the geometrical expansion leads to:

$$\ln\frac{A}{A_h} = \frac{1}{2}\mathrm{cosec}\Delta \int_0^{\Delta} \sin(\Delta-\phi) \left[ \sin\phi \partial^2_{\theta\theta} \frac{\delta C}{C_0} - \cos\phi \partial_{\phi} \frac{\delta C}{C_0} \right] d\phi \quad (8.4)$$

where $A_h$ is the amplitude corresponding to a homogeneous sphere. This is for the amplitude the equivalent to (8.1) for the phase, giving the amplitude anomaly as a great circle integral over local velocity anomalies. Note that the second order transverse derivative has a symmetrical weight maximum at the center of the path whereas the derivative along the path is important near the source.

## 3. Examples

For illustration we will consider deviations from the great circle tomographic approximation, observed on long-period records of the Geoscope array. Their interpretation is made by ray tracing on the global laterally heterogeneous models of the Earth of Nakanishi and Anderson (1984) and M84C of Woodhouse and Dziewonski (1984). Fig.1 displays the geographical distribution of Rayleigh phase velocity anomalies at a period of 200 s for these two models. The ray-tracing was performed with a version of the program RAY81 by Pšenčík (1983) for 2-D cartesian coordinates. Its velocity and derivatives representation has been replaced by an analytical representation of phase and group velocities with spherical harmonics. The Mercator and associated transformations for velocities, and integration of the group delay along the ray (Jobert, 1986a) have been included.

### 3.1 Polarization anomalies

Reports of such anomalies for mantle waves are made by Fukao and Kobayashi (1983). Evidence of anomalous polarization of mantle waves is reported by G.Roult and al.(1986)

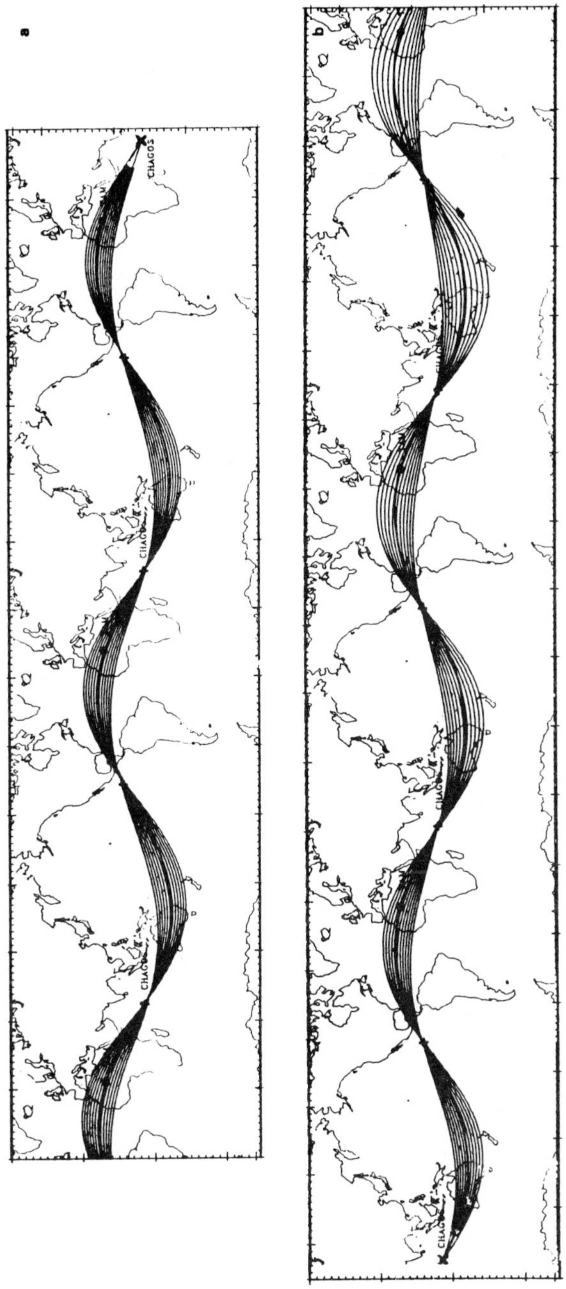


*Figure 6* - Ray tracing on M84C model (T=100s) for the Chagos Islands earthquake to the Geoscope station TAM - a) $R_1, R_3, R_5$ b) $R_2, R_4, R_6$.

on the records of the Akita-Oki (Honshu) earthquake of May 26, 1983 at the Geoscope station PAF. Fig.2 presents the horizontal components respectively rotated with azimuths N58°E and N48°E. 48° is the theoretical azimuth, still $R_1$ appears on the transverse component and $G_1$ and $G_2$ waves on the longitudinal component. The separation for $R_1$ is much better for an azimuth of 58°, which is an evidence of off-great circle propagation, with an anomalous deviation of the path towards the East of about 10°. Indeed the ray-tracing performed on the model of Nakanishi and Anderson at 166 s displays anomalous refractions of the paths in a region of high transverse gradient of the phase velocity, N.E. and S.W. of the epicenter (Fig.1). At PAF (Fig.3) the paths for odd order arrivals are rotated towards the East relatively to the great circle path by 5° for $R_1$ , and the deviation increases by 5° at each orbit, with an opposite effect for even order arrivals. This is in agreement with the observation for $R_1$ .

### 3.2 Time anomalies

Differences in average one-orbit phase velocities for opposite directions of propagation, obtained by analyzing separately odd and even order mantle arrivals, are reported by G.Roult and al. (1986). Examples of such observations are displayed in Fig.4. It is seen that in some cases (Akita-Oki (Honshu) and Costa-Rica recorded at PAF) relative differences less than 1% appear for the average one-orbit velocities at periods shorter than 200 s, and they increase with frequency. These anomalies are fairly well predicted by ray tracing on the model of Nakanishi and Anderson for the third example. Indeed on the ray tracing of Fig.3, it is seen that in the region of minor arc propagation, for odd order arrivals, the ray paths are shifted more and more, with increasing order, into a region of higher velocity to the East, with the opposite effect for even order arrivals. This explains the average one-orbit velocities higher for odd order arrivals than for even order arrivals.

The ray tracing of Fig.3 shows also that at periods shorter than 200 s the paths of mantle waves can deviate by as much as 10° from the regions predicted by great circle propagation, thus introducing mislocation effects in the localization of velocity anomalies by great circle tomography.

### 3.3 Amplitude anomalies

Such observed anomalies, with an amplitude ratio reaching 2 for $R_2$ and $R_3$ are interpreted by Lay and Kanamori (1985). In Fig.5 is shown the vertical record at the Geoscope station TAM of the Chagos Islands earthquake of November 30, 1983. The epicentral distance is 71.4°. In this figure are displayed the successive Rayleigh arrivals up to order 6. The effect of attenuation along the path, due to the anelasticity in the mantle, should appear as an amplitude decrease regular with increasing order of arrivals (an exponential law of distance). Yet, it is seen that $R_3$ is as important as $R_2$ , $R_5$ as $R_4$. This is qualitatively predicted by the ray tracing on Fig.6, performed on model M84C at a period of 100 s: whereas the geometrical spreading at TAM does not change appreciably with increasing order for odd arrivals, for even arrivals a widening of the pattern of rays occurs, increasing the geometrical spreading at TAM. It seems to increase at each orbit while the wave is passing over Australia, where indeed is situated a positive velocity anomaly (Fig.1) for this model.

# Chapter 13

# Waveform tomography

*G. Nolet*

Why use delay times only? Obviously, a seismogram contains much more information than the arrival time of the first wavelet, and we should wish to include the whole seismic time series in the data set, and use this in an imaging technique. There are two obstacles to such an approach. The first is the enormity of the numerical problem. The second is that it is often difficult or impossible to develop a manageable approximation to the solution of the wave equation, especially one that is adequate enough to model the response of complicated Earth models. This chapter deals with the first question in general, and with the second for the case of low frequency seismic waves with relatively low horizontal phase velocity.

## 1. Introduction

Seismic tomography with body wave delay times has some obvious disadvantages when we try to model near-surface structure. Except in a few highly seismic areas, the ray coverage is strongly determined by the station coverage, which is often not very dense. Usually we find that the dominant ray direction is near the vertical, so that the vertical resolution is impaired. The oceans are not covered by seismic stations, so that the oceanic Upper Mantle cannot be probed by methods of body wave tomography.

The existence of a Low Velocity Layer (LVL) in the upper Mantle of the Earth is another complicating factor, since depth and extent of the LVL is highly influential on the ray geometry in the background model. If the LVL in the background model is not like the one in the real Earth, strong nonlinear effects may corrupt the tomographic experiment.



*G. Nolet (ed.), Seismic Tomography, 301–322.*

Finally, delay times of anything but the first arrival are difficult to determine. This is an important disadvantage when we wish to study the fine structure of the Upper Mantle discontinuities, or when our interest is in the S-wave structure. It is not surprising that some of the most detailed laterally homogeneous Upper Mantle models have been constructed by fitting complete waveforms of groups of body wave arrivals even if only in a qualitative way (e.g., Grand and Helmberger, 1984; Leven, 1985).

Surface waves have provided us so far with most of the information on the Upper Mantle, especially in oceanic regions. The classical method is to search for an event/station or station/station pair within one (relatively) homogeneous geophysical province, and determine the phase shift of a surface wave travelling along the connecting path (for reviews, see Knopoff, 1972, or Kovach, 1979). This method works with fundamental modes of seismic surface waves, which do not penetrate very deep unless their frequencies are far below 10 mHz. For this special, broad band instruments are available that give reliable recordings of the very low frequencies (Wielandt and Knopoff, 1982). The alternative is to measure the higher modes of surface waves with an array of stations (Nolet, 1977), but this, again, presupposes an often unrealistic degree of lateral homogeneity over large distances.

An approach that bears strong resemblance with the method of body wave tomography is to use *regionally varying* surface wave phase velocities. This is essentially a geometrical optics approximation, in which one assumes that the surface wave travels with a *local* phase velocity, which is equal to the phase velocity that the wave would have had if the Earth were laterally homogeneous with the same structure as the local structure (Backus, 1964). Madariaga (1972) provided a theoretical basis for this approximation based on the theory of normal modes. Very often, however, the conditions for the validity of the 'ray approximation' to surface waves are not valid in the real Earth (Madariaga and Aki, 1972). Ray theory fails if the length scale of the heterogeneities are of the same order as the wavelength of the surface wave. Scattering from these heterogeneities has to be treated with different methods that are still in a very experimental stage and do seem to require larger seismic station coverage than is now available (Snieder, chapter 14).

Despite the theoretical objections to the approximations involved, ray approximations have been used to infer regional surface wave phase velocities from world circling surface waves in many cases, first by Toksoz and Anderson (1966). Essentially two approaches can be followed: the first one is to use a priori geophysical knowledge to divide the Earth's surface into a number of geophysical 'provinces', and determine the phase velocity as a function of frequency for each of these. The second is to divide the Earth into cells, and solve the resulting large underdetermined problem with algebraic methods similar to those used in body wave tomography. An example of both strategies is given by Nataf et al. (1986).

## 2. Waveform fitting

A seismogram contains much more information than just the arrival time of the first arriving P-wave and the phase velocities of the fundamental modes of surface waves. These types of data derive their prominence in global seismology mostly from the ease with which they can be measured and modelled. Nevertheless, the extra information provided by later arrivals of body waves, diffracted waves and higher modes of surface waves is badly needed often to compensate for the inadequate coverage of the Earth by seismic stations and earthquakes.

Seismologists engaged in the investigation of seismic sources have recognized for some time that synthetic seismograms of groups of arrivals (such as P, pP, sP) are a very powerful diagnostic tool. The complicated multiple arrivals of seismic waves at epicentral distances between 17° and 25° can be modeled with synthetic seismograms to produce detailed models of the 400- and 660-km discontinuities. This kind of waveform fitting is generally done by trial and error, although very recently more algorithmic procedures have been developed, either using linearizations (Shaw and Orcutt, 1985; Chapman and Orcutt, 1985) or using fully nonlinear searches with Monte Carlo techniques (Cary and Chapman, 1986).

Trial and error waveform fitting is inadequate for the matching of longer wavetrains that propagate in complicated structures. Lerner-Lam and Jordan (1983) developed a linearized waveform-fitting technique to fit higher Rayleigh modes. However, the linearizations involved in this technique impose an upper limit to the frequencies that can be handled. This is due to the oscillatory nature of seismic time signals. As a simple illustration, consider a time signal consisting of a number of plane waves with differing frequencies of the form $\cos[kx-\omega t]$. The wavenumber $k$ is related to the model velocity $c$ as $k=\omega/c$. Perturbations $\delta k$ in the wavenumber are (approximately) related to model perturbations in a linear way. However, the resulting time signal contains terms $\delta kx$ in the harmonic, and therefore is strongly nonlinear, especially when $x$ is large.

In this chapter I shall describe a formalism for waveformfitting by stating the problem in a form, suitable for attack by methods of nonlinear optimization (Gill et al., 1981). The general form of an optimization problem is the minimization of some nonlinear *objective function* , denoted by $F(\mathbf{p})$. $\mathbf{p}$ is a vector of model parameters. Although we shall treat $\mathbf{p}$ in this chapter as a vector of finite dimension, this is no essential restriction, and models (or rather model functions) of infinite dimensions can be treated in a very much similar fashion.

## 3. General statement of the problem

To model the observed seismograms we must decide upon a single measure of the goodness of fit between one or more observed seismograms, and the corresponding time series predicted by an Earth model and the earthquake source parameters. We shall denote this measure by the objective function $F(\mathbf{p})$, where $\mathbf{p}$ is an M-dimensional vector of all parameters.

The waveform tomography problem is then reduced to an optimization problem: the minimization of $F(\mathbf{p})$. Care must be taken in the definition of $F(\mathbf{p})$. The simplest definition is:

$$F_1(\mathbf{p}) = \frac{1}{q}\sum_{i=1}^{N}\int_0^{T_i} |\Psi_i(\mathbf{p},t) - s_i(t)|^q \, dt \tag{1}$$

where $N$ is the total number of time series available, $T_i$ is a (sufficiently large) time span, $\Psi_i$ is an operator that generates, from a model **p**, the synthetic seismogram corresponding to the i-th time series, $s_i(t)$ is the observed seismogram and $q$ defines the norm (usually $q$=2). In (1), a weighting of each contributing time series may be applied to compensate for different variances of the time signals $s_i(t)$, and this may in general be sufficient for seismic data of a quite uniform nature, such as obtained in reflection seismics.

The simple definition (1) has distinct shortcomings in global seismology. First of all the operator $\Psi_i$ usually involves approximations in order to keep the calculational effort manageable. We shall wish to apply frequency filters and time windows to exclude as much of the unmodelled signal from the data as we can. The WKBJ method for example (Chapman and Orcut, 1986) does not in general model reverberations or surface waves. Mode summation algorithms, on the other hand, do not fit all body wave arrivals adequately.

Secondly, amplitudes of some waves may dominate the signal to a degree that is not proportional to their information content. For example, in one of our numerical tests we had difficulties fitting the SS and SSS waves from a shallow focus earthquake using (1), simply because the energy of the fundamental mode made up more than 90% of the total energy in the time series. Paradoxically, this fundamental mode energy is so high because the energy of this mode is trapped near the surface, and contains very little information on deeper structure. Thus, we shall in general wish to weigh different parts of the signal in a different way.

If we include a filtering and windowing operator $\boldsymbol{R}$ in (1) , which operates on the data as well as on the synthetics, we get:

$$F_2(\mathbf{p}) = \frac{1}{q}\sum_{i=1}^{N}\int_0^{T_i} |\boldsymbol{R}\Psi_i(\mathbf{p},t) - \boldsymbol{R}s_i(t)|^q \, dt \tag{2}$$

As with so many problems in geophysics, the inversion is likely to be unstable because $F$ may be insensitive to particular parameters $p_j$ or to combinations of such parameters. To maintain stability in that case we may add a measure of the model size to $F$, and define the objective function (for the case $q$=2) as:

$$F_3(\mathbf{p}) = \frac{1}{2}\sum_{i=1}^{N}\int_0^{T_i} |\boldsymbol{R}\Psi_i(\mathbf{p},t) - \boldsymbol{R}s_i(t)|^2 dt + \frac{1}{2}\gamma\mathbf{p}^T\mathbf{C}\mathbf{p} \tag{3}$$

where $\gamma$ governs the trade off between model norm and data fit, and **C** scales the model parameters, ideally to unit a priori variance. Because the stabilizing term is in a quadratic form, there is no loss of generality in assuming **C** symmetric. In many practical applications it will suffice to choose **C** diagonal.

The solution to the inverse problem is the vector **p** that minimizes the selected $F$. Standard methods are available to solve such multi-parameter optimization problems (see, e.g., Gill et al., 1981, or Adby and Dempster, 1974), but in general these lead to very large amounts of computer time or computer memory requirements. Only a few methods are able to handle problems of the size encountered in waveform tomography.

## 4. Optimization

In this section I discuss various optimization strategies that are in theory capable of handling the type of problem under consideration. It should be kept in mind that each type of wavefrom inversion comes with its own specific requirements, and that a method that works with one type of data may perform inefficient on another. Very little testing of these procedures on actual data has been done so far, and the field of actual application offers a large potential for further research.

The choice of the Euclidean norm ($q$=2) greatly simplifies mathematics and is therefore often preferred. For the derivatives of $F$ as defined in (3) with respect to the model parameters, we find:

$$\mathbf{g}(\mathbf{p}) \equiv \nabla F(\mathbf{p}) = \sum_{i=1}^{N} \int_{0}^{T_i} \boldsymbol{R} \nabla \Psi_i [\boldsymbol{R} \Psi_i(\mathbf{p},t) - \boldsymbol{R} s_i(t)] dt \; + \gamma \mathbf{C}\mathbf{p} \tag{4}$$

where we implicitly assumed that $\boldsymbol{R}$ is independent of the model. For the Hessian matrix, the matrix of second derviatives $\partial^2 F(\mathbf{p})/\partial p_i \partial p_j$ we find:

$$\mathbf{H}(\mathbf{p}) \equiv \nabla \mathbf{g}(\mathbf{p}) = \sum_{i=1}^{N} \int_{0}^{T_i} \left[ \boldsymbol{R} \nabla \Psi_i (\boldsymbol{R} \nabla \Psi_i)^T + \boldsymbol{R} \nabla\nabla \Psi_i [\boldsymbol{R} \Psi_i - \boldsymbol{R} s_i] \right] dt \; + \gamma \mathbf{C} \tag{5}$$

Nonlinear optimization algorithms work in an iterative way. A starting model $\mathbf{p}$ is updated with a correction $\Delta\mathbf{p}$ to give a new model $\mathbf{p}+\Delta\mathbf{p}$. This new model is then taken as a starting model in the next iteration. Suppose that at some stage we have arrived at model $\mathbf{p}$. With a simple Taylor expansion we may now find an approximation for the step $\Delta\mathbf{p}$ that should bring us close to the minimum $\mathbf{g}(\mathbf{p})=0$:

$$F(\mathbf{p}+\Delta\mathbf{p}) \approx F(\mathbf{p}) + \mathbf{g}(\mathbf{p})^T \Delta\mathbf{p} + \frac{1}{2} \Delta\mathbf{p}^T \mathbf{H}(\mathbf{p}) \Delta\mathbf{p} \tag{6}$$

differentiating with respect to $\Delta\mathbf{p}$ gives:

$$\mathbf{g}(\mathbf{p}+\Delta\mathbf{p}) \approx \mathbf{g}(\mathbf{p}) + \mathbf{H}(\mathbf{p})\Delta\mathbf{p}$$

So that $\Delta\mathbf{p}$ can be found by setting $\mathbf{g}(\mathbf{p}+\Delta\mathbf{p})=\mathbf{0}$ and solving:

$$\mathbf{H}(\mathbf{p})\Delta\mathbf{p} = -\,\mathbf{g}(\mathbf{p}) \tag{7}$$

In general, the exact calculation of $\mathbf{H}(\mathbf{p})$ is an unattainable goal, and we have to resort to approximate methods again.

In the classical Gauss-Newton method, the term involving $\nabla\nabla\Psi_i$ is ignored in (5), and the calculation of $\nabla\Psi_i$ suffices to construct $\mathbf{H}(\mathbf{p})$. When iterative methods are used to solve (7), such as the LSQR algorithm (Paige and Saunders, 1982 - see also chapter 3), there is in principle no need to store the $N \times N$ matrix $\mathbf{H}(\mathbf{p})$, but we pay for this by having to

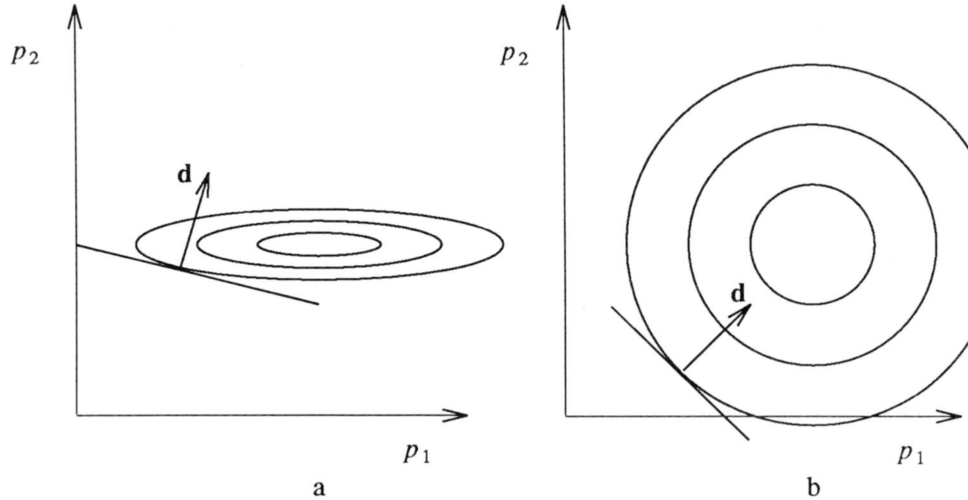


*Figure 1:* Isolines for two different configurations for the minimum of $F(\mathbf{p})$ in the case of a two-dimensional $\mathbf{p}$, and the steepest descent direction $\mathbf{d}$.

recalculate the time integral for every iteration. Although this could be done for very short time series, this method looses its attraction for longer time series or larger dimensions of $\mathbf{p}$. The Gauss-Newton algorithm is also unattractive in seismological applications for another reason: the justification for ignoring the $\nabla\nabla\Psi_i$ term in (5) is that the residual $\boldsymbol{R}\Psi_i - \boldsymbol{R}s_i$ is expected to be small with respect to the gradient. Our experience with seismological data so far indicates that the residual may remain quite large even when we approach the minimum and $\nabla\Psi_i$ is small, due to the high noise level.

An alternative approach is to search for the minimum along one direction, and repeat this with new search directions until the residual has reached an acceptable level, or any other stopping criterion is met. These methods offer the big advantage that memory requirements are minimal.

The *method of steepest descent* defines the search direction $\mathbf{d}$ as a vector in parameter space anti-parallel to the local gradient of the objective function:

$$\mathbf{d}^i = -\,\mathbf{g}(\mathbf{p}^i) \tag{8}$$

where $i$ is the iteration number. Steepest descent methods have an intuitive appeal (a blind man walking in the mountains would use a steepest descent path to find his way to the beach) but their convergence is slow . The reason is that the vector $\mathbf{d}$ rarely points into the direction of the minimum. This is shown in figure 1a for a two-parameter model and ellipsoidal isolines for the objective function. The result is that the steepest descent method follows an indirect, zig-zag path towards the minimum. Only if the isolines are circular (figure 1b) does $\mathbf{d}$ point in the direction of the minimum. In multidimensional optimization, figure 1a corresponds to the case that the eigenvalues of $\mathbf{H}(\mathbf{p})$ are widely different, whereas they are equal in the 'circular' case of figure 1b. We may therefore expect that descent methods converge faster when $\mathbf{H(p)}$ has one, or only a few, groups of closely spaced eigenvalues (see Gill et al., 1981, page 147). Knowing $\mathbf{H(p)}$ or its estimate, this could be accomplished by preconditioning, i.e. by a scaling of the parameters $p_i$. In the more common case that $\mathbf{H(p)}$ is unknown, we have no better option than to scale the parameters intuitively to roughly equal magnitude, e.g. as a percentage of the estimated model value.

Another strategy to speed up the convergence is to use *conjugate directions* for subsequent iterations, i.e.:

$$\mathbf{d}^{iT}\mathbf{H}(\mathbf{p}_{\min})\mathbf{d}^{j} = 0 \quad \text{if} \quad i \neq j \tag{9}$$

Eigenvectors of $\mathbf{H}(\mathbf{p}_{\min})$ are orthogonal, and therefore conjugate, but there are other - nonorthogonal - sets of conjugate vectors that are easier to calculate. Conjugate vectors have the nice property that, for a quadratic objective function, a model change along the direction given by one of these vectors does not impair the function decrease which has been obtained along other directions. This is easily seen by expanding the parameter correction $\Delta\mathbf{p}$ and using the Taylor series for $F(\mathbf{p})$ around the minumum location $\mathbf{p}_{\min}$. Suppose we start from $\mathbf{p}_0$:

$$\mathbf{p}_0 = \mathbf{p}_{\min} + \Delta\mathbf{p} \tag{10}$$

where we may expand:

$$\Delta\mathbf{p} = \sum_i \mu_i \mathbf{d}^i \tag{11}$$

Using a second order Taylor expansion around $\mathbf{p}_{\min}$:

$$F(\mathbf{p}_0) = F(\mathbf{p}_{\min}+\Delta\mathbf{p}) = F(\mathbf{p}_{\min}) + \tfrac{1}{2}\sum_{ij} \mu_i \mu_j \, \mathbf{d}^{iT}\mathbf{H}(\mathbf{p}_{\min})\mathbf{d}^{j}$$

Because of (9), the complicated sum with cross terms over different direction vectors reduces to a simple sum:

$$F(\mathbf{p}_{\min}+\Delta\mathbf{p}) = F(\mathbf{p}_{\min}) + \tfrac{1}{2}\sum_{i} \mu_i^2 \, \mathbf{d}^{iT}\mathbf{H}(\mathbf{p}_{\min})\mathbf{d}^{i} \tag{12}$$

We now minimize in subsequent directions $\mathbf{d}^i$, slowly reducing $\Delta\mathbf{p}$, and we see that each new direction subtracts a positive amount from the objective function which is determined by that direction only.

Conjugate directions may be found by a simple recursive scheme (see Fletcher and Reeves, 1964, or any of the books on optimization):

$$\mathbf{d}^{i+1} = -\,\mathbf{g}^{i+1} + \beta_i \mathbf{d}^i \tag{13}$$

where

$$\beta_i = \frac{|\mathbf{g}^{i+1}|^2}{|\mathbf{g}^{i}|^2}\,, \quad \beta_0 = 0 \tag{14}$$

and $\mathbf{g}$ denotes the gradient of $F$ for the i-th iteration model $\mathbf{p}^i$.

Williamson (1986) and Kennett and Williamson (1987) advocate the use of a *subspace method* for the case of delay-time tomography, but it may be worthwhile to investigate this method for waveform tomography as well. In the subspace method $\Delta\mathbf{p}$ is the optimum step within a subspace of dimension $m$, spanned by basis vectors $\mathbf{a}^j$:

$$\Delta\mathbf{p} = \sum_{j=1}^{m} \alpha_j \mathbf{a}^j \tag{15}$$

Equation (7) now reduces to $m$ equations for the coefficients $\alpha_j$:

$$(\mathbf{a}^j)^T \mathbf{g} + (\mathbf{a}^j)^T \mathbf{H}(\mathbf{p}) \sum_{k=1}^{m} \alpha_k \mathbf{a}^k = 0 \quad (j=1,...,m) \tag{16}$$

Williamson chooses $m$=2 with

$$\mathbf{a}^1 = \mathbf{g} \quad \text{and} \quad \mathbf{a}^2 = \mathbf{H}(\mathbf{p})\mathbf{a}^1 - \frac{(\mathbf{a}^1)^T \mathbf{H}(\mathbf{p})\mathbf{a}^1}{(\mathbf{a}^1)^T \mathbf{a}^1} \mathbf{a}^1 \tag{17}$$

Thus, the subspace method again requires an estimate of $\mathbf{H}(\mathbf{p})$. Because of the errors involved in the Gauss-Newton approximation, it may be worthwhile to adopt the solution of (16) as a *search direction* $\mathbf{d}$, rather than as a fixed-length step. We then search for the $\mu$ that minimizes $F(\mathbf{p}+\Delta\mathbf{p})$ for $\Delta\mathbf{p}=\mu\sum\alpha_j \mathbf{a}^j$, i.e. along $\mathbf{d}$, with $\mu\approx 1$.

Whatever the method employed to find a good search direction, we use a simple and practical technique for locating the minimum along $\mathbf{d}$. We calculate $F$ in 3 or more points along $\mathbf{d}$ until we have passed a minimum, and then locate the minimum approximately with quadratic interpolation. We test whether this new point is indeed a point where the value of $F$ is less than in the other points. Although in theory this does not guarantee convergence to the minimum, it works well in practice.

Another potentially useful method is the Broyden-Fletcher-Goldfarb-Shanno or BFGS update. This belongs to the class of quasi-Newton methods which exploit the fact that an approximation to the curvature of a nonlinear function can be computed without explicitly forming the Hessian matrix. In the BFGS method, we assume $\mathbf{H}(\mathbf{p})\approx\mathbf{H}$, i.e. constant, and we try to find an approximation $\mathbf{H}^{i+1}$ for the $i$+1-th iteration. Gill et al. (1981) give full details on the BFGS method. Here we merely list the result:

$$\mathbf{H}^{k+1} = \mathbf{H}^k - \frac{\mathbf{H}^k \Delta\mathbf{p}^k (\Delta\mathbf{p}^k)^T \mathbf{H}^k}{(\Delta\mathbf{p}^k)^T \mathbf{H}^k \Delta\mathbf{p}^k} + \frac{\Delta\mathbf{g}^k (\Delta\mathbf{g}^k)^T}{(\Delta\mathbf{g}^k)^T \Delta\mathbf{p}^k} \tag{18}$$

where

$$\Delta\mathbf{p}^k = \mathbf{p}^{k+1} - \mathbf{p}^k \quad , \quad \Delta\mathbf{g}^k = \mathbf{g}^{k+1} - \mathbf{g}^k$$

With the step direction computed from $\mathbf{H}^i \mathbf{d}^i = -\mathbf{g}^i$ so that $\mathbf{H}^k \Delta\mathbf{p}^k = -\mu_i \mathbf{g}^k$, we find:

$$\mathbf{H}^{k+1} = \mathbf{H}^k + \frac{\mathbf{g}^k (\mathbf{g}^k)^T}{(\mathbf{d}^k)^T \mathbf{g}^k} + \frac{\Delta\mathbf{g}^k (\Delta\mathbf{g}^k)^T}{\mu_k (\Delta\mathbf{g}^k)^T \mathbf{d}^k} \tag{19}$$

At first sight, this does seem to result in very large memory requirements when the number of parameters $M$ is large, since any initial sparsity in the Hessian for individual seismograms is lost in the sum over all data. Individual Hessian matrices are sparse with a suitably chosen model parametrization, since our data consist, in general, of path integrals. $\nabla\Psi_i(\mathbf{p},t)$ will have many zero elements if we parametrize our model into nonoverlapping cells for example. If $\nabla\Psi_i(\mathbf{p},t)$ contains $n_i$ nonzero elements, it may prove convenient to write $\mathbf{H}^k$ as a sum of convex "element functions" (Griewanck and Toint, 1982):

$$\mathbf{H}^k = \sum_{i=1}^{N} \mathbf{H}_i^k \tag{20}$$

where $\mathbf{H}_i^k$ denote $M \times M$ matrices with $n_i \times n_i$ nonzero entries. If we use the sparse vector:

$$\mathbf{g}_i = \int_0^{T_i} R\nabla\Psi_i [R\Psi_i(\mathbf{p},t) - Rs_i(t)]dt \tag{21}$$

we may apply the BFGS update formula (19) separately for each pair $(\mathbf{H}_i^k, \mathbf{g}_i^k)$, before using (20) to find a search direction from:

$$\left[\sum_{i=1}^{N} \mathbf{H}_i^k\right] \mathbf{d}^k = -\sum_{i=1}^{N} \mathbf{g}_i^k \tag{22}$$

Note that we set the damping parameter $\gamma=0$ to retain sparsity of $\mathbf{g}_i$. To guard against possible numerical instabilities in solving (22) the damped version of LSQR should probably be used. The method eases the storage requirements. For example, if $N$=200, $M=10^4$ and $n_i$=200, the (symmetric) Hessian contains $\pm 5\times 10^7$ elements, but the actual storage requirement for the element matrices is merely $2\times 10^4$. This scheme lends itself particularly well to parallel computing.

Since the Hessian is only approximated, a few iterations of a sparse matrix solver such as LSQR should suffice to give $\Delta\mathbf{p}^k$ to acceptable accuracy. Again, it seems wise to use this as a direction $\mathbf{d}$ only, and find the true location of the minimum along $\mathbf{d}$ by fitting a quadratic through 3 points of $F(\mathbf{p})$.

## 5. Fitting the spectrum

Instead of (3), we may fit the signals in the frequency domain by minimizing:

$$F(\mathbf{p}) = \frac{1}{2}\sum_{i=1}^{N} \int_0^{\Omega_i} |R\Psi_i(p,\omega) - Rs_i(\omega)|^2 d\omega + \frac{1}{2}\gamma \mathbf{p}^T \mathbf{C}\mathbf{p} \tag{23}$$

For those algorithms that compute the seismogram in the frequency domain this method has the advantage that it saves $N$ FFT's with each evaluation of $F(\mathbf{p})$ or its gradient. That is, if $\boldsymbol{R}$ is a multiplicative operator in the frequency domain, such as a high- or a low-pass filter. But if $\boldsymbol{R}$ also involves windowing in the time domain (and it usually does), the use of a Fourier transform is usually more efficient than a convolution in the frequency domain and we might use time-domain fitting with (3) as well. A hybrid approach will be suggested in section 7.

## 6. The forward problem for broad-band signals

We may look upon a seismogram either as a collection of localized waves, progressing in time, or as a record of harmonic reverberations in the Earth. The first viewpoint is connected with the ray theoretical approximation. It works well with early arrivals of body waves. The second approach takes the correct, if somewhat laborious, view that (small) displacements in the Earth result from a summation of normal modes. This works with the S-wave and surface wave part of the seismograms, for frequencies below about 0.1 Hz in teleseismic applications, although higher frequencies are feasible for more local events in not too heterogeneous regions (e.g. Panza and Suhadolc, 1986). Hybrid formulations are possible as well (Felsen, 1984) and might be useful in special applications.

In this chapter we are more concerned with the view that a seismogram can be approximated by summing a finite number of modes at each frequency component. Snieder and Nolet (1987) model the spectrum of a far field seismic wave as:

$$\Psi_R(\mathbf{p}\,|\,\mathbf{r},\omega) = -\frac{1}{2}\sum_n \frac{[\hat{\mathbf{r}}U_n(r) + i\,\hat{\boldsymbol{\theta}}\lambda_n(\omega)V_n(r)]}{u_n(\omega)\sqrt{\sin\theta}} A_{Rn}(\phi)e^{i\lambda_n\theta - \omega\theta/2Q_n u_n} \tag{24}$$

for Rayleigh waves, and for Love waves:

$$\Psi_L(\mathbf{p}\,|\,\mathbf{r},\omega) = \frac{1}{2}\sum_n \frac{i\,\hat{\boldsymbol{\phi}}\lambda_n(\omega)W_n(r)}{u_n(\omega)\sqrt{\sin\theta}} A_{Ln}(\phi)e^{i\lambda_n\theta - \omega\theta/2Q_n u_n} \tag{25}$$

where $\lambda_n(\omega)$ is the wavenumber and $u_n(\omega)$ is the group velocity, both scaled with the Earth's radius $a$: $\lambda_n(\omega)=ak_n(\omega)$ and $u_n(\omega)=(\partial\lambda_n/\partial\omega)^{-1}$. We note that $\lambda_n(\omega)$ is related to the angular order $l$ of the wave by $\lambda_n(\omega)=\sqrt{l(l+1)}$. The amplitude of the surface wave is linearly related to the components $M_i$ of the moment tensor:

$$A_{R,Ln}(\phi) = (\lambda_n/2\pi)^{1/2}\sum_{k=1}^{6} B_{kn}^{R,L} M_k \tag{26}$$

with the convention: $M_1=M_{rr}, M_2=M_{r\theta}, M_3=M_{r\phi}, M_4=M_{\theta\theta}, M_5=M_{\phi\phi}, M_6=M_{\theta\phi}$, and:

$$\begin{aligned}
B_1^R &= \partial_r U_s\, e^{-i\pi/4} \\
B_2^R &= -\lambda_n[\partial_r V_s + r_s^{-1}(U_s - V_s)]\cos\zeta_s\, e^{i\pi/4} \\
B_3^R &= -\lambda_n[\partial_r V_s + r_s^{-1}(U_s - V_s)]\sin\zeta_s\, e^{i\pi/4} \\
B_4^R &= \tfrac{1}{2}r_s^{-1}(2U_s - \lambda_n^2 V_s)\, e^{-i\pi/4} - \tfrac{1}{2}\lambda_n^2 r_s^{-1} V_s\cos 2\zeta_s\, e^{-i\pi/4} \\
B_5^R &= \tfrac{1}{2}r_s^{-1}(2U_s - \lambda_n^2 V_s)\, e^{-i\pi/4} + \tfrac{1}{2}\lambda_n^2 r_s^{-1} V_s\cos 2\zeta_s\, e^{-i\pi/4} \\
B_6^R &= -\lambda_n^2 r_s^{-1} V_s \sin 2\zeta_s\, e^{-i\pi/4}
\end{aligned} \tag{27}$$

$$\begin{aligned}
B_1^L &= 0 \\
B_2^L &= -\lambda_n[\partial_r W_s - r_s^{-1}W_s]\sin\zeta_s\, e^{i\pi/4} \\
B_3^L &= \lambda_n[\partial_r W_s - r_s^{-1}W_s]\cos\zeta_s\, e^{i\pi/4}
\end{aligned} \tag{28}$$

$$B_4^L = -\tfrac{1}{2}\lambda_n^2 r_s^{-1} W_s \sin 2\zeta_s \, e^{-i\pi/4}$$
$$B_5^L = -B_4^L$$
$$B_6^L = \lambda_n^2 r_s^{-1} W_s \cos 2\zeta_s \, e^{-i\pi/4}$$

Where $\zeta_s$ denotes the source-to-station azimuth, measured counterclockwise from South. In (24)-(28) we assumed a Fourier convention $s(\omega)=\int s(t)e^{+i\omega t}dt$. $r_s$ denotes the source location, and the normalization used is

$$\omega^2 \int_0^a \rho[U_n^2+\lambda_n^2 V_n^2] r^2 dr = 1 , \qquad \omega^2 \int_0^a \rho\lambda_n^2 W_n^2 r^2 dr = 1 \tag{29}$$

In the following we shall use a shorthand notation, and use for one of the components of the displacement spectrum (the j-th datum):

$$\Psi_j(\omega) = \sum_n \Psi_{nj}(\omega) = \sum_n A_{nj}(\omega) e^{i\chi_{nj}(\omega)-\alpha_{nj}(\omega)} \tag{30}$$

To accomodate slight variations in phase velocity along the wave path, we replace the simple expressions for $\chi_{nj}$ and $\alpha_{nj}$ by integrals along the surface ray path. This approximation, which is widely applied but may not be valid for strong gradients in velocity or for certain wavelengths (see section 1) is essentially a phase integral, or WKBJ approximation:

$$\chi_{nj}(\omega) = \int_{S_j} [k_n(\omega) + \delta k_n(\omega,\theta,\phi)] ds \tag{31}$$

$$\alpha_{nj}(\omega) = \frac{1}{2} \int_{S_j} [k_n(\omega) + \delta k_n(\omega,\theta,\phi)][Q_n^{-1}(\omega) + \delta Q_n^{-1}(\omega,\theta,\phi)] ds \tag{32}$$

where $S_j$ denotes the surface ray path, $\delta k_n(\omega,\theta,\phi)$ and $\delta Q_n^{-1}(\omega,\theta,\phi)$ are the *local* perturbations in wavenumber and quality factor of the mode, respectively, as a function of colatitude $\theta$ and longitude $\phi$.

To a very good approximation, $\delta k_n(\omega,\theta,\phi)$ and $\delta Q_n^{-1}(\omega,\theta,\phi)$ are linearly related to the local perturbations in the starting model. The theory is well known (e.g. Takeuchi and Saito, 1972; Nolet, 1981) and here we merely list the resulting formulae:

$$\delta k_n(\omega,\theta,\phi) = \int_0^a \left\{ \left[\frac{\partial k_n}{\partial \alpha}\right]_{\beta,\rho} \delta\alpha + \left[\frac{\partial k_n}{\partial \beta}\right]_{\alpha,\rho} \delta\beta + \left[\frac{\partial k_n}{\partial \rho}\right]_{\alpha,\beta} \delta\rho \right\} dr$$

$$\begin{aligned} \left[\frac{\partial k_n}{\partial \alpha}\right]_{\beta,\rho} &= D_1^{L,R}\, 2\alpha\rho \\ \left[\frac{\partial k_n}{\partial \beta}\right]_{\alpha,\rho} &= (D_2^{L,R} - 2D_1^{L,R})\, 2\beta\rho \\ \left[\frac{\partial k_n}{\partial \rho}\right]_{\alpha,\beta} &= (D_1^{L,R} - 2D_2^{L,R})\beta^2 + D_2^{L,R}\alpha^2 + D_3^{L,R} \end{aligned} \tag{33}$$

where, for Love and Rayleigh waves, respectively:

$$D_1^L = 0$$

$$D_2^L = -\frac{\omega}{2V_g}[r^2(\partial_r W_n - \frac{1}{r}W_n) + (\lambda_n^2 - 2)W_n^2 \tag{34}$$

$$D_3^L = \frac{\omega^3}{2V_g} r^2 W_n^2$$

$$D_1^R = -\frac{\omega}{2V_g}[r^2(\partial_r U_n)^2 + 2r\,\partial_r U_n(2U_n - \lambda_n^2 V_n) + (2U_n - \lambda_n^2 V_n)^2]$$

$$D_2^R = -\frac{\omega}{2V_g}[2r^2(\partial_r U_n)^2 + (2U_n - \lambda_n^2 V_n)^2 + \lambda_n^2(r\,\partial_r V_n + U_n - V_n)^2 + \lambda_n^2(\lambda_n^2 - 2)V_n^2] \tag{35}$$

$$D_3^R = \frac{\omega^3}{2V_g}[U_n^2 + \lambda_n^2 V_n^2]$$

and where $V_g$ denotes group velocity $(\partial k_n(\omega)/\partial\omega)^{-1}$

Assuming zero bulk loss, the spatial quality factor $Q_n^{-1}(\omega)$ of the mode is linearly related to the intrinsic $Q_\mu^{-1}(r)$ of the Earth:

$$Q_n^{-1}(\omega) = -\frac{1}{k_n(\omega)}\int_0^a \beta(r)\left[\frac{\partial k_n}{\partial\beta}\right]_{\alpha,\rho} Q_\mu^{-1}(r) r^2 dr \tag{36}$$

so that

$$\delta Q_n^{-1}(\omega,\theta,\phi) = \frac{1}{k_n(\omega)}\int_0^a \left[\frac{\partial k_n}{\partial\beta}\right]_{\alpha,\rho} [\beta\delta Q_\mu^{-1}(r,\theta,\phi) + \delta\beta(r,\theta,\phi) Q_\mu^{-1}(r)] r^2 dr \tag{37}$$

The term with $\delta\beta$ dissappears when we have a perfectly elastic starting model. If we define $\delta Q_\mu^{-1}$ rather than $\delta Q_\mu$ as our model, inversion for $Q_n^{-1}(\omega)$ is linear in $Q_\mu^{-1}(\mathbf{r})$. Similar expressions can be found for deviations from isotropy (Montagner and Nataf, 1986; Romanowicz and Snieder, 1987).

The linear expressions (31) and (32) enable us to calculate $\chi_{nj}(\omega)$ and $\alpha_{nj}(\omega)$ in a very efficient manner. We only need to solve the eigenvalue problem for the $k_n(\omega)$ once, and store the partial derivatives with respect to model perturbations. Thus, if we are willing to ignore perturbations in the amplitudes $A_n$ due to perturbations in the elastic structure, it is a very simple matter to calculate the spectrum of the synthetic signal. We do ignore relevant information when we ignore perturbations in $A_n$ (Levshin, 1985) and it may be worthwhile to investigate the applicability of linearized theories for the effects of focussing and defocussing that have recently been developed (Woodhouse and Wong, 1986; Snieder and Romanowicz, 1987).

If necessary, we may repeat the optimization after location of a minimum using new eigenvectors. In that case we have a two-level iterative process. The outer level iteration also remedies possible shortcomings in the linearized calculation of $\chi_{nj}(\omega)$ and $\alpha_{nj}(\omega)$ if the starting model needs large perturbations to fit the data. By retaining the essential nonlinearity of the problem also in the inner level iterations, our starting model does not have to be close to the real Earth. Some striking examples of the power of this strategy have been presented in Nolet et al. (1986).

## 7. The gradient

For the calculation of the gradient (4) or (21) we need $\nabla\Psi_i$. In the frequency domain, we find for elastic variables and the density (again ignoring variations in $A_{nj}(\omega)$):

$$\frac{\partial\Psi_j(\omega)}{\partial p_i} = \sum_n \left[ i\frac{\partial\chi_{nj}(\omega)}{\partial p_i} - \frac{\partial\alpha_{nj}(\omega)}{\partial p_i} \right] \Psi_{nj}(\omega) \quad (p_i = \alpha,\beta,\rho) \tag{38a}$$

and for variations with respect to moment tensor elements or source depth:

$$\frac{\partial\Psi_j(\omega)}{\partial p_i} = \sum_n \frac{\partial A_{nj}(\omega)}{\partial p_i} e^{i\chi_{nj}(\omega)-\alpha_{nj}(\omega)} \tag{38b}$$

where

$$\frac{\partial\chi_{nj}(\omega)}{\partial p_i} = \int_{S_j} \left[ \frac{\partial k_n}{\partial p_i} \right] ds \tag{39}$$

$$\frac{\partial\alpha_{nj}(\omega)}{\partial p_i} = \int_{S_j} [k_n(\omega)\frac{\partial Q_n^{-1}(\omega,\theta,\phi)}{\partial p_i} + Q_n^{-1}(\omega)\frac{\partial k_n(\omega,\theta,\phi)}{\partial p_i}]ds \tag{40}$$

The gradient (38) is easy to calculate, since (39) and (40) can be computed once before the whole optimization exercise is started, whereas the derivatives of $A_{nj}$ need only be recalculated if $p_i$ is the source depth level $r_s$.

For the gradient in the time domain, (38) must be Fourier transformed:

$$\frac{\partial\Psi_i(\mathbf{p},t)}{\partial p_i} = \frac{1}{2\pi}\int_{-\infty}^{\infty} \frac{\partial\Psi_i(\omega)}{\partial p_i} e^{-i\omega t} d\omega \tag{41}$$

This is obviously a major extra computational effort, since an FFT is now needed for each of the components $p_i$. An alternative worth considering is to determine the search direction by calculating the gradient directly in the frequency domain, i.e. from (23). Since, for the derivative of $ff^*$, where $f$ is a complex quantity, $(ff^*)'=2\mathrm{Re}(ff')$ we have:

$$\mathbf{g} = \sum_{j=1}^{N} \int_0^{\Omega_i} \mathrm{Re}\,[\boldsymbol{R}_\omega \nabla\Psi_j(\boldsymbol{R}_\omega\Psi_j - \boldsymbol{R}_\omega s_j)^*]d\omega \tag{42}$$

or, with a multiplicative 'filter' $\mathbf{R}_\omega$:

$$g_i = \mathrm{Re}\sum_{j=1}^{N}\sum_{k=1}^{K}\left[ \sum_n \frac{\partial A_{nj}(\omega_k)}{\partial p_i} + A_{nj}(\omega_k)\left[ i\frac{\partial\chi_{nj}(\omega_k)}{\partial p_i} - \frac{\partial\alpha_{nj}(\omega_k)}{\partial p_i} \right]\right] \times$$

$$\times e^{i\chi_{nj}(\omega_k)-\alpha_{nj}(\omega_k)}[\Psi_j(\omega_k) - s_j(\omega_k)]\ |\boldsymbol{R}_\omega(\omega_k)|^2\Delta\omega_k \tag{43}$$

A very flexible way to compute the gradient in the time domain is by finite differencing $F(\mathbf{p})$. The advantage of this is that any type of operator $\boldsymbol{R}$ can easily be accommodated, and the programming is simplified tremendously. The disadvantage is the very large

**Table 1: Earthquake parameters**

| Date | Time | Epicentre | Depth | Region |
|---|---|---|---|---|
| May 26, 1983 | 2:59:58.8 | 40.5N 139.1E | 16 | W Coast Honshu |
| Dec 22, 1983 | 4:11:29.3 | 11.8N 13.5W | 11 | NW Africa |

computing time required, since the evaluation of one component of the gradient now requires the calculation of one full set of synthetics if a two-point difference formula is used, and even two sets if we calculate **g** from 3 points (two on either side of the central point).

## 8. A case history

We shall illustrate some aspects of the theory outlined above with a few numerical tests that we performed on broad band seismograms from the two earthquakes listed in table 1. The seismographs all belong to the NARS network (Dost et al., 1984; Nolet et al., 1986). Ray paths are illustrated in figure 2. Both events, and the stations of the NARS array, are approximately located on one great circle. This allows us to construct a 2-D cross section of the upper mantle along this great circle.

In the initial tests we used only the recordings from the Honshu earthquake. One of the first factors that we investigated was how well the objective function $F(\mathbf{p})$ can be approximated by a quadratic, since the assumption of a constant Hessian matrix underlies the convergence proofs of all of the optimization strategies in section 4 that are feasible for models with many parameters.

The various pitfalls are probably best illustrated in figure 3, which depicts $F(\mathbf{p})$, calculated as the time domain fit (3), in a 2-dimensional subspace of the adopted parameterspace. $p_1$ is, in this case the (constant) perturbation of the S velocity over the whole surface ray path for a depth region comprising the lithosphere *and* the low velocity layer (33-240 km depth). $p_2$ is like $p_1$ but involves the lithosphere (33-120 km) only. Thus, $p_1$ and $p_2$ are not completely independent and the resulting trade-off is visible as the band-like structure in figure 3. Figure 3a gives $F(\mathbf{p})$ without damping ($\gamma$=0); in figure 3b, $\gamma$ has been raised to a value that brings the global minimum (the black dot) about halfway between the starting model and the global minimum of the undamped $F(\mathbf{p})$.

The isolines in figure 3 are not ellipsoidal, except perhaps in the direct neighbourhood of the global minimum. In fact, there is a series of local minima located in the left of the diagrams that are closer to the starting point "s" than is the global minimum. Had the starting model been located in the point (-1%,-1%), the optimization would end up in one of these local minima. In the time domain one expects local minima to occur whenever a full $2\pi$ phase shift occurs over a dominant part of the wavetrain. The difference in velocity in the two solutions is quite drastic: in fact $p_1$ changes sign!

This situation is not solved in the damped version (figure 3b), and it is obvious that more drastic measures are needed, such as using a strong low-pass filter or even a reduction

*Figure 2* Surface ray paths for the Honshu and NW Africa earthquakes (square symbols) to the NARS array in Europe.

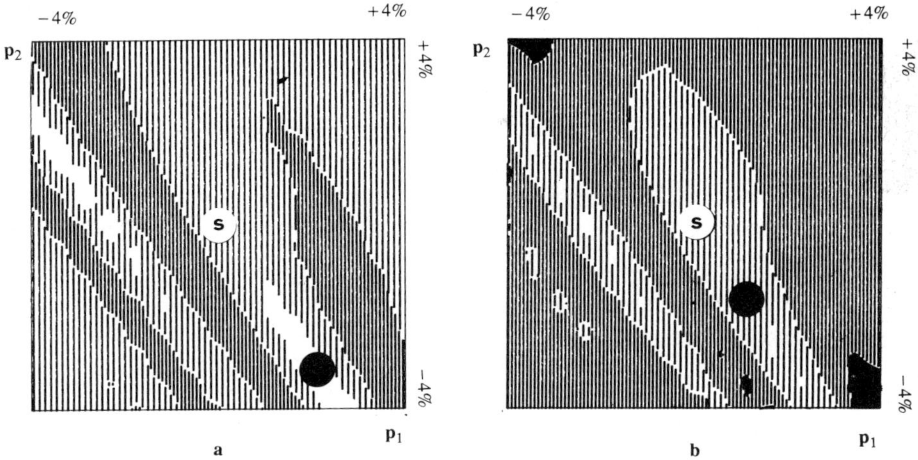


*Figure 3* Isolines for $F(\mathbf{p})$ with real data: (a) undamped, (b) damped case. For details see text.

of the time signal to its envelope. These operations are symbolically denoted by $\boldsymbol{R}$. In a later stage of the optimization, when we have approached the global minimum, the influence of $\boldsymbol{R}$ can be relaxed.

Local minima will still exist if we choose $p_1$ and $p_2$ as nonoverlapping regions in depth, for instance by redefining $p_1$ as the velocity in the LVL (120-240 km). Sometimes a smoothing of $F(\mathbf{p})$ can be obtained by working with transformed parameters: $\mathbf{p}' = \mathbf{Sp}$, a strategy called *multiple gridding* by Nolet et al. (1986). $\mathbf{S}$ could for instance be used to

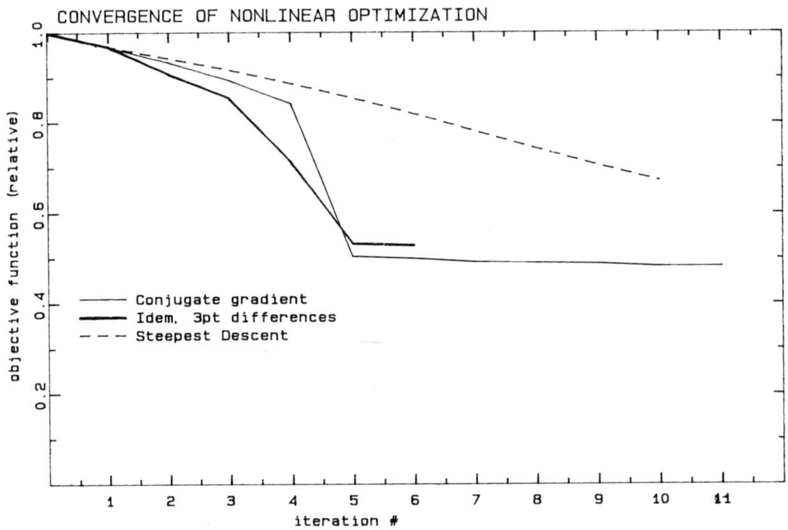


*Figure 4* Convergence behaviour of various methods for nonlinear optimization.

eliminate certain parameters, connected with the more detailed aspects of the model, in the early stages of the inversion. However, the influence of this is often dissappointing. Again figure 3 serves as an example: if we eliminate $p_2$ and invert for $p_1$ only, we will end up very close to the starting model and little improvement in data fit will result. Only when **S** is redefined such that $p_1'$ follows the "valley" roughly corresponding to $p_1'=p_1-p_2$, will a significant improvement in convergence occur and we avoid the region with local minima (but imagine what happens with $p_1'=p_1+p_2$!). In general we lack the knowledge to define **S** in such an optimal way. In practice we use **S** to select a few independent parameters for which we perform the inversion first. These are changes in source origin time and fault plane parameters, source depth and Earth structure between the source(s) and the first station of the network over which we try to monitor local variations in Earth structure.

Using the same data set, we have compared the convergence behaviour of the methods of steepest descent (8) and conjugate gradients (13). The behaviour shown in figure 4 is typical for the tests that we have made. Steepest descent optimization results in a steady but slow convergence to a minimum of $F(\mathbf{p})$. The conjugate gradient method gives only a slight improvement in the first few iterations, but then suddenly finds a model with a much better fit, after which the convergence slows down. We have calculated $\mathbf{g}(\mathbf{p})$ with a simple two-point difference, but find that the inaccuracy of this has no serious effects. A run using the 3-point difference formula gives comparable convergence, but requires twice as much computing time. However, in later runs with many parameters we noticed that the error in a two-point difference formula may play us tricks. In that case we found that for those elements of **g** that are close to the minimum, but have a large second derivative, 3-point differences may be necessary to obtain sufficient accuracy in the orthogonalization (13).

After these preliminaries, we have tested the algorithm on a realistic data set consisting of both events in table 1. We allowed the structure to vary locally from a background

**Table 2: background model**

| Depth(km) | $V_s$ (m/s) |
|---|---|
| 29 | 4321 |
| 40 | 4400 |
| 60 | 4500 |
| 300 | 4500 |
| 400 | 4770 |
| 400 | 4933 |
| 500 | 5224 |
| 600 | 5516 |
| 671 | 5945 |

model listed in table 2. The initial fit of the synthetics to the observations is shown in figure 5. We used a summation over 15 Rayleigh modes with a limiting phase velocity of 11 km/s to construct these synthetics. The data have been low-passed at a period of about 30 seconds and the same filter has been applied to the synthetics. To match the synthetics to the data, we performed a multi-stage nonlinear inversion.

In the first stage, we limited the data set to 4 seismograms: NE01 and NE14 for the NW Africa event, NE02 and NE11 for the Honshu event. The upper mantle was divided in three different regions: one region spanning the distance between the first and the last station of the array (Western Europe), one between the northernmost station (NE01) and Honshu, one between the southernmost station (NE14) and the epicentre of the NW Africa event. Within each of these regions the S-velocity in the mantle was allowed to be perturbed on a number of grid points at the Moho and at depths of 140, 400, 670 and 1370 km. Linear interpolation was used to determine the velocity perturbation in between these points. In addition we allowed perturbations in upper and lower crust. The main purpose of this inversion was to determine the average structure over the Eurasian and African parts, which have a large influence on the waveforms. Unfortunately, in the absence of any digital broad band stations in between the NARS array and these events, we have no information about the lateral heterogeneity in these regions, and we shall treat the structure there merely as a correction factor, rather than attach any physical importance to it.

After some experimenting with different starting models, we found that the model in table 2 gave the most satisfactory starting point for nonlinear inversion. The remarkable feature of the model is the absence of a low velocity layer, which is not required for the Eurasian path data (where the inversion raised $\beta$ to 4569 m/s at the 140 km grid point) nor for the West African path, where $\beta$ was changed to 4643 m/s at 140 km depth. From higher mode results in the same area (Dost, 1986) we know that Western Europe has an LVZ - one of the motivations for this research is that we wish to know the lateral extent of this zone.

It was not difficult to obtain a good fit of the fundamental Rayleigh mode at this stage, but after this the convergence slowed down appreciably, and a satisfactory fit to the S and SS waves could not be obtained within a reasonable number of iterations (~10). We solved this by application of a window to the signals, which reduced the fundamental mode parts

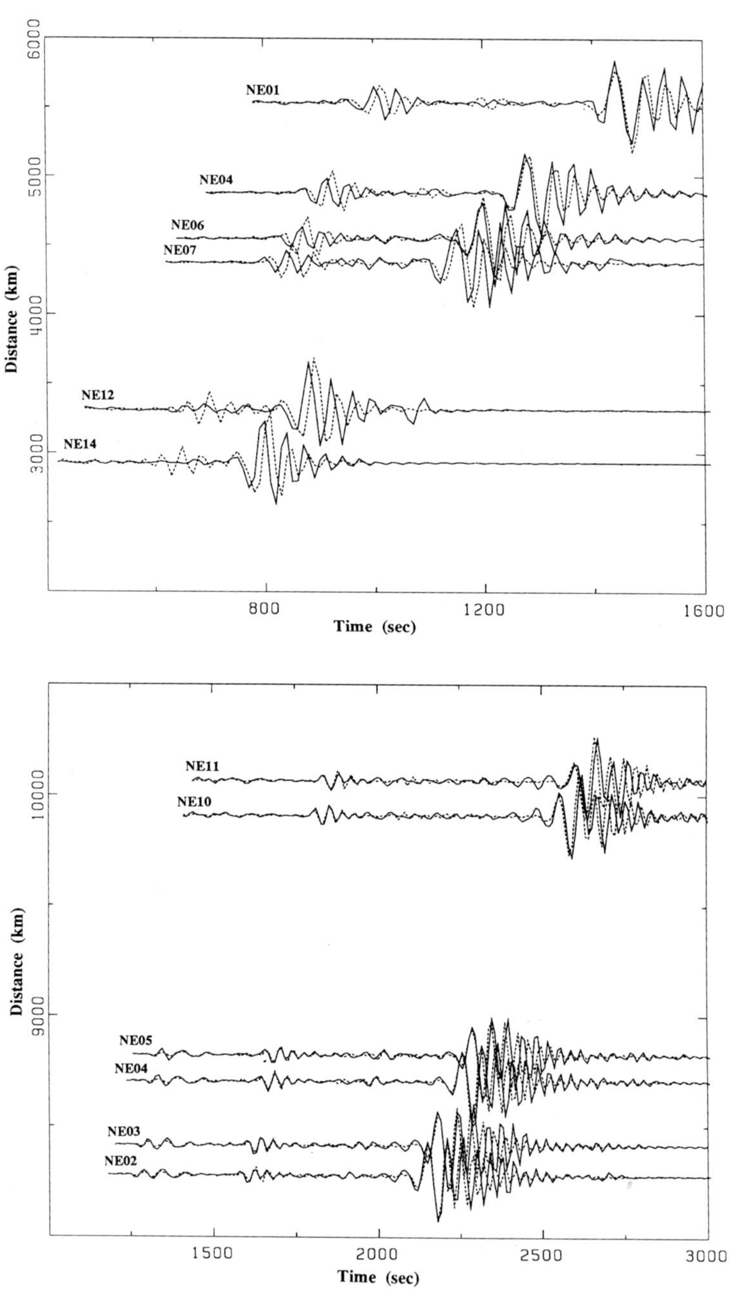


*Figure 5* Comparison of observed vertical seismograms (broken line) with computed synthetics for the laterally homogeneous model listed in table 2. Top: NW Africa event, bottom: Honshu event.

of the seismogram by a factor of 50%. With this reduced weighting of the dominant surface wave part, the model quickly converged to a reasonable fit for the S wave arrivals, with one notable exception: the model predicts a far too large amplitude for the S wave in station NE14 for the NW Africa event.

In the second stage, we developed the lateral heterogeneity in a bilinear form, using basis functions $g_i(x)$ and $h_j(z)$ for our adopted coordinate system of epicentral distance $x$ and depth $z$:

$$\delta\beta(x,z) = \sum_{ij} p_{ij} g_i(x) h_j(z) \tag{44}$$

The basis functions have a triangular shape as a function of $x$ and $z$ respectively. For the horizontal functions, the location of the maxima of the triangle functions is indicated with the filled triangles in the map in figure 7. The location of the top of one triangle always coincides with a base point of the next. This way, each pair of basis functions allows us to interpolate linearly along the $x$-coordinate. We did the same for the depth dependent functions: The $h_j(z)$ where chosen to allow linear interpolation along the z-coordinate, with pivots at the same depths as in the first inversion stage. We did not vary the depth of the Moho, and possible variations in Moho depth will map as velocity variations. We also added the remaining seismograms to the data set.

The final outcome is shown in figure 7, the data fit in figure 6. There is still a misfit in the amplitude of the S waves of the NW Africa event in stations NE12 and NE14 (Spain). For the Honshu event, the beginning of the fundamental Rayleigh mode is not well matched. We suspect that effects of anelasticity play a role in both mismatches, but further research is in order to confirm this.

The model shows that the low velocity layer is confined to the central part of the array. The northern and southern boundary are placed within the grid points of the horizontal basis functions, and evidently a finer horizontal discretization (with higher frequencies in the data and more stations to provide better resolution) is needed to define these boundaries more precisely. There is a qualitative agreement of the location of the northern and southern boundaries of this LVZ with the LVZ as determined from P-delays by Spakman (personal communication, 1976). There is disagreement with the laterally homogeneous model WEPL1 obtained by Dost (1986): the minimum $\beta$ in our model is 4389 m/s at a depth of 140 km. This is not far from Dost's average value at this depth (4340 m/s), but the LVZ in WEPL1 has a sharp minimum at 200 km depth (4243 m/s) and a steep gradient starting near 250 km. In comparison, our model has a much more smeared-out minimum. The difference is even larger with respect to a more recent model WEPL2 (Dost, in preparation) which also satisfies S wave arrival times for European events. We attribute this difference to the relatively coarse parametrization with depth that we were forced to impose, in order to keep the number of parameters within acceptable limits.

All our calculations where done on a Gould PN6080 minicomputer. The longest runs took a few hours turn-around time. Unfortunately, we have no precise measurements of the actual CPU time consumed.

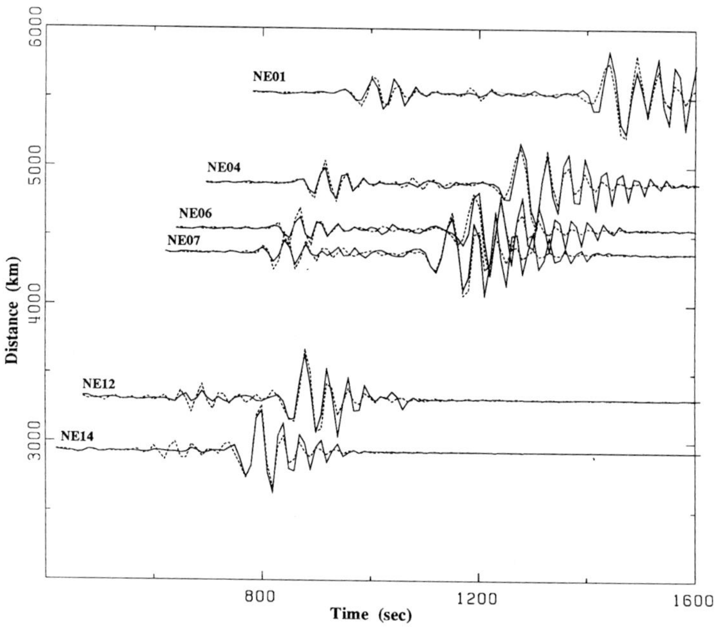


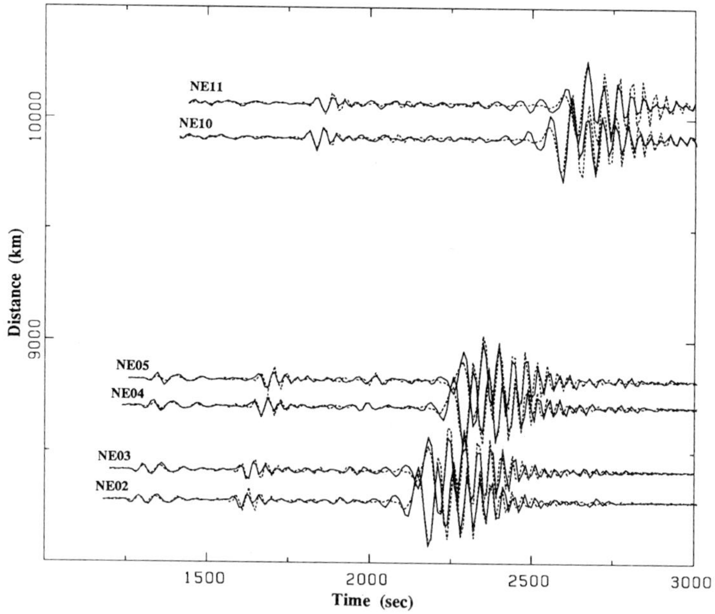


*Figure 6* Comparison of observed vertical seismograms (broken line) with computed synthetics for the laterally heterogeneous model shown in figure 7. Top: NW Africa event, bottom: Honshu event.

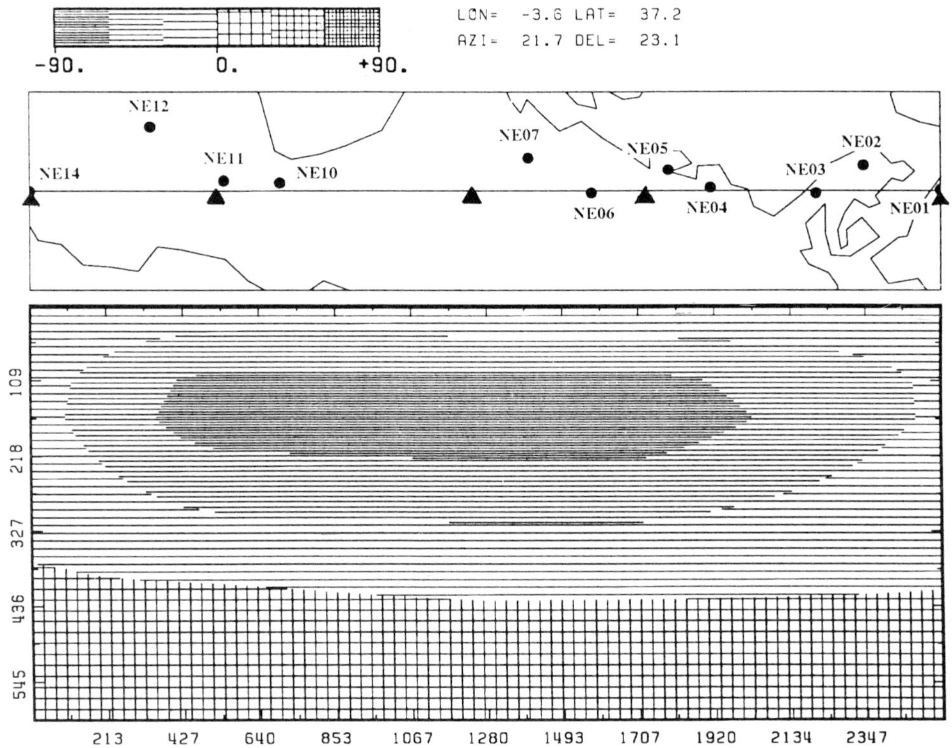


*Figure 7* Laterally heterogeneous model for the S-velocity under Western Europe, along a transect from NARS stations NE01 to NE14. Deviations (in m/s) are with respect to the model listed in table 2. The numbers along the axes denote depth and horizontal distance in km.

From the results of this excercise, we conclude that the methods of nonlinear optimization provide a useful approach to model building in seismology, especially at intermediate frequencies, where effects of nonlinearity are large but where a laterally homogeneous starting model is still effective. The fitting of complete seismograms with periods as small as 30 seconds constitutes a significant advance with respect to earlier, low frequency inversions.

Further research is needed in several areas: analytical calculation of $\boldsymbol{R}\nabla\Psi$, even for complicated operators $\boldsymbol{R}$, would increase the computation speed by an order of magnitude or more, thus allowing for more complicated model parametrization. The possibility to use the sparsity of the Hessian matrix, and use partitioned BFGS updates may lead to the fastest convergence available and should be explored.

Although in this test, the synthetics have been computed using summation of surface wave modes, the methods outlined in this paper should be applicable to any type of synthetic seismogram.

*Acknowledgments.* I benefitted from discussions with Roel Snieder, who also helped in getting some of the software working correctly. Berend Scheffers' help in processing data and doing most of the calculations for the test example is greatly appreciated. The NARS project is financed by the Earth Science Branch (AWON) of the Netherlands Organization for the Advancement of Pure Research (ZWO).

*Chapter 14*

# Surface wave holography

*R. Snieder*

## 1. Introduction

Surface waves have proven to be very useful in determining the properties of the Earth's crust and mantle. The traditional surface wave analysis consists of two steps. First, from surface wave recordings, dispersion data (phase velocities or group velocities) are retrieved for each source receiver pair (Dziewonski and Hales, 1972; Nolet, 1977). Next, the information for different frequencies and many source receiver pairs is combined to yield an image of the Earth's interior (e.g. Woodhouse and Dziewonski, 1984; Montagner, 1986; Nataf et al., 1986). These methods implicitly use ray theory by resorting to the "great circle theorem" (Backus, 1964; Jordan, 1978; Dahlen, 1979). This theorem states that for a sufficiently smooth medium the surface wave data are only influenced by the structure under the great circle joining the source and the receiver. The great circle theorem is acceptable provided the inhomogeneity varies little on the scale of the wavelength of the surface waves.

It turns out, however, that this condition is often violated in realistic situations. A Rayleigh wave with a period of 20 seconds has a horizontal wavelength of about 70 kilometers. It is well known that, especially in continents, the lateral heterogeneity on this scale can be considerable. In fact, the models constructed from surface wave data using the great circle theorem sometimes vary strongly on a distance of one wavelength (Panza et al., 1980). In that case the constructed model is inconsistent with the ray theory used for producing the model. It is clear that in these situations one has to resort to a more complete wave theory which takes surface wave scattering and reflection into account. Since these effects are most sensitive to the horizontal *gradient* in the Earth's structure, scattered



*G. Nolet (ed.), Seismic Tomography, 323–337.*

surface waves could provide valuable independent information on the structure of the Earth.

Surface wave scattering and reflection can be treated analytically in two dimensions (Kennett, 1984), but for three dimensional surface wave scattering no analytical solutions are available. In that case one either has to use numerical methods, or make some simplifying assumptions. The Born approximation has been used successfully for describing surface wave scattering in three dimensions (Snieder, 1986ab). A brief outline of this theory is presented in section 2. The Born approximation gives a linear relation between the scattered waves and the heterogeneity. This situation is closely analogous to the wave theories forming the basis of modern migration schemes in exploration geophysics (Clayton and Stolt, 1981; Tarantola, 1984ab; Bleistein et al., 1985; Bleistein and Gray, 1985; Ikelle et al., 1986).

It is therefore not surprising that an inversion scheme using scattered surface waves can be formulated along similar lines. In section 3 it is shown that this scheme can be derived using a least squares criterion, as in Tarantola (1984ab). Without making additional simplifying assumptions the resulting inversion scheme isn't very manageable. It is shown in section 4 how some simplifications result in a workable scheme for reconstructing an inhomogeneity using scattered surface waves. The resulting reconstruction method is similar to holographic techniques used in optics.

In order to check if the method works with real data, a field experiment was conducted on a tidal flat, where surface waves were reflected by a dam. The results for this inversion are presented in section 5. A field experiment, as presented here, is an ideal tool for testing the feasibility of abstract mathematical inversion schemes.

In this chapter the summation convention is used throughout for vector and tensor indices. The dot product which is used is defined by

$$[\mathbf{p}.\mathbf{q}] \equiv p_i{}^* q_i\,. \tag{1.1}$$

## 2. Linearized theory for surface wave scattering

The equation of motion combined with the equations for linear elasticity lead to the following expression for the displacement field in the frequency domain

$$L_{ij} u_j = F_i \tag{2.1}$$

where the differential operator L is defined by

$$L_{ij} = -\rho\omega^2\delta_{ij} - \partial_n c_{inmj} \partial_m \tag{2.2}$$

and $\mathbf{F}$ is the point force which excites the wavefield.

Now suppose that the elastic medium (i.e., the density and the elasticity tensor) can be decomposed as follows:

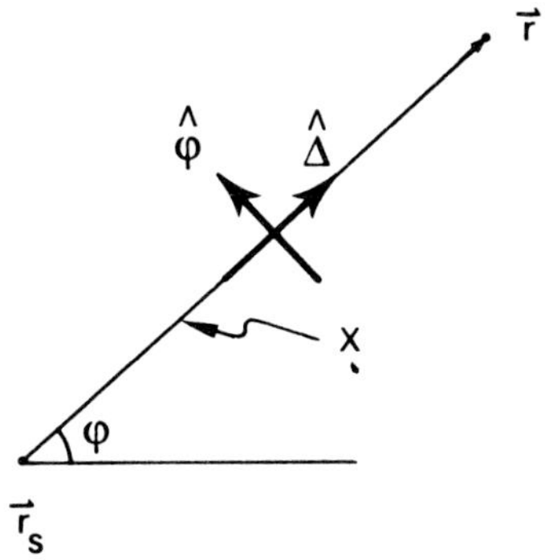


*Figure 1* Definition of the geometric variables for the direct wave.

$$\rho(x,y,z) = \rho^0(z) + \rho^1(x,y,z) \tag{2.3.a}$$

$$\underline{c}(x,y,z) = \underline{c}^0(z) + \underline{c}^1(x,y,z). \tag{2.3.b}$$

This means that the medium is viewed here as a laterally homogeneous reference medium, with heterogeneities superposed on it. This decomposition suggests the following decomposition of the displacement

$$\mathbf{u} = \mathbf{u}^0 + \mathbf{u}^1. \tag{2.4}$$

$\mathbf{u}^0$ is the displacement in the laterally homogeneous reference medium, this term is usually called the direct wave. $\mathbf{u}^1$ describes the effect of inhomogeneities, this term is usually labelled the scattered wave.

In order to derive expressions for $\mathbf{u}^0$ and $\mathbf{u}^1$ it is convenient to introduce the surface wave polarization vectors. For Love waves the polarization vector is

$$\mathbf{p}^\nu(z,\phi) = l_1^\nu(z)\,\hat{\phi} \tag{2.5.a}$$

and for Rayleigh waves

$$\mathbf{p}^\nu(z,\phi) = r_1^\nu(z)\,\hat{\Delta} + ir_2^\nu(z)\,\hat{\mathbf{z}}. \tag{2.5.b}$$

In this chapter Greek indices are used to label the surface wave modes. A summation over these indices indicates a summation over both Love waves and Rayleigh waves, thus treating both kinds of waves in an unified way. The unit vectors $\hat{\Delta}$, $\hat{\phi}$ and $\hat{\mathbf{z}}$ point respectively in the radial, transverse and down direction, see figure 1. The functions $l_1^\nu(z)$, $r_1^\nu(z)$ and $r_2^\nu(z)$ are the surface wave eigenfunctions as defined in Aki and Richards (1980). These eigenfunctions are assumed to be normalized according to:

$$8c_\nu U_\nu I_1^\nu = 1 \tag{2.6}$$

$U_\nu$ and $c_\nu$ are the group and phase velocity of the mode under consideration, and $I_1^\nu$ is the kinetic energy integral (Aki and Richards, 1980).

The far field surface wave Green's function of the laterally homogeneous reference medium can conveniently be expressed as a dyad of the polarization vectors. As shown in

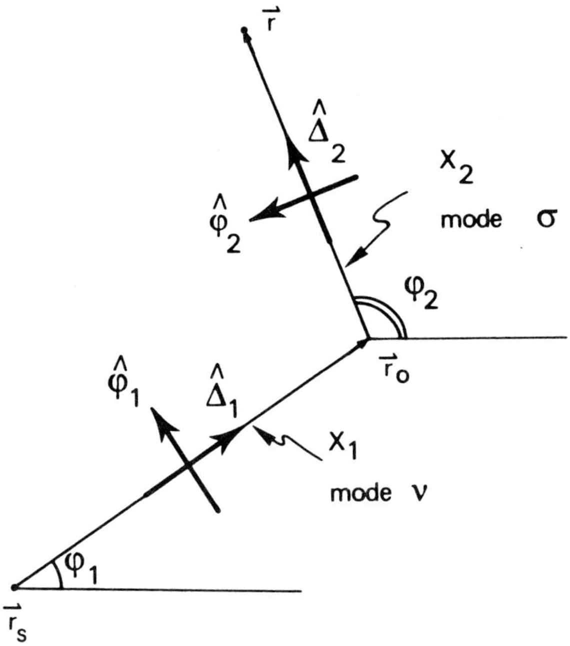


*Figure 2*. Definition of the geometric variables for the scattered wave.

Snieder (1986a), this leads to the following far field expression for the direct wave in the frequency domain

$$\mathbf{u}^0(\mathbf{r}) = \sum_\nu \mathbf{p}^\nu(z,\phi) \frac{\exp i\,(k_\nu X + \frac{\pi}{4})}{(\frac{\pi}{2} k_\nu X)^{1/2}} [\mathbf{p}^\nu(z_s,\phi).\mathbf{F}] \tag{2.7}$$

see figure 1 for the definition of the geometric variables. It is assumed here that the wavefield is excited by a point force $\mathbf{F}$ at location $\mathbf{r}_s$. Note that the direct wave is written as a superposition of modes (Love waves and Rayleigh waves), and that the modes don't interact with each other. Using the Born approximation, one can show that for sufficiently weak scatterers the scattered wave in the frequency domain is given by

$$\mathbf{u}^1(\mathbf{r}) = \sum_{\sigma,\nu} \iint \mathbf{p}^\sigma(z,\phi_2) \frac{\exp i\,(k_\sigma X_2 + \frac{\pi}{4})}{(\frac{\pi}{2} k_\sigma X_2)^{1/2}} V^{\sigma\nu}(x_0,y_0) \frac{\exp i\,(k_\nu X_1 + \frac{\pi}{4})}{(\frac{\pi}{2} k_\nu X_1)^{1/2}} [\mathbf{p}^\nu(z_s,\phi_1).\mathbf{F}]\, dx_0 dy_0 \tag{2.8}$$

see figure 2 for the definition of geometric variables. This expression is derived in Snieder (1986a) for buried scatterers in an isotropic medium. It is shown in Snieder (1986b) that scattering due to surface topography can also be described by (2.8). Reading (2.8) from right to left, one follows the "life history" of the scattered wave. At the source (in $\mathbf{r}_s$), mode

$\nu$ is excited by the projection of the point force $\mathbf{F}$ on the polarization vector $\mathbf{p}^\nu$. Then, a propagation to the scatterer occurs. This gives a phase shift and amplitude decay due to geometrical spreading, described by the term $\exp i(k_\nu X_1+\frac{\pi}{4})/(\frac{\pi}{2}k_\nu X_1)^{1/2}$ . At the scatterer (in $\mathbf{r}_0$), scattering and mode conversion occurs. This is described by the interaction terms $V^{\sigma\nu}$. This term gives the coupling between the incoming mode $\nu$, and the outgoing mode $\sigma$. After this, the mode $\sigma$ propagates to the receiver, which is shown by another propagator term. Finally, the oscillation at the receiver (in $\mathbf{r}$) is described by the polarization vector $\mathbf{p}^\sigma$. An integration over the scatterer, and a summation over all outgoing and incoming modes $(\sigma,\nu)$ superposes the different parts of the scattered waves $\mathbf{u}^1$.

The interaction terms $V^{\sigma\nu}$ are a linear function of the perturbations in the density $(\rho^1)$, the Lame parameters ($\lambda^1$ and $\mu^1$), and the surface topography $h$. It can explicitly be seen that in (2.8) a single scattering approximation is used, since the interaction terms appear only once. For buried inhomogeneities the interaction terms are given in Snieder (1986a), while the interaction terms due to surface topography are derived in Snieder (1986b). For example, the Love wave-Love wave interaction for buried heterogeneities is given by

$$V_{LL}^{\sigma\nu} = \int \left[ (l_1^\sigma l_1^\nu \rho^1 \omega^2 - (\partial_z l_1^\sigma)(\partial_z l_1^\nu)\mu^1) \cos\phi - k_\sigma k_\nu l_1^\sigma l_1^\nu \mu^1 \cos 2\phi \right] dz. \qquad (2.9)$$

In this expression $\phi=\phi_2-\phi_1$ is the scattering angle, and $k_\nu$ is the wavenumber of mode $\nu$. The interaction terms are a very simple function of the scattering angle $\phi$.

At this point we can already conclude that in inversions using scattered surface waves, we can only obtain information of the scatterers through the interaction terms $V^{\sigma\nu}$. Information at different frequencies (and possibly different modes) is needed to obtain the depth dependence of the inhomogeneities. The dependence of $V^{\sigma\nu}$ on the scattering angle can in principle be used to unravel the contributions from the density and the Lame parameters.

The theory is presented here for a point force excitation in a plane geometry. The excitation by a moment tensor is discussed in Snieder (1986a), and the formulation of this theory in a spherical geometry is shown in Snieder and Nolet (1987). In both cases only minor changes in the theory have to be made.

## 3. A formalism for surface wave holography

Scattered surface waves can be used to map the inhomogeneities in the Earth. The theory in the previous section is linear(ized), therefore least squares inversion techniques can conveniently be used for this. Least squares inversion for variables depending continuously on one or more space variables has been discussed in detail by Tarantola and Valette (1982). Suppose we want to find the following model vector

$$\mathbf{m}(\mathbf{r}) = \begin{pmatrix} \rho^1(\mathbf{r}) \\ \lambda^1(\mathbf{r}) \\ \mu^1(\mathbf{r}) \end{pmatrix} \qquad (3.1)$$

and suppose we describe the a-priori knowledge of the heterogeneity with the vector $\mathbf{m}_0(\mathbf{r})$.

Let the vector **u** denote all available data in the time domain. With "data" we mean here the difference between the recorded signals, and the synthetics produced by the a-priori model $\mathbf{m}_0(\mathbf{r})$. We shall assume here that the a-priori model is zero ($\mathbf{m}_0(\mathbf{r})=0$). This means that the data (**u**) consist of the difference between the recorded signals, and the synthetic seismograms of the laterally homogeneous reference medium.

The inversion scheme of Tarantola and Valette (1982) requires the a-priori covariances of the model ($C_m(\mathbf{r},\mathbf{r}')$), and of the data ($C_u$). If the a-priori cross covariances between the model and the data ($C_{um}$) are assumed to vanish, the least squares solution of the model is given by (Tarantola, 1984a):

$$\mathbf{m} = \mathbf{M}^{-1}\, \mathbf{C}_m \mathbf{G}^T\, \mathbf{C}_u^{-1}\, \mathbf{u} \tag{3.2}$$

with

$$\mathbf{M} = \mathbf{C}_m\, \mathbf{G}^T\, \mathbf{C}_u^{-1}\, \mathbf{G} + \mathbf{I} \tag{3.3}$$

and $G$ is the gradient of the data with respect to the model parameters.

In principle, (3.2) can be used to compute the model $\mathbf{m}(\mathbf{r})$ at every point in three dimensional space. In practice one shouldn't be too optimistic about a straightforward use of (3.2), since three different kinds of inversion are implied in (3.2):

[1] The surface wave energy should be focussed in the horizontal directions on the scatterers.

[2] The contribution of the three parameters $\rho^1$, $\mu^1$ and $\lambda^1$ should be unraveled.

[3] The depth dependence of these parameters should be reconstructed.

It shall be clear that with band limited, noisy data for a limited range of scattering angles, the goals [2] and [3] can never be fully reached. As a simplification it is therefore appropriate to expand the depth dependence of $\rho^1$, $\mu^1$ and $\lambda^1$ in a suitably chosen set of basis functions $\mathbf{b}_p(z)$. The subscripts p and q are used throughout this chapter to denote these basisfunctions. The basisfunctions are used to parameterize the depth dependence of the heterogeneity, and to separate the contributions from $\rho^1$, $\mu^1$ and $\lambda^1$. From now on, we assume that the inhomogeneity can be decomposed as follows

$$\mathbf{m}(\mathbf{r}) = \sum_p h_p(\mathbf{x})\, \mathbf{b}_p(z) \tag{3.4}$$

and the aim of the inversion is to reconstruct the fields $h_p(\mathbf{x})$. The vector **x** shall be used to denote the horizontal components of **r** ($\mathbf{x}\equiv\mathbf{r}-(\mathbf{r}.\hat{\mathbf{z}})\hat{\mathbf{z}}$) this convention will be followed throughout this chapter.

In order to obtain a workable formalism, more notation needs to be introduced. A superscript "rs" shall be used to denote the source receiver pair which is considered, thus $u^{rs}(t)$ is the time signal of the recorded scattered surface wave for source "s" and receiver "r". Furthermore, let the synthetic seismogram for source receiver pair "rs", basis function $\mathbf{b}_p(z)$ and a scatterer at location **x** be denoted by $s_p^{rs}(\mathbf{x},t)$. Since the theory is linear, this synthetic seismogram is precisely the contribution of source receiver pair "rs" to the gradient ($G$) of the data at location **x** and basisfunction p.

Now let us assume that the data are uncorrelated, but that the autocorrelation of different seismograms may be different

$$C_u^{rs,r's'}(t,t') = \delta_{r,r'}\,\delta_{s,s'}\,\delta(t-t')\sigma_{rs}^2 \tag{3.5}$$

Inserting this in (3.2-3), and working out the implied operator products yields

$$h_p(\mathbf{x}) = \sum_{p_1,p_2} \int d^2x_1 \int d^2x_2\, M_{pp_1}^{-1}(\mathbf{x},\mathbf{x}_1)\, C_{m,p_1p_2}(\mathbf{x}_1,\mathbf{x}_2)\, H_{p_2}(\mathbf{x}_2) \tag{3.6}$$

where

$$H_p(\mathbf{x}) = \sum_{rs} \frac{1}{\sigma_{rs}^2} \int s_p^{rs}(\mathbf{x},t)\, u^{rs}(t)\, dt \tag{3.7}$$

and

$$M_{pq}(\mathbf{x},\mathbf{x}') = \sum_{p_1} \int d^2x_1\, C_{m,pp_1}(\mathbf{x},\mathbf{x}_1) \sum_{rs} \frac{1}{\sigma_{rs}^2} \int s_{p_1}^{rs}(\mathbf{x}_1,t)\, s_q^{rs}(\mathbf{x}',t)\, dt + \delta_{pq}\,\delta(\mathbf{x}-\mathbf{x}'). \tag{3.8}$$

It can be seen from (3.6) that the inversion consist of three steps. The data ($u^{rs}(t)$) enter the inversion through the "holography term" $H_p(\mathbf{x})$. After this, an integration with the model covariance ($C_m$) is performed. Finally, a contraction with the inverse operator $M^{-1}$ completes the inversion. Now let us focus on the holography term (3.7).

This term can be interpreted most easily by converting (3.7) to a frequency integral using Parseval's theorem (Butkov, 1968). Inserting (2.8) for the synthetic seismogram $s_p^{rs}(\mathbf{x},\omega)$ we get

$$H_p(\mathbf{x}) = \frac{1}{2\pi} \sum_{rs} \int d\omega \sum_{\sigma,\nu} [\mathbf{u}^{rs}.\mathbf{p}^{\sigma}(z_r)] \frac{\exp i(k_\sigma X_2 + \frac{\pi}{4})}{(\frac{\pi}{2}k_\sigma X_2)^{1/2}} V_p^{\sigma\nu}(\mathbf{x}) \frac{\exp i(k_\nu X_1 + \frac{\pi}{4})}{(\frac{\pi}{2}k_\nu X_1)^{1/2}} [\mathbf{p}^{\nu}(z_s)\cdot\mathbf{F}] \tag{3.9}$$

It is understood that all quantities at the right hand side are evaluated in the frequency domain, and that the geometric variables are to be considered for each source receiver pair separately. The interaction terms $V_p^{\sigma\nu}(\mathbf{x})$ are for scattering (and conversion) by basis function $\mathbf{b}_p(z)$ at location $\mathbf{x}$. Equation (3.9) can be interpreted by considering the terms on the left and on the right of the interaction matrix. The term $\exp i(k_\nu X_1 + \frac{\pi}{4})/(\frac{\pi}{2}k_\nu X_1)^{1/2}$ $[\mathbf{p}^{\nu}(z_s).\mathbf{F}]$ describes the waves excited by the point force $\mathbf{F}$, which travel to the scatterer. In optics this term would be called "the illumination", since this term describes how much energy emanating from the source reaches the scatterer. The term $[\mathbf{u}^{rs}.\mathbf{p}^{\sigma}(z_r)] \exp i(k_\sigma X_2 + \frac{\pi}{4})/(\frac{\pi}{2}k_\sigma X_2)^{1/2}$ can be interpreted as the backpropagation of the data $\mathbf{u}^{rs}$, into the medium. This can most easily be understood by noting the symmetry in (3.9) in the excitation $\mathbf{F}$ and the data $\mathbf{u}^{rs}$. The holographic term (3.9) depends on the correlation between the illumination and the backpropagated signal. A summation over all source receiver pairs completes this term. This procedure is similar to holographic techniques in optics, where an image is reconstructed using the interference between the

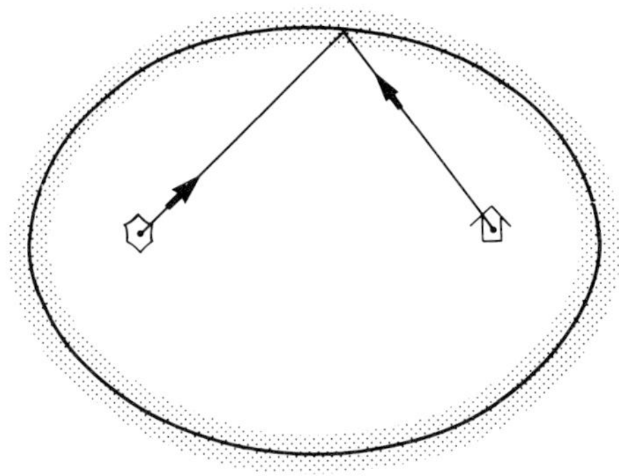

*Figure 3.* Ellipsoidal area over which the contribution of one source-receiver pair to the holographic term (3.7) is spread out in the absence of mode conversions.

illumination, and the light which has (back)propagated from the hologram to the area of reconstruction.

This holographic reconstruction procedure amounts to smearing out the recorded scattered energy over ellipses, or egg shaped curves in the medium. For instance, if mode conversions are absent, the recorded scattered wave for one source receiver pair is smeared out over an ellipse with the source and the receiver as focal points (figure 3). Using many different source receiver pairs, these ellipses are superposed to reconstruct the heterogeneity. Virtually all migration schemes used in exploration seismics use the same principle (either explicitly or implicitly). Insufficient data, or an inadequate reference model for the propagation leads to an imperfect reconstruction, producing the characteristic "smiles" in migrated seismic sections (Berkhout, 1984, chapter 5).

After applying the holographic operator in (3.6), an integration with the model covariance $C_m$ is to be applied. This covariance operator makes it possible to impose a-priori knowledge on the spatial scale of variation in the medium. The integration over $\mathbf{x}_2$ with this operator implies a smoothing of the holographic image. One should be careful not to apply too much smoothing. The reason for this is that the scattering effects are most sensitive to the horizontal gradients of the inhomogeneities. Smoothing scatterers over one wavelength of the surface waves eliminates virtually all scattering effects. Therefore it is crucial to allow sufficiently horizontal abrupt variations of the inhomogeneities.

The last step in the inversion (3.6) entails the inversion of the operator $M$ (3.8). After discretizing the model in cells, this inversion amounts to inverting a huge matrix. The matrix is in general very large, since the cell size should be much smaller than a wavelength. In order to do an inversion on a continental scale using surface waves with periods less than 100 seconds, several thousands of cells are required. A direct inversion of such a matrix is not feasible, but iterative techniques such as steepest descent, or conjugate gradients can be used for this, see chapter 1.2. Alternatively, one can complete the reconstruction (3.6) by making strongly restricting assumptions on the matrix $M$, which allows for a more convenient, but less accurate inversion of this matrix.

## 4. A simplified reconstruction procedure

In this section a simplified version of the reconstruction (3.6) is proposed. It is assumed that the heterogeneity can be described by one basisfunction $\mathbf{b}_p(z)$ and the subscript "p" is therefore dropped. Furthermore, it is assumed that the heterogeneity has a zero correlation length

$$C_m(\mathbf{x},\mathbf{x}') = \sigma_m^2 \, \delta(\mathbf{x}-\mathbf{x}') \tag{4.1}$$

and that all data have the same covariance $\sigma_u^2$. Lastly, and this is the most restricting assumption, we ignore the off-diagonal elements of the operator $M(\mathbf{x},\mathbf{x}')$. In this approximation

$$h(\mathbf{x}) = M^{-1}(\mathbf{x},\mathbf{x}) \, H(\mathbf{x}) \tag{4.2}$$

with

$$M(\mathbf{x},\mathbf{x}) = 1 + \frac{\sigma_m^2}{\sigma_u^2} \sum_{rs} \int s^{rs}(\mathbf{x},t)^2 \, dt. \tag{4.3}$$

Assuming the operator $M$ to be diagonal means that one assumes that for each point $\mathbf{x}$, all the scattered waves for all source receiver pairs are generated by a single scatterer at location $\mathbf{x}$. This assumption clearly breaks down when different scatterers collectively generate scattered waves for all source receiver pairs. In that case (4.2-3) cannot be expected to give results which are quantitatively correct. However, it is shown in section 5 that this simplifying assumption is able to produce qualitatively meaningful results. In fact, many migration schemes used in exploration geophysics implicitly use this assumption. (As an alternative, the system (3.2-3) could be solved iteratively, as shown in Tarantola (1984ab). In that case the substitution (4.3) specifies a preconditioning parameter for the iterative inversion (Tarantola, 1984c), and the final model is insensitive to the choice of this parameter. An explicit inversion of the operator M can then be avoided.)

Combining (3.7) and (4.2-3) the image reconstruction is in this approximation

$$h(\mathbf{x}) = \frac{\sum_{rs} \int s^{rs}(\mathbf{x},t) \, u^{rs}(t) \, dt}{\frac{\sigma_u^2}{\sigma_m^2} + \sum_{rs} \int s^{rs}(\mathbf{x},t)^2 \, dt} \tag{4.4}$$

The numerator is simply the holographic term. The denominator contains two terms. The autocorrelation of the synthetic seismograms in the denominator serves to normalize the reconstructed heterogeneity. The $\sigma_u^2/\sigma_m^2$ term serves to suppress the contaminating influence of noise.

It is shown in Snieder (1986ab) that the radiation pattern for surface wave scattering usually has one or more nodes. For one source receiver pair, near a node of the radiation pattern, the autocorrelation of the synthetic seismograms in the denominator approaches zero faster than crosscorrelation in the numerator. This might lead to a numerical instability. The regularization term $\sigma_u^2/\sigma_m^2$ damps this instability.

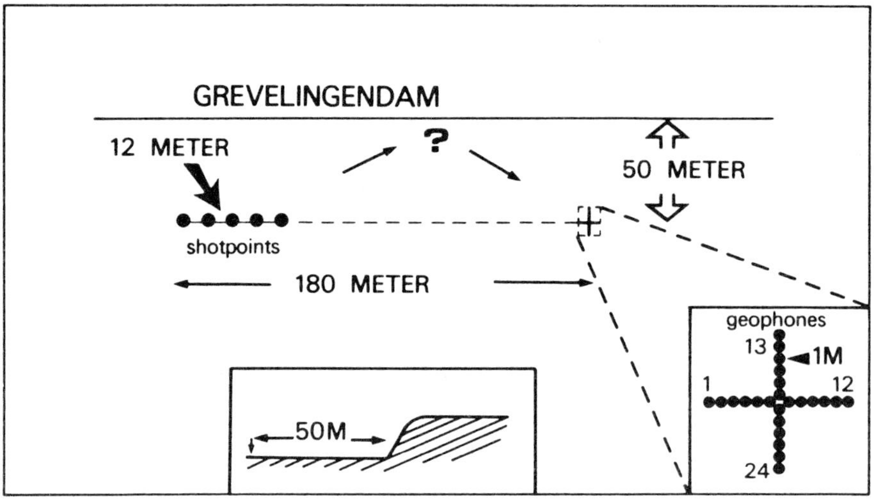


*Figure 4.* Layout of the field experiment.

## 5. A field experiment for image reconstruction with scattered surface waves

A field experiment was carried out in order the test the feasibility of surface wave holography. Surface wave measurements were done on a tidal flat in the Netherlands. A cross shaped array of 24 (10 Hz.) geophones was placed 50 meters from a concrete dam (the "Grevelingendam"). A weight drop source (of 30 kg.) was used to generate surface waves at several locations 50 meters from the dam, see figure 4. A description of the field equipment is given by Doornenbal and Helbig (1983). The reference model $(\rho_0(z), \mu_0(z), \lambda_0(z))$ used in the inversion was determined using standard surface wave dispersion analysis, using the fundamental Rayleigh modes and five higher modes (Gabriels et al., 1987).

An example of the geophone records for one shotpoint is shown in figure 5. Note the relatively strong higher mode signal before the arrival of the fundamental mode. It can be seen that the direct fundamental mode arrives simultaneously at the geophones on the transverse leg of the array (geophone 13-24), confirming that this wave propagates parallel to the dam. After this, the scattered fundamental mode arrives. On both the parallel (geophone 1-12), and the transverse (geophone 13-24) leg of the array this wave has a slanted lineup, indicating that this part of the signal comes from the direction of the dam. In this inversion the signal was muted until just after the arrival of the direct fundamental mode, so that only the scattered fundamental Rayleigh mode was used in the inversion.

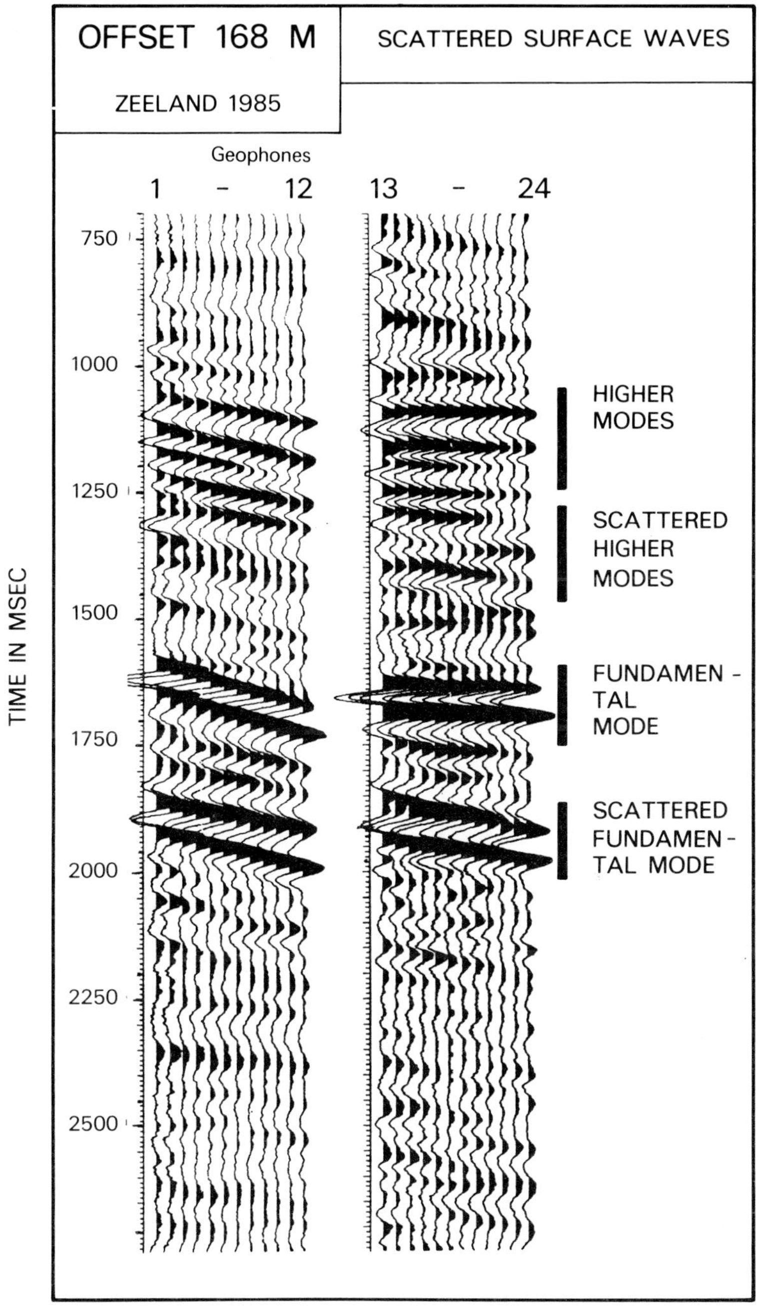


Figure 5. Field record for a shotpoint 168 meters from the geophone array.

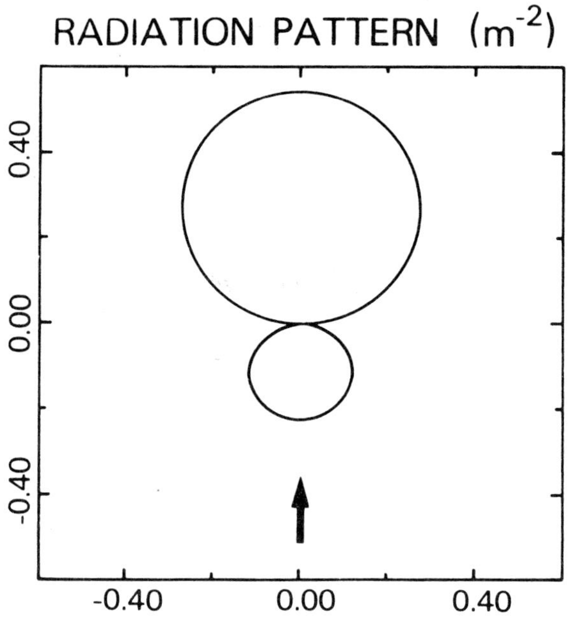


*Figure 6.* Radiation pattern for the basis function employed for the scattering of the fundamental Rayleigh mode to itself. The direction of the incoming wave is indicated by an arrow. The numbers indicate the scattering amplitude / $m^2$.

(The Love wave contribution to the data and the synthetic seismograms is zero, because a vertical force excites only Rayleigh waves, and vertical component geophones don't register Love waves.)

The sediments composing the tidal flat have a shear wave velocity of 100-300 m/sec (depending on depth), and a density of approximately 1500 $kg/m^3$. In the dam, shear wave velocities of several kilometers per second are possible, and the density can be as large as 2500 $kg/m^3$. It will be clear that the dam cannot be considered a "small perturbation", so that we cannot expect to obtain quantitatively correct information. However, the geometry of the scatterer isn't favourable to multiple scattering, which explains why this linear reconstruction technique can be employed.

As a basis function, a constant relative shear wave velocity perturbation of 500%, and a constant relative density perturbation of 25% was assumed down to a depth of 12 meters. The radiation pattern for fundamental mode Rayleigh wave scattering is shown in figure 6. Note that the radiation pattern has a node for a scattering angle of approximately 90 degrees.

The image reconstruction was performed with a straightforward numerical implementation of (4.4). The synthetic seismograms $s^{rs}(\mathbf{x},t)$ were computed in the frequency domain using (2.8), and then Fourier transformed. Imaging experiments were

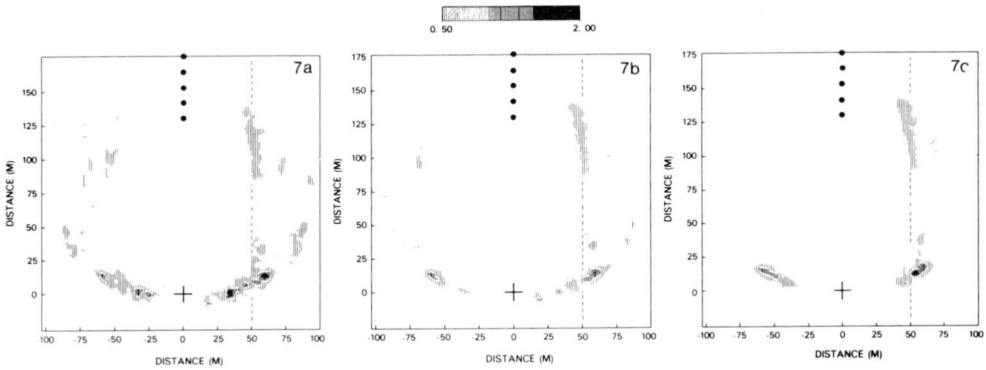


*Figure 7a*. Envelope of the reconstructed image $h(\mathbf{x})$ in the undamped case ($\sigma_M=0$), using only 4 geophones of the array. The true location of the edge of the dam is shown by the vertical dashed line. The shotpoints and the geophone array are marked with dots and a cross.
*Figure 7b*. As figure 7a, using only 8 geophones of the array.
*Figure 7c*. As figure 7c, using all 24 geophones of the array.

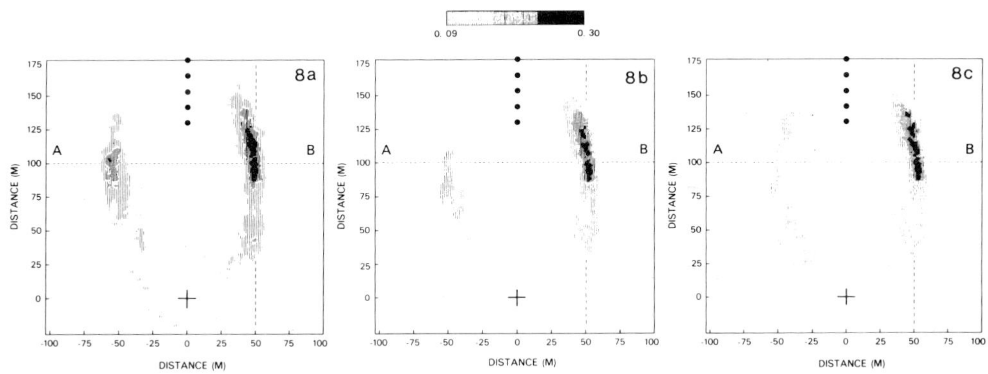


*Figure 8a*. Envelope of the reconstructed image $h(\mathbf{x})$ in the damped case ($\sigma_M\neq0$), using only 4 geophones of the array. The true location of the edge of the dam is shown by the vertical dashed line. The shotpoints and the geophone array are marked with dots and a cross.
*Figure 8b*. As figure 8a, using only 8 geophones of the array.
*Figure 8c*. As figure 8c, using all 24 geophones of the array.

performed for geophone spacings of 6 meters (using 4 geophones), 3 meters (using 8 geophones) and 1 meter (using all geophones). (The dominant wavelength of the fundamental Rayleigh mode is 6 m.) In all cases five shotpoints were used in the inversion. The reconstructed inhomogeneity is a highly oscillatory function of the space variables, since the reconstructed inhomogeneity $h(\mathbf{x})$ consists of the temporal correlation of two dispersed wavetrains. In the results presented here, the envelope of the function $h(\mathbf{x})$ is

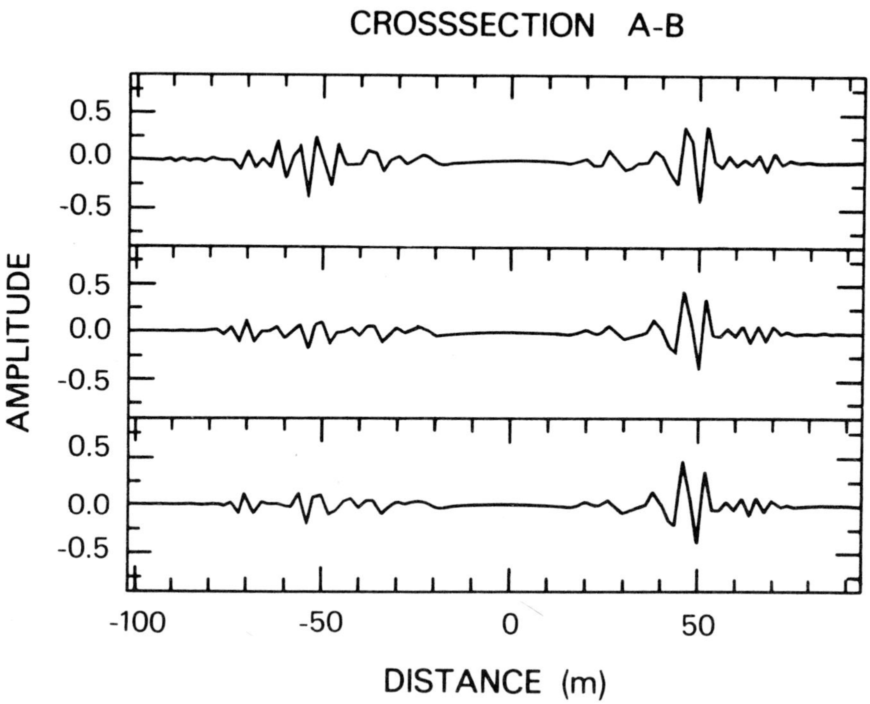


*Figure 9*. Cross sections of the reconstructed image $h(\mathbf{x})$ along the line AB for the solution in figure 8a (top panel), figure 8b (middle panel) and figure 8c (bottom panel).

therefore shown. In the figures 7a,b,c the reconstructed image is shown in the undamped case ($\sigma_u=0$) for different geophone spacings. The dam is not reconstructed very well, and the reconstructed heterogeneity is dominated by a sickle shaped body near the geophone array. This is caused by the fact that for the basis function employed here, the radiation pattern has a node near 90 degrees, see figure 6. Therefore, all the points near the circle with the source and the receiver as antipodal points produce a scattered wave $s^{rs}(\mathbf{x},t)$ with very small amplitude. Since the denominator in (4.4) goes faster to zero with $s^{rs}(\mathbf{x},t)$ than the numerator, this leads to an unrealistic inhomogeneity where these circles for different source receiver pairs overlap. This happens close to the geophone array. Taking more geophones into account gives some improvement, but the result isn't very good.

If the damping is nonzero ($\sigma_u \neq 0$), the results are considerably better, as can be seen in the figures 8abc. The sickle-shaped "ghost heterogeneity" has disappeared, and in all cases a clear image of the dam is visible at the correct location. In all cases a mirror image of the dam (at the left side of the shotpoint-geophone line) is visible, but if more source receiver pairs are taken into account this mirror image weakens. The reason for this is that only the geophones on the transverse leg of the array contribute to a determination between "left and right" for the incoming waves. Taking more geophones into account leads to a better determination of the direction of the incoming wave.

Note that with a geophone spacing comparable to the dominant wavelength (as in figure 8a), the inhomogeneity can still be reconstructed. This is fortunate, because in global seismology the station density is usually so small that the stations are more than a wavelength apart. Apparently, spatial aliasing effects don't affect the reconstruction strongly.

Cross sections of the field $h(\mathbf{x})$ along the line AB in figures 8abc are shown in figure 9 for the three geophone spacings employed. Note the oscillatory character of the reconstructed image, which is a by-product of the correlation technique used here. The image of the dam can clearly be seen at 50 meters. The mirror image of the dam is also visible, but it can be seen that using more geophones leads to a weakening of this mirror image. Unfortunately, it is not possible to determine the sign of the heterogeneity from figure 9. In reality, the inhomogeneity is certainly positive because both the shear wave velocity and the density are much higher in the dam than in the tidal flat. Due to the oscillatory character of the reconstructed image this cannot be determined from figure 9. This experiment has shown the feasibility of locating lateral heterogeneities in the Earth using scattered surface waves. Application of this technique to seismological data recorded with the NARS array (Dost et al., 1984) is currently in progress.

*Acknowledgements.* I am much indebted to Guust Nolet, both for proposing the field experiment as well as for his continuous interest and advice. Wout Brouwer helped developing the software for the data analysis. K. Helbig and Johan Tempels from the Department of Exploration Geophysics of the University of Utrecht kindly lent us their field equipment and provided technical assistance.

# Chapter 15

# Tomographic imaging of seismic sources

*Larry J. Ruff*

One area of active research in seismology is the determination and interpretation of spatial variations in the earthquake rupture process. The equations of elastodynamics connect the spatial-temporal distribution of fault slip to the radiated seismic waves. The observed waveforms are generated by the space-time history of fault slip (the moment rate density function) via linear integral equations. Source time functions can be deconvolved from far-field P waves; these observed source time functions are the Radon transform of the moment rate density function. Unfortunately, the moment rate density function cannot be recovered by direct application of the inverse Radon transform to the observations. Far-field source time functions sample only a small range of the transform variable, rather than the range of $-\infty$ to $+\infty$ as required by the inverse Radon transform. Zero-frequency quantities of the moment rate density function, which are proportional to the fault displacements, are consequently not resolved by the observations. To invert the source time functions, the inverse Radon kernel must be modified to properly treat information near zero-frequency and to prevent incoherent high-frequency noise from appearing in the model images. A series expansion provides the theoretical framework to construct the optimal inverse Radon kernel. An example that uses a synthetic data set demonstrates various aspects of the inverse Radon kernel and seismic source tomographic imaging.

## 1. Introduction

Tomographic imaging is typically regarded as the determination of some physical property that varies in two or three spatial dimensions. Seismic source tomographic imaging determines a function that varies with time in addition to spatial variations. Although seismic source and structural tomographic imaging involve different physical quantities, in



*G. Nolet (ed.), Seismic Tomography, 339–366.*

both cases the data are projections of a model function and the inverse problems share many of the same difficulties. Since an extensive literature already exists on structural tomographic imaging, I will focus on those aspects that are particularly relevant to seismic source imaging; this includes a description of the general circumstances for which the seismic source variations are related by the Radon transform to the observed source time functions.

Seismic waves are generated by a source and propagate through the earth to be recorded by seismographs. There are some seismic phases and frequency bands for which the propagational aspects are rather stable and easily modeled. For example, long-period P waves and very long-period surface waves are routinely used to study the seismic source. For these phases, all the propagational effects as well as the seismographic system response are completely characterized by a Green's function for a "unit impulse" earthquake. In the following development, it is assumed that this Green's function is easily calculated and "almost" correct over a certain frequency band. With this assumption, the seismic source imaging problem is separated from the earth structure perturbation problems that occupy other contributions in this volume.

In many applications of tomographic imaging the "raypaths" along which the model projections occur are well-approximated by straight lines; the tomographic imaging problem can then be written as a Radon transform (see Deans, 1983). The inverse problem then consists of finding a suitable form of the inverse Radon transform. In this treatment of seismic source imaging, only the forms that reduce to a Radon transform between the model and data are developed. We will see that the Radon transform represents the first-order connection between spatial variations in faulting and the far-field seismic waves (see Figure 1). It is only recently that this connection has been used to invert data; this formalism will undoubtedly be employed in future studies.

In the following sections, the equations that lead to seismic source tomographic imaging will be schematically derived. Properties of the seismic source function are then discussed and various important integral quantities of the model function are derived. The inverse Radon transform is given an abbreviated treatment, and then two "problem areas" of seismic source imaging are considered: (1) the zero frequency problem, and (2) the high-frequency problem. These problems are remedied by appropriate choice of the inverse Radon kernel.

## 2. The forward problem

Seismic sources are disturbances within the solid or fluid earth with a time scale short enough to generate detectable seismic waves. Henceforth, our attention is restricted to earthquakes; the most common and largest internal transient sources. An earthquake possesses a finite spatial extent in addition to a finite temporal duration. The spatial length scale varies over many orders of magnitude from less than 0.1 km for small earthquakes to over 1000 km for some of the largest earthquakes (e.g. the 1960 Great Chilean earthquake, see Plafker and Savage, 1970) Faulting duration varies from less than a second to more than one minute. There are significant differences between earthquakes other than

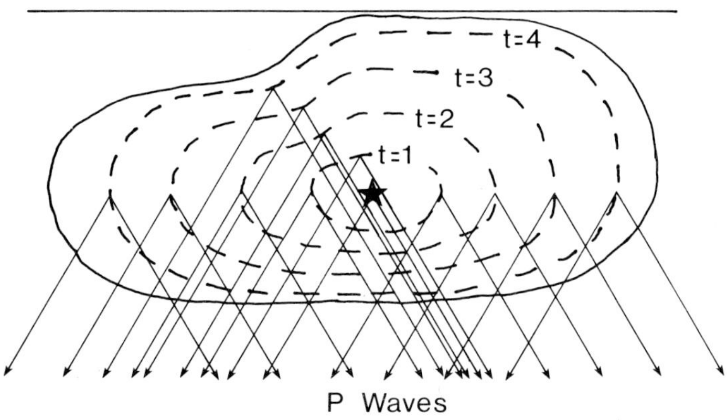


*Figure 1* Schematic diagram of seismic source imaging. The rupture starts at the hypocenter (star) and the propagating rupture front initiates fault slip as it sweeps across the fault area, bounded by the heavy solid line. The dashed contours show the rupture front location for increasing time (arbitrary time units). The solid lines are the raypaths for P waves radiated by fault slip at the rupture front. A few raypaths for two stations at opposite azimuths are shown. The P waves sample the fault slip history in both time and space.

variations in rupture length and duration. For example, two earthquakes might have approximately the same overall rupture duration and fault area, yet possess dramatically different time histories and spatial distributions of fault slip. The time dependence of the faulting history is directly displayed in P wave seismograms (see Figure 2). This variability in the temporal character of earthquakes presumably indicates variability in the spatial character as well.

**Table 1: Source parameters for earthquakes listed in figs. 2 and 3.**

| Region | Date | O.T. hr:mn:s | Epicenter Lat° | Fault Lon° | Area $L \times W$ $km^2$ | $M_w$ |
|---|---|---|---|---|---|---|
| Colombia | 12 Dec 79 | 7:59:04 | 1.6N | 79.3W | 200×100 | 8.2 |
| Japan | 16 May 68 | 0:48:57 | 40.9N | 143.4E | 250×80 | 8.2 |
| Kurile Is. | 11 Aug 69 | 21:27:41 | 43.6N | 147.2E | 200×80 | 8.2 |
| Peru | 17 Oct 66 | 21:41:57 | 10.7S | 78.6W | 100×120 | 8.2 |
| Philippines | 16 Aug 76 | 16:11:05 | 6.2N | 124.1E | 160×80 | 8.1 |
| Solomon Is. | 26 Jul 71 | 1:23:21 | 4.9S | 153.2E | 300×70 | 8.1 |
| Peru | 3 Oct 74 | 14:21:29 | 12.2S | 77.6W | 250×60 | 8.1 |

The point source approximation is one of the more useful expressions in seismology. The seismic source is reduced to a point location, though it may still have a complicated temporal history. The point source approximation is valid for a source dimension smaller than the wavelength of the seismic waves of interest. For example, a magnitude ~7 earthquake might rupture a fault area of $10 \times 10 km^2$, while very long period surface waves, say a period of T~250 s, have a wavelength on the order of ~1000 km. Hence a magnitude 7 earthquake is well-approximated by a point source for very long period surface wave

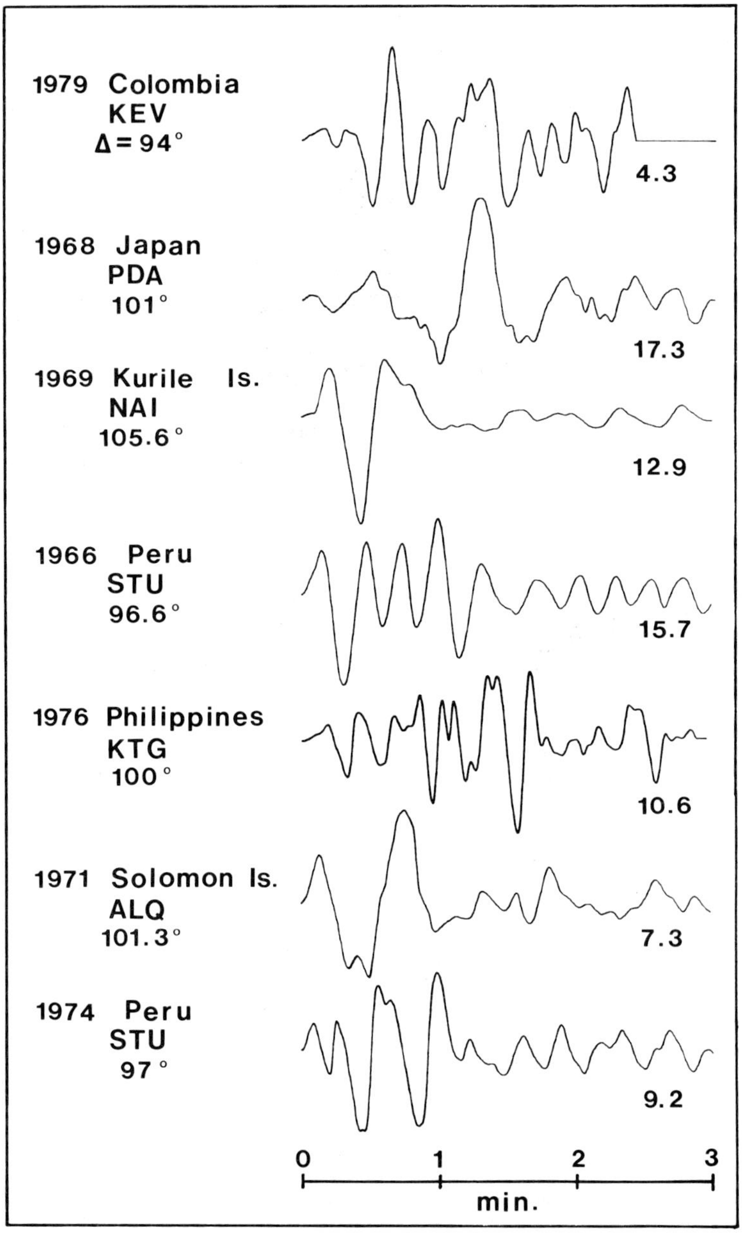


*Figure 2* Representative long-period WWSSN P wave seismograms from the seven largest underthrusting earthquakes, 1966 to present. See Table 1 for earthquake parameters. Station codes and epicentral distances are listed for each seismogram. The numbers to the right are the maximum amplitude ranges of the traces in cm, corrected to an instrument magnification of 1500. The time history of moment release shows great variability.

studies. On the other hand, P waves with a period of ~1 s would have a ~6 to ~10 km wavelength, thus the point source expression is not as useful for short period P wave seismograms. To interpret P waves at periods of a few seconds radiated by magnitude 8 earthquakes (these earthquakes have a fault dimension of 100 km or more), we need to use the theoretical framework for a finite fault. The most general equations for a finite earthquake source are of little use for interpreting seismograms; we must apply a sequence of approximations to obtain a more useful connection between the source and observations.

## 2.1 Dislocation sources and far-field seismic waves

The subsequent development of the equations for seismic waves radiated by earthquake sources uses a rather abbreviated and compact notation so that the fundamental relationships are readily seen. A comprehensive development of the equations for elastodynamics with seismic sources is given in Aki and Richards (1980); in particular their Chapters 2,3,4,7 and 14 give detailed expressions for the seismic source and various Green's functions. Aki and Richards derive a quite general expression for the elastodynamic displacement field that arises from a displacement discontinuity across an internal surface, R, equation (3.19) in Aki and Richards:

$$u_n(t) = \iint m_{pq}(t) * G_{np,q}(t) d\Sigma \tag{1}$$

where: * denotes convolution with respect to time, $t$; the indices refer to spatial components and range from 1 to 3 with the convention of summation over repeated indices; $u_n(t)$ is the $n^{th}$ component of displacement at the observation site; $m_{pq}(t)$ represents the moment density tensor; the Green's function is $G_{np}(t)$ which gives the $n^{th}$ displacement component for a point impulse force acting in the pth direction; and $G_{np,q}(t)$ is then the derivative of the Green's function with respect to the $q^{th}$ spatial direction. The seismic source is completely characterized by $m_{pq}(t)$ over the dislocation surface $\Sigma$, and in general each component of the moment density tensor varies across the dislocation surface. The Green's function varies across $\Sigma$ since the source-receiver distance and geometry change over $\Sigma$.

For a general displacement discontinuity, the moment density tensor is

$$m_{pq}(t,\Sigma) = c_{ijpq} v_j(\Sigma) D_i(t,\Sigma) \ , \tag{2}$$

where: $c_{ijpq}$ represents the elastic constants; $v_j(\Sigma)$ is the $j^{th}$ component of the unit vector normal to the dislocation surface; and $D_i(t,\Sigma)$ is the $i^{th}$ component of the displacement discontinuity. We will now place some restrictions on the moment density tensor. First, since most earthquakes are a shear dislocation source, the isotropic part of the moment density tensor is set to zero ($v_1 D_1 = 0$). This condition, combined with that for an isotropic elastic medium, reduce equation (2) to

$$m_{pq}(t,\Sigma) = \mu\left[ v_p(\Sigma) D_q(t,\Sigma) + v_q(\Sigma) D_p(t,\Sigma)\right] \tag{3}$$

where: $\mu$ is the shear modulus of the elastic medium. Note that $m_{pq}$ has units of <force><length>/<area>, or <moment>/<area>, hence the term moment density. We now place two further restrictions on the earthquake source: assume that the fault surface is planar and that the displacement direction is constant, i.e. $v_j$ does not vary across the fault surface, and $D_i(t,\Sigma) = w_i D(t,\Sigma)$ where $w_i$ is one component of the unit vector for displacement direction. In other words, the faulting geometry remains constant over time and space. It is convenient to place a local Cartesian coordinate system $(x,y)$ on the fault surface with the origin located at the hypocenter, the point of rupture initiation. The moment density tensor now becomes,

$$m_{pq}(t,x,y) = \mu(v_p w_q + v_q w_p) D(t,x,y) = \tilde{m}_{pq} \mu D(t,x,y) \tag{4}$$

where $\tilde{m}_{pq}$ is the "unit" non-dimensional moment tensor that specifies the faulting geometry. With these modifications to the source description, equation (1) becomes

$$u_n(t) = \tilde{m}_{pq} \mu \iint G_{np,q}(t,x,y) * D(t,x,y)\, dx\, dy \quad . \tag{5}$$

For further analysis, some properties of the Green's function must now be specified. Henceforth, our attention is restricted to far-field traveling waves; these conditions are appropriate for a wide class of seismic phases including body and surface waves. Under these circumstances, the principle variation in the Green's function is the travel time shift due to variations in epicentral distance and hypocentral depth. Although the amplitude of the Green's function changes with increasing distance, as does the waveshape, these changes are small compared to the temporal shift of the entire waveform as the travel time varies. This approximation is valid for relatively small perturbations to epicentral distance and relatively long periods: the far-field approximation. The limitations of this approximation for P waves generated by shallow finite faults become rather complicated, and these details will not be considered here (see Ruff (1987)). To apply the above stated approximation, write the Green's function as

$$G_{np}(t) = g_{np}\, h(t-T) \;, \tag{6}$$

where: $h(t)$ is the basic waveshape recorded for a particular source-receiver geometry, and $T$ is the travel time. Equation (5) requires that the derivative with respect to the $q^{th}$ direction be applied to the Green's function. With our far-field approximation, the spatial dependence of $g_{np}$ and $h(t)$ can be ignored: the spatial dependence of the travel time is the most important variation. Use of the chain rule and only keeping the dominant term gives

$$G_{np,q}(t) = -g_{np} T_{,q} \dot{h}(t-T) = g_{npq} \dot{h}(t-T) \tag{7}$$

where: $\dot{h}(t)$ is the derivative of $h$ with respect to the argument; and $g_{npq}$ is the modified form of the Green's function factor evaluated for a particular source-receiver geometry. Equation (5) can now be written as

$$u_n(t) = \tilde{m}_{pq}\, g_{npq} \iint \dot{h}(t-T) * \mu D(t,x,y)\, dx\, dy \quad . \tag{8}$$

One of the well-known properties of convolution allows the time derivative to be switched to the other function,

$$u_n(t) = \tilde{m}_{pq}\, g_{npq} \iint h(t-T) * \mu \dot{D}(t,x,y)\, dx\, dy \tag{9}$$

Completion of the fault integral requires some further specification of the functions that appear within the fault integral.

*2.1.1 Point source.* It is instructive to derive the point source approximation. To obtain this expression, assume that the observed seismograms record a low-pass filtered version of the ground displacements. The Green's function only needs to contain the longer periods and therefore will be essentially constant across a sufficiently small fault surface. The entire Green's function can then be taken outside the fault integral. In the context of our previous approximations, the point source approximation means that we ignore variations in travel time across the fault surface. Hence, $h(t-T)$ can be taken out of the fault integral,

$$u_n(t) = \tilde{m}_{pq}\, g_{npq}\, h(t-T) * \iint \mu \dot{D}(t,x,y)\, dx\, dy \tag{10}$$

By taking the entire Green's function out of the fault integral the displacements at all observation sites depend only on the integral of the displacement rate function,

$$\dot{M}(t) = \mu \iint \dot{D}(t,x,y)\,dx\,dy = \mu A\,\dot{\mathbf{D}}(t) \quad , \tag{11}$$

where the fault averaged displacement rate is $\dot{\mathbf{D}}(t)$, and $A$ is the fault area. With the point source assumption, the source description consists of a normalized "geometric" part ($\tilde{m}_{pq}$) and a scalar function ($\dot{M}(t)$) that gives the time history and size of the earthquake. The static seismic moment is:

$$M_0 = \int \dot{M}(t)\,dt = \mu A \int \dot{\mathbf{D}}(t)\,dt = \mu A D_0 \tag{12}$$

where $D_0$ is the fault averaged static displacement.

The point source expression has been extensively used by seismologists to interpret earthquake seismograms. Note that seismic waves are "generated" by the moment rate function rather than the moment function. With appropriate estimates for the point source Green's function ($g_{npq}\,h(t-T)$) and the fault geometry ($\tilde{m}_{pq}$) in equation (10), the moment rate function can be deconvolved from the observed seismograms at different stations (see Kikuchi and Kanamori (1982) and Ruff and Kanamori (1983) for examples). $\dot{M}(t)$ is commonly referred to as the source time function and is equivalent to $f(t)$ in our notation. Figure 3 displays the source time functions deconvolved from the seismograms in Figure 2. Since the obscuring effects of the Green's functions are removed, variations in the moment release between earthquakes are quite clear. The static seismic moment is recovered by the integral of any one source time function.

## 2.2 Finite fault source

The point source expression does not retain any information on the spatial variation of $\dot{D}(t,x,y)$ over the fault. This is a direct consequence of taking the entire Green's function out of the fault integral. As the fault dimension increases, or the wave period decreases, we are no longer able to make this approximation. In a more optimistic tone, the far-field seismic waves now contain information on the spatial variation of displacement rate across the finite fault.

To develop the finite fault expressions, we return to equation (9) and carefully consider the variations in travel time over the fault. Rewrite equation (9):

$$u_n(t) = \tilde{m}_{pq}\,g_{npq} \iint h(t-T(x,y)) * \mu \dot{D}(t,x,y)\,dx\,dy \quad , \tag{13}$$

where the $(x,y)$ dependence now appears in the travel time. The travel time is composed of two parts: the hypocentral time, $T_0$, evaluated at ($x$=0,$y$=0); and the variation in travel time, $dT(x,y)$, for $x$ and $y$ variations from the hypocentral values. Hence $h(t-T)$ becomes

$$h(t-T(x,y)) = h(t-T_0-dT(x,y)) = h(t-T_0) * \delta(t-dT(x,y)) \quad , \tag{14}$$

where $\delta(t)$ is the Dirac delta function. It is then possible to extract the Green's function, evaluated at the hypocentral location, and leave the travel time variation within the fault integral,

$$u_n(t) = \tilde{m}_{pq}\,g_{npq}\,h(t-T_0) * \iint \delta(t-dT(x,y)) * \mu \dot{D}(t,x,y)\,dx\,dy \quad . \tag{15}$$

A key assumption that we now use is that a good estimate of the faulting geometry is known. (Various long-period moment tensor inversion schemes produce reliable estimates,

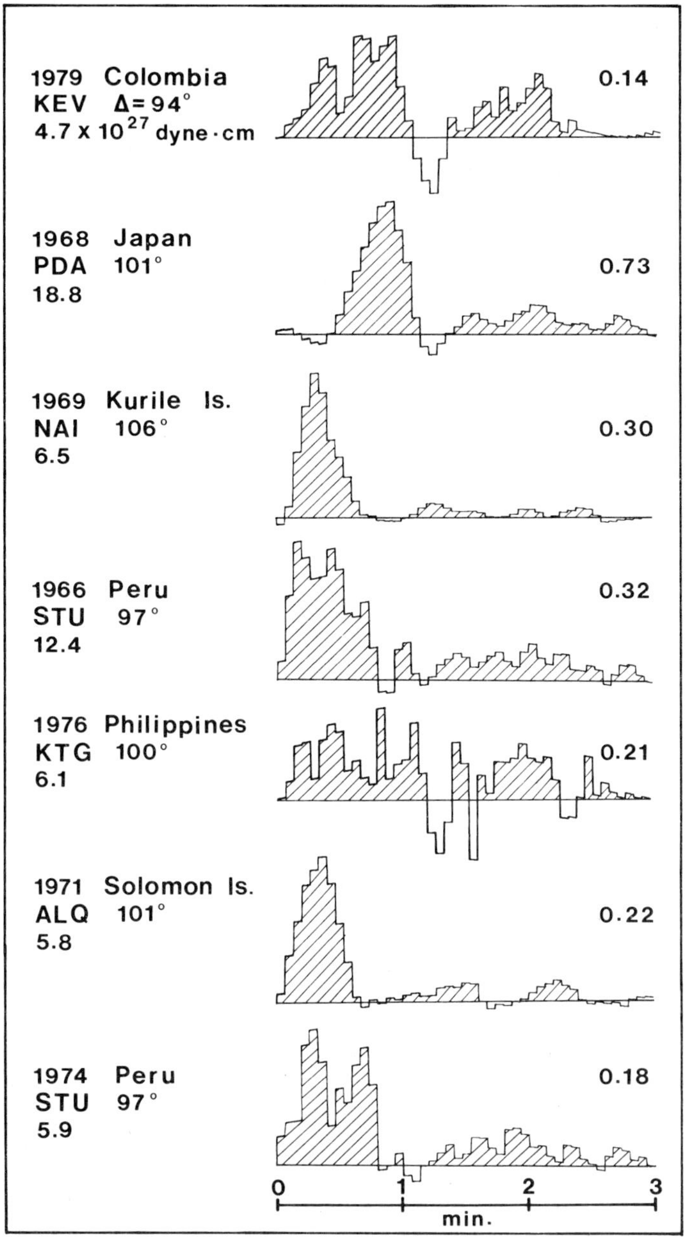


*Figure 3* Representative source time functions for the large underthrusting earthquakes, deconvolved from the seismograms in Figure 2. The single-station seismic moment estimates are listed on the left, the numbers to the right are the maximum ranges of the source time functions, in units of 1027 dyne-cm/s. The rupture character of these earthquakes varies from simple single-pulse events to complicated multiple event sequences.

see Dziewonski et al., 1981; Dziewonski and Woodhouse, 1983; and Kanamori and Given, 1981). Thus, $\tilde{m}_{pq}\, g_{npq}\, h\,(t-T_0)$ is known and can be replaced by $\mathbf{G}_n\,(t-T_0)$,

$$u_n\,(t) = \mathbf{G}_n\,(t-T_0)*\iint \delta(t-dT\,(x\,,y))*\,\mu\dot{D}\,(t\,,x\,,y)\,dx\;dy = \mathbf{G}_n\,(t-T_0)*\dot{M}\,(t) \qquad (16)$$

The fault integral is written as the moment rate function in the latter line to emphasize the parallel between equations (16) and (10). The deconvolved source time function, $f\,(t)$ , as prescribed by equation (16) is,

$$f\,(t) = \iint \delta(t-dT\,(x\,,y))*\,\mu\dot{D}\,(t\,,x\,,y)\,dx\;dy \qquad (17)$$

For a point source, the travel time variations $dT\,(x\,,y)$ are set to zero. For a finite fault, the travel time variations cause the source time functions to vary for different stations, i.e. different source-receiver geometries. The travel time variations must be written in terms of the source-receiver geometry. Since equation (16) is an approximation that is correct to first order, a consistent approximation is to use a first order expression for $dT\,(x\,,y)$. The first order terms of the Taylor series expansion for the travel time variations are,

$$dT\,(x\,,y) \approx (dT\,/dx\,)_0 x + (dT\,/dy\,)_0 y \quad , \qquad (18)$$

where the partial derivatives are evaluated at the hypocentral location. The derivatives depend on two factors: (1) the geometry of the seismic ray to the particular station relative to the fault coordinate direction, and (2) the wave slowness (inverse phase velocity) of the wave. If the seismic raypath to a station is perpendicular to a fault coordinate direction, the derivative is zero. The maximum absolute value of the derivative corresponds to a raypath along the coordinate direction: this maximum value equals the reciprocal of the seismic wave velocity. There is a strong upper bound on the absolute value of $(dT\,/dx\,)$ and $(dT\,/dy\,)$ since waves cannot propagate slower than the seismic wave velocity. Each station will in general have different values of the travel time derivatives, but they are easily calculated given the station-source epicentral distance and azimuth (see, e.g., Ruff (1987) for explicit formulas). These derivatives are referred to as the "directivity parameters", and we use the symbols $\Gamma$ and $\Lambda$ for the $x$ and $y$ travel time partial derivatives, respectively. The dependence of the source time functions on the source-receiver geometry and wave-type is now explicitly stated:

$$f\,(t\,,\Gamma,\Lambda) = \iint \delta(t-\Gamma x-\Lambda y\,)*\,\mu\dot{D}\,(t\,,x\,,y)\,dx\;dy \quad . \qquad (19)$$

The moment rate density function is defined as: $m\,(t\,,x\,,y)=\mu\dot{D}\,(t\,,x\,,y)$. Convolution with the delta function simply transfers the directivity time shifts to the argument of the moment rate density,

$$f\,(t\,,\Gamma,\Lambda) = \iint m\,(t-\Gamma x-\Lambda y\,,x\,,y)\,dx\;dy \quad . \qquad (20)$$

The observed source time functions are projections of the moment rate density function, where the projection slopes are given by $\Gamma$ and $\Lambda$ (see Chapman (1981) and chapter 2 for discussion of slant stack integrals in seismology). To complete this development, note that the limits of integration can be extended to $\pm\infty$ for both $x$ and $y$ since the moment rate density is zero beyond the faulted area. (Note: the integration limits for all following integrals are from $-\infty$ to $+\infty$ unless specified otherwise). Thus, equation (20) shows that

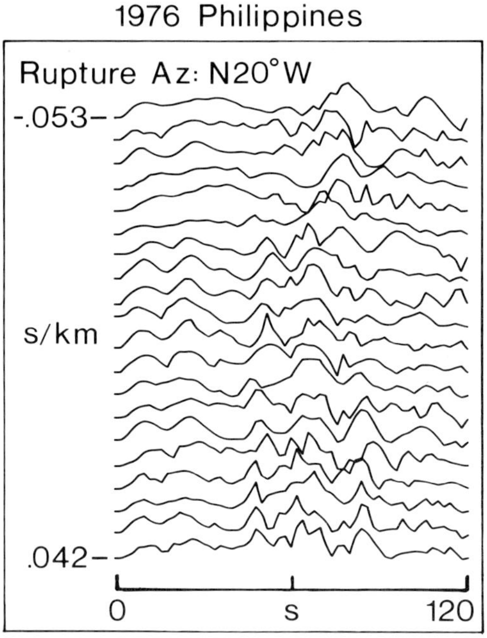


*Figure 4* Twenty source time functions for the 1976 Philippines earthquake. These time functions are deconvolved from WWSSN seismograms (see Beck & Ruff, 1985). The time functions are ordered top to bottom as a function of the horizontal directivity parameter:$\Gamma$. The rupture azimuth of N20°W is the best estimate from Beck & Ruff (1985). While certain features in the time functions are coherent, observed source time functions contain appreciable noise.

$f(t,\Gamma,\Lambda)$ is the three-dimensional Radon transform of $m(t,x,y)$. The ubiquitous Radon transform makes another appearance.

*2.2.1 Ribbon fault source: Two-dimensional Radon transform.* The rupture area of many earthquakes is elongated in one dimension. For example, large strike-slip earthquakes rupture vertical fault planes that may have a horizontal length of more than one hundred kilometers, with a fault width (depth) of only ten kilometers. These earthquakes can be regarded as a finite source only along one coordinate direction. It is also possible that the data set may be capable of resolving spatial variations along only one fault coordinate. This restricted faulting model is referred to as a ribbon fault (Ruff, 1983). The simplest development of the ribbon fault is to start with equation (19). Suppose that the $x$ coordinate coincides with the direction of fault elongation. Due to the limited range in $y$ (or equivalently, with $\Lambda$), the maximum directivity time shift due to the $(\Lambda y)$ term is much smaller than the maximum value of the $(\Gamma x)$ term. By ignoring the $(\Lambda y)$ term, the source time functions do not depend on $\Lambda$, hence equation (19) becomes,

$$f(t,\Gamma) = \iint \delta(t-\Gamma x) * \mu\dot{D}(t,x,y)\,dx\,dy \quad . \tag{21}$$

The integral with respect to y is simply the product of the average displacement rate and the fault width ($W$ ) as a function of $x$ , and this can be written as

$$f(t,\Gamma) = \int \delta(t-\Gamma x) * \mu W\dot{D}(t,x)\,dx \quad , \tag{22}$$

where: $\dot{D}(t,x)$ is the displacement rate, averaged with respect to the $y$ coordinate. The moment rate *line* density, as opposed to the *surface* density, is $m(t,x)=\mu W\dot{D}(t,x)$. With this definition, equation (22) becomes

$$f(t,\Gamma) = \int m(t-\Gamma x,x)\,dx \quad . \tag{23}$$

The observed source time functions are generated by a two-dimensional Radon transform of the moment rate (line) density function. All seismic source imaging applications in Ruff (1984), Schwartz and Ruff (1985), Schell and Ruff (1986), and Ruff (1987) are based on the ribbon fault formulation. Thus, both the two- and three-dimensional inverse Radon transforms are important. Most of the difficulties in seismic source imaging are present in both cases, hence our effort shall be concentrated on the two-dimensional case.

Let us review the two main conditions that lead to equations (20) and (23): (1) use of far-field traveling waves, and (2) use of first-order quantities in the travel time expansion. Perhaps the best perspective is to view this development as the first step toward "looking inside" the earthquake rupture process. While the point source expression represents the zero$^{th}$ order of resolution on faulting variations, retention of the directivity time shifts within the fault integral produces the first order resolution of the rupture process. Spatial variations of the moment release are now reflected by variations in the different source time functions (see Figure 4). This greatly increases the data processing requirements: many seismograms need to be collected to deconvolve many source time functions. To recover the seismic source process, "merely" invert the Radon transform, i.e. apply the inverse Radon transform to the observed source time functions. Unfortunately, this inversion is not so straightforward. Before proceeding to the inverse problem, it is important to discuss some properties of the moment rate density function.

## 2.3 Properties of the moment rate density function

Most seismologists accept a few basic concepts concerning the rupture process of earthquakes. One important concept is that a rupture front initiates fault slip. In particular, the dislocation nucleates at a certain point (the hypocenter) and a rupture front sweeps across the fault area with a velocity somewhat less than the elastic wave velocity. While the rupture front initiates the dislocation slip, the cessation of slip is influenced by many factors. It is possible that only a small fraction of total rupture area is slipping at any one time. We usually envision the rupture front to have a smooth and regular shape, such as a line or a circular arc, though it could have a highly irregular shape if the failure properties of the fault zone display strong spatial heterogeneity (see Mikumo and Miyatake (1978)).

Another important concept is that the fault slip at any one location does not reverse during the earthquake. The fault displacement function increases from zero to the final displacement value in an ever-increasing fashion. Since the moment rate density function depends on the displacement rate, $m(t,x,y)$ can only take on positive or zero values, i.e. it is a one-sided function. Early studies of body waves would generate synthetic seismograms using earthquake source time functions with simple geometric shapes, e.g. trapezoids, triangles, and various exponential functions with one or two adjustable parameters (see Langston and Helmberger (1975), Kanamori and Stewart (1976) for examples). All of these simple shapes are one-sided functions. Given an adequate source time function, the static seismic moment is simply the integral of the source time function. Of course, this is still true if the directivity time shifts are retained in the definition of the source time function. To demonstrate this, integrate equation (19) over all time for a fixed $\Gamma$ and $\Lambda$,

$$\int f(t,\Gamma,\Lambda)\,dt = \iiint \mu\dot{D}(t-\Gamma x-\Lambda y,x,y)\,dx\,dy\,dt \tag{24}$$

The integral over $t$ on the right-hand side of the equation yields $\mu D_0(x,y)$, the product of the shear modulus and the static displacement at fault location $(x,y)$. The integral of static displacement over the fault surface is then equal to the product of the shear modulus, the fault area, and the average static displacement: this is the seismic moment, $M_0 = \mu A D_0$. The seismic moment is obtained by integrating a single source time function, regardless of whether the interpretive framework is the point source or finite source expression.

Source time functions described by the ribbon fault formulation can also be integrated:

$$\int f(t,\Gamma)\,dt = \iint \mu W\dot{D}(t-\Gamma x,x)\,dx dt = \mu W \int D_0(x) dx \ , \tag{25}$$

where $D_0(x)$ is the static displacement as a function of $x$. The $D_0(x)$ function is of fundamental seismological and geological interest. For example, if a large earthquake occurs along a plate boundary, $D_0(x)$ represents the coseismic plate boundary slip. Static fault displacements can occasionally be measured by direct field observations for large strike-slip earthquakes, thus $D_0(x)$ is an important comparative function. It is also plays a key role in seismological models of fault plane heterogeneity, i.e. asperities and barriers. The above relations are now transformed into the frequency domain.

*2.3.1 Frequency domain representation.* Since it will later be convenient to discuss several aspects of the inverse problem in the frequency domain, I now define various Fourier transform pairs. Although all of these relations are well known to seismologists, it is still useful to introduce the basic notation and definitions. The Fourier transform pairs exist for $m(t,x,y)$ with respect to all three variables since the moment rate density function is time and space limited and the integral of $m(t,x,y)$ is finite. In addition, $m(t,x,y)$ is in reality a rather smooth function with respect to all variables due to physical constraints on fault slip. The Fourier transform of $m(t,x,y)$ with respect to time is then

$$\underline{m}(\omega,x,y) = \int m(t,x,y) e^{-i\omega t} dt \ , \tag{26}$$

and the inverse Fourier transform is given by,

$$m(t,x,y) = (1/2\pi)\int \underline{m}(\omega,x,y) e^{+i\omega t}\, d\omega \ , \tag{27}$$

where this integral can be interpreted as the principal value if necessary. For the ribbon fault model, the Fourier transform of $m(t,x)$ with respect to time is $\underline{m}(\omega,x)$.

The Fourier transform can also be applied to the spatial coordinates, for example

$$\overline{m}(t,k) = \int m(t,x)e^{-ikx}\,dx \quad , \tag{28}$$

and the inverse transform is

$$m(t,x) = (1/2\pi)\int \overline{m}(t,k)e^{+ikx}\,dk \quad . \tag{29}$$

The double Fourier transform of $m(t,x)$ is,

$$\overline{\underline{m}}(\omega,k) = \iint m(t,x)e^{-i\omega t}e^{-ikx}\,dtdx \quad . \tag{30}$$

In a similar fashion, we can define the transform of $m(t,x,y)$ with respect to y as $\hat{m}(t,x,l)$, where $l$ is the transform variable. The triple transform of $m(t,x,y)$ is $\hat{\overline{\underline{m}}}(x,k,l)$ and follows from the above definitions.

Various integrals of the moment rate density function were discussed in the previous section. These integrals correspond to the zero frequency components of the time transformed functions. In particular, note that with $\omega$=0 in equation (26): $\underline{m}(0,x,y)=\mu D_0(x,y)$ and similarly $\underline{m}(0,x) = \mu W D_0(x)$. Now let $\omega = 0$ in equation (30):

$$\overline{\underline{m}}(0,k) = \int \mu W D_0(x)e^{-ikx}\,dx \tag{31}$$

and $D_0(x)$ is then given by the inverse spatial transform,

$$D_0(x) = (1/2\pi\mu W)\int \overline{\underline{m}}(0,k)e^{ikx}\,dk \quad . \tag{32}$$

Thus, $D_0(x)$ is determined if $\overline{\underline{m}}(0,k)$ is known. Note that with $k$=0 in equation (31): $\overline{\underline{m}} = M_0$. The seismic moment is thus obtained from the value of the doubly transformed moment rate density function at the origin, $\omega = 0$ and $k = 0$.

The Fourier transform pair for a source time function also exists:

$$\underline{f}(\omega,\Gamma,\Lambda) = \int f(t,\Gamma,\Lambda)e^{-i\omega t}\,dt \tag{33a}$$

$$f(t,\Gamma,\Lambda) = (1/2\pi)\int \underline{f}(\omega,\Gamma,\Lambda)e^{i\omega t}\,d\omega \tag{33b}$$

and $\underline{f}(\omega,\Gamma)$ is defined in a similar manner. Note that if $\omega = 0$ in equation (33a), the Fourier transform again reduces to an integral over time (see equation 24), hence

$$\underline{f}(0,\Gamma,\Lambda) = M_0 \tag{34}$$

It is important to remember that the observed source time functions are deconvolved from seismograms and consequently are band-limited versions of the actual source time functions. Observed source time functions contain relatively more noise at the higher frequencies, partly due to the inadequacy of the assumed Green's functions at high frequencies. Although the zero-frequency value of the source time function is not recorded by the seismograph, it is possible to reconstruct the zero-frequency value by spectral extrapolation based on the assumption that source time functions are one-sided.

In conclusion, the integral of the moment rate density function over time is proportional to the static displacement along the fault. Furthermore, the integral of the moment rate density function over time and space is the static seismic moment, the fundamental seismological measure of earthquake size.

## 3. The inverse problem

The properties of the Radon and inverse Radon transform pair have been extensively described elsewhere (see Herman, 1980; Deans, 1983; Chapman, chapter 2). It is possible to simply write down the inverse Radon transform; source process imaging is one of the few geophysical inverse problems with an analytical inverse formula (see Chapman, 1978 and 1981, for other examples and discussion). However, application of the inverse Radon transform is beset with difficulties. Due to the properties of the source time functions, we cannot perform all of the steps as prescribed by the inverse Radon transform formula and this results in a rather poor estimate of the moment rate density function. The data set inadequacies can be divided into three specific problems: (i) limited range in the Radon transform variables, (ii) discrete sampling of the transform variables, (iii) each individual source time function is a noisy bandpass filtered version of the source time function. Most aspects of the latter two problems are treated in the tomography literature. The first problem is particularly severe for seismic source imaging. This problem is noted by Aki and Richards (1980). Menke (1985) points out that truncating the back-projection integral causes sidelobes in the point-spread function for the model reconstruction, and Ruff (1984) shows an example for a typical case of source imaging. While the above problems have been recognized, the zero-frequency problem in seismic imaging has not been specifically addressed before and will receive considerable attention in the following sections.

Deans (1983) and other earlier workers derive the connection between Radon and Fourier transforms. The basic form of the inverse Radon transform can be easily derived using this connection. Since the Fourier transform pairs for the source time functions $\underline{f}(\omega,\Gamma)$ and the model function $\underline{\overline{m}}(\omega,k)$ exist and were described in the previous section, this simple approach to the inverse Radon transform will be followed. Since I am merely recopying formulae that have received mathematical scrutiny in previous papers, I shall eliminate excessive terminology and condition appraisal. For example, I will routinely switch the order of integration. My focus will be on those aspects that cause particular problems for seismic source imaging. It is worth mentioning at the outset that the usual procedure for deriving the inverse Radon transform makes physically unrealizable assumptions with respect to observations of the seismic source. The main goal of this paper is to elucidate the properties of the optimal inverse transform for seismic source imaging, within the framework of the inverse Radon transform.

### 3.1 The two-dimensional inverse Radon transform

The prescription for the inverse Radon transform derivation is as follows: the first step is to take the Fourier transform of $f(t,\Gamma)$ with respect to time; then simply identify the resulting integral as a spatial Fourier transform; write down the inverse spatial Fourier transform; change the variable of integration; and finally take the inverse Fourier transform with respect to time. Although changing the variable of integration would appear to be an innocent step, this leads to many unfortunate properties of the inverse Radon transform.

The Fourier transform of equation (23) is,

$$\underline{f}(\omega,\Gamma) = \iint m(t-\Gamma x, x) e^{-i\omega t} dt dx = \int \underline{m}(\omega,x) e^{-i\omega\Gamma x} dx \quad , \tag{35}$$

where the exponential factor in the latter line results from the shift property of the Fourier transform. The integral in equation (35) now looks like the spatial Fourier transform of $\underline{m}(\omega,x)$ (see equation (28)) if we identify ($\omega\Gamma$) as k. Thus we have

$$\underline{f}(\omega,\Gamma) = \overline{\underline{m}}(\omega,k) = \overline{\underline{m}}(\omega,\omega\Gamma) \quad . \tag{36}$$

Equation (36) directly displays the connection between model and data; we shall return to this equation for a more careful analysis. However, let us now proceed in a purely formal manner. Since the inverse Fourier transform of $\overline{m}$ exists, we can immediately write down

$$\begin{aligned} \underline{m}(\omega,x) &= c\int \overline{\underline{m}}(\omega,k) e^{ikx} dk \\ &= c\int \overline{\underline{m}}(\omega,\omega\Gamma) e^{i\omega\Gamma x} d(\omega\Gamma) \\ &= c\int \underline{f}(\omega,\Gamma) e^{i\omega\Gamma x} d(\omega\Gamma) \end{aligned} \tag{37}$$

where equation (36) is used to substitute $\underline{f}$ for $\overline{\underline{m}}$ , and $c$ replaces $(1/2\pi)$ . Now change the integration variable from ($\omega\Gamma$) to $\Gamma$: $d(\omega\Gamma)=\omega d\Gamma$. The tricky part is changing the limits of integration. If $\omega>0$, then ($\omega\Gamma$) will run from $-\infty$ to $\infty$ as $\Gamma$ runs from $-\infty$ to $\infty$. (Note: This is not physically possible due to the bounds on maximum wave slowness). If $\omega<0$, then ($\omega\Gamma$) will run from $-\infty$ to $\infty$ *if* $\Gamma$ varies from $+\infty$ to $-\infty$. Now, if $\omega=0$, how must $\Gamma$ vary so that ($\omega\Gamma$) runs from $-\infty$ to $\infty$? This is too difficult a question to answer at this point, so let us proceed with this case excluded:

$$\begin{aligned} \omega>0: \quad & \underline{m}(\omega,x) = c\int \omega \underline{f}(\omega,\Gamma) e^{i\omega\Gamma x} d\Gamma \\ \omega<0: \quad & \underline{m}(\omega,x) = c\int (-\omega) \underline{f}(\omega,\Gamma) e^{i\omega\Gamma x} d\Gamma \end{aligned} \tag{38}$$

where the integration limits are switched for the latter case. Since ($\omega$) is positive for $\omega>0$ and ($-\omega$) is positive for $\omega<0$ , the two cases in equation (38) can be combined into a single equation by taking the absolute value of $\omega$ ,

$$\omega\neq 0: \quad \underline{m}(\omega,x) = c\int |\omega| \underline{f}(\omega,\Gamma) e^{i\omega\Gamma x} d\Gamma \tag{39}$$

To finish this derivation, we now take the inverse Fourier transform of equation (39), momentarily ignoring the fact that equation (39) is not defined for $\omega=0$,

$$\begin{aligned} (\omega\neq 0) \quad m(t,x) &= c^2 \iint |\omega| \underline{f}(\omega,\Gamma) e^{i\omega\Gamma x} d\Gamma e^{i\omega t} d\omega \\ &= c\int [c\int |\omega| \underline{f}(\omega,\Gamma) e^{i\omega(t+\Gamma x)} d\omega] d\Gamma \\ &= c\int I(t) * f(t+\Gamma x, \Gamma) d\Gamma \end{aligned} \tag{40}$$

where: the time shift and convolution properties of the inverse Fourier transform are used; and the inverse Radon kernel, $I(t)$ , is the inverse Fourier transform of $|\omega|$. Equation (40) is the "filter-and-back project" form of the inverse Radon transform. Back-projection refers to the action of the integral over $\Gamma$, note that it is a projection integral just as $f(t,\Gamma)$ is a projection of the model function.

Before we return to the zero-frequency problem, the same prescription is used to quickly derive the three-dimensional inverse Radon transform.

### 3.2 The three-dimensional inverse Radon transform

The Fourier transform of equation (20) is,

$$\begin{aligned} \underline{f}(\omega,\Gamma,\Lambda) &= \iiint m(t-\Gamma x-\Lambda x, x, y)e^{-i\omega t}\,dtdxdy \\ &= \iint \underline{m}(\omega,x,y)e^{-i\omega\Gamma x}e^{-i\omega\Gamma y}\,dxdy \end{aligned} \tag{41}$$

The double integral in equation (41) is then the double spatial Fourier transform of $\underline{m}$:

$$\begin{aligned} \underline{f}(\omega,\Gamma,\Lambda) &= \hat{\underline{m}}(\omega,k,l) \\ &= \hat{\underline{m}}(\omega,(\omega\Gamma),(\omega\Lambda)) \ . \end{aligned} \tag{42}$$

It is then possible to retrieve $\underline{m}(\omega,x,y)$ by application of the double inverse spatial Fourier transform,

$$\underline{m}(\omega,x,t) = c^2 \iint \underline{f}(\omega,\Gamma,\Lambda)e^{i\omega\Gamma x}e^{i\omega\Lambda y}\,d(\omega\Gamma)d(\omega\Lambda) \tag{43}$$

As before, change the integration variables from $(\omega\Gamma)$ and $(\omega\Lambda)$ to $\Gamma$ and $\Lambda$. These changes result in an integrand factor of $(|\omega|.|\omega|)$, which can be replaced with $\omega^2$:

$$\omega \neq 0: \quad \underline{m}(\omega,x,y) = c^2 \iint \omega^2 \underline{f}(\omega,\Gamma,\Lambda)e^{i\omega(\Gamma x+\Lambda y)}d\Gamma d\Lambda \ . \tag{44}$$

Application of the inverse temporal Fourier transform yields,

$$\omega \neq 0: \quad m(t,x,y) = c^2 \iint I_2(t) * f(t+\Gamma x+\Lambda y,\Gamma,\Lambda)d\Gamma d\Lambda \ .$$

$I_2(t)$ is the inverse Fourier transform of $\omega^2$, which can also be written as $(-\partial^2/\partial t^2)$.

The inverse Radon transforms for both the two and three-dimensional cases show the same basic form, namely convolution with the IRK (Inverse Radon Kernel) function and back-projection. The IRK is a rather irregular function (the inverse Fourier transform exists in the sense of the principal value), but has essentially the same properties for both the two and three-dimensional cases. The main properties of the IRK are that it is a purely real and even one-sided function in the frequency domain, and it is zero at zero-frequency and "blows up" for high frequencies; only the "strength" of the high frequency behavior varies for the two and three-dimensional cases. Given the similarity between the two and three-dimensional inverse Radon transforms, subsequent analysis only considers the two-dimensional case.

### 3.3 Zero-frequency

There is one serious problem with the inversion formula of equation (40): it is not defined for $\omega=0$. This problem is not serious for many tomographic applications as the main interest is in a sharp image rather than adjustments to the baseline level of the image. However, various zero-frequency quantities are important in seismic source imaging, e.g

$\underline{\overline{m}}(0,0)$ is the seismic moment and $\underline{m}(0,x)$ gives the static fault displacements. It is shown below that the zero-frequency problems curiously result from the redundancy of zero-frequency information in the observed source time functions. Correction of this effect then leads to the instabilities present at the high-frequencies.

Since equation (37) also displays the zero-frequency problem with $\omega$ appearing in the integration variable, we must return to equation (36) to explore the zero-frequency problem. The simplest procedure is to examine the doubly transformed function $\underline{\overline{m}}(\omega,k)$ in $(\omega,k)$ space and use integration paths. Note that $\underline{\overline{m}}(\omega,k)$ is a complex function with real and imaginary parts, this is implied in the following discussion and figure. Rewriting the basic definition of the inverse double Fourier transform,

$$m(t,x) = c^2 \iint \underline{\overline{m}}(\omega,k) e^{ikx} e^{i\omega t} dk d\omega = c \int \underline{m}(\omega,x) e^{i\omega t} d\omega \;, \tag{46}$$

we see that $m(t,x)$ is obtained by the double integral over the product of $\underline{\overline{m}}(\omega,k)$ with the weighting function $(e^{ikx} e^{i\omega t})$. This double integral is repeated many times for different combinations of $t$ and $x$ . The double integration is taken with respect to the Cartesian coordinates, $\omega$ and $k$ . Partial inversion of the double integral is obtained by integrating along certain paths in the $(\omega,k)$ plane. For example, $\underline{m}(\omega,x)$ is obtained by integrating along paths parallel to the $k$ axis (Figure 5).

Other coordinates can be used to integrate a function over the $(\omega,k)$ plane. Suppose that the integration path is a straight line that passes through the origin in the $(\omega,k)$ plane with some slope, $\Gamma$. The entire $(\omega,k)$ plane can be covered by changing the slope of the integration path. Although these radial integration paths can in general be described with cylindrical coordinates, this description is *not* appropriate for seismic source imaging due to the physical inhomogeneity of the $\omega$ and k variables. It is possible to keep $\omega$ as one integration variable, $k$ then depends on $\omega$ as follows: $k=\Gamma\omega$. Since all integration paths pass through the origin, the integrand is redundantly sampled around the origin as $\Gamma$ varies, in contrast to the Cartesian integration (i.e. eq. (46)) which evenly samples the integrand over the whole $(\omega,k)$ plane.

A different perspective is to view equation (46) as a Riemann volume integral where an individual contribution is the product of the integrand and the surface area element, $(d\omega \cdot dk)$. If the integration is reparameterized in terms of $\Gamma$, then the surface area element changes from $(d\omega \cdot dk)$ to $(d\omega \cdot \omega d\Gamma)$, which should be written as $(d\omega \cdot |\omega| d\Gamma)$ when considering the full range of $\omega$ . The $|\omega|$ factor accounts for the convergence of the $\Gamma$ integration paths toward the $(\omega,k)$ origin. In other words, $|\omega|$ downweights the contributions near the $(\omega,k)$ origin for each integration path such that the integral over all $\Gamma$ integration paths faithfully recovers $m(t,x)$. Thus there is a delicate balance between the $|\omega|$ factor and the integral over all $\Gamma$ paths. It certainly seems that this change in integration coordinates offers no advantages. However, we have no choice in the matter as this change is imposed by the fact that the observed source time functions are generated by the Radon transform of $m(t,x)$. With the substitution of $k=\omega\Gamma$ in equation (36), $\underline{m}(\omega,\omega\Gamma)$ can then be identified as $\underline{f}(\omega,\Gamma)$. Thus, a source time function corresponds to a particular $\Gamma$ integration path.

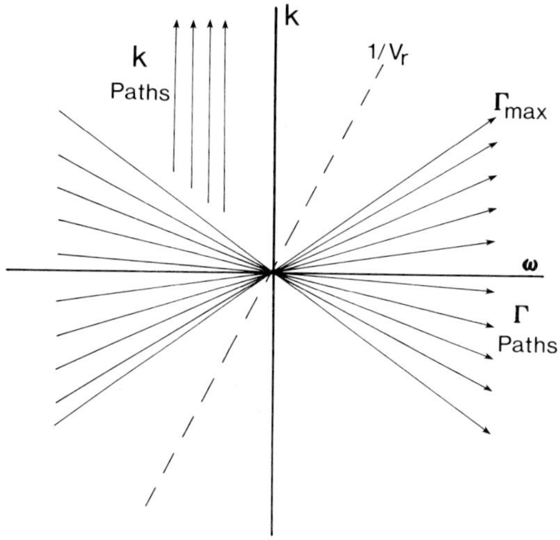


*Figure 5* Paths in the $(\omega,k)$ domain for the ribbon fault model. The moment rate density function, $m(t,x)$, is recovered by taking the double inverse Fourier transform with respect to $\omega$ and $k$. Integration along all paths parallel to the $k$-axis (as indicated in the upper-left quadrant) would recover $\underline{m}(\omega,x)$. A source time function corresponds to integration along a path that passes through the $(\omega,k)$ origin with slope $\Gamma$. Observed source time functions cover a fan-shaped protion of the $(\omega,k)$ plane between slopes $\Gamma_{max}$ and $-\Gamma_{max}$. If the moment release of an earthquake is concentrated along a rupture front, then a ridge of maximum amplitude trends across the $(\omega,k)$ plane with a slope of $1/v_r$ (dashed line).

To cover the entire $(\omega,k)$ plane, a $\Gamma$ integration path must coincide with the $k$ axis. Integration along this zero-frequency path, which corresponds to the limit $\Gamma=\infty$ or $\Gamma=-\infty$, produces $\underline{m}(0,x)$: the static fault displacement function. To recover $\underline{m}(0,x)$, the double limit, $\omega \rightarrow 0$ and $\Gamma \rightarrow \infty$, must be specified so that the product $(\omega\Gamma=k)$ varies from $-\infty$ to $+\infty$. These double limit difficulties arise because $\Gamma$ is the transform variable rather than the "angle" of the integration path; the latter choice is commonly used in structural tomography analysis and this avoids some of the apparent difficulties with zero-frequency and large aperture angles. For the seismic source imaging problem, the use of $\Gamma$ as the transform variable is due to the role of wave slowness and the inhomogeneity of time and spatial coordinates. In practice, the double limit at $\Gamma \rightarrow \infty$ is of no consequence since the integration path for $\Gamma=\infty$ is *not* physically realizable. The maximum value for $|\Gamma|$ is much less than 1 sec/km, with .05 sec/km a typical value for teleseismic P waves. Thus, while the formula for the inverse Radon transform requires that $\Gamma$ ranges from $-\infty$ to $+\infty$ to completely recover $m(t,x)$, in fact $\Gamma$ only ranges over -.05 to +.05 sec/km. Even if source time functions are determined for an infinite number of stations, only a "fan-shaped" portion of the $(\omega,k)$ plane is sampled (also see Menke, 1985). We must conclude that some aspects of the moment release function cannot be recovered. In particular, the static fault displacements cannot be directly determined since the equivalent integration path is not observed.

There is one physically realizable integration path in the $(\omega,k)$ plane that merits some discussion: the integration path for $\Gamma=0$. This path corresponds to $\overline{m}(t,0)$. A glance at equation (28) shows that: $\overline{m}(t,0)=\mu WL\dot{\mathbf{D}}(t)=\dot{M}(t)$, i.e. $\overline{m}(t,0)$ is the equivalent point source moment rate. As source time functions for non-zero $\Gamma$ are added to the data set, some information on the spatial distribution of displacement is acquired.

*3.3.1 Example: rupture fronts.* The double inverse Fourier transform for $m(t,x)$ requires knowledge of $\overline{\underline{m}}(\omega,k)$ over a large rectangular region, but we directly observe $\overline{\underline{m}}$ in only a fan-shaped region. Recovery of the zero-frequency function $(\overline{\underline{m}}(0,k))$ thus requires extrapolation in the $(\omega,k)$ plane. Unfortunately, this extrapolation is rather difficult.

If a rupture front plays a significant role in the moment release, $\overline{\underline{m}}(\omega,k)$ is characterized by a ridge of large spectral amplitude that passes through the $(\omega,k)$ origin and lies between the observable sector and the $k$ axis (see Figure 5). To see this for a particular example, let us specify a rupture model where the moment release occurs along a rupture line that trends across $(t,x)$ space with a rupture velocity of $v_r$,

$$m(t,x)=H(x)\delta(t-x/v_r) \tag{47}$$

where: $H(x)$ is a one-sided function that is proportional to the static displacement as a function of $x$ , and $H(x)=0$ for $x$ outside the faulted region. The double Fourier transform of $m(t,x)$ is then,

$$\overline{\underline{m}}(\omega,k)=\overline{H}(k)*\delta(k-\omega/v_r)=\overline{H}(k-\omega/v_r)\ . \tag{48}$$

Thus, a rupture line in $(t,x)$ space transforms into a rupture line in $(\omega,k)$ space; the slope of this rupture line is $(1/v_r)$ . Since the rupture velocity is less than the phase velocity of P or S waves, $(1/v_r)>\Gamma_{max}$. The rupture line lies in the unobserved sector of $(\omega,k)$ space between the $k$ axis and $\Gamma_{max}$ . Given that $\overline{H}(k')$ achieves maximum spectral amplitude at $k'=0$ , a ridge of maximum spectral amplitude will follow the rupture line. Extrapolation from $\Gamma_{max}$ to the $k$ axis must cross this ridge in spectral amplitude to some smaller value on the $k$ axis, a difficult proposition in general.

For the specific rupture model given above, it is possible to actually recover $\underline{m}(0,x)$ from information in the observable sector of the $(\omega,k)$ plane. A source time function observed at a station for directivity parameter $\Gamma$ is given by equation (36), and substitution from equation (48) gives

$$\underline{f}(\omega,\Gamma)=\overline{\underline{m}}(\omega,\omega\Gamma)=\overline{H}(\omega\Gamma-\omega/v_r)=\overline{H}(\omega(\Gamma-1/v_r))\ . \tag{49}$$

Thus, the spectrum for each source time function is a "stretched" version of $\overline{H}(k)$, with the stretch factor equal to $(\Gamma-1/v_r)$. Since $\Gamma$ is always less than $1/v_r$, increasing $\omega$ corresponds to decreasing $k$, though symmetry properties of $\overline{H}(k)$ can be used to switch this direction. Given one source time function, $H(x)$ can be recovered if $v_r$ is known. Comparison of source time functions at different $\Gamma$ allows a good estimate of $v_r$. Resolving the details of $H(x)$ depends on the total range of $k$ for which $\overline{H}(k)$ is known, or by equation (49) the total range in $\omega$. In other words, high spatial resolution of $\underline{m}(0,x)$ stems directly from the availability of high frequencies in the observed source time functions: e.g. for $\Gamma=0$, we have $|k|_{max}=|\omega|_{max}/v_r$. In practice, the spatial resolution is related to the finite time of slip duration at any point on the fault. If the rise time is $\tau$, instead of the zero duration implied by equation (47), then equation (49) will only be valid for a frequency range out to the corner frequency, which is approximately $(\pi/\tau)$. Thus, the spatial resolution for a single source time function is $|k|_{max} \approx \pi/v_r\tau$. With $v_r$=2 km/s and a $\tau$ of more than 3 seconds, then it is not possible to resolve spatial wavelengths smaller than about 10 km with $\underline{f}(\omega,0)$. Of course, the simultaneous use of many source time functions might improve the spatial resolution. Since the high-frequency information is crucial for spatial resolution, the best approach for data inversion is to let data coherency determine the high-frequency cut-off, rather than some arbitrary *a priori* notion (for an example of such a misguided, yet sincere, approach see Menke (1985)). The optimal inversion formula for seismic source imaging must properly treat the zero-frequency information and also maximize the inclusion of *coherent* high-frequencies to enhance image sharpness.

## 4. Optimal inverse Radon kernel for seismic source imaging

This section is concerned with the problem of model recovery based on the observed source time functions. The analytical prescription for the two-dimensional inverse Radon transform is given by equations (39) and (40) for the frequency and time domains (except for the case of $\omega$=0). The $\Gamma$ integral in equations (39) and (40) cannot be completed since we only have discrete observations of $f(t,\Gamma)$ within a certain range of $\Gamma$. To counter some detrimental effects of this limited sampling of the $\Gamma$ integral, the IRK function is modified in the vicinity of zero-frequency. Also, there will be a high-frequency limit to the IRK. The problems at zero-frequency and high-frequency are treated in separate sections, though the results will be combined into a single description of the inverse Radon kernel.

The restricted integration range of $\Gamma$ can be introduced into equation (39) by changing the integration limits from $\pm\infty$ to $\pm\Gamma_{max}$, or equivalently the integrand is modified by adding a window function factor. However, integration over a fan-shaped region is still not the correct specification since the source time functions are recorded at a finite number of stations, thus a sampling function must be applied to the $\Gamma$ integral:

$$[\sum \delta(\Gamma-\Gamma_j)\,\Delta(\Gamma)] \tag{50}$$

where: a total of $N$ source time functions are observed at discrete values of $\Gamma$, not necessarily equally spaced; and the integration interval weight is defined as $\Delta(\Gamma_j)=\Delta\Gamma_j=(\Gamma_{j+1}-\Gamma_{j-1})/2$ (other definitions are possible, but will not be pursued here). Inclusion of the sampling function changes equation (39) to,

$$\underline{m}_e(\omega,x) = c \int \underline{I}(\omega)\,[\sum \delta(\Gamma-\Gamma_j)\,\Delta(\Gamma)]\,\underline{f}(\omega,\Gamma)\,e^{i\omega\Gamma x}\,d\Gamma \tag{51}$$

where $\underline{m}_e$ is the model estimate, and $\underline{I}(\omega)$ is the IRK function. The $\Gamma$ integral in equation (51) is now easily solved,

$$\underline{m}_e(\omega,x) = c\,\underline{I}(\omega)\,[\sum (\Delta\Gamma_j \underline{f}(\omega,\Gamma_j) e^{i\omega\Gamma_j x})] \ . \tag{52}$$

Application of the inverse Fourier transform to equation (52) gives

$$m_e(t,x) = c \sum \Delta\Gamma_j I(t) * f(t+\Gamma_j x,\Gamma_j) \ . \tag{53}$$

Equation (53) is the sampled inverse transform for source imaging. We shall investigate the properties of $I(t)$ such that $m_e(t,x)$ is a "good" estimate.

### 4.1 IRK at zero-frequency

The IRK function in equation (39) is $|\omega|$. If we let $\underline{I}(0)=0$ in equation (52), the zero-frequency value for each "filtered" source time function is zero, and the sum over a finite number of source time functions still gives a model value of zero (i.e. $m_e(0,x)=0$). In other words, the balance between $|\omega|$ and the infinite range of $\Gamma$ is lost due to the limited range and discretization of $\Gamma$.

The exponential factor in equation (52) can be expanded in the vicinity of zero-frequency. Since $\Gamma_j$ and $x$ are bounded to some finite range (total fault length is $L$), the argument of the exponential function in equation (52) goes to zero as $\omega \to 0$. Thus, for sufficiently small $\omega$, equation (52) is,

$$\underline{m}_e(\omega \to 0,x) = c\,\underline{I}(\omega) \sum \Delta\Gamma_j\,\underline{f}(\omega,\Gamma_j)\,[1 + i\,\omega\Gamma_j x - \ldots]$$

$$\underline{m}_e(0,x) = c\,\underline{I}(0) \sum \Delta\Gamma_j\,\underline{f}(0,\Gamma_j) \tag{54}$$

There is no $x$ dependence in the right-hand side of equation (54). This behavior is expected since for $\omega \to 0$ the only model value sampled in the $(\omega,k)$ plane is at (0,0). Recall that $\overline{\underline{m}}(0,0)=M_0$, and this value is also given by the integral of $\underline{m}(0,x)$ over $x$. Thus, if equation (54) is integrated over the fault length,

$$M_0 = \int_L \underline{m}_e(0,x)\,dx = \int_L c\,\underline{I}(0) \sum \Delta\Gamma_j\,\underline{f}(0,\Gamma_j)\,dx\ . \tag{55}$$

Since the integrand on the right-hand side of (55) is constant with respect to $x$, the integral over $x$ simply gives the fault length, $L$ :

$$c\,\underline{I}(0)\,[\sum \Delta\Gamma_j\,\underline{f}(0,\Gamma_j)]\,L = M_0\ . \tag{56}$$

Note that for $f(0,\Gamma_j)=M_0$ (i.e. all source time functions are normalized to the same moment), and the total weighted range of $\Gamma$ is defined as $\Pi = \sum \Delta\Gamma_j$, then (56) is

$$\underline{I}(0)\,[c\,\Pi M_0 L\,] = M_0\ . \tag{57}$$

Since the factor in brackets is a non-zero number, the zero-frequency value of IRK is determined:

$$\underline{I}(0) = 1/(c\,\Pi L)\ . \tag{58}$$

Equation (58) shows that the "ideal" analytical form of the IRK, $\underline{I}(\omega)=|\omega|$, must be modified in the vicinity of zero-frequency (Figure 6). Use of the above formula ensures that the resultant model image has a net seismic moment of $M_0$. $\underline{I}(0)$ depends on the nature of the data set ($\Pi$) and the specification of the fault length ($L$).

As the frequency increases, $\underline{I}(\omega)$ should retain the constant value in equation (58) until $\omega > 1/(c\,\Pi L)$. This cross-over frequency corresponds to a period around 100 s when imaging large earthquakes. At shorter periods, $\underline{I}(\omega)$ should indeed increase with increasing frequency to compensate for the diverging radial integration paths.

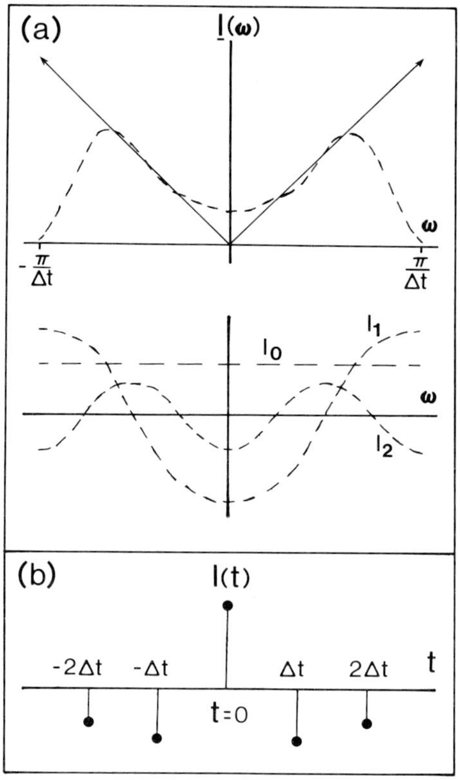


*Figure 6* The IRK (Inverse Radon Kernel) in frequency (a) and time (b) domains. The analytical form of the IRK, $\underline{I}(\omega)=|\omega|$ shown as the solid line in (a), must be modified near zero-frequency and at high-frequency. This modified IRK (e.g. the dashed line in top graph) can be constructed from a Fourier cosine series in the frequency domain (lower graph in (a)). This representation produces a discrete IRK in the time domain (b).

## 4.2 IRK at high-frequency

$\underline{I}(\omega)$ cannot increase as $|\omega|$ indefinitely. Since the source time functions are discretized, we eventually encounter the highest allowed frequencies. Ruff (1984) considered an IRK that increased as $|\omega|$ out to the Nyquist frequency (see Figure 6). Further experience with seismic source imaging shows that there are severe problems with this ultimate high-frequency cut-off: seismic noise can overwhelm the coherent signal in the source time functions before the Nyquist frequency is reached. Since these highest frequencies receive the largest weight from the IRK, the model images can be filled with large spurious high-frequency pulses. At what frequency do source time functions become incoherent with respect to fault plane moment release? This cut-off frequency depends on many factors and will vary for different earthquakes and data sets. Indeed, the cut-off must be considered as one of the key variables to be determined by the imaging process. In short, we cannot determine the high-frequency coherence limit unless we attempt to invert the data.

The IRK can be constructed in a stepwise manner, where the addition of new components is determined by data coherency. One approach is to start with low-frequencies and then admit progressively higher frequency components. The high-frequency contributions will eventually be incoherent and therefore will not significantly improve the match between the observed and synthetic source time functions. Several different algorithms can be devised to implement this approach, and in fact several of the iterative techniques that are used to solve linear systems can be viewed as the numerical application of the above strategy. Deans (1983) gives many references to these iterative methods. The iterative method employed by Ruff (1987) is one example that has been used in seismic source imaging. Rather than delve into specific numerical methods, I will present a series expansion for the IRK that serves as a useful theoretical framework for many different numerical applications.

### 4.3 Example of IRK expansion

One simple expansion for the IRK is a Fourier series in the frequency domain. This expansion makes explicit, and implicit, use of the discrete Fourier transform. Since $\underline{I}(\omega)$ is a purely real and even function, we can use a Fourier cosine series to represent $\underline{I}(\omega)$ over the frequency range of interest.

Let the sample interval of the source time function be $\Delta t$. Then we can write a Fourier series expansion for $\underline{I}(\omega)$ as,

$$-\pi/\Delta t \leq \omega \leq \pi/\Delta t: \quad \underline{I}(\omega) = I_0 + \sum I_n \cos(n\,\Delta t\,\omega) \quad . \tag{59}$$

The above series expansion in frequency domain can easily be transformed to a series in time domain. Application of the inverse Fourier transform to equation (59) gives,

$$I(t) = I_0 \delta(t) + \sum I_n/2\,[\delta(t+n\,\Delta t) + \delta(t-n\,\Delta t)] \quad . \tag{60}$$

Thus, the coefficients $(I_0, I_1, \ldots)$ are the weights attached to the time domain sampled IRK which is then convolved with the source time functions. Note that the weight of the central sample at $t$=0 is proportional to $I_0$, $I_1$ gives the weight of the two samples on either side of the central sample, $I_2$ yields the weight of the next two adjacent samples, and so on. Adding higher order cosine terms to the IRK in the frequency domain is equivalent to the symmetric broadening of the IRK in the time domain. A three-point IRK, such as the one used in Ruff (1984), corresponds to inclusion of $I_0$ and $I_1$, the "fundamental" mode. There is a major difference between our present formulation and that of Ruff (1984) even for the three-point IRK function; namely, Ruff (1984) used the condition that $\underline{I}(0)$=0, while we now realize that for an incomplete discrete sampling of $\overline{\underline{m}}(\omega,k)$ , $\underline{I}(0)$>0. Also, equations (59) and (60) provide the framework for customizing the IRK for different applications: the coefficients should be determined by data set properties rather than preconceived notions. In particular, high-frequency noise requires that we include more Fourier terms into $\underline{I}(\omega)$ to move the highest weight to a smaller frequency. For a specified application, it is possible to solve for the $I_n$ coefficients by the minimization of some norm, for example least-squares fit to data. However, the applications of a modified IRK that are shown in the next section

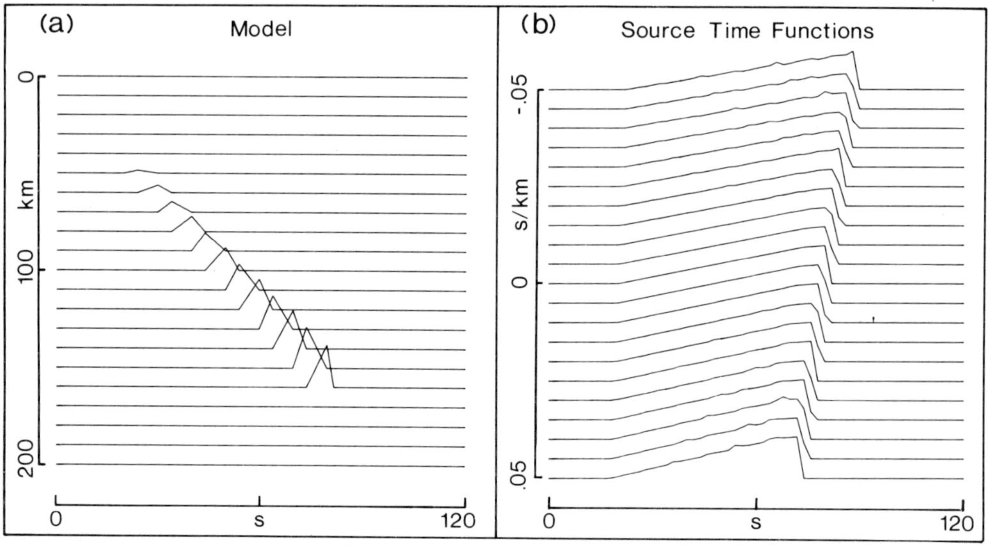


*Figure 7* Ribbon fault model simulation: Unilateral rupture front with sawtooth time history. The moment rate density function is displayed in (a). The rupture front velocity is 2 km/s. The moment rate density function is zero except along the rupture line between a time-space location of (20 s, 40 km) and (80 s, 160 km). The "observed" source time functions in (b) result from slant-stacking the model with values of Γ between -.05 and .05 s/km.

use the iterative method of Ruff(1987).

### 4.4 Examples of seismic source imaging

Although the main goal of this paper is to present the equations that lead to seismic source imaging and discuss some properties of the inverse transform, it is nonetheless interesting to look at a few simple examples of source imaging. In particular, some advantages of the optimal IRK over the three-point IRK of Ruff (1984) will become apparent.

We will use a synthetic data set of source time functions that are generated for a known model function. The ribbon fault formulation is used. The rupture model uses a unilateral constant-velocity (2 km/s) rupture front that produces a sawtooth-shaped time function. The model function is shown in Figure 7a, and the subsequent source time functions are shown in Figure 7b. The model image is composed of time traces, 120 s in duration, spaced at 10 km from the epicenter (0 km) to 200 km. The synthetic source time functions are generated by applying the slant-stack integral to this model image. Twenty-one source time functions are generated for evenly spaced values of Γ between +0.05 and -0.05 s/km. The time interval is 2 s for both the source time functions and the model images. The data set is almost noise-free: there is noise at periods close to the sampling interval. Two different specifications of the IRK have been used: (1) the three-point IRK of Ruff (1984), and (2) an optimal IRK based on the iterative method of Ruff (1987) where the zero-frequency and high-frequency behavior are dynamically determined.

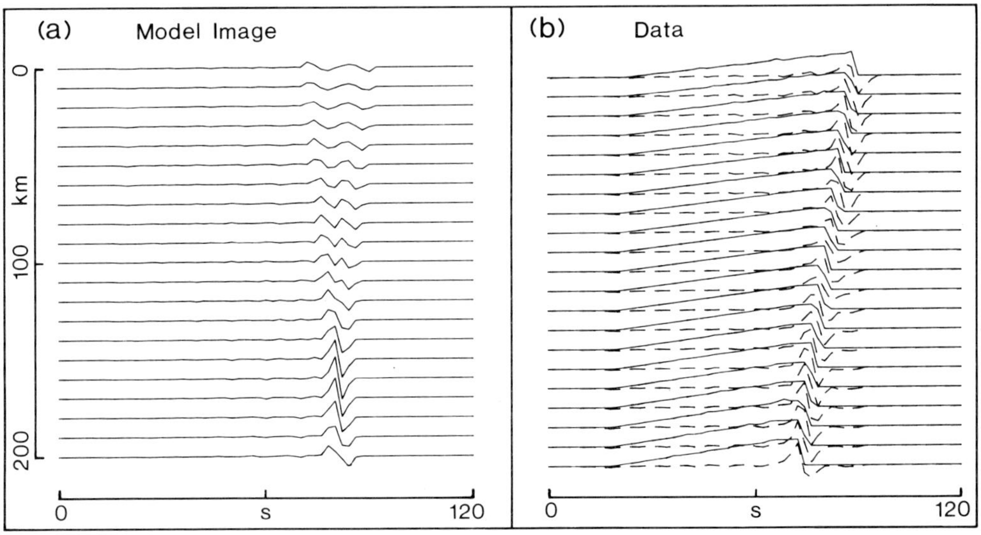


*Figure 8* Model image and synthetic source time functions for the three-point IRK. The three-point IRK is convolved with the "observed" functions in Figure 7b, and the results are then back-projected to produce the model image in (a). This model image is then slant-stacked to produce the synthetic time functions (dashed traces in (b)) which are plotted with the "observed" time functions (solid traces, same as in Figure 7b). Note that long-period components are completely lacking in the model image and synthetic time functions.

The three-point IRK is convolved with the source time functions, and these "filtered" time functions are then back-projected to produce the model image in Figure 8a. Compare this model image to the true model in Figure 7a: while the abrupt truncation of moment release at $x=160$ km is located to ±20 km, the long-period part of the moment release is not present in the IRK image. This fact is made more obvious by slant-stacking the model image in Figure 8a to produce a set of source time functions, plotted as the dashed traces in Figure 8b. The match to the "observed" source time functions is rather poor. Indeed, the synthetic time functions are proportional to the second-differences of the "observed" functions. Note that rupture initiation starts off as a ramp, and this feature is missed by the three-point IRK imaging. It is clearly evident from Figure 8 that three-point IRK imaging emphasizes jumps in the moment release.

A more optimal IRK can be found with the iterative method of Ruff (1987). Given the numerical implementation of this iterative method, the final IRK function is not explicitly known. The first iteration uses only the $I_0$ term from equation (60), i.e. the first model image is the simple back-projection of the observed time functions, scaled to minimize the error between observed and synthetic time functions. The next iteration then significantly broadens the time-domain form of the IRK as several new components are activated. After a few iterations, the optimal IRK has been found in terms of the primary criterion: best least-squares match between the synthetic and observed source time functions. The model images and time functions are shown in Figure 9 for the first and last iterations. Note the

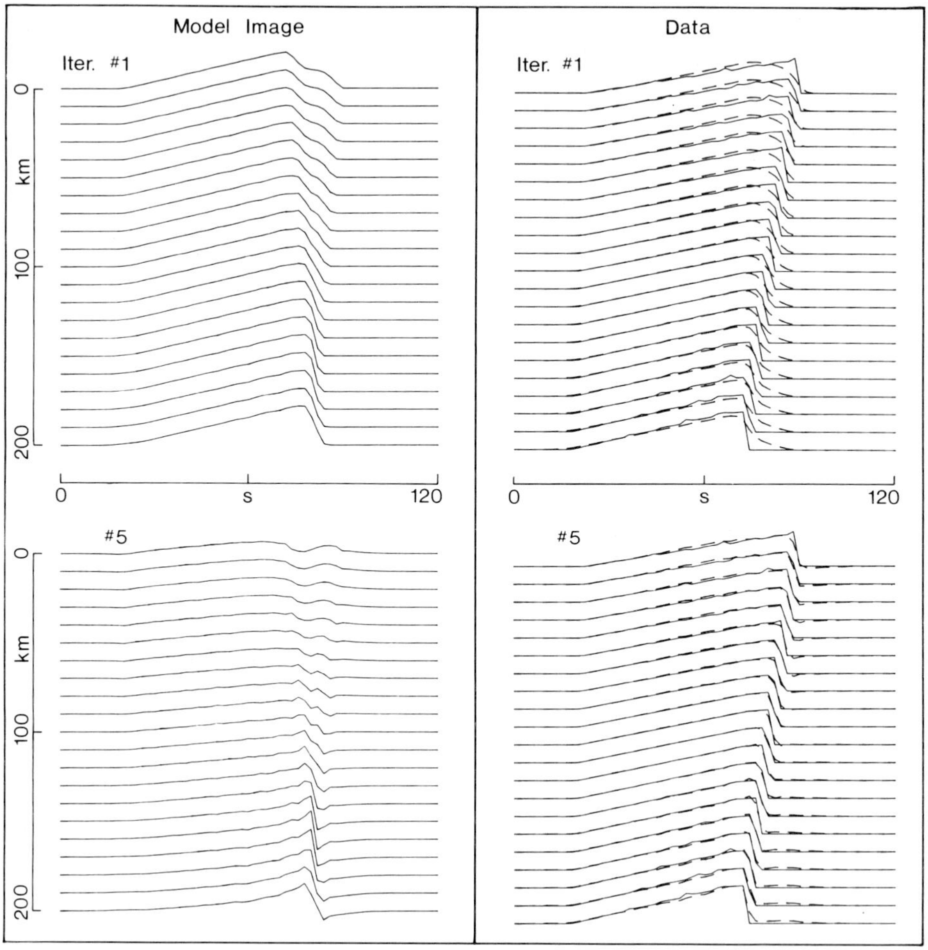


*Figure 9* Model image and synthetic source time functions for an "optimal" IRK. The iterative procedure of Ruff (1987) is used to generate progressively refined model images. The first and fifth iterations are shown at top and bottom. Simple back-projection of the "observed" time functions produces the model image for the first iteration: this is equivalent to a one-point IRK. The synthetic source time functions for this model image are plotted on the right (dashed traces) with the "observed" time functions. Successive iterations admit optimal combinations of the discrete components for the IRK. By the fifth iteration, the sawtooth truncation is well-fit in the time functions and this feature is localized in the model image. While the long-period components are present, they occur throughout the model image.

pronounced differences between the model image in Figure 8 and the first iteration in Figure 9. While the three-point IRK over-emphasizes the high-frequencies, the "one-point" IRK used for the first iteration in Figure 9 over-emphasizes the low-frequencies. The synthetics generated from the first iteration match the low-frequency characteristics of the data, however the rupture truncation is not yet sharp enough. After just five iterations, both the low- and high-frequency features of the observed time functions are well-matched.

Note that the high-frequency character of the final model image in Figure 9 bears some resemblance to Figure 8a. The optimal IRK is preferred to the three-point IRK even at the highest frequencies since the data match in Figure 9 is vastly superior to that in Figure 8.

Regarding the low-frequency aspects of the model image in Figure 9; though the necessary zero-frequency components are present such that the sawtooth time functions are properly matched, these components are distributed over most of the model image. The static moment release is *not* properly localized. This lack of resolution follows quite directly from the zero-frequency problems that were previously discussed. Although we will not delve further into this subject, Ruff(1987) advocates and employs the explicit incorporation of *a priori* information on the moment rate density function into seismic source tomographic imaging.

## 5. Conclusions

Many researchers in earthquake seismology are trying to "look inside" the earthquake rupture process. Although several different approaches are available, it is certainly important to rigorously pursue the angle of tomographic imaging for the moment rate density function based on far-field waves. In addition to the strong observational basis, there are substantial theoretical advantages of source imaging based on far-field waves: the model and data are connected by linear integral equations; and for a wide variety of circumstances, this connection is the Radon transform.

Seismic source imaging cannot make direct use of the analytical inverse Radon transform due to severe inadequacies of the observed waveforms. At the same time, seismologists want to know certain zero-frequency quantities of the model that are not directly observed. In this paper, I have established the theoretical framework to discuss these problems, but certainly have not solved them. Ultimately, new information must be added to the problem, use of a priori conditions is one approach. Another approach would be to add near-field waves to the data set. Although this addition destroys the analytical linear transform connection between model and data, an iterative method could easily incorporate the non-linear near-field data in the latter iterations.

Does seismic source tomographic imaging offer anything to experienced tomographers? At the present, seismic source imaging uses the well-established concepts, theory, and techniques of tomography. Seismic source imaging is a new physical system that falls within the tomography aegis. There is at least one significant difference between seismic source imaging and typical structural imaging problems: seismic source imaging uses observations of time-varying seismic waves to determine model properties in both time and space. Although some tomographers might be merely amused by the mixture of space and time coordinates, this aspect of seismic source imaging could lead to new insights for other problems. The emphasis on zero-frequency quantities and the multiplicity of seismic wave types present a rich physical and mathematical structure that is waiting to be explored.

*Acknowledgments.* The research program for earthquake studies at the University of Michigan has been supported by grants from the National Science Foundation

(EAR8407786) and the Shell Companies Foundation to LJR.

# References

Abramowitz, M. and I. A. Stegun, 1964.*Handbook of Mathematical Functions*, National Bureau of Standards, Washington, D.C.

Adby, P.R. and M.A.H. Dempster, 1974. *Introduction to Optimization methods*, Chapman and Hall, London.

Aggarwal, Y.P., Sykes, L.R., Armbruster, J., and Sbar, M.L., 1973, Premonitory changes in seismic velocities and prediction of earthquakes. Nature, **241**, 101-104.

Aki, K., Christoffersen, A. and Husebye, E.S., 1977. Determination of the three-dimensional seismic structure of the lithosphere, J.Geophys.Res., **82**, 277-296.

Aki, K., De Fazio, T., Reasenberg, P., and Nur, A., 1970, An active experiment with earthquake fault for an estimation of the in situ stress. Bull. Seism. Soc. Am., **60**, 1315.

Aki, K. and P. Richards (1980). *Quantitative Seismology: Theory and Methods*. 932 pp., W.H. Freeman and Company, San Francisco.

Allen,R.V., 1982. Automatic phase pickers: their present use and future prospects, Bull. Seism. Soc. Am., **72**, S225-S242.

Alsop, L.E., 1966. Transmission and reflection of Love waves at a vertical discontinuity. J. Geophys. Res., **71**, 3969-3984.

Anderson, D.L., and Whitcomb, J.H., 1973, The dilatancy-diffusion model of earthquake prediction, in Proceedings of the conference on tectonic problems of the San Andreas fault system. Stanford University Publ., Univ. Ser., Geol. Sci., **13**, 417-426.

Anderson, O. L., 1986. Properties of iron at the Earth's core conditions, Geophys. J. R. astr. Soc.,**84**, 561-579.

Anderssen, R.S. and J.R.Cleary, 1980. Estimation of *PKP* times from ISC data, Phys. Earth Planet. Inter., **23**, 207-214.

Babich, V.M. and Buldyrev, V.S., 1972. *Asimptoticheskie Metody v zadachakh Difraktsii Korotkikh Voln*. Nauka, Moscow.

Babich, V.M. and Rusakova, N.Ya., 1962, O rasprostranenii voln Releya po pover- khnosti neodnorodnogo uprugogo tela proizvol'noi formy. Jurn. Vych. Mat. Mat. Fiz., **2**, 652-665. English transl., 1963. Comput. Math. and Math. Phys., **2**, 719-735.

Babich, V.M., 1961a. O rasprostranenii voln Releya vdol' poverkhnosti odnorodnogo uprugogo tela proizvol'noi formy. Dokl. Akad. Nauk USSR, **137**, 1263-1266.

Babich, V.M., 1961b. Luchevoi metod vychisleniya intensivnosti volnovykh frontov v sluchae uprugoi neodnorodnoi anizotropnoi sred'i. Mat. Vopr. Teor. Rasprostr. Voln, Inst. Steklov Akad. Nauk, **5**, 36-46.

Babich, V.M., Chikhachev B.A. and Yanovskaya, T.B., 1976. Poverkhnostnye volny v vertikal'no-neordnorodnom uprugom poluprostranstve so slaboi gorizontal'noi neodnorodnostyu. Izv. Akad. Nauk, Fiz. Zemli, 24-31.

Babich, V.M. and Kirpichnikova, N.Ya., 1974. *The boundary-layer method in diffraction problems*. (In Russian) Leningrad Univ. Press. English transl. Springer Verlag 1980.

Babuska, V., J. Plomerova and J. Sileny, 1984. Spatial variations of P residuals and deep structure of the European lithosphere, Geophys. J. R. astr. Soc., **79**, 363-383.

Backus, G.E., 1962. The propagation of short elastic surface waves on a slowly rotating Earth. Bull. seism. Soc. Am., **52**, 823-846.

Backus, G.E., 1964. Geographical interpretation of measurements of average phase velocities of surface waves over great circular and great semi-circular paths, Bull. Seismol. Soc. Am., **54**, 571-610.

Backus, G. and J.F. Gilbert, 1970. Uniqueness on the inversion of inaccurate gross Earth data, Phil. Trans. Roy. Soc. Lond., **A266**, 123.

Bates, C.C., T.F. Gaskell and R.B. Price, 1982. *Geophysics in the Affairs of Man*, Pergamon Press, Oxford, 492pp.

Bates, R.H.J. & McKinnon, G.C., 1979. Towards improving images in ultrasonic transmission tomography, Australian Phys. Sci. Med., **2**, 134-140.

Beck, S.L. and L.J. Ruff (1985). The rupture process of the 1976 Mindanao earthquake. J. Geophys. Res., **90**, 6773-6782.

Berkhout, A.J., 1984, *Seismic migration, imaging of acoustic energy by wave field extrapolation, B. Practical aspects*, Elsevier, Amsterdam.

Beylkin, G., 1982. Generalized Radon transform and it applications, Ph. D. thesis, New York University.

Beylkin, G., 1983. Inversion of the generalized Radon transform, Proc. SPIE, **413**, 32-39.

Beylkin, G., 1984. The inversion problem and applications of the generalized Radon transform, Comm. Pure Appl. Math., **27**, 579-599.

Beylkin, G., 1985. Imaging of discontinuities in the inverse scattering problem by inversion of a causal generalized Radon transform, J. Math. Phys., **26**, 99-108.

Björck, A., 1985. A bidiagonalization algorithm for solving ill-posed systems of linear equations. Univ. of Linköping (Sweden), Dept. of Math., Report LiTH-MAT-R-80-33, revised 1985-06-27.

Björck, A., Elfving, T., 1979. Accelerated projection methods for computing pseudoinverse solutions of systems of linear equations. BIT, **19**, 145-163.

Blair, D.P., 1984. Rise times of attenuated seismic pulses detected in both empty and fluid-filled cylindrical boreholes, Geophysics, **49**, 398-410.

Bleistein, N., Cohen, J.K. & Hagin, F.G., 1985, Computational and asymptotic aspects of velocity inversion, Geophys., **50**, 1253-1265.

Bleistein, N. & Gray, S.H., 1985, An extension of the Born inversion method to a depth dependent reference profile, Geophys. Prosp., **33**, 999-1022.

Boatwright, J., 1979. The Radon transform and the inversion of body-wave pulse shapes, Earthquake Notes, **50**, 4, 31-32.

Bodoky, T., Hermann, L. & Dianiska, L., 1985. Processing of the in-seam seismic transmission measurements, presented at the 47th Annual EAEG Meeting in Budapest, Hungary.

Bois, P., La Porte, M., Lavergne, M. & Thomas, G., 1972. Well-to-well seismic measurements, Geophysics, **37**, 471-480.

Bolt, B.A., 1960. The revision of earthquake epicentres, focal depths and origin times using a high-speed computer, Geophys. J. R. astr. Soc., **3**, 433-440.

Bonner, B.P., 1974, Shear wave birefringence in dilating granite. Geophys. Res. Lett., **1**, 217.

Bracewell, R. N., 1956. Strip integration in radio astronomy, Aust. J. Phys., **9**, 198-217.

Bregman, N. D., Chapman, C. H. and Bailey, R. C., 1986. Crosshole seismic tomography using raytracing, SEG 56th Annual Meeting, extended abstract S13.7, 560-563.

Bretherton, F.P., 1968. Propagation in slowly varying waveguides. Proc. R. Soc., **A302**, 555-576.

Brooks, R. A., Weiss, G. H. and Talbert, A. J., 1978. A new approach to interpolation in computed tomography, J. Comp. Ass. Tomog., **2**, 577-585.

Buchbinder, G.G.R., 1972. Travel times and velocities in the outer core from *PKmKP*, Earth Planet.Sci.Lett., **14**, 161-168.

Buland, R., 1984. Residual statistics, Terra Cognita, **4**, 268.

Buland, R., 1987. Uniform reduction error analysis, preprint.

Bullen, K.E. and B.A.Bolt, 1985. *An introduction to the theory of seismology*, 4th ed., 499 pp, Cambridge Univ. Press, Cambridge.

Bulletin of the International Seismological Centre: Catalogue of Events and Associated Observations (Years 1964-1982), Vols. **1-19**, International Seismological Centre, Newbury, Berkshire, England.

Burmakov, Ju.A., A.V. Treussov and L.P. Vinnik, 1984. Determination of three dimensional velocity structure from observations of refracted body waves, Geophys. J. R. astr. Soc., 79, 285-292.

Butkov, E., 1968, *Mathematical Physics*, Addison-Wesley, Reading Ma.

Červený V., 1982. Direct and inverse kinematic problem for inhomogeneous anisotropic media, Contr. Geophys. Inst. Slov. Acad. Sci., **13**, 127-133.

Červený, V., 1985. The application of ray tracing to the numerical modelling of seismic wave fields in complex structures, *Seismic Shear Waves*, Part A: Theory, pp.1-124, Geophysical Press, London.

Červený, V., 1986a. Seismic ray theory, *Encyclopedia of Geophysics*, Van Nostrand Reinhold, Stroudsburg, in press.

Červený, V., 1986b. *Seismic ray theory*, Lecture Notes, Internat. Centre for Theoretical Physics, Miramare, Trieste.

Červený V. and Firbas, P., 1984. Numerical modelling and inversion of travel times of seismic body waves in inhomogeneous anisotropic media, Geophys.J.R.astr.Soc., **76**, 41-56.

Červený, V. and Hron, F., 1980. The ray series method and dynamic ray tracing systems for 3-D inhomogeneous media, Bull. Seism. Soc. Am., **70**, 47-77.

Červený, V., Klimeš, L. and Pšenčík, I., 1984. Paraxial ray approximations in the computation of seismic wavefield in inhomogeneous media, Geophys.J.R.astr.Soc., **79**, 89-104.

Červený, V., Klimeš, L., and Pšenčík, I., 1986. *Paraxial boundary value ray tracing*, Manuscript (in Czech).

Červený, V., Molotkov, I.A. and Pšenčík, I., 1977. *Ray Method in Seismology*, Universita Karlova, Praha.

Červený, V., Popov, M. and Pšenčík, I., 1982. Computation of wave fields in inhomogeneous media - Gaussian beam approach. Geophys. J. R. astr. Soc., **70**, 109-128.

Červený, V. and Pšenčík, I., 1981. *2D seismic ray package*. Research report. Charles University, Prague.

Červený, V. and Pšenčík, I., 1983. Gaussian beams in two-dimensional elastic inhomogeneous media. Geophys. J. R. astr. Soc., **72**, 417-433.

Capon, J., 1970. Analysis of Rayleigh wave multipath propagation at LASA. Bull. seism. Soc. Am., **69**, 1701-1731.

Capon, J., 1971. Comparison of Love- and Rayleigh- waves multipath propagation at LASA. Bull. seism. Soc. Am., **61**, 1327.

Cary P.W. and C.H. Chapman, 1986. Non-linear bayesian inversion of marine seismic data, Terra Cognita, **6**, 305.

Chander, R., 1975. On tracing rays with specified end points, J.Geophys., **41**, 173-177.

Chapman, C. H., 1978. A new method for computing synthetic seismograms, Geophys. J. R. astr. Soc., **54**, 481-518.

Chapman, C.H. 1981. Generalized Radon transforms and slant stacks. Geophys. J. R. astr. Soc., **66**, 445-453.

Chapman, C. H. and Cary, P. W., 1986. The circular harmonic Radon transform, Inverse Problems, **2**, 23-49.

Chapman, C.H. and J.A. Orcut, 1985. Least-squares fitting of marine seismic refraction data, Geophys. J. R. astr. Soc., **82**, 339-374.

Chapman, C.H. and J.A. Orcut, 1986. The computation of body wave synthetic seismograms in laterally homogeneous media, Rev. Geophys., **23**, 105-164.

Chapman, C. H. and Phinney, R. A., 1972. Diffracted seismic signals and their numerical solution. In: *Methods of Computational Physics*, **12**, ed. B. A. Bolt, Academic Press Inc., New York and London.

Chin, R.C.Y., G. Hedstrom and L. Thigpen, 1984. Numerical methods in seismology, J. Comp. Phys., **54**, 18-56.

Chou, C. and J. Booker, 1979. A Backus-Gilbert approach to inversion of travel time data for three-dimensional velocity structure, Geophys. J. R. astr. Soc., **59**, 325-344

Choy, G.L. and V.F.Cormier, 1983. The structure of the inner core inferred from short period and broadband GDSN data, Geophys.J.R. astr.Soc., **72**, 1-21.

Christensen, N.I., and Wang, H.F., 1985, The influence of pore pressure and confining pressure on dynamic elastic properties of Berea sandstone. Geophysics, **50**, 207-213.

Claerbout, J.F., 1971, Toward a unified theory of reflector mapping, Geophysics, **36**, 467-481.

Claerbout, J.F., 1985. *Imaging the Earth's Interior*, Blackwell, Oxford, 398pp.

Claerbout, J.F. and F. Muir, 1973. Robust modelling with erratic data, Geophys., **38**, 826-844.

Clark, H.E. and E.S. Medina, 1976. *Tsunami seismic system*, USGS Albuquerque Seism. Lab. Rept 77-777.

Clayton, R.W. and R.P.Comer, 1983. A tomographic analysis of mantle heterogeneities from body wave travel times (abstract), Eos Trans.AGU, **64**, 776.

Clayton, R.W. and R.P. Comer, 1984. A tomographic analysis of mantle heterogeneities, Terra Cognita, **4**, 282-283.

Clayton, R.W. and A.M. Dziewonski, 1984. Lateral variations in lower mantle velocities determined from travel time data (abstract), Eos Trans. AGU, **65**, 271.

Clayton, R.W. & Stolt, R.H., 1981, A Born-WKBJ inversion method for acoustic reflection data, Geophys., **46**, 1559-1567.

Cline, A.K., 1981. FITPACK - software package for curve and surface fitting employing splines under tension, Dept. of Comp.Sci., Univ. of Texas, Austin.

Cohen, J. K., Hagin, F. G. and Bleistein, N., 1986. Three dimensional Born inversion with an arbitrary reference, Geophysics, **51**, 1552-1558.

Comer, R.P., 1984. Rapid seismic ray tracing in spherically symmetric Earth via interpolation of rays, Bull. Seism. Soc. Am., **74**, 479-492.

Comer, R.P. and R.W. Clayton, 1984. Tomographic reconstruction of velocity heterogeneity in the Earth's mantle (abstract), Eos Trans. AGU, **65**, 236.

Cormack, A. M., 1963. Representation of a function by its line integrals, with some radiological applications. I, J. Appl. Phys., **34**, 2772-2727.

Cormack, A. M., 1964. Representation of a function by its line integrals, with some radiological applications. II, J. Appl. Phys., **35**, 2908-2913.

Cormier, V., 1984. The polarization of S waves in a heterogeneous isotropic Earth model, J.Geophys., **56**, 95-113.

Cormier, V., 1986. An application of the propagator matrix of dynamic ray tracing: The focussing and defocussing of body waves by three-dimensional velocity structure in the source region, Geophys.J.R.astr.Soc., in press.

Cosma, C., 1986. *Crosshole investigations - short- and medium-range survey by seismic tomography*, Stripa Project Internal Report, SKBKBS, Stockholm.

Costagna, J.P., Batzle, M.L., and Eastwood, R.L., 1985, Relationship between compressional wave and shear wave velocities in clastic silicate rocks. Geophysics, **50**, 551-570.

Cottin, J.-F., Deletie, P., Jacquet-Francillon, H., Lakshmanan, J., Lemoine, Y. & Sanchez, M., 1986. Curved ray seismic tomography: application to the Grand Etang Dam (Reunion Island), First Break, **4:7**, 25-30.

Courant, R. and Hilbert, D., 1966. *Methods of mathematical physics*, II. Intersc. publ.

Crampin, S., 1985, Evaluation of anisotropy by shear wave splitting. Geophysics, **50**, 142.

Crosson, R.S., 1976. Crustal structure modelling of earthquake data. 1.Simultaneous least squares estimation of hypocenters and velocity parameters. J.Geophys.Res., **81**, 3036-3046.

Crowther, R. A., DeRosier, D. J. and Klug, A., 1970. The reconstruction of a three-dimensional structure from projections and its application to electron microscopy, Proc. R. Soc. A, **317**, 319-340.

Dahlen, F.A., 1979, The spectra of unresolved split normal mode multiplets, Geophys. J. R. astr. Soc., **58**, 1-33.

De Martini, D.C., Beard, D.C., Danburg, J.S., and Robinson, J.H., 1976, Variation of seismic velocities in sandstones and limestones with lithology and pore fluid at simulated in situ conditions, Proceedings EGPC Exploration seminar.

DeVilbiss, J., 1980, Wave dispersion and absorption in partially saturated rocks. Stanford University Ph.D. dissertation.

Deans, S.R., 1977. *The Radon transform: some remarks and formulas for two dimensions*, Lawrence Berkeley Laboratory Report LBL-5691, Berkeley.

Deans, S.R., 1983. *The Radon Transform and Some of Its Applications*. 289 pp., John Wiley & Sons, New York.

Del Pino, E. & Nur, A., 1985. Seismic wave polarization applied to geophysical tomography, presented at the 55th Annual SEG Meeting in Washington, U.S.A.

Devaney, A.J., 1984. Geophysical diffraction tomography, IEEE Trans. Geosci. Remote Sensing, **22**, 3-13.

Dines, K., & Lytle, J., 1979. Computerized geophysical tomography, Proc. IEEE, **67**, 1065-1073, 1679.

Docherty, P., 1985. *A fast ray tracing routine for laterally inhomogeneous media*, Report CWP-018, Centre for Wave Phenomena, Colorado School of Mines, Golden.

Domenico, S.N., 1984, Rock lithology and porosity determination from shear and compressional wave velocity. Geophysics, **49**, 1188-1195.

Doornenbal, J.B. & Helbig, K., 1983, High resolution reflection seismics on a tidal flat in the Dutch delta - Acquisition, Processing and Interpretation, First Break, **9**, 9-20.

Dorbath, C., L. Dorbath, F.D. Fairhead and G.W. Stuart, 1986. A teleseismic delay time study across the Central African Shear zone in the the Adamawa region of Cameroon, West Africa, Geophys. J. R. astr. Soc., **86**, 751-766.

Dost, B., 1986. Preliminary results from higher-mode surface-wave measurements in western Europe using the NARS array, Tectonophys., **128**, 289-301.

Dost, B., A. van Wettum and G. Nolet, 1984. The NARS array, Geol. Mijnb., **63**, 381-386.

Doyen, P., Journel, A., and Nur, A., 1982, Porosity mapping in petroleum reservoirs using seismic data. Extended abstract, Society of Exploration Geophysicists Annual Meeting, p. 328-330.

Dziewonski, A.M., 1984. Mapping the lower mantle: Determination of lateral heterogeneity in *P* velocity up to degree and order 6, J.Geophys.Res., **89**, 5929-5952.

Dziewonski, A.M., 1984. Science plan for global digital network, EOS, **65-16**, 245.

Dziewonski, A.M. and Anderson, D.L., 1981. Preliminary reference Earth model. Phys. Earth Planet. Inter., **25**, 297-356.

Dziewonski, A.M. and Anderson, D.L., 1983. Travel times and station corrections for *P* waves at teleseismic distances, J. Geophys. Res., **88**, 3295-3314.

Dziewonski, A.M., T.A. Chou, and J.H. Woodhouse (1981). Determination of earthquake source parameters from waveform data for studies of global and regional seismicity. J. Geophys. Res., **86**, 2825-2852.

Dziewonski, A.M. and F. Gilbert, 1976. The effect of small aspherical perturbations on travel times and a re-examination of the corrections for ellipticity, Geophys. J. R. astr. Soc., **44**, 7-16.

Dziewonski, A.M., B.H. Hager and R.J. O'Connell, 1975. Large-scale heterogeneities in the lower mantle, J. Geophys. Res., **82**, 239-255.

Dziewonski, A.M. and Hales, A.L., 1972, Numerical analysis of dispersed seismic waves, *Methods in Comp. Phys.*, **11**, 39-85, Academic Press, New York.

Dziewonski, A.M. and J.H. Woodhouse (1983). Studies of the seismic source using normal mode theory. in: *Earthquakes: Observation, Theory, and Interpretation*, edited by H. Kanamori and E. Boschi, pp. 45-137, Elsevier North-Holland, New York.

Ein-Gal, M., 1974. *The shadow transform: an approach to cross-sectional imaging*, Technical Report, SEL-74-050, Information Systems Laboratory, Stanford University.

Ellsworth, W.L. and R.Y. Koyagani, 1977. Three-dimensional crust and upper mantle structure of the Kilanea volcano, Hawaii, J. Geophys. Res., **82**, 5379-5394.

Endo, E.T., Ward, P.L., Harlow, D.H., Allen R.V. and J.P. Eaton, 1974. A prototype global volcano surveillance system monitoring seismic activity and tilt, Bull. Volcanol., **38**, 315-344.

Evans, J.R. and S.S. Allen, 1983. A teleseismic-specific detection algorithm for single short period traces, Bull. Seismol. Soc. Am., **71** 1173-1186.

Evernden, J.E., 1953. Direction of approach of Rayleigh waves and related problems, Part I. Bull. seism. Soc. Am., **43**, 335.

Evernden, J.E., 1954. Direction of approach of Rayleigh waves and related problems, Part II. Bull. seism. Soc. Am., **44**, 159.

Fawcett, J.A., 1983. *I. Three dimensional ray-tracing and ray-inversion in layered media, II. Inverse scattering and curved ray tomography with applications to seismology*, Ph. D. thesis, California Institute of Technology, Pasadena.

Fawcett, J.A. and R.W. Clayton, 1984. Tomographic reconstruction of velocity anomalies, Bull. Seism. Soc. Am., **74**, 2201-2219.

Fehler, M. & Pearson, C., 1984. Crosshole seismic surveys: Applications for studying subsurface fracture systems at a hot dry rock geothermal site, Geophysics, **49**, 37-45.

Felsen, L.B., 1984 (ed.). *Hybrid formulation of wave propagation and scattering*, Nijhoff Publ., Dordrecht, 432pp.

Fessenden, R.A., 1917. Method and apparatus for locating ore-bodies, U.S. patent 1.240,328.

Firbas, P., 1981. Inversion of travel time data for laterally heterogeneous velocity structure - linearization approach. Geophys.J.R.astr.Soc., **64**, 189-198.

Firbas, P., 1984a. Two-dimensional inverse kinematic problem - application to the Saudi Arabia refraction profile. In: Proc. of the 1980 workshop of IASPEI on the Earth's interior on the seismic modelling of laterally varying structures. U.S.Geological Survey Circular 937, (Mooney W.D. and Prodehl C. - editors), Alexandria, 14-21.

Firbas, P., 1984b. Travel time curves for complex inhomogeneous slightly anisotropic media. Stud. geoph. et geod., **28**, 393-406.

Firbas, P., 1984c. Interpretation of the test profile Zurich. Workshop proc.: Interpretation of seismic wave propagation in laterally heterogeneous structures, (Finlayson, D.M. and Ansorge, J. - editors) Report 258 of Bureau of Mineral Resources, Geology & Geophysics, Australian Government Publishing Service, Canberra, 166-170.

Firbas, P., 1984d. Inverse kinematic problem for inhomogeneous media - linearization approach. Ph.D. thesis. Charles University Prague-Geofyzika Brno. 217 pp. (In Czech).

Fletcher, R. and Reeves, C.M., 1964. Function minimization by conjugate gradients, Computer J., **7**, 149-154.

Firbas, P. and Skorkovská M., 1986. Velocity distribution modelling in laterally heterogeneous media (Program for a desk-top computer). Applied Geophysics **20**, Prague, ISSN 0036-5319, 139-153.

Fletcher, R., 1980, *Practical methods of optimization*, **1**, Unconstrained optimization, Wiley.

Franklin, J.N., 1970. Well posed stochastic extention of ill-posed linear problems. J.Math.Anal.Appl., **31**, 682-716.

Fukao, Y. and Kobayashi, M., 1983. Phase and group velocities and Q of mantle Love and Rayleigh waves of the first two modes and their azimuthal dependence for the 1983 Kurile Islands earthquake. Phys. Earth Planet. Inter., **4**, 35.

Gabriels, P., Snieder, R. & Nolet, G., 1987, In situ measurements of shear velocity in sediments using higher mode Rayleigh waves, Geophys. Prosp., in press.

Gauthier, O., Virieux, J. and Tarantola, A., 1986, Two-dimensional nonlinear inversion of seismic waveforms, Numerical results, Geophysics, **51**, 1387-1403.

Gel'fand, I.M., Graev, M.I. and Vilenkin, N.Ya., 1966. *Generalized Functions*, **5**, Integral Geometry and Representation Theory, Academic Press, New York.

Gel'fand, I.M. and Shilov, G.E., 1964. *Generalized Functions*, **1**, Properties and Operations, Academic Press, New York.

Gerver, M. and Markuschevich, V., 1966. Determination of the seismic wave velocity from the travel-time curve, Geophys. J. R. astr. Soc., **11**, 165-173.

Giardini, D., X. Li and J. H. Woodhouse, 1986. Heterogeneous models of the Earth from modal splitting functions of low order multiplets (abstract), Terra Cognita, **6**, 291.

Giardini, D., X. Li and J. H. Woodhouse, 1987. Three dimensional structure of the Earth from Splitting in free oscillation spectra, Nature (in press).

Gilbert, J.F., 1972. Ranking an winnowing gross Earth data for inversion and resolution, Geophys. J. R. astr. Soc., **43**, 125, 1971.

Gilbert, J.F. and A.M. Dziewonski, 1975. An application of normal mode theory to the retrieval of structural parameters and source mechanisms from seismic spectra, Phil. Trans. Roy. Soc. Lond., **278**, 187.

Gill, P.E., W. Murray and M.H. Wright, *Practical Optimization*, Acad. Press, London, 401pp.

Giloi, W.K., 1978. *Interactive computer graphics*. Prentice Hall.

Gjevik, B., 1973. A variational method for Love waves in nonhorizontally layered structures. Bull. seism. Soc. Am., **63**, 1013-1023.

Gjöystdal, H., Reinhardsen, J.E. and Ursin, B., 1984. Travel time and wavefront curvature calculations in three-dimensional inhomogeneous layered media with curved interfaces, Geophysics, **49**, 1466-1494.

Glot, J.P., Gresta, S., Patane, G. and G. Poupinet, 1986. Earthquake activity during the 1983 Etna eruption, Bull. Volcanol., to be published.

Godfrey, R., Muir, F., and Rocca, F., 1980, Modeling seismic impedance with Markov chains, Geophysics, **45**, 1351-1372.

Goforth, T. and E. Herrin, 1981. An automatic seismic signal detection algorithm based on the Walsh transform, Bull. Seismol. Soc. Am., **71**, 1351-1360.

Golub, G.H., Van Loan, C.F., 1983. *Matrix computations*. North Oxford Academic, Oxford.

Grand, S.P. and D.V. Halmberger, 1984. Upper Mantle shear structure beneath the Northwest Atlantic Ocean, J. Geophys. Res., **89**, 11465-11475.

Grasso, J.R., M. Cuer and G. Pascal, 1983. Use of two inverse techniques. Application to a local structure in the New Hebrides island arc, Geophys. J. R. astr. Soc., **75**, 437-472.

Gregersen, S. and Alsop, L.E., 1974. Amplitudes of horizontally refracted Love waves. Bull. seism. Soc. Am., **64**, 535-553.

Griewanck, A. and Ph. L. Toint, 1982. Partitioned variable metric updates for large structured optimization problems, Numer. Math., **39**, 119-137.

Gustavsson, M., Ivansson S., Moren, P., & Pihl, J., 1986. Seismic borehole tomography - measurement system and field studies, Proc. IEEE, **74**, 339-346.

Gutenberg, B., 1953. Travel times of longitudinal waves from surface foci, Proc.Natl.Acad.Sci., **39**, 849-853.

Haddon, R.A.W. and E.S. Husebye, 1978. Joint interpretation of P-wave time and amplitude anomalies in term of lithospheric heterogeneities, Geophys. J. R. astr. Soc., **55**, 19-43.

Hager, B.H., 1984. Subducted slabs and the geoid: constraints on mantle rheology and flow, J.Geophys.Res., **89**, 6003-6015.

Hager, B.H., Clayton, R.W., Richards, M.A., Comer, R.P., Dziewonsky, A.M., 1985. Lower mantle heterogeneity, dynamic topography and the geoid. Nature, **313**, 541-545.

Hammarstrom, M., Ivansson, S., Moren, P., & Pihl, J., 1986. *Crosshole investigations - results from seismic borehole tomography*, Stripa Project Internal Report, SKBKBS, Stockholm.

Han, D., Nur, A., and Morgan, D., 1986, Effects of porosity and clay content on wave velocities in sandstones. Geophysics, **51**, 2093-2107.

Hansen, E.W., 1979. *Image reconstruction from projections using circular harmonic expansion*, Ph. D. thesis, Stanford University.

Hansen, E.W., 1981. Theory of circular harmonic image reconstruction, J. Opt. Soc. Am., **71**, 304-308.

Hashida, T. and K. Shimazaki, 1985. Seismic tomography: 3-D image of upper mantle attenuation beneath the Kanto district, Japan, Earth Plan. Sc. Lett., **75**, 403-409.

Hawkins, W. G., 1983. Mathematics of computed tomography, Ph. D. thesis, University of Arizona.

Herman, G.T. (1980). *Image Reconstruction from Projections*. 316 pp., Academic Press, New York.

Hermann, L., Dianiska, L. & Verboci, J., 1982. Curved ray algebraic reconstruction technique applied in mining geophysics, Geophys. Trans. of the Eotvos Lorand Geophys. Inst. of Hungary, **28**, 33-46.

Herrera, I., 1964. On a method to obtain a Green's function for a multilayered halfspace. Bull. seism. Soc. Am., **54**, 1087-1096.

Herrin, E.and J.Taggart, 1962. Regional variations in $P_n$ velocity and their effect on the location of epicenters, Bull.Seism.Soc.Am., **52**, 1037-1046.

Hestenes, M.R., Stiefel, E., 1952. Methods of conjugate gradients for solving linear systems. J. Res. N. B. S., **49**, 409-436.

Hudson, J.A., 1981. A parabolic approximation for surface waves. Geophys. J. R. astr. Soc., **67**, 755-770.

Hudson, J.A., and Heritage, J.R., The use of Born approximation in seismic scattering problems, Geophys. J. R. astr. Soc., **66**, 221-240, 1981.

*IRIS - Incorporated Research Institutions for Seismology*, 1985. The program for Seismic Studies of the Continental Lithosphere (PASSCAL), Washington,DC,USA.

Ikelle, L.T., Diet, J.P. & Tarantola, A., 1986, Linearized inversion of multioffset seismic reflection data in the ω-k domain, Geophys., **51**, 1266-1276.

Ito, H., DeVilbiss, J., and Nur, A., 1979, Compressional and shear waves in saturated rock during water-steam transition. J. Geophys. Res., **84**.

Its, E.N. and T.B. Yanovskaya, 1985. Propagation of surface waves in a halfspacewith vertical, inclined and curved interfaces, Wave Motion, **7**, 79-94.

Ivansson, S., 1983. Remark on an earlier proposed iterative tomographic algorithm, Geophys. J. R. astr. Soc., **75**, 855-860.

Ivansson, S., 1985. A study of methods for tomographic velocity estimation in the presence of low-velocity zones, Geophysics, **50**, 969-988.

Ivansson, S., 1986. Seismic borehole tomography - theory and computational methods, Proc. IEEE, **74**, 328-338, 1166.

Ivansson, S., 1986. Some remarks concerning seismic reflection tomography and velocity analysis, Geophys. J. R. astr. Soc., **87**, 539-557.

Iyer, H.M., 1975. Anomalous delays of teleseismic P-waves in Yellowstone National Park, Nature, **253**, 425-427.

Jackson, D.D., 1972. Interpretation of inaccurate, insufficient and inconsistent data. Geophys.J.R.astr.Soc., **28**, 97-109.

Jackson, D.D., 1979. The use of a priori data to resolve non-uniqueness in linear inversion, Geophys. J. R. astr. Soc., **57**, 137-158.

Jeffreys, H., 1932. An alternative to the rejection of observations, Proc. Roy. Soc. Lond., **187**, 78-87.

Jeffreys, H., 1936. On travel times in seismology, Bur. Centr. Seism. Trav. S., **14**, 3-86 (reprinted in *The collected papers of Sir Harold Jeffreys*, **2**, Gordon and Breach, London, 1973).

Jeffreys, H. and Bullen, K.E., 1958. *Seismological tables*, British Association for the Advancement of Science London.

Jobert, G., 1976. Matrix methods for generally stratified media. Geophys. J. R. astr. Soc., **47**, 351-362.

Jobert, N. 1976. Propagation of surface waves on an ellipsoidal Earth. Pure appl. Geophys., **114**, 797-804.

Jobert, N. 1986a. Mantle wave propagation anomalies on laterally heterogeneous global models of the Earth by gaussian beam synthesis. Ann. Geophys., **4**, 261-270.

Jobert, N. 1986b. Mantle wave deviations from "pure-path" propagation on aspherical models of the Earth, by Gaussian beam waveform synthesis. Phys. Earth Planet. Int.,in press.

Jobert, N. and Jobert, G., 1983. An application of ray theory to the propagation of waves along a laterally heterogeneous spherical surface. Geophys. Res. Lett., **10**, 1148-1151.

Jones, T.D., 1986, Pore fluids and frequency-dependent wave propagation in rocks. Geophysics, **51**, 1939-1953.

Jones, T.D., and Nur, A., 1984, The nature of seismic reflections from deep crustal fault zones. J. Geophys. Res., **89**, 3153-3171.

Jordan, T.H., 1975. The continental tectosphere, Rev. Geophys. Space Phys., **13**, 1-12.

Jordan, T.H., 1978, A procedure for estimating lateral variations from low frequency eigenspectra data, Geophys. J. R. astr. Soc., **54**, 571-610.

Julian, B.R., 1970. *Ray tracing in arbitrarily heterogeneous media*. Techn. Not. 1970-45 Lincoln Lab. M.I.T. Cambridge U.S.A.

Julian, B.R. and Gubbins, D., 1977. Three-dimensional seismic ray tracing, J.Geophys., **43**, 95-113.

Julian, B.R., and Sengupta, M.K., 1973. Seismic travel time evidence for lateral inhomogeneity in the deep mantle, Nature, **242**, 443-447.

Kaczmarz, S., 1937. Angenäherte Auflösung von Systemen linearer Gleichungen. Bull. Acad. Polon. Sci. Lett. A.,

35, 355-357.

Kanamori, H. and Given, G.W., (1981). Use of long-period surface waves for rapid determination of earthquake source parameters. Phys. Earth Planet. Int., **27**, 8-31.

Kanamori, H. and Stewart, G.S., (1976). M sub 0de of strain release along the Gibbs fracture zone, mid-Atlantic ridge. Phys. Earth Planet. Int., **11**, 312-332.

Karal, J.B. and Keller, F.C., 1960. Surface wave excitation and propagation. J. Appl. Phys., **31**, 1039-1046.

Keenan, J.H., Keyes, F.G., Hill, P.G., and Moore, J.G., 1969, *Steam Tables*. John Wiley, New York.

Keilis-Borok, V.I. (ed.), 1986. Seismic surface waves in a laterally inhomogeneous Earth (in Russian), Nauka, Moscow, 278pp.

Keller, H.B. and Perozzi, P.J., 1983. Fast seismic ray tracing, SIAM J.Appl.Math., **43**, 981-992.

Kennett, B.L.N., 1984. Guided wave propagation in laterally varying media - I.Theoretical development. Geophys. J. R. astr. Soc., **79**, 235-255.

Kennett, B.L.N. and Mykkeltveit, S., 1984.Guided wave propagation in laterally varying media - II. Lg waves in north-western Europe. Geophys. J. R. astr. Soc., **79**, 257-268.

Kennett, B.L.N. and Williamson, P., 1987. Subspace methods for large-scale nonlinear inversion, in: N.J. Vlaar, G. Nolet, M.J.R. Wortel and S.A.P.L. Cloetingh (eds.) *Mathematical Geophysics*, Reidel, Dordrecht.

Kershaw, D., 1970. The determination of the density distribution of a gas flowing in a pipe from mean density measurements, J. Inst. Math. Appl., **6**, 111-114.

Kikuchi, M. and H. Kanamori (1982). Inversion of complex body waves. Bull. Seism. Soc. Am., **72**, 491-506.

King, M.S., 1977, Acoustic velocities and electrical properties of frozen sandstones and shales. Canadian Journal Earth Sciences, **14**, 1004-1013.

Kirpichnikova, N.Ya., 1969. Rayleigh waves concentrated near a ray on the surface of an inhomogeneous elastic body. In Mat. Probl. Wave Propag. Theory. II. Steklov Mat. Inst., Nauka, Leningrad, **15**, 49-62, in Russian. Engl. Transl. Consultants Bureau, New York, 1971.

Klug, A., Crick, F. H. C. and Wyckoff, H. W., 1958. Diffraction by helical structures, Acta Crystallogr., **11**, 199-213.

Knopoff L., 1972. Observation and inversion of surface wave dispersion, Tectonophys., **13**, 497-519.

Knopoff, L. and Hudson, J.A., 1964. Transmission of Love waves past a continental margin. J. Geophys. Res., **69**, 1649-1653.

Koch, M., 1985. A numerical study on the determination of the 3-D structure of the lithosphere by linear and non-linear inversion of teleseismic travel times, Geophys. J.R.astr.Soc., **80**, 73-93.

Kormendi, A., Bodoky, T., Hermann, L., Dianiska, L., & Kalman, T., 1986. Seismic measurements for safety in coal mines - case histories, Geophys. Prosp., in press.

Kovach R.L., 1979. Seismic surface waves and crustal and Upper Mantle structure, Rev. Geophys. Space Phys., **16**, 1-13.

Kowallis, B., Jones, L.E.A., and Wang, H.F., 1984, Velocity-porosity-clay content; systematics of poorly consolidated sandstones. J. Geophys. Res., **89**, 10355-10364.

Kravtsov,Y.A. and Orlov,Y.I., 1980. Limits of applicability of the method of geometric optics and related problems, Soviet Phys. Usp., **23**, 750-762.

Lager, D. & Lytle, J., 1976. Determining a subsurface electromagnetic profile from high-frequency measurements by applying reconstruction-technique algorithms, Radio Science, **12**, 249-260.

Lakshminarayanan, A. V., 1975. *Reconstruction from divergent ray data*, Dept. of Computer Science Technical Report Number 92, SUNY - Buffalo, Amherst.

Lanczos, C., 1950. An iteration method for the solution of the eigenvalue problem of linear differential and integral operators. J. Res. N. B. S., **45**, 255-282.

Langan, R.T., Lerche, I. and Cutler, R.T., 1985. Tracing of rays through heterogeneous media: An accurate and efficient procedure, Geophysics, **50**, 1456-1465.

Langston, C.A. and D.V. Helmberger (1975). A procedure for modeling shallow dislocation sources. Geophys. J. R. astr. Soc., **42**, 117-130.

Lavrentiev, M. M., Romanov, V. G. and Vasiliev, V. G., 1970. *Multidimensional Inverse Problems for Differential Equations*, **167**, Lecture Notes in Mathematics, Springer-Verlag, Berlin.

Lay, and Kanamori, 1985. Geometric effects of global heterogeneity on long period surface wave propagation. J. Geophys. Res., **90**, 605-621.

Leary, P.C & Henyey, T.L., 1985. Anisotropy and fracture zones about a geothermal well from P-wave velocity profiles, Geophysics, **50**, 25-36.

Lee, W.H.K. and Stewart, S.W., 1981. *Principles and Applications of Microearthquake Networks*, Academic Press, New York.

Lerner-Lam A.L. and T.H. Jordan, 1983. Earth structure from fundamental and higher mode waveform analysis, Geophys. J. R. astr. Soc., **75**, 759.

Leven, J.H., 1985. The application of synthetic seismograms to the interpretation of the upper mantle P-wave velocity structure in northern Australia, Phys. Earth Plan. Int., **38**, 9-27.

Levshin, A.L., 1985. Effects of lateral inhomogeneities on surface wave amplitude measurements, Ann. Geophysicae, **3**, 511-518.

Levshin, A.L and L. Ratnikova, 1984. Apparent anisotropy in inhomogeneous media, Geophys. J. R. astr. Soc., **76**, 65-69.

Li, X.D., Giardini, D. and J.H. Woodhouse, 1985. The interpretation of modal splitting functions in terms of aspherical Earth structure (abstract), Eos Trans.AGU, **66**, 300.

Li, X.D., Giardini, D. and J.H. Woodhouse, 1986. Heterogeneous models of the earth using modal splitting functions of low order normal modes: 2) Earth models (abstract), Eos Trans. AGU, **67**, 307.

Love, A.E.H., 1927. *A treatise on the mathematical theory of elasticity*, 4th ed., Cambridge University Press, Cambridge, 643 pp.

Lynn, H.B., and Thomsen, L.A., 1986, Shear wave exploration along the principal axes. Extended abstract, 56th Annual Meeting Society of Economic Geologists.

Lysmer, J. and Drake, L.A., 1971. The propagation of Love waves across nonhorizontally layered structures. Bull. seism. Soc. Am., **61**, 1233-1251.

Lytle, J. & Dines, K., 1980. Iterative ray-tracing between boreholes for underground image reconstruction, IEEE Trans. Geosci. Remote Sensing, **18**, 234-240.

Madariaga, R., 1972. Toroidal free oscillations of the laterally heterogeneous Earth, Geophys. J. R. astr. Soc., **27**, 81-100.

Madariaga, R. and K. Aki, 1972. Spectral splitting of toroidal free oscillations due to lateral heterogeneity of the Earth's structure, J. Geophys. Res., **77**, 4421-4431.

Maguire, P.K.H., D.J. Francis and D.N. Whitcombe, 1985. Determination of three-dimensional seismic structure of the crust and upper mantle in the Central Midlands of England, Geophys. J. R. astr. Soc., **83**, 347-362.

Mao, N.H., and Sweeney, J.J., 1986, Estimation of in-situ stresses from ultrasonic measurements. Society of Petroleum Engineers Formation Evaluation 532.

Mason, I.M., 1981. Algebraic reconstruction of a two-dimensional velocity inhomogeneity in the High Hazles seam of Thoresby colliery, Geophysics, **46**, 298-308.

Masters, G., T.H. Jordan, P. G. Silver, and F. Gilbert, 1982. Aspherical earth structure from fundamental spheroidal mode data, Nature, **298**, 609-613.

Masters, G. and F. Gilbert, 1981. Structure of the inner core inferred from observations of its spheroidal shear modes, Geophys. Res. Lett., **8**, 569-571.

Matthews, J. and R.L. Walker, 1973. *Mathematical Methods of Physics, 2nd ed.*, W.A.Benjamin Inc, Menlo Park, 501pp.

Mc Garr, A., 1969.Amplitude variations of Rayleigh waves - horizontal refractions. Bull. seism. Soc. Am., **59**, 1307-1334.

McCann, D.M., Baria, R., Jackson, P.D. & Green, A.S.P., 1986. Application of crosshole seismic measurements in site investigation surveys, Geophysics, **51**, 914-929.

Melrose, E. and A. Vernucci, 1983. Data broadcast to microterminal via satellite using **spread-spectrum** techniques, Space Communication and Broadcasting **1**, 211-218.

Menke, W., 1977. Lateral inhomogeneities in P velocity under the Tarbella array of the Lesser Himalayas of Pakistan, Bull. Seism. Soc. Am., **67**, 725-734.

Menke, W., 1984. *Geophysical data analysis: Discrete inverse theory*. Academic Press.

Menke, W., 1985. Imaging fault slip using teleseismic waveforms: Analysis of a typical incomplete tomography problem. Geophys. J. R. astr. Soc., **81**, 197-204.

Meyer, R.P. and R.F. Mereu, 1983. *Proceedings of the Workshop on Portable seismograph development*, International Association of Seismology and Physics of the Earth's Interior, Commission on Controlled Source Seismology, University of Wisconsin,Madison,WI,USA.

Mikumo, T. and T. Miyatake (1978). Dynamical rupture process on a three-dimensional fault with non-uniform frictions and near-field seismic waves. Geophys. J. R. astr. Soc., **54**, 417-438.

Minerbo, G. N. and Sanderson, J. G., 1977. *Reconstruction of a source from a few (2 or 3) projections*, Informal

Report LA-6747-MS, Los Alamos Scientific Laboratory, Los Alamos.

Mitchell, B.J., C.C. Cheng and W. Stauder, 1977. A three dimensional velocity model of the lithosphere beneath the New Madrid seismic zone, Bull. Seism. Soc. Am., **62**, 1061-1074.

Montagner, J.-P., 1985. Seismic anisotropy of the Pacific ocean inferred from long-period surface wave dispersion, Phys. Earth Plan. Int., **28**, 28-50.

Montagner, J.-P., 1986, Regional three-dimensional structures using long-period surface waves, Ann. Geophys., **B4**, 283-294.

Montagner, J.-P. and H.-C. Nataf, 1986. A simple method for inverting the azimuthal anisotropy of surface waves, J. Geophys. Res., **91**, 511-520, 1986.

Morelli, A. and A.M. Dziewonski, 1985. Stability of aspherical models of the lower mantle (abstract), Eos Trans. AGU, **66**, 975.

Morelli, A. and A.M. Dziewonski, 1986. 3D structure of the Earth's core inferred from travel time residuals (abstract), Eos Trans. AGU, **67**, 311.

Morelli, A. and A.M. Dziewonski, 1987. Topography of the core-mantle boundary and lateral homogeneity of the liquid core, Nature (in press).

Morelli, A., A.M. Dziewonski and J.H. Woodhouse, 1986. Anisotropy of the inner core inferred from *PKIKP* travel times, Geophys. Res. Lett. (in press).

Murphy, W.F., III, 1982, Effects of partial water saturation on attenuation in Massilon sandstone and Vycor porous glass. Journal Acoustical Society, **71**, 1458-1468.

Nakanishi, I. and Anderson, D.L., 1982. Worldwide distribution of group velocity of mantle Rayleigh waves as determined by spherical harmonic inversion, Bull. Seismol. Soc. Am., **72**, 1185-1194.

Nakanishi, I. and Anderson, D.L., 1983. Measurements of mantle wave velocities and inversion for lateral heterogeneity and anisotropy, 1, Analysis of great circle phase velocities, J.Geophys.Res., **88**, 10267-102684.

Nakanishi, I. and Anderson, D.L., 1984. Measurements of mantle wave velocities and inversion for lateral heteogeneity and anisotropy. II. Analysis by single-station method. Geophys. J. R. astr. Soc., **78**, 573-617.

Nakanishi, I. and D. Suetsugu, 1986. Resolution matrix calculated by a tomographic inversion method, J. Phys. Earth, **34**, 95-99.

Nakanishi, I. and K. Yamaguchi, 1986. A numerical experiment on nonlinear image reconstruction from first-arrival times for two-dimensional island arc structure, J. Phys. Earth., **34**, 195-201.

Nataf, H.-C., I. Nakanishi and D.L. Anderson, 1986. Measurements of mantle wave velocities and inversion for lateral heterogeneities and anisotropy, 3, Inversion, J. Geophys. Res., **91**, 7261-7308.

Nercessian, A., A. Hirn and A. Tarantola, 1984. Three-dimensional seismic transmission prospecting of the Mont Dore volcano, France, Geophys. J. R. astr. Soc., **76**, 307-316.

Nersesov, I.L., Semenov, A.N., and Simbireva, I.G., 1969, Space-time distribution of the travel time ratios of transverse and longitudinal waves in the Garm area, in: *The Physical Basis of Foreshocks*, Akad. Nauk SSSR Publication.

Neumann-Denzau, G. and J. Behrens, 1984. Inversion of seismic data using tomographic reconstruction techniques for investigation of laterally inhomogeneous media, Geophys. J. R. astr. Soc., **79**, 305-316.

New, B.M., 1985a. *The seismic investigation of rock properties at the Carwynnen Test Mine*, Department of the Environment report no DOERW85 16, U.K.

New, B.M., 1985b. An example of tomographic and Fourier microcomputer processing of seismic records, Q.J. Eng. Geol. London, **18**, 335-344.

Nolet, G., 1977. The upper mantle under Western Europe inferred from the dispersion of Rayleigh modes, J. Geophys., **43**, 265-286.

Nolet, G., 1978. Simultaneous inversion of seismic data, Geophys. J. R. astr. Soc., **55**, 679-691.

Nolet, G., 1981. Linearized inversion of (teleseismic) data, *in: The solution of the inverse problem in geophysical interpretation*, ed. R. Cassinis, Plenum Press, New York, 9-38.

Nolet, G., 1984. Damped least squares methods to solve large tomographic systems and determine the resolving power, Terra Cognita, **4**, 285.

Nolet, G., 1985. Solving or resolving inadequate and noisy tomographic systems. J. Comp. Phys., **61**, 463-482.

Nolet, G. and G.F. Panza, 1976. Array analysis ofseismic surface waves: limits and possibilities, Pure Appl. Geophys., **114**, 776-790.

Nolet, G., B. Romanowicz, R.Kind and E. Wielandt, 1986. *Orfeus Science Plan*, Reidel, Dordrecht, 45pp.

Nolet G., J. van Trier and R. Huisman, 1986. A formalism for nonlinear inversion of seismic surface waves, Geophys. Res. Lett., **13**, 26-29.

Nomofilov, V.E., 1978. O rasprostranenii kvazistacionarn'ix voln Releya b neodnorodnoi anizotropnoi uprugoi srede. Mat. Vopr. Teor. Rasprostr. Voln, Inst. Steklov Akad.Nauk, **10**, 234-245.

Nordqvist, A., 1986. *Application of ultrasonic crosshole seismics for hard rock conditions*, Licentiate Thesis, University of Lulea, Sweden.

Nur, A., 1971, Effects of stress on velocity anisotropy in rocks with cracks. J. Geophys. Res., **76**, 2022-

Nur, A., 1972, Dilatancy, pore fluids, and premonitory variations of $t_s/t_p$ travel times. Bull. Seism. Soc. Am., **62**, 1217-1222.

Nur, A., 1982, Seismic imaging in enhanced recovery. Society of Petroleum Engineers Third Joint Symposium on Enhanced Oil Recovery, Tulsa, Oklahoma.

Nur, A., and Simmons, G., 1969a, Stress-induced velocity anisotropy in rock, an experimental study. J. Geophys. Res., **74**, 6667.

Nur, A., and Simmons, G., 1969b, The effect of saturation on velocity in low porosity rocks. Earth Planetary Science Letters, **7**, 183.

Paige, C.C., Saunders, M.A., 1982. LSQR: An algorithm for sparse linear equations and sparse least squares. ACM Trans. Math. Softw. **8**, 43-71.

Panza, G.F., Mueller, S. & Calcagnile, G., 1980, The gross features of the lithosphere-asthenosphere system in Europe from seismic surface waves and body waves, Pure Appl. Geophys., **118**, 1209-1213.

Park, J., 1986. Synthetic seismograms from coupled free oscillations: effects of lateral structure and rotation. J. Geophys. Res., **91**, 6441-6464.

Parlett, B.N., 1980. *The symmetric eigenvalue problem*. Prentice Hall, Englewood Cliffs, N.J..

Paterson, M.S., and Weiss, L.E., 1961, Symmetry concepts in the structural analysis of deformed rocks. Bull. Geol. Soc. Am., **72**, 841.

Patton, H., 1980. Crust and upper mantle structure of the Eurasian continent from phase velocity and Q of surface waves. Rev. Geophys. Space Phys., **18**, 605-625.

Paulsson, B.N.P., Cook, N.G.W. & McEvilly, T.V., 1985. Elastic-wave velocities and attenuation in an underground granitic repository for nuclear waste, Geophysics, **50**, 551-570.

Pavlis, G.L. and Booker, J.R., 1980. The mixed discrete-continuous inverse problem: application to the simultaneous determination of earthquakes and velocity structure. J.Geophys.Res., **85**, B9, 4801-4810.

Pereyra, V., Lee, W.H.K. and Keller, H.B., 1980. Solving two-point seismic ray tracing problem in a heterogeneous medium, Part I: A general adaptive finite difference method, Bull. Seismol. Soc. Am., **70**, 79-99.

Perry, R. M., 1975. Reconstructing a function by circular harmonic analysis of its line integrals, Image Processing for 2-D and 3-D Reconstruction from Projections: Theory and Practice in Medicine and the Physical Sciences, August 4-7, Stanford University, Technical Digest (Washington, DC: Optical Society of America), **ThA6-1**, 6-4.

Peterson, J.E., Paulsson, B.N.P & McEvilly, T.V., 1985. Applications of algebraic reconstruction techniques to crosshole seismic data, Geophysics, **50**, 1566-1580.

Petrowsky, I.G., 1945. On the propagation velocity of discontinuities of the displacement derivatives on the surface of an unhomogeneous elastic body of arbitrary form. Compt. Rend. Acad. Sc. URSS, **47**, 255-258.

Pickett, G.R., 1963, Acoustic character logs and their application in formation evaluation. Journal Petrology Technology, **15**, 659-667.

Plafker, G. and J.C. Savage (1970). Mechanism of the Chilean earthquakes of May 21 and 22, 1960. Geol. Soc. Am. Bull., **81**, 1001-1030.

Popov, M.M. and Pšenčík, I., 1978a. Ray amplitudes in inhomogeneous media with curved interfaces, in Geof. sb., **24**, pp.118-129, Academia, Praha.

Popov, M.M. and Pšenčík, I., 1978b. Computation of ray amplitudes in inhomogeneous media with curved interfaces, Studia geoph. et geod., **28**, 248-258.

Poupinet, G. ,1979. On the relation between P-wave travel time residuals and the age of continental plates, Earth Planet. Sci. Lett., **43**, 149-161.

Poupinet, G. and J. P. Glot, 1982. A low power seismic event detector transmitting data via ARGOS satellite data collecting system, EOS, **63**, 1266.

Poupinet, G., R. Pillet and A. Souriau, 1983. Possible heterogeneity of the Earth's core deduced from *PKIKP* travel times, Nature, **305**, 204-206.

Press, F., 1956. Determination of crustal structure from phase velocity of Rayleigh waves. Bull. geol. Soc. Am., **67**, 1647-1658.

Prothero, W.A., 1984. Ocean bottom seismometer technology, EOS, **65-13**, 113-116.

Pšenčík, I., 1983. Programs for the computation of kinematic and dynamic parameters of seismic waves in 2-D laterally heterogeneous media with curved interfaces. Programs interpr. seism. observ., **3** . Nauka. Leningrad. (in Russian).

Radcliff, R., Balanis, C. & Hill, H., 1984. A stable geotomography technique for refractive media, IEEE Trans. Geosci. Remote Sensing, **22**, 698-703.

Radon, J., 1917. Uber die Bestimmung von Funktionen durch ihre Integralwerte langs gewisser mannigfaltigkeiten, Ber. Verh. Sachs. Akad. Wiss., **69**, 262-267.

Rafavich, F., Kendall, C.H.St.C., and Todd, T.P., 1984, The relationship between acoustic properties and the petrographic character of carbonate rocks. Geophysics, **49**, 1622-1636.

Raymer, D.S., Hunt, E.R., and Gardner, J.S., 1980, An improved sonic transit time-to-porosity transform. Society Professional Well Log Analysis, **21**, Paper P.

Reasenberg, P., W. Ellsworth and A. Walter, 1980. Teleseismic evidence for a low-velocity body under the Coso Geothermal Area, J. Geophys. Res., **85**, 2471-2483.

Ritzwoller, M., G. Masters and F. Gilbert, 1986. Observations of anomalous splitting and their interpretation in terms of aspherical structure, J. Geophys. Res., **91**, 10203-10228.

Robert, C.W., 1975, *CRC Handbook of Chemistry and Physics*. F-18.

Robertson, J.D., 1983, Carbonate porosity from S/P traveltime ratios. 53rd Annual International SEG Meeting, Las Vegas, Nevada.

Romanov, V.G., 1972. *Někotoryje obratnyje zadaci dlja uravněnij giperboličeskogo tipa*. Nauka. Novosibirsk.

Romanov, V.G., 1974. *Integral Geometry and Inverse Problems for Hyperbolic Equations*, **26**, Springer Tracts in Natural Philosophy, Springer-Verlag, Berlin.

Romanowicz, B.A., 1979. Seismic structure of the upper mantle beneath the United States by three dimensional inversion of body wave arrival times, Geophys. J. R. astr. Soc., **57**, 479-506.

Romanowicz, B.A., 1980. A study of large scale variations of P velocity in the upper mantle beneath western Europe, Geophys. J. R. astr. Soc., **63**, 217-232.

Romanowicz, B., 1986. Multiplet-multiplet coupling due to lateral heterogeneity: asymptotic effects on the amplitude and frequency of the Earth's normal modes. Geophys. J.R. astr. Soc., in press.

Romanowicz, B. and R. Snieder, 1987. A new formalism for the effect of lateral heterogeneity on normal modes and surface waves: - II: General anisotropic perturbations, subm. to Geophys. J. R. astr. Soc.

Roult, G., Romanowicz, B. and Jobert, N., 1986. Observations of departures from classical approximations on very long period Geoscope records. Ann. Geophys., **4**, 241-250.

Ruff, L.J., 1983. Fault asperities inferred from seismic body waves. in *Earthquakes: Observation, Theory, and Interpretation*, edited by H. Kanamori and E. Boschi, pp. 251-276, Elsevier North-Holland, New York.

Ruff, L.J., 1984. Tomographic imaging of the earthquake rupture process. Geophys. Res. Lett., **11**, 629-632.

Ruff, L.J., 1987. Inversion of P waves for images of the earthquake rupture process: Theory and application. J. Geophys. Res., in press.

Ruff, L.J. and H. Kanamori (1983). The rupture process and asperity distribution of three great earthquakes from long-period diffracted P waves. Phys. Earth Planet. Int., **31**, 202-230.

Santo, T., 1966.Lateral variation of Rayleigh wave dispersion character. Part III. Pageoph, **63**, 40.

Sato, 1958. Attenuation, dispersion and the wave-guide of the G-wave. Bull. seism. Soc. Am., **48**, 231-251.

Scales, J.A., A. Gersztenkorn and S. Treitel, 1987. Fast $l_p$ solutions of large, sparse, linear systems: Application to seismic travel time tomography, J. Comp. Phys., in press.

Scales, L.E., 1985, *Introduction to nonlinear optimization*, Macmillan.

Schell, M.M. and L.J. Ruff (1986). Southeastern Alaska tectonics: Source process of the large 1972 Sitka earthquake. EOS, **67**, 304-305 (abstract).

Schultz, P. S. and Claerbout, J. F., 1978. Velocity estimation and downward continuation by wavefront synthesis, Geophysics, **43**, 691-714.

Schwartz, S. and Lay,T., 1985.Comparison of long-period surface wave amplitude and phase anomalies for two models of global lateral heterogeneity. Geophys. Res. Lett., **12**, 321-234.

Schwartz, S.Y. and L.J. Ruff (1985). The 1968 Tokachi-Oki and the 1969 Kurile Islands earthquakes: Variability in the rupture process. J. Geophys. Res., **90**, 8613-8626.

Semenov, A.N., 1969, Variations in the travel time of transverse and longitudinal waves before violent earthquakes. Bulletin Academy of Science USSR, Physics Solid Earth, **3**, 245-248.

Shaw, P.R. and J.A. Orcut, 1985. Waveform inversion of seismic refraction data and applications to young Pacific crust, Geophys. J. R. astr. Soc., **82**, 375-414.

Shepp, L. A. and Logan, J. A., 1974. The Fourier reconstruction of a head section, IEEE Trans. Nucl. Sci., **NS-21**, 21-43.

Silvey, S.D., 1970. *Statistical Inference.* Penguin, Harmondsworth.

Sluis, A. van der, van der Vorst, H.A., 1986. The rate of convergence of conjugate gradients. Numer. Math., **48**, 543-560.

Snieder, R., 1986a, 3D Linearized scattering of surface waves and a formalism for surface wave holography, Geophys. J. R. astr. Soc., **84**, 581-606.

Snieder, R., 1986b, The influence of topography on the propagation and scattering of surface waves, Phys. Earth Plan. Int., **44**, 226-241.

Snieder, R. and Romanowicz, B., 1987. A new formalism for the effect of lateral heterogeneity on normal modes and surface waves - I: isotropic perturbations, perturbations of interfaces and gravitational perturbations, submitted to Geophys. J. R. astr. Soc.

Snieder, R. and Nolet, G., 1987,Linearized scattering of surface waves on a spherical Earth, J. Geophys, in press.

Sobel, P.A. and von Seggern, D.H., 1978. Application of surface wave ray tracing. Bull. seism. Soc. Am., **68**, 1359-1380.

Sommerfeld, A., 1947. *Partielle Differentialgleichungen der Physik.* Vorlesungen ueber theoretische Physik Band 4. Akademische Verlags- Gesellschaft Geest und Portig KG, Leipzig.

Soonawala, N.M., 1984. An overview of the geophysics activity within the Canadian Nuclear Fuel Waste Management Program, Geoexploration, **22**, 149-168.

Spencer, C., and Gubbins, D., 1980. Travel-time inversion for simultaneous earthquake location and velocity structure determination in laterally varying media. Geophys.J.R.astr.Soc., **63**, 95-116.

Spetzler, H., and Anderson, D.L., 1968, The effect of temperature and partial melting on velocity and attenuation in a simple binary mixture. J. Geophys. Res., **73**, 6051.

Stevenson, D.J., 1986. Limits on lateral density and velocity variations in the Earth's outer core, Geophys.J.R.astr.Soc. (in press).

Strichartz, R., 1982. Radon inversion - variations on a theme, Amer. Math. Monthly, **89**, 377-384.

Takeuchi H. and M. Saito, 1972. Seismic surface waves, *in: Methods in Computational Physics*, ed. B.A.Bolt, Acad. Press, London, 217pp.

Tanimoto, T. and D.L Anderson, 1984. Lateral heterogeneity and azimuthal anisotropy of the upper mantle: Love and Rayleigh waves 100-250 sec, J. Geophys. Res., **90**, 1842-1858, 1985.

Tarantola, A., 1984a, Inversion of seismic reflection data in the acoustic appriximation, Geophys., **49**, 1259-1266.

Tarantola, A., 1984c, The seismic reflection inverse problem, p.104-181, *in Inverse problems of acoustic and elastic waves*, ed. F. Santosa, Y.H. Pao, W.W. Symes and C. Holland, SIAM Philadelphia.

Tarantola, A., 1986a, A strategy for nonlinear elastic inversion of seismic reflection data, Geophysics (october 1986, in press).

Tarantola, A., 1986b, Linearized inversion of seismic reflection data, Geophys. Prosp., **32**, 998-1015.

Tarantola, A., 1987, *Inverse problem theory. Methods for data fitting and model parameter estimation.*, Elsevier, in press.

Tarantola, A. and A. Nercessian, 1984. Three-dimensional inversion without blocks, Geophys. J. R. astr. Soc., **76**, 299-306.

Tarantola, A. and Valette, B., 1982, Generalized nonlinear inverse problems solved using the least squares criterion, Reviews of Geophys. and Space Phys., **20**, No. 2, 219-232.

Taylor, S.R. and M.N. Toksoz, 1979. Three-dimensional crust and upper mantle structure of the northeastern United States, J. Geophys. Res., **84**, 7627-7644.

Thill, R.E., Willard, R.J., and Bur, T.R., 1969, Correlation of longitudinal velocity variation with rock fabric. J. Geophys. Res., **74**, 4897.

Thomson, C.J. and Chapman, C.H., 1985. An introduction to the Maslov asymptotic method, Geophys.J.R.astr.Soc., **83**, 143-168.

Thomson, C.J. and Gubbins, D., 1982. Three dimensional lithospheric modelling at NORSAR: linearity of the method and amplitude variations from the anomalies, Geophys. J. R. astr. Soc., **71**, 1-36.

Timur, A., 1968, Velocity of compressional waves in porous media at permafrost temperatures. Geophysics, **33**, 584.

Tocher, D., 1957, Anisotropy in rocks under simple compression. American Geophysical Union Transactions, **38**, 89.

Todd, T., and Simmons, G., 1972, Effect of pore pressure on the velocity of compressional waves in low-porosity rocks. J. Geophys. Res., **77**, 3731-3743.

Toksöz, M.N. and D.L. Anderson, 1966. Phase velocities of long period surface waves and structure of the Upper Mantle, J. Geophys. Res., **71**, 1649-1667.

Tosaya, C., 1982, Acoustical properties of clay-bearing rocks. Stanford University Ph.D. dissertation.

Tosaya, C. and Nur, A., 1982, Effects of diagenesis and clays on compressional velocities in rocks. Geophys. Res. Lett., **9**, 5-8.

Tosaya, C., Nur, A., Aronstam, P., and DaPrat, G., 1984, Monitoring of thermal EOR fronts by seismic methods. SPE California Regional Meeting, preprint.

Valette, B., 1986. About the influence of prestress upon adiabatic perturbations of the Earth. Geophys. J. R. astr. Soc., **85**, 179-208.

Vazquez, A., Aranda, R. & Benhumea, M., 1985. Cave detection using a tomographic algorithm, presented at the 47th Annual EAEG Meeting in Budapest, Hungary.

Verly, J. G., 1981. Circular and extended circular harmonic transforms and their relevance to image reconstruction from line integrals, J. Opt. Soc. Am., **71**, 825-835.

Walder, J., and Nur, A., 1984, Porosity reduction and crustal pore pressure development. J. Geophys. Res., **89**, 11,539-11,548.

Wang, Z., and Nur, A.M., 1986, Effect of temperature on wave velocities in sands and sandstones with heavy hydrocarbons. Society of Petroleum Engineers 61st Annual Technical Conference and Exhibition, New Orleans, Louisiana.

Webster, W.J., Miller, W.H., Whitley, R., Allenby R.J. and R.T. Dennison, 1981. A seismic signal processor suitable for use with NOAA—GOES satellite data collection system, IEEE Trans. on Geoscience and Remote Sensing, **GE 19-2**, 91-94.

Wesson, R.L., 1971. Travel time inversion for laterally inhomogeneous crustal models, Bull. Seismol. Soc. Am., **61**, 729-746.

Whitcomb, J.H., Garmany, J.D., and Anderson, D.L., 1973, Earthquake prediction: Variation of seismic velocities before the San Fernando earthquake. Science, **180**, 632.

Wielandt E. and L. Knopoff, 1982. Dispersion of very long-period Rayleigh waves along the East Pacific Rise: Evidence for S wave velocity anomalies to 450 km depth, J. Geophys. Res., **87**, 8631-8641.

Wiggins, R.A., 1972. General linear inverse problem - Implication of surface waves and free oscillations for Earth structure, Rev.Phys.Space Phys., **10**, 251-285.

Williamson P.R., 1986. Tomographic inversion of travel time data in reflection seismology, Ph.D. thesis, Cambridge (UK), 1986.

Winkler, K.W. and Nur, A., 1979, Pore fluids and seismic attenuation in rocks. Geophys. Res. Lett., **6**, 1-4.

Winkler, K.W. and Nur, A., 1982, Seismic attenuation: Effects of pore fluids and frictional sliding. Geophysics, **47**, 1-15.

Wong, J., Hurley, P. and West, G.F., 1983. Crosshole seismology and seismic imaging in crystalline rocks, Geophys. Res. Lett., **10**, 686-689.

Wong, J., Hurley, P. and West, G.F., 1984. Crosshole audiofrequency seismology in granitic rocks using piezoelectric transducers as sources and detectors, Geoexploration, **22**, 261-279.

Wong, J., Hurley, P. and West, G.F., 1985. Investigation of subsurface geolocigal structure at the Underground Research Laboratory with crosshole seismic scanning, in Proceedings of the 17th information meeting of the Nuclear Fuel Waste Management Program: **TR-229**, 593-608, Canada.

Woodhouse, J.H., 1974. Surface waves in a laterally varying layered structure. Geophys. J. R. astr. Soc., **37**,461-490.

Woodhouse, J.H. and Dziewonski, A.M., 1984. Mapping the upper mantle: Three dimensional modeling of the Earth structure by inversion of seismic waveforms, J. Geophys. Res., **89**,5953-5986.

Woodhouse, J.H. and Dziewonski, A.M., 1986. Three dimensional mantle models based on mantle wave and long period body wave data (abstract), Eos Trans AGU, **67**, 307.

Woodhouse, J.H., Giardini, D. and X. Li, 1986. Evidence for inner core anisotropy from splitting in free oscillation spectra, Geophys.Res.Lett. (in press).

Woodhouse, J.H. and Girnius, T.P., 1982. Surface waves and free oscillations in a regionalized earth model. Geophys. J. R. astr. Soc, **68**, 653-674.

Woodhouse, J.H. and Wong, Y.K., 1986. Amplitude, phase and path anomalies of mantle waves. Geophys. J. R. astr. Soc., **87**, 753-774.

World Meteorological Organization, 1982. ***Satellites in Meteorology, Oceanography and Hydrology***, WMO publication number 585, Geneva, CH.

World Meteorological Organization, 1985. *Information on meteorological satellite programmes operated by Members and organizations*, WMO publication number 411, Geneva,CH.

Wyllie, M.R.J., Gregory, A.R., and Gardner, G.H.F., 1958, An experimental investigation of factors affecting elastic wave velocities in porous media. Geophysics, **23**, 459-493.

Wyllie, M.R.J., Gregory, A.R., and Gardner, L.W., 1956, Elastic wave velocities in heterogeneous and porous media. Geophysics, **21**, 41-70.

Yanovskaya, T.B., 1984. Solution of the inverse problem of seismology for laterally inhomogeneous media, Geophys. J. R. astr. Soc., **79**, 293-304.

Yanovskaya, T.B., R. Maaz, P.G. Ditmar and H. Neunhofer, 1987. A method for joint interpretation of the phase and group surface wave velocities for estimation of lateral variations of the Earth's structure, Phys. Earth. Plan. Int., in press.

Yomogida, K., 1985. Gaussian beams for surface waves in laterally slowly varying media. Geophys. J. R. astr. Soc., **82**, 511-533.

Yomogida, K. and Aki, K., 1985. Waveform synthesis of surface waves in a laterally heterogeneous Earth by the Gaussian beam method. J. Geophys. Res., **98**, 7665-7688.

Yomogida, K. and Aki, K., 1986. Amplitude and phase data inversions for phase velocity anomalies in the Pacific Ocean Basin. Subm. to Geophys. J. R. astr. Soc. .

# Index

acoustic birefringence 210
acoustic waves 3,86
adjoint operator 136
algorithm, conjugate gradients 71ff
-, Dines-Lyttle 60
-, LSQR 75ff
-, ray tracing 99ff
-, SIRT 58ff
alkenes 223
amplitude anomalies 164,213,300
anelasticity 319
angular order 310
anisotropy 23,175,178,207ff,288
-, and cracks 207
-, and stress 207,208
-, cylindrical 271
-, dependence on symmetry 214
-, of inner core 267
ARGOS 243
ART 49,58,139,171
asperities 350
attenuation 213,218,219

back projection 139,353
back propagation 329
barriers 350
BFGS update 308
bicharacteristics 278
borehole effects 165
Born approximation 146,147,326
brightspots 213
bulk modulus 4

cave detection 165
cell parametrization 13,111,112,171
-, vs. bilinear elements 171
-, vs. harmonic expansion 261
Central Limit theorem 12
central slice theorem 29
characteristics 102
Chebyshev polynomial 33,34
Circular Harmonic Decomposition (CHD) 32
clay content 204
CMB 252
coal seam 182
coefficient of influence 139
column-scaling 52
common mid point data 41
confining pressure 210

conjugate directions 307
conjugate gradients 19,70,71ff,137,315
coordinate quantization 255
coordinates, ray-centered 122,288ff
-, scaling of 280ff
-, transformation 35,290,291
core-mantle boundary 252,262
corner frequency 358
correlation length 140,331
covariance matrix 15,261
covariance operators 135ff
crosshole tomography 159ff

dam investigation 165
Data Collection Platforms (DCP) 242ff
data errors 12,54,85ff,191,266
data, selection 254
data, telemetry 239ff
delay time 8
-, bias 97
-, distribution 11,257
-, errors 12,85ff
-, maximum 12
-, outliers 11,12,14
diffracted waves 85ff
diffraction tomography 179
Dines-Lyttle algorithm 60
directivity parameters 347
discontinuity, depth of 262
dislocation sources 343
dispersion 4,276,288
displacement spectrum 311
displacement, static 350
dyadic Green's function 325
dynamic ray tracing 121ff,291

Earth Flattening Transformation (EFT) 35
effective pressure 212
Eicosene 221
Eikonal equation 4,7,100,114,117,279
elasticity tensor 3
ellipsoidal surface, ray tracing 291
Euclidean norm 14,50,305
Evinson wave 183

Fan-Beam Geometry 34
far-field approximation 344
fault detection 165,181
fault integral 344
Fermat's principle 8,9,22,102,163,190,288,293
Filtered Back Projection (FBP) 30,353
finite fault source 345
focusing 22,300,312
fracture 159
fracture detection 184ff,231
Fréchet derivatives 146
fundamental matrix 125
fundamental mode 276

gas cap movement 236
gas pockets 213
Gauss Markov theorem 52,54,
Gaussian beams 133,285ff
Gaussian statistics 195
Gauss-Newton method 305
generalized inverse 54,194,262
geometrical spreading 284
geophones 180
geostationary satellites 242
geothermal exploration 159,218
GDSN 276
GEOSCOPE 276,278
GMS 242
GOES 242
granite 209,315
great circle approximation 293,311,323
Green's function 143,343ff
-, reciprocity 144
group velocity 20,310

Hamilton-Jacobi equation 279,282
Hessian matrix 305
-, sparse 308
higher modes 21,276
-, interaction terms 327
Hillbert transform 30
holography 327ff
Hooke's law 3
Huygens' principle 10
hydrocarbons 220ff
hypocenter 343

IDA 276,278
impedance, P wave 148
-, S wave 150
imcompressibility 4
initial value ray tracing 106
INTELSAT 242
International Seismological Centre (ISC) 22,254
inverse scattering 25
inversion, stability 304
IRK (Inverse Radon Kernel) 30ff,354ff
IRLS 14

Jeffreys-Bullen model 22,256

Kaczmarz's method 58
Krylov subspace 70

Lagrangian 283
Lame's parameter 6
Lanczos method 70,73ff
Lanczos vectors 74,75
LASA 240
least squares 34,49ff,135ff,170,194,262,304,327
LITHOSCOPE project 241
local minima 191,314
Love wave 275,310,325
low velocity layer (LVZ) 39ff,301,317
LSQR 18,19,71,75ff,171,305
-, compared to SIRT 82
-, convergence 78
-, regularization 76

Maslov method 131
matrix, norm 51
-, nullspace 51
-, range 51
-, propagator 123
-, sparse 50
-, symplectic 125
maximum likelihood estimate 19
METEOSAT 242
Mercator transformation 290
migration 26,153,324,330
mineral exploration 165
minimum norm solution 16,51
mode conversion 327
mode summation 304,310ff
model, cell size 16
-, covariance 16,135,260
-, initial 193
-, interpolation 131
- norm 17,195
-, parameter choice 13,140,171,261
-, resolution 19,151,172,260
-, scaling 16,262
-, smoothness 17,131
-, spherical harmonic 13,260

moment density tensor 343ff
moment rate function 349ff
moment tensor 310
multiple gridding 315

NARS array 314,337
nonlinear inverse problems 22,136ff,305ff
normal equations 14,51
NORSAR 240

objective function 303
ocean bottom seismometer 245
oil-saturated sands 225ff
OMEGA receiver 246
operator, adjoint 136
-, covariance 138
-, roughening 17
-, smoothing 22
-, transpose 136,148
optimization 137,305ff
ore body 180
overburden pressure 210

P wave 7,101,340
parabolic approximation 281
paraxial rays 124,285ff
PASSCAL project 239
path elimination 177
*PcP* 252
permafrost zones 220
permeability 239
phase integral 277
phase transitions 214
phase velocity 20
piezoelectric transducers 184
*PKIKP* 252
$PKP_{BC}$ 252
plane wave points 123
point source approximation 341,344ff
Poisson's ratio 211
polarization anomalies 298
polarization vectors 123,325
pore pressure 210
porosity 204,229
-, and clay content 206
-, and $V_p$ 204ff
-, mapping 229
preconditioning 307,331
projection experiments 24
projection method 49,69ff
- method, compared to SIRT 82
propagator matrices 123

QR algorithm 77
quadratic slowness 108
quality factor Q 23,159,312

radiation pattern 334
radioactive waste 165,166
Radon transform 25ff,340ff
ray 4
-, bending 175,177
-, coordinates 10,121,285ff
-, parameter 5
-, propagator 125
-, width 10
ray approximation 85ff,278,302
-, conditions of validity 102,279
Rayleigh wave 275,310,325
ray tracing 99ff,177,192,287ff
-, analytic 107
-, bending 132
-, cartesian coordinates 102
-, continuation 133
-, dynamic 121ff,291
-, in cells 111
-, initial condition 102
-, shooting 132
-, spherical coordinates 116
-, surface waves 275ff
ray tracing equation, ellipsoidal surface 291
-, isotropic dispersion relation 287
-, plane surface 289
-, spherical surface 289
-, transformation-, 290
reciprocity 144
reference model 9,193
reflected waves 116,118
refraction profiles 189ff
regularization 17,55,331
relocation 256
representation theorem 10,143,149
reservoir description 203ff
resolution 19,151,172,260
resolution matrix 20,173
ribbon fault source 348ff
Richardson iteration 70
Ritz polynomial 77
Ritz values 76ff
rock properties 203ff
row-scaling 52
rupture front 349,357
rupture velocity 357

S wave 7,101,122,310,317

sandstone 204,216,223
satellite transmission 239ff
saturation 210
scaling 16,141,262
scattering 324ff
search direction 137,306,308
sedimentary rocks 206
sediments 334
eismic profiles 189ff
seismic source imaging ,358,362,365
seismic sources 339ff
shadow zone 89
shales 204,209
shear modulus 343
shifting 52
signal duration 160
Singular value decomposition 53,194
SIRT 58ff,171,188
-, associated matrix 63
-, compare to LSQR 82
-, convergence 61
-, errors 64
-, regularization 63
-, statistical properties 63
slant stacking 26
slowness 15
-, quadratic 104
-, vector 101,103ff
smiles 329
smoothing 330
Snell's law 5,292
-, for curved interfaces 105ff
solution, approximation 56
-, errors 19
-, iteration error 65
-, maximum likelihood 136
-, non-uniqueness 15
-, perturbation error 65
-, regularization error 56
-, resolution 19,151,172,260
source time function 345
-, resolution 358
spectrum, fitting 309
spherical surface, ray tracing 289
spherical harmonics 13,259
spline approximation 193
SS waves 317
starting model 9
steamflooding 232
steepest ascent 136
steepest descent 306,315
stress determination 231
stress monitoring 165,181
Strichartz set 167
Stripa Mine 166
subroutine *pstomo* 18
subspace method 308
summary earthquakes 256
surface rays, theory of 278ff
surface wave 20,340
-, amplitude 310
-, far-field spectrum 310
-, fundamental modes 302
-, higher modes 302
-, holography 327ff
-, incidence on vertical surface 292
-, ray theory 278
-, scattering 302,324ff

tar-satuated sands 225ff,
tau-p domain 26
telescopic points 123
thermal fronts 232
transpose operator 136
triplications 192

uniform reduction, method of 11
Upper Mantle 314

velocities, pure path 276
-, and clay content 205
-, and phrase transitions 214
-, and porosity 205
-, and saturation 212
-, and temperature 220ff
-, in hydrocarbons 220ff
$V_p - V_s$ ratio 206
Volterra equation 25,38,41

water flooding 236
wave equation, acoustic 4,86
-, elastodynamic 7,142
-, exact solution 87
wave fronts 278
wave polarization 165
waveform fitting 140ff,302ff
wavefront 4,101
weighting functions 141
Wielandt effect 10
WKBJ 278,304,311

Young's modulus 211